Nanobiomaterials
Research Trends and Applications

Editors

Thandapani Gomathi
Department of Chemistry
D.K.M. College for Women
Vellore, Tamil Nadu, India

P.N. Sudha
Department of Chemistry
D.K.M. College for Women
Vellore, Tamil Nadu, India

Sabu Thomas
Department of Chemical Sciences
Mahatma Gandhi University
Kottayam, Kerala, India

CRC **CRC Press**
Taylor & Francis Group
Boca Raton   London   New York

CRC Press is an imprint of the
Taylor & Francis Group, an **informa** business

A SCIENCE PUBLISHERS BOOK

First edition published 2024
by CRC Press
2385 NW Executive Center Drive, Suite 320, Boca Raton FL 33431

and by CRC Press
4 Park Square, Milton Park, Abingdon, Oxon, OX14 4RN

© 2024 Taylor & Francis Group, LLC

CRC Press is an imprint of Taylor & Francis Group, LLC

Library of Congress Cataloging-in-Publication Data (applied for)

ISBN: 978-0-367-17485-9 (hbk)
ISBN: 978-1-032-48026-8 (pbk)
ISBN: 978-0-429-05703-8 (ebk)

DOI: 10.1201/9780429057038

Typeset in Times New Roman
by Radiant Productions

Dedication

Dedicated to my dear parents Mr. P. Thandapani and Mrs. T. Manimegalai

By
Dr. T. Gomathi

Preface

In several fields of nanoscience, there is an ongoing interchange of ideas between the biological and physical sciences. Nanotechnology creates nanosized materials for use in biology and other fields by using biomimetic or bio-inspired processes. In this approach, nanoscale materials act as the "building blocks" for the creation of the "bridge," which is how nanobiotechnology bridges the gap between nano and bio. *Nanobiomaterials: Research Trends and Applications* cover a wide range of topics that can be used to illustrate the precious value of biopolymers in prospective applications in several disciplines, including drug delivery, tissue engineering, agriculture, electrochemical field, biosolar cells and wastewater treatment applications. As a contemporary collection, this book examines the extremely beneficial biodegradable and biocompatible materials. It will give an overview of various successful uses of nanobiomaterials. These nanobiomaterials are suitable for a variety of industrial applications since they can be employed directly or chemically altered to take advantage of their various chemical sites.

This book examines innovative methods for creating nanobiomaterials, characterisation of nanobiomaterials and its application for better agricultural practises, protracted drought conditions, and recent developments in biomedicine. Anyone who is interested in nanobiotechnology and nanobiomaterials can benefit from it.

Contents

Part I
Preparation, Classification and Physico-chemical Properties

1

Nanobiomaterials
An Overview

Thandapani Gomathi,[1,]* *P.N. Sudha*[1] and *Sabu Thomas*[2]

1. Introduction

Nanomaterials are very small molecules in the nanometer scale range. Nanomaterials are the foundations of nanoscience and nanotechnology. The development of nanomaterial has attracted great interest in the past few years. In 1959, Feynman described a process of nanotechnology that permits one to individually manipulate atoms and molecules throughout high precision instruments. This technique might be applied to style and build systems at the nanoscale level, atom by atom (Feynman 1959, Nanotech 2009), and its applications in many areas of wide interest like health, industry, pharmacy, etc., seem to be unlimited. The term "nanotechnology" was first applied by Drexler (Drexler 1981, 1986).

Nanotechnology refers to the world of the knowledge that designs and produces structures, devices and systems by manipulating atoms and molecules at the nanoscale level (Savage et al. 2007). The field of nanotechnology is an interdisciplinary area and it is one of the most popular areas of current research and development basically in all disciplines. According to Drexler, "Nanotechnology is the principle of manipulation of the structure of matter at the molecular level. It entails the ability to build molecular systems with atom-by-atom precision, yielding a variety of nanomachines (Stander and Theodore 2011, Waseem et al. 2012)." Nanoparticles are microscopic particles smaller than 100 nanometers (Medina et al. 2007). Nanoparticles have unusual properties which make their use in nanomedicine advantageous (Jos et al. 2009). Nowadays most of the nanoparticles are obtained from transition metals, metal oxides, carbon and silicon. Moreover, the nanoparticles can be prepared using biopolymers.

The use of nanomaterials in the biological fields is referred to as nanobiomaterials. Nanoparticles utilized in biology are grouped into three categories: organic, inorganic and mixed (organic/inorganic) (Holban et al. 2016). In recent years, many nanoparticles are developed for diverse applications in medicine, including infectious diseases. Nanobiomaterials are utilized for various biological and biomedical applications such as drug and gene delivery, biosensors, bio-imaging, tissue engineering, bio-electronics, and for antimicrobial activities (Vo-Dinh 2007). Shuttle systems

[1] Department of Chemistry, D.K.M. College for Women, Affiliated to Thiruvalluvar University, Vellore, Tamil Nadu, India.
[2] Mahatma Gandhi University, Kottayam, Kerala, India.
* Corresponding author: drgoms1@gmail.com

are commonly used for the delivery and stabilization of bioactive drugs and antimicrobial molecules, ensuring not just their specificity, but also controlled release (Holban et al. 2016). Coating medical devices is of an excellent advantage within the communicable disease field, e.g., nano modified surfaces and devices proved to be very efficient to scale back microbial attachment and biofilm formation (Holban et al. 2014, Lara et al. 2015).

The use of nanomaterial as biosensors has currently a huge impact and a quick development within the usage of smart nanobiomaterials. Biosensors are accurate and offer a price effective approach for the detection of pathogenic infectious agents in natural environment, and also clinical specimens (Veigas et al. 2014). Nanodiagnostics was first introduced by Mirkin et al., in 1996 where the authors published the utilization of gold (Au)-nanoparticles to permit anthrax detection (Bailey et al. 2003).

The material chemistry and properties are not defined solely by the term "biomaterial." The name biomaterial was confined based on the interaction of material, either natural or synthetic, with a biological environment. Thus, "nanobiomaterials" can be of natural and synthetic origin. The aim of this review is to highlight and discuss the recent progress and applications of nanobiomaterials.

2. Natural Materials and their Derivatives

Naturally occurring materials are often found everywhere in nature and only with recent advances in instrumentation and metrology equipment are researchers starting to locate, isolate, characterize and classify the vast range of their structural and chemical changes. Also, natural nanomaterials are often created due to the by-products of combustion reactions or are produced purposefully through engineering to perform a specialized function. These materials can have different physical and chemical properties from their bulk-form peers.

2.1 Polymeric Nanobiomaterials

Polysaccharides are biodegradable and biocompatible materials that have a high content of functional groups, which include hydroxyl, amino, and carboxylic acid groups. Polysaccharides include chitin, chitosan, and cellulose, which are plentiful in nature, renewable, resorbable, biocompatible, and quite low-cost, which can be utilized in drug delivery, cell-encapsulating biomaterials, and tissue engineering (Wen and Oh 2014, Heidari et al. 2016a). The chitin nanoparticles were also found to possess antibacterial property and its ferromagnetic performance allows for its possible application in drug-tracking systems (Jayakumar et al. 2010a, Dev et al. 2010).

2.1.1 Alginate

Alginates are natural polymers consisting of linear copolymers of an anionic and hydrophilic polysaccharide containing β-(1-4)-linked d-mannuronic acid and β-(1-4)-linked 1-guluronic acid units. Commonly, alginate is one of the most affluent biosynthesized natural materials and is derived primarily from two sources: marine plants, i.e., brown seaweed (40% of dry matter) such as species of *ascophyllum, durvillaea, escklonia, luminaria, lessonia, macrocystis, sargassum, turbinaria* and bacteria (from Antimicrobial Textiles 2016). Alginate is soluble in water and its solubility is governed by pH of the solvent, ionic strength and presence of gelling ions. It has been reported that alginate chelates with divalent ions such as sodium, calcium, magnesium, strontium and barium ions to form hydrogels and crosslinking glutaraldehyde will improve stability (Elçin 1995, Flores-Maltos et al. 2011). Calcium chloride ($CaCl_2$) is one of the most frequently used agents to ionically cross-link alginate.

Alginate has a broad range of applications as a biomaterial and especially as the supporting matrix or delivery system and tissue repair and regeneration. Due to its outstanding properties in terms of biocompatibility, biodegradability, non-antigenicity and chelating ability, alginate has been widely used in a variety of biomedical applications and tissue engineering. Alginate is now largely used to prepare nanoparticles via complexation method, and w/o emulsification coupled with the gelation of the alginate emulsion droplet.

2.1.2 Chitosan

Chitin is the most abundant natural polymer next to the cellulose and is similar to cellulose in many respects. Chitin contains 2-acetamido-2-deoxy-β-D-glucose through a β (1→4) linkage. The most abundant source of chitin is the shell of crab and shrimp. The deacetylated form of chitin is chitosan. Chitosan was discovered in 1859 by Professor C. Rouget. Chitosan contains 2-acetamido-2-deoxy-β-D-glucopyranose and 2-amino-2-deoxy-β-D-glucopyranose residues (Bhatnagar and Sillanpää 2009). Chitosan is the most important derivative of chitin (Shukla et al. 2013), which is more suitable for the biological applications compared to chitin owing to its superior solubility in organic solvents and water (Mima et al. 1983, Heidari et al. 2016b). It is well known that some of the structural characteristics of chitin/chitosan and its derivatives such as degree of acetylation (DA), degree of substitution (DS) and molecular weight (MW) greatly influence various properties such as solubility, and physiological activities (Hirano and Nagano 1989, Suzuki et al. 1990). Chitosan is a unique biopolymer than chitin since it can be modified with different functional groups to control hydrophobic, cationic, and anionic behaviors. The most interesting properties of chitosan come from the presence of primary amines along with its backbone. In addition to highly beneficial physicochemical characteristics, these structures also give these polysaccharides the ability to interact specifically with cells and biomolecules (Shukla et al. 2013).

It is simply processed into nanofibers (Jayakumar et al. 2010b), nanoparticles (Anitha et al. 2009) and chitosan-based nanobiomaterials possess superior chemical and physical properties including high surface area, porosity, mechanical strength, conductivity, and photoluminescence compared to pure chitosan. Also, chitosan is amenable to molding into porous scaffolds with controllable characteristics. Moreover, chitosan surfaces support the attachment and the subsequent proliferation and growth of different types of cells, which have been attributed to the high cationic charge density of chitosan (Mohammed et al. 2017, Bumgardner et al. 2007). However, owing to the insolubility in a neutral and basic media, its uses are restricted.

Chitosan nanofibers have many different applications for the development of wound dressing (Jayakumar et al. 2010b, Croisier and Jérôme 2013), which can be prepared from trifluoroacetic acid and dichloromethane mixtures (Czaja et al. 2007, Klemm et al. 2001).

2.1.3 Cellulose

Cellulose is the most abundant form of living terrestrial biomass and finds numerous usages in modern industry. It is a long-chain polymer with repeating units of D glucose, a simple sugar that appears in almost pure form in cotton fiber such as wood, plant leaves, and stalks. It is found in combination with other materials such as hemicelluloses and lignin. Some bacteria are also found to synthesize cellulose. In nature, cellulose is a fibrous, tough, and water-insoluble substance with chemical formula $(C_6H_{10}O_5)n$. Each cellulose fibril is composed of d-glucose monomer solely linked with β-1,4 glucosidic bonds. The cellulose content of cotton fibers, hardwood stems, softwood stems, rice and wheat straw is 95–98%, 40–55%, 45–50%, and 35–45%, respectively.

Cellulose macro and nanofibers have attracted much attention because of the high strength and stiffness, biodegradability, non-toxicity, facile chemical modification, and renewability, and their production and use in composite developments. Fabrication of cellulose nanofibers is a relatively new research area. Cellulose nanofibers may be used as filler in composite materials owing to the enhanced mechanical and biodegradation behavior of composites. Application of cellulose nanofibers as polymer filler is also a novel research area (Siró and Plackett 2010). Although cellulose has numerous applications and uses, in the present scenario the large volumes of cellulosic waste materials which are originating from agricultural or industrial activities are often regarded as worthless.

2.1.4 Lignin

Lignin is the second most abundant natural polymer after cellulose (Li 2012). It plays an important role in plants, providing rigidity to strengthen the structures of cell walls and resistance to microbial attack (Laurichesse and Avérous 2014). Lignin is an amorphous, three-dimensional polymer that is different from both carbohydrates and proteins in composition and properties. As plants mature, their cell walls increase in lignin concentration leading to a tough, stringy texture (Santosh 2012). Lignin is a fiber that is not sugar, but rather a saccharide consisting of long chains of phenolic resin alcohols connected along an oversized advanced molecule.

Lignin has degradable property and in common practice, hydrogenation and oxidation are the two most common techniques used to degrade lignin (Upton and Kasko 2016, Jia et al. 2011). Lignin has many other properties such as high thermal stability (Argyropoulos et al. 2014), antioxidant, biodegradability (Hon 1996) and antimicrobial behavior, adhesive properties and relative abundance (Bode et al. 2001). Lignin shows the additives properties, dust dispersant and blending properties (Mahmood et al. 2016). The most common basic structural unit of lignin is a phenyl propanoid. Lignin forms by oxidative condensation (C—C and C—O bond formation) between substituted group such as —OH, —OCH$_3$ (or) =O group and phenyl propanoid units (Agrios 2005, Windeisen and Wegener 2014). Depending on botanical origin, lignin has different chemical structure such as H-lignin (Coumaryl alcohol), G-lignin (Coniferyl alcohol), and S-lignin (Sinapyl alcohol). Generally, softwood has G-lignin and hardwoods have G and S-lignin (Figure 1).

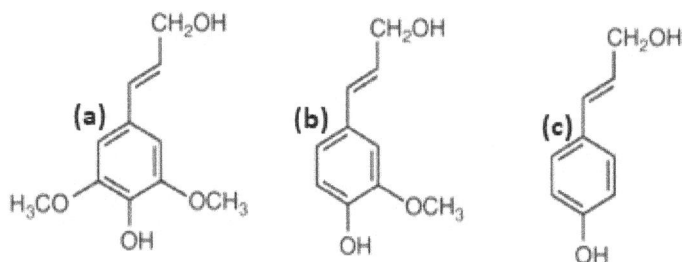

Figure 1: Different chemical structure of lignin (a) S-lignin, (b) G-lignin and (c) H-lignin.

2.1.5 Starch

Starch is the second largest biomass produced on Earth. Starch consists mainly of amylose and amylopectin. Starch is a mixture of a linear α-1,4-glucan (amylose) and a branched glucan (amylopectin), also containing 1,4,6-bonded glucose units. Generally, the weight ratio for amylopectin:amylose is about 75:25, but high-amylopectin starches can be obtained by genetic modification of corn or potato. Depending on botanical origin and genetic background, starch has different chemical structures such as branch chain lengths, phosphate derivatives and different functional properties. This structural and functional diversity makes these starches suitable for different applications. High-amylose starch produces strong films which are suitable for making biodegradable plastics and small granule starch is used to polyethylene film filler (Jane 1995).

Most starches in the native form present limitations such as high viscosity, susceptibility to retrogradation, limited digestibility for some, and limited solubility for others. For this reason, most starch used in food or industrial applications is first modified (Daniel et al. 2007).

2.2 Metallic Nanobiomaterials

Recently, metallic nanobiomaterials are used in the drug delivery, nanofabrication, as a catalyst, etc., in the nanotechnology field and certain achievements have been made in the metal nanomaterials as heterogeneous catalysts (Dutta et al. 2015, Yasukawa et al. 2015, Zhang et al. 2015a, 2015b). These catalysts have a very high catalytic activity and selectivity for specific reactions.

Nanocatalysts are used to catalyze certain chemical reactions, which include oxidation reaction (Dutta et al. 2015), reduction reaction (Yasukawa et al. 2015), coupling reaction (Zhang et al. 2015a, 2015b) and electrochemical reaction (Zhou et al. 2003, Tripkovic et al. 2002, Choi et al. 2006, Shen et al. 2015, Xiao et al. 2015). Nanomaterials have attracted much attention because they are a bridge between atoms and bulk materials. In addition, they have some special properties such as surface and interface effect, small size effect, quantum size effect and macroscopic quantum tunnel effect.

2.2.1 Gold Nanoparticles

Gold nanoparticles are synthesised by various methods such as physical, chemical, biosynthesis and seeding growth method (Madu et al. 2011, Ya et al. 2011, Haustrup et al. 2007) (Figure 2).

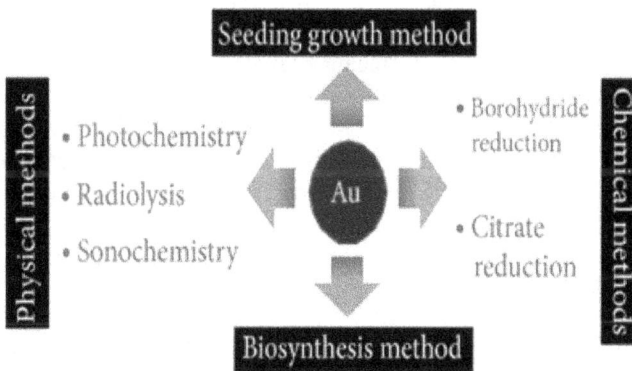

Figure 2: Synthesis of AuNPs by different methods.

In bionanotechnology, gold nanoparticles (AuNPs) have been widely used because of their unique properties and multiple surface functionalities. AuNPs are extensively used in colorimetric biosensors (Podsiadlo et al. 2008, Chen et al. 2007), drug delivery (Paciotti et al. 2004, Javier et al. 2008, Aaron et al. 2007), cancer imaging (Lee et al. 2008, El-Sayed et al. 2005, von Maltzahn et al. 2009, Cheng et al. 2008) and cancer therapies (Kim et al. 2002, Sun and Xia 2004) due to their oxide-free surface, bioconjugation properties, good biocompatibility and unique optical properties (Dekany 2011, Sunita et al. 2011, Sónia and Carabineiro 2010).

2.2.2 Silver Nanoparticles

Madhumathi et al. (2010) have synthesized novel α-chitin/nanosilver composite, which was found to possess outstanding antibacterial activity against *Staphylococcus aureus* and *Escherichia coli*, combined with good blood-clotting ability and wound healing. Similarly, Kumar et al. (2010) have developed and characterized β-chitin/nanosilver composite, which was found to be antibacterial, whole blood clotting, swelling, cytotoxicity and wound healing. Furthermore, β-chitin/nanosilver composite bioscaffolds were assessed for their cell adhesion behavior using epithelial cells, which are encouraging matrices offering good cell attachment apart from their antibacterial activity and is an ideal for wound healing.

2.2.3 Iron Oxide Nanoparticles

Magnetic materials, especially iron oxides nanoparticles, are known since ancient times to have many spectacular properties, but in the last decade the properties that they possess at nanometric scale have been the starting point of great potential applications such as drug delivery, magnetic cell separation, tumor labeling and cell labeling. The most common forms of iron oxides, magnetite and maghemite (Fe_3O_4, $-Fe_2O_3$) are studied due to the outstanding properties they exhibit at nanometric scale (high specific surface area, super paramagnetism, etc.) (Laurent et al. 2008, Mornet et al. 2004, Prodan et al. 2013, Katz et al. 2004, Laurent et al. 2011).

2.2.4 Other Metal Oxides

Hydrophobic nature is particularly important in textiles because cottons and synthetic textiles build up static charges due to moisture. Recently, nanometric oxides like TiO_2, SiO_2, and ZnO and their hybrids were found to be successful in producing hydrophobic, eco-friendly functional coatings (Fei et al. 2006, Wang et al. 2006, Xu and Xie 2003). In addition to the antimicrobial, antifungal, and antistatic characteristics, they also offer UV-blocking property (Wang et al. 2004, Vigneshwaran et al. 2006, Farouk et al. 2009). UV shielding sunscreen surface coatings on textiles help to protect human skin from sun burn and skin-cancer.

Moreover unlike TiO_2 nanoparticles, the ZnO nanocrystals can be easily grown by varieties of chemical techniques (Sanoop et al. 2012). Nano TiO_2/Ag, ZnO/SiO_2/polyester hybrids, and ZnO-starch composites are being tested as UV shielding textile coatings. ZnO/TiO_2 nanocomposites are better candidate for UV active self-cleaning surface coatings.

Concerning the photoactivity, TiO_2 is more effective only when the TiO_2 precursor coatings are heat treated at 400°C (Xu et al. 2010, Inagaki et al. 2003). Functional nano-ZnO has advantages over TiO_2 because ZnO can block UV in the entire ranges (UV-A, UV-B, and UV-C). It also exhibits antibacterial property in neutral pH even with low quantity of ZnO.

2.3 Ceramic Nanobiomaterials

Ceramics are also unique biomaterials used for repairing and regenerating several parts of the human body (Vallet-Regi 2014). Modern ceramic substrates and packages are sophisticated combinations of glasses, ceramics, and metals that can form compact cost-effective solutions for a variety of applications such as analytical tools, nanoimaging, nanomaterials and nano-devices, novel therapeutics and drug delivery systems, clinical, regulatory and toxicological issues (Balasubramanian et al. 2017). Among the varieties of nanomaterials, nanostructured ceramics, cements and coatings are being considered for major applications in orthopaedic and dental treatments.

Biocompatible ceramics, also known as bioceramics, include both macro and nano materials mainly used for bone, teeth and other medical applications (Juhasz and Best 2012). By understanding the potential biomedical applications of bioceramics, the major insights for the future developments can be provided.

2.3.1 Properties of Ceramic Nanobiomaterials

➢ Ceramics are compounds between metallic and non-metallic elements; they are most frequently oxides, phosphates, nitrides, and carbides.

➢ These materials are typically insulative to electricity and heat, and are highly resistant to harsh chemical environments than metals and polymers.

➢ With regard to their mechanical behavior, ceramics are very hard and brittle.

➢ At nanoscale also, ceramic materials exhibit higher hardness, excellent heat and corrosion resistance, and electrical insulation properties (Xing-Jie et al. 2013).

➢ In addition to these properties, fine ceramics (also known as "advanced ceramics") have many advanced mechanical, electrical, electronic, magnetic, optical, chemical and biochemical characteristics.

➢ The bioceramics have good biocompatibility, osteoconductivity, osteoinductivity, biodegradability, resorbability, and hydrophilicity.

2.3.2 Types of Ceramic Nanobiomaterials

Based on their inherent properties (Doremus 2002),

a) Bioactive ceramics (CaP, HAP, Bioactive Glass (BAG), and Glass Ceramics (GC)) which form direct chemical bonds with bone or even soft tissues of living systems

b) Bioresorbable ceramics (TCP) that actively participate in the metabolic process of an organism

c) Bioinert high strength ceramics (alumina and zirconia)

Based on their applications,

a) Cardiovascular biomaterials

b) Dental biomaterials

c) Orthopedic biomaterials

d) Biomaterials to promote tissue generation

2.3.2.1 Hydroxyapatite (HAP)

The hydroxyapatite has a few favorable bioactive and osteoconductive properties which help in rapid bone formation, with a strong biological fixation to bony tissues, dentistry and orthopaedics (Roeder et al. 2008, Mendelson et al. 2010). It also has very low mechanical strength and fracture toughness, which is an obstacle to its applications in load-bearing areas (Choi et al. 1998).

HAP can incorporate the drug molecules either physically or chemically so that the drug retains interact until it reaches to the target site. It could also gradually degrade and then deliver the drug in a controlled manner over time (Uskokovic et al. 2011).

Due to its outstanding properties, the HAPs are employed in a variety of applications. They are (Kantharia et al. 2014),

• Bone tissue engineering, bone void fillers for orthopaedic, traumatology, spine, maxillofacial and dental surgery, orthopedic and dental implant coating,

• Repair of mechanical furcation perforations and apical barrier formation,

• Desensitizing agent in post teeth bleaching,

• Early carious lesions treatment and drug and gene delivery,

• Osteoblast and dental adhesion (Balasundarama et al. 2006, Shojai et al. 2010),

• Repair of dental enamel (Li et al. 2008),

• Controlled-release carrier of bone morphogenetic protein (Xie et al. 2010),

• Intracellular bio-imaging (Wu et al. 2011),

• Photocatalytic applications (Yang et al. 2010),

• Antibacterial applications (Lin et al. 2007),

• Drug Delivery Systems.

2.3.2.2 Alumina Ceramics

Technical Ceramics mainly includes aluminum oxides. It has very outstanding physical stability due to its high melting point (2050°C), and the highest hardness among oxides. It is strong and heat-resistant. It is the most widely used as the best-known fine ceramic material. The major applications of aluminum oxide ceramics include

- Dentistry
- Anthroplasty
- Treatment of hand and elbow fractures and
- Antimicrobial activities (Seil and Webster 2012).

2.3.2.3 Zirconia Ceramics

Zirconia Ceramics include Zirconium, Zirconium dioxide ceramics and Zirconia ceramics. Zirconium dioxide nanoparticles appear in the form of a white powder. It is non-magnetic, highly resistant against acids, has high thermal stability, and excellent compatibility with bones and surrounding connective tissue. The zirconium dioxide nanoparticles are non-toxic to organisms. The field of dentistry makes use of its special properties for manufacturing corona frames and bridge frames, tooth root studs, and metal-free dental implants.

The major applications include

- Coatings for metallic orthopaedic implants (Wang et al. 2010)
- Stabilizing hydroxyapatite (Evis et al. 2006)
- Dentistry (Puigdollers et al. 2016).

2.3.2.4 Silicon

Adding a material such as silica can also improve the bioactivity and biocompatibility of chitin. Madhumathi et al. (2010) have reported that the prepared chitin/nanosilica composite has very good bioactivity, swelling ability, and cytotoxicity *in vitro* studies, which is beneficial for bone tissue engineering.

2.3.2.4.1 Silicon Substituted HAP

Silicon (Si) substitution in the crystal structures of calcium phosphate (CaP) ceramics such as hydroxyapatite (HAP) and tricalcium phosphate (TCP) generates materials with superior biological performance. Silica is an essential trace element required for healthy bone and connective tissues.

Silica has direct effects on the physiological processes in the skeletal tissue (Pietak et al. 2007, Porter et al. 2003, Bohner 2009). The two main applications of silica-based materials in medicine and biotechnology are seen in bone-repairing devices and for drug delivery systems (Vallet-Regí and Balas 2008, Thian et al. 2007, Cerruti 2012).

2.3.2.5 Other Ceramics

Recent studies include the development of bioactive glass ceramics for bone regeneration. Development of sintered Na-containing bioactive glasses, borate-based bioactive glasses, and those doped with trace elements like Cu, Zn and Sr have been employed for bone tissue engineering (Chen et al. 2012).

Currently, composites of polymers and ceramics are being developed with the aim to increase the mechanical scaffold stability and to improve tissue interaction. These ceramic nanobiomaterials are mainly used in bone tissue engineering, dentistry and hip joint replacements (Gangadoo et al. 2015). Magnetic oxides are also used in association with ceramic nanomaterials for therapeutic drug, gene and radionuclide delivery, cancer therapy and as contrast enhancing agents (Pankhurst et al. 2003).

The technical ceramics are divided into oxides and non-oxides like aluminium oxide, ceramics, carbide ceramics, nitride ceramics, oxide ceramics, silicon carbide ceramics, silicon nitride ceramics, and zirconium ceramics dioxide. These ceramics can be used to make surgical implements and implants, such as hip joint replacements that can stay in place for up to 20 years.

3. Interaction of Bionanomaterials with Biological Systems

Nanoparticles (NPs) have unique properties with its size, shape and surface morphology that may be useful in a diverse range of applications, and consequently they have attracted significant interest due to its controlling physical, chemical, biological, optical and electronic properties. Particularly in the bio-medical field, the use of nano vaccines and nano drugs is being intensively researched.

Although the knowledge about bio-compatibility and risks of exposure of nanomaterials to human systems is limited. Human exposure to nanoparticles can be inadvertent, as in occupational exposure, or intentional, as in the usage of consumer products with nanotechnology. There is an increasing number of studies that demonstrate adverse effects of nanomaterials in *in-vitro* cellular systems. But it is unclear whether the available data can be reliably extrapolated to predict the adverse effects of nanotechnology for humans. Any toxic effects of nanomaterials will be specific to the type of base material, size, shape and coatings. Hence, there is a need to understand the molecular mechanisms of nanoparticles-to-biological system interaction. In a biological medium, NPs may interact with bio-molecules such as proteins, nucleic acids, lipids and even biological metabolites due to their nano-size and large surface-to-mass ratio.

From the biological viewpoint, the majority of microorganisms such as bacteria and virus appears in different shapes. Qi et al. (2004) and Ma and Lim (2003) investigated and reported that the higher antibacterial activity and antifungal activity of nanochitosan was exhibited by the interaction of negatively charged surface of the bacterial cell wall with the polycationic nanochitosan (NH_3^+). This interaction hindered the growth of the microorganism, which then changed their metabolism and led to cell death (Leceta et al. 2013, Zheng and Zhu et al. 2003); therefore, it suggested that surface charge is an important parameter for the interaction of nanoparticles with biological systems.

The particle size is the prime factor for all biological applications. It determines the half-life of drug clearance in tissues (Jin et al. 2014). The remarkable factors in the competence of most of the drug delivery system are particle size and surface area. When the particles size of the nanoparticles gets smaller, the surface area of the nanoparticles gets larger, which results in the adherence of the drug molecules to the surface of the particles. Shape also plays a major role in not only nanoparticle pharmacokinetics, but also intravascular transport, binding, and accumulation at the site of a tumor. Hence, the basic ideas about nano-bio interactions will be helpful in engineering the nanostructured material for the specific design of carrier to ensure the highest possible delivery efficiency.

Of particular importance is the adsorption of proteins on the nanoparticle surface. The formation of nanoparticle-protein complexes is commonly referred to as the nanoparticle-protein corona (NP-PC). A number of consequences of protein adsorption on the NP surface can be speculated. Overall, the NP-PC can influence the biological reactivity of the NP (Casals et al. 2010, Cedervall et al. 2007).

4. Interaction of Nanoparticles with Protein

NPs interact with cells via the protein (Shruti et al. 2013).

(A) Uptake of large sized NP-protein complexes, agglomerates of NP may be ingested by specialized cells such as macrophages and neutrophils via phagocytosis. It involves folding of the plasma membrane over the NP complex to form the phagosome.

(B) Non-specific uptake of extracellular fluid containing aggregates of NP may also be taken up by cells via macropinocytosis which involves ruffling of the plasma membrane to form vesicles which ultimately fuse to form lysosomes.

(C) Endocytosis of NP complexes may also be directed by specific receptors involving formation of caveolae that are plasma membrane indentations consisting of cholesterol binding proteins called caveolins.

(D) Clathrin-coated vesicles.

(E) Apart from these, other endocytic mechanisms, independent of clathrin or caveolae, may also facilitate uptake of NP.

The NP-PC dictates the overall biological reactivity of the otherwise inorganic NP surface. Understanding the dynamics of this complex interaction can thus provide useful insights into cytotoxic, inflammatory potential and other key properties of the novel materials that can be explored for developing safer and value added nanomaterials for future applications.

5. Applications

Nanobiomaterials exhibit distinctive characteristics, including mechanical, electrical, and optical properties, which make them suitable for a variety of biological applications. Because of their versatility, they are poised to play a central role in nanobiotechnology and make significant contributions to biomedical research and health care (Patra et al. 2018).

The properties and behaviors of nanobiomaterials therefore allow the diagnosis, monitoring, treatment and prevention of diseases, such as cardiovascular diseases, cancer, musculoskeletal and inflammatory conditions, neurodegenerative and psychiatric diseases, diabetes and infectious diseases. Due to the specific properties such as solubility (for otherwise insoluble drugs), carriers for hydrophobic entities, multifunctional capability, active and passive targeting, ligands (size exclusion) and reduced toxicity, nanobiomaterials have the potential to detect diseases (as imaging tools), deliver treatments and allow prevention in new ways. In addition, some nanobiomaterials imparted with electrical, optical, magnetic, or antimicrobial properties can satisfy the requirement for the regeneration of some special tissues such as cardiac, neural, and skin tissues (Hasan et al. 2018).

Ceramic nanobiomaterials, especially calcium phosphate, are used mainly as bone substitutes or scaffolds due to their good biocompatibility and compositional similarity with the inorganic components of human bone (Ratner et al. 2004). The major applications of calcium phosphates are found to be:

a) Bone Grafts, Bone Reconstruction and Repair: Bone-grafting is usually required to stimulate bone-healing. In addition, spinal fusions, filling defects following removal of bone tumors and several congenital diseases may require bone grafting (Schulz-Siegmund et al. 2011, Banerjee 2011, Pietrzak 2008).

b) Bioactive coatings for orthopaedic implants (Zhang et al. 2014, Catledge et al. 2002).

c) Controlled Drug Release: Ceramics can be effective carriers of bioactive peptide or bone cell seeds and are therefore potentially useful in tissue engineering and drug delivery.

d) Gene delivery and gene therapy (Roy et al. 2003, Maitra 2005).

e) Dentistry.

f) Bone tissue regeneration.

6. Commercial Prospects

Nanobiomaterials used in commercial products, especially ceramic nanobiomaterials, have excellent applications in several areas of health and medicine, which have a promising future in all commercial applications of nanotechnology.

7. Future Challenges

Natural based and biologically derived nanobiomaterials are still in their limits but upward strides are being made to advance and enhance protocols for clinical applications and continued research in both biomedical engineering and biochemical engineering will be needed to understand the potential of this type of nanobiomaterials for medicine and surgery.

Moreover, process development is needed to enable reliable, cost-effective, scaled-up fabrication of bioderived polymers with desired physical, mechanical, chemical, and biological properties. Many types of naturally derived materials are on the horizon for clinical medicine. Synthesizing new polymers using monomers obtained from agricultural resources is one avenue for future innovation. Agricultural resources such as cellulose may also provide useful starting materials for implantable medical devices. Moreover, additional polymers derived from microbial production are under exploration.

8. Conclusion

In summary, the nanobiomaterials are recent trends in the nanotechnology field, and their types are discussed with their appropriate examples. In all the above discussions, ceramics have played a vital role and their applications are more. Therefore, this review discussed more about the different nanobiomaterials and their applications in various fields. There is still room for new contributions and learning through study.

References

Aaron, J., N. Nitin, K. Travis, Sonia Kumar, Tom Collier, Sun Young Park, Miguel José-Yacamán, Lezlee Coghlan, Michele Follen, Rebecca Richards-Kortum and Konstantin Sokolov. 2007. Plasmon resonance coupling of metal nanoparticles for molecular imaging of carcinogenesis *in vivo*. Journal of Biomedical Optics 12: 034007.

Amit Bhatnagar and Mika Sillanpää. 2009. Applications of chitin- and chitosan-derivatives for the detoxification of water and wastewater—A short review. Adv. Colloid Interface Sci. 152: 26–38.

Anitha, A., V.V. Divya Rani, R. Krishna, V. Sreeja, N. Selvamurugan, S.V. Nair, H. Tamura and R. Jayakumar. 2009. Synthesis, characterization, cytotoxicity and antibacterial studies of chitosan, O-carboxymethyl and N, O-carboxymethyl chitosan nanoparticles. Carbohydr. Polym. 78: 672–7.

Argyropoulos, D.S., H. Sadeghifar, C. Cui and S. Sen. 2014. Synthesis and characterization of poly(arylene ether sulfone) kraft lignin heat stable copolymers. Chem. Eng. 2: 264–271.

Bailey, R.C., J.M. Nam, C.A. Mirkin and J.T. Hupp. 2003. Real-time multicolor DNA detection with chemoresponsive diffraction gratings and nanoparticle probes. J. Am. Chem. Soc. 125: 13541–13547.

Balasubramanian, S., B. Gurumurthy and A. Balasubramanian. 2017. Research biomedical applications of ceramic nanomaterials: A review. Int. J. Pharm. Sci. Res. 8: 4950–59. Doi: 10.13040/IJPSR.0975-8232.8(12).4950-59.

Balasundarama, G., M. Satoa and T.J. Webstera. 2006. Using hydroxyapatite nanoparticles and decreased crystallinity to promote osteoblast adhesion similar to functionalizing with RGD. Biomaterials 27: 2798–2805.

Banerjee, R. 2011. Nanobiomaterials for Ocular Applications. Nanobiomaterials Handbook, CRC Publication. 12.

Bin Fei, Zhaoxiang Deng, John H. Xin, Yihe Zhang and Geoffrey Pang. 2006. Room temperature synthesis of rutile nanorods and their applications on cloth. Nanotechnology 17: 1927.

Bode, H., K. Kerkhoff and D. Jendrossek. 2001. Bacterial degradation of natural and synthetic rubber. Biomacromolecules 2: 295–303.

Bohner, M. 2009. Silicon-substituted calcium phosphates—A critical view. Biomaterials 30: 6403–6406.

Bumgardner, J.D., J.L. Ong and Y. Yang. 2007. The effect of cross-linking of chitosan microspheres with genipin on protein release. Carbohydr. Poly. 68: 561–567.

Casals, E., T. Pfaller, A. Duschl, G.J. Oostingh and V. Puntes. 2010. Time evolution of the nanoparticle protein Corona. ACS Nano. 4: 3623–3632.

Catledge, S.A., M.D. Fries, Y.K. Vohra, W.R. Lacefield, J.E. Lemons, S. Woodard and R. Venugopalan. 2002. Nanostructured ceramics for biomedical implants. J. Nanosci. Nanotechnol. 2: 293–312.

Cedervall, T., I. Lynch, M. Foy, T. Berggad, S. Donnelly, G. Cagney, S. Linse and K. Dawson. 2007. Detailed identification of plasma proteins adsorbed on copolymer nanoparticles. Angew. Chem. Int. Ed. 46: 5754–5756.

Cerruti, M. 2012. Surface characterization of silicate bioceramics-Review. Phil. Trans. R. Soc. A 370: 1281–1312.

Chen, Q., C. Zhu and G.A. Thouas. 2012. Progress and challenges in biomaterials used for bone tissue engineering: Bioactive glasses and elastomeric composites. Progress in Biomaterials 1: 22.

Cheng, Y., A.C. Samia, J.D. Meyers, I. Panagopoulos, B. Fei and C. Burda. 2008. Highly efficient drug delivery with gold nanoparticle vectors for *in vivo* photodynamic therapy of cancer. Journal of the American Chemical Society 130: 10643–10647.

Choi, J.-H., K.-J. Jeong, Y.J. Dong, J.H. Han, T.-H. Lim, Jae-S. Lee and Y.-E. Sung. 2006. Electro-oxidation of methanol and formic acid on PtRu and PtAu for direct liquid fuel cells. J. Power Sources 163: 71–75.

Choi, J.W., Y.M. Kong, H.E. Kim and I.S. Lee. 1998. Reinforcement of hydroxyapatite bioceramic by addition of Ni_3Al and Al_2O_3. Journal of the American Ceramic Society 81: 1743–1748.

Croisier, F. and C. Jérôme. 2013. Chitosan-based biomaterials for tissue engineering. Eur. Polym. J. 49: 780–92.

Czaja, W.K., David J. Young, Marek Kawecki and R. Malcolm Brown. 2007. The future prospects of microbial cellulose in biomedical applications. Biomacromolecules 8: 1–12.

Daniel, J.R., R.L. Whistler and H. Roper. 2007. Starch. *In*: Wiley (ed.). Ullmann's Encyclopedia of Industrial Chemistry (VCH Verlag GmbH & Co, 2000).

Dev, A. Jithin, C. Mohan, V. Sreeja, H. Tamura, G.R. Patzke, F. Hussain, S. Weyeneth, S.V. Nair and R. Jayakumar. 2010. Novel carboxymethyl chitin nanoparticles for cancer drug delivery applications. Carbohydr. Polym. 79: 1073–1079.

Doremus, Rh. 1992. Bioceramics. Journal of Materials Science 27: 285–297.

Drexler, K.E. 1981. Molecular engineering: An approach to the development of general capabilities for molecular manipulation. Proc. Natl. Acad. Sci. USA 78: 5275–5278.

Drexler, K.E. 1986. Engines of Creation: The Coming Era of Nanotechnology; Doubleday: New York City, NY, USA.

Dutta, D.K., B.J. Borah and P.P. Sarmah. 2015. Recent advances in metal nanoparticles stabilization into nanopores of montmorillonite and their catalytic applications for fine chemicals synthesis. Catal. Rev. 57: 257–305.

El-Sayed, I.H., X. Huang and M.A. El-Sayed. 2005. Surface plasmon resonance scattering and absorption of anti-EGFR antibody conjugated gold nanoparticles in cancer diagnostics: Applications in oral cancer. Nano Letters 5: 829–834.

Evis, Z., M. Sato and T.J. Webster. 2006. Increased osteoblast adhesion on nanograined hydroxyapatite and partially stabilized zirconia composites. J. Biomed. Mater. Res. A 78: 500–507.

Farouk, A., T. Textor, E. Schollmeyer, A. Tarbuk and A.M. Grancacic. 2009. Sol-gel derived inorganic-organic hybrid polymers filled with ZnO nanoparticles as ultraviolet protection finish for textiles. Autex Res. J. 9: 114.

Feynman, R.P. 1959. Plenty of room at the bottom. Am. Phy. Soc. Available online: http://www.pa.msu.edu/~yang/RFeynman_plentySpace.pdf (accessed on 30 June 2016).

Gangadoo, S., W. Andrew, T. Robinson and J. Chapman. 2015. From replacement to regeneration: Are bionanomaterials the emerging prospect for therapy of defective joints and bones? J. Biotechnol. Biomaterials 5: 2.

George N. Agrios. 2005. How pathogens attack plants. Plant Pathology (Fifth Edition), 175–205.

Hasan, A., M. Morshed, A. Memic, S. Hassan, T. Webster and H. Marei. 2018. Nanoparticles in tissue engineering: Applications, challenges and prospects. International Journal of Nanomedicine 13: 5637–5655.

Haustrup, N. and G.M. O'Connor. 2011. Nanoparticle generation during laser ablation and laser-induced liquefaction. Physics Procedia 12: 46–53.

Heidari, F., Mehdi Razavi, Mohammad E. Bahrololoom, Reza Bazargan-Lari, Daryoosh Vashaee, Hari Kotturi and Lobat Tayebi. 2016. Mechanical properties of natural chitosan/hydroxyapatite/magnetite nanocomposites for tissue engineering applications. Mater. Sci. Eng. C 65: 338–44.

Heidari, F., Mehdi Razavi, Mohammad E. Bahrololoom, Mohammadreza Tahriri, Morteza Rasoulianborroujeni, Hari Koturi and Lobat Tayebi. 2016. Preparation of natural chitosan from shrimp shell with different deacetylation degree. Mater. Res. Innovations, 1–5.

Hirano, S. and N. Nagano. 1989. Effects of chitosan, pectic acid, lysozyme, and chitinase on the growth of several phytopathogens. Agric. Biol. Chem. 53: 3065–6.

Holban, A.M., M.C. Gestal and A.M. Grumezescu. 2014. New molecular strategies for reducing implantable medical devices associated infections. Current Med. Chem. 21: 3375–3382.

Holban, A.M., M.C. Gestal and A.M. Grumezescu. 2016. Control of biofilm-associated infections by signaling molecules and nanoparticles. Int. J. Pharm. 510: 409–18.

Hon, D.N.S. 1996. Chemical Modification of Lignucellulosic Material, Library of congress Cataloging in Publication Data, ed., 129–150.

Imre DÉKÁNY. 2011. Titanium dioxide and gold nanoparticle for environmental and biological application. Annals of Faculty Engineering Hunedoara; Tome IX, Fascicule 1: 161–166.

Inagaki, M., Y. Hirose, T. Matsunaga, T. Tsumura and M. Toyoda. 2003. Carbon coating of anatase-type TiO2 through their precipitation in PVA aqueous solution. Carbon 41: 2619.

Jane, J. 1995. Starch properties, modifications, and applications. Journal of Macromolecular Science, Part A: Pure and Applied Chemistry 32: 751–757.

Javier, D.J., N. Nitin, M. Levy, A. Ellington and R. Richards-Kortum. 2008. Aptamer-targeted gold nanoparticles as molecular specific contrast agents for reflectance imaging. Bioconjugate Chemistry 19: 1309–1312.

Jayakumar, R., Deepthy Menon, K. Manzoor, S.V. Nair and H. Tamura. 2010a. Biomedical applications of chitin and chitosan based nanomaterials—A short review. Carbohydr. Polym. 82: 227–32.

Jayakumar, R., M. Prabaharan, S.V. Nair and H. Tamura. 2010b. Novel chitin and chitosan nanofibers in biomedical applications. Biotechnol. Adv. 28: 142–50.

Jia, S., B.J. Cox, X. Guo, Z.C. Zhang and J.G. Ekerdt. 2011. Hydrolytic cleavage of beta-O-4 ether bonds of lignin model compounds in an ionic liquid with metal chlorides. Ind. Eng. Chem. Res. 50: 849–855.

Jin, R., B. Lin, D. Li and H. Ai. 2014. Super paramagnetic iron oxide nanoparticles for MR imaging and therapy: Design considerations and clinical applications. Current Opinion in Pharmacology 18: 18–27.

Jos, A., S. Pichardo, M. Puerto, E. Sanchez, A. Grilo and A.M. Camean. 2009. Cytotoxicity of carboxylic acid functionalized single wall carbon nanotubes on the human intestinal cell line caco-2. Toxicol. *In Vitro* 23: 1491–1496.

Juhasz, J.A. and S.M. Best. 2012. Bioactive ceramics: Processing, structures and properties. J. Mater. Sci. 47: 610–624.

Kantharia, N., S. Naik, S. Apte, M. Kheur, S. Kheur and B. Kale. 2014. Nano-hydroxyapatite and its contemporary applications. J. Dent. Res. Sci. Develop. 1: 15.

Katz, E. and I. Willner. 2004. Integrated nanoparticle-biomolecule hybrid systems: synthesis, properties, and applications. Angewandte Chemie 43: 6042–6108.

Kim, F., J.H. Song and P. Yang. 2002. Photochemical synthesis of gold nanorods. Journal of American Chemical Society 124: 14316–14317.

Klemm, D., Dieter Schumann, Ulrike Udhardt and Silvia Marsch. 2001. Bacterial synthesized cellulose—artificial blood vessels for microsurgery. Prog. Polym. Sci. 26: 1561–603.

Kumar, P.T., Abhilash, S., Koyakutty Manzoor, Nair, S.V., Tamura Hiroshi and Jayakumar Rakshana. 2010. Preparation and characterization of novel β-Chitin/nano silver composite scaffolds for wound dressing applications. Carbohydrate Polymers 80: 761–767. 10.1016/j.carbpol.2009.12.024.

Lara, H.H., D.G. Romero-Urbina, C. Pierce, J.L. Lopez-Ribot, M.J. Arellano-Jimenez and M. Jose-Yacaman. 2015. Effect of silver nanoparticles on candida albicans biofilms: An ultrastructural study. J. Nanobiotechnol. 13: 91.

Laurent, S., D. Forge, M. Port, Alain Roch, Caroline Robic, Luce Vander Elst and Robert N. Muller. 2008. Magnetic iron oxide nanoparticles: Synthesis, stabilization, vectorization, physic chemical characterizations and biological applications. Chemical Reviews 108: 2064–2110.

Laurent, S., S. Dutz, U.O. Häfeli and M. Mahmoudi. 2011. Magnetic fluid hyperthermia: Focus on superparamagnetic iron oxide nanoparticles. Advances in Colloid and Interface Science 166: 8–23.

Laurichesse, S. and L. Avérous. 2014. Chemical modification of lignins: Towards biobased polymers. Progress in Polymer Science 39(7): 1266–1290.

Lee, S., E.J. Cha, K. Park, Seung-Young Lee, Jin-Ki Hong, In-Cheol Sun, Sang Yoon Kim, Kuiwon Choi, Ick Chan Kwon, Kwangmeyung Kim and Cheol-Hee Ahn. 2008. A near-infrared-fluorescence quenched gold-nanoparticle imaging probe for *in vivo* drug screening and protease activity determination. Angewandte Chemie 47: 2804–2807.

Leo Stander and Louis Theodore. 2011. Environmental implications of nanotechnology—An update. Int. J. Environ. Res. Public Health 8: 470–479.

Li, L., H. Pan, J. Tao, X. Xu, C. Mao, X. Gub and R. Tang. 2008. Repair of enamel by using hydroxyapatite nanoparticles as the building blocks. J. Mater. Chem. 18: 4079–4084.

Li, Z. 2012. Research on renewable biomass resource-lignin. Journal of Nanjing Forestry University (Natural Science Edition) 36(1): 1–7.

Lian-Ying Zheng and Jiang-Feng Zhu. 2003. Study on antimicrobial activity of chitosan with different molecular weights. Carbohydrate Polymers 54(4): 527–530.

Lin, Y., Z. Yang and J. Cheng. 2007. Preparation, characterization and antibacterial property of cerium substituted hydroxyapatite nanoparticles. Journal of Rare Earths 25: 452–456.

Madhumathi, K., P.T. Sudheesh Kumar, S. Abhilash, V. Sreeja, H. Tamura, K. Manzoor, S.V. Nair and R. Jayakumar. 2010. Development of novel chitin/nanosilver composite scaffolds for wound dressing applications. J. Mater. Sci.: Mater. Med. 21: 807–813.

Madu, A.N., P.C. Njoku, G.N. Iwuoha and U.M. Agbasi. 2011. Synthesis and characterization of gold nanoparticles using 1-alkyl, 3-methyl imidazolium based ionic liquids. International Journal of Physical Sciences 6: 635–640.

Mahmood, N., Z. Yuan, J. Schmidt and C.C. Xu. 2016. Depolymerization of lignins and their applications for the preparation of polyols and rigid polyurethane foams: A review. Renew. Sust. Energy Rev. 60: 317–329.

Maitra, A. 2005. Calcium phosphate nanoparticles-second generation nonviral vectors in gene therapy. Expert Rev. Mol. Diagn. 5: 893–905.

Medina, C., M.J. Santos-Martinez, A. Radomski, O.I. Corrigan and M.W. Radomski. 2007. Nanoparticles: Pharmacological and toxicological significance. Br. J. Pharmacol. 150: 552–558.

Mendelson, B.C., S.R. Jacobson, A.M. Lavoipierre and R.J. Huggins. 2010. The fate of porous hydroxyapatite granules used in facial skeletal augmentation. Aesthetic Plastic Surgery 34: 455–461.

Mima, S., Masaru Miya, Reikichi Iwamoto and Susumu Yoshikawa. 1983. Highly deacetylated chitosan and its properties. J. Appl. Polym. Sci. 28: 1909–1917.

Mirkin, C.A., R.L. Letsinger, R.C. Mucic and J.J. Storhoff. 1996. A DNA-based method for rationally assembling nanoparticles into macroscopic materials. Nature 382: 607–609.

Mohammed, M.A., J.T.M. Syeda, K.M. Wasan and E.K. Wasan. 2017. An overview of chitosan nanoparticles and its application in non-parenteral drug delivery. Pharmaceutics 9: 53.

Mornet, S., S. Vasseur, F. Grasset and E. Duguet. 2004. Magnetic nanoparticle design for medical diagnosis and therapy. Journal of Materials Chemistry 14: 2161–2175.

Paciotti, G.F., L. Myer, D. Weinreich, Dan Goia, Nicolae Pavel, Richard E. McLaughlin and Lawrence Tamarkin. 2004. Colloidal gold: A novel nanoparticle vector for tumor directed drug delivery. Drug Delivery 11: 169–183.

Pankhurst, Q.A., J. Connolly, S.K. Jones and J. Dobson. 2003. Applications of magnetic nanoparticles in biomedicine. J. Phys. D Appl. Phys. 36: R167–181.

Patra, J.K., G. Das, L.F. Fraceto, E.V.R. Campos, M. del, P. Rodriguez-Torres, L.S. Acosta-Torres and H.-S. Shin. 2018. Nano based drug delivery systems: Recent developments and future prospects. Journal of Nanobiotechnology 16: 2–33.

Pietak, A.M., J.W. Reid, M.J. Stott and M. Sayer. 2007. Silicon substitution in the calcium phosphate bioceramics—Review. Biomaterials 28: 4023–4032.

Pietrzak, W.S. 2008. Musculoskeletal tissue regeneration: Biological materials and methods. New Jersey: Humana Press, 161–162.

Plenty of room' revisited. 2009. Nat. Nanotech. 4: 781.

Porter, A.E., N. Patel, J.N. Skepper, S.M. Best and W. Bonfield. 2003. Comparison of *in vivo* dissolution processes in hydroxyapatite and silicon-substituted hydroxyapatite bioceramics. Biomaterials 24: 4609–4620.

Prodan, A.M., S.L. Iconaru, C.M. Chifiriuc, Coralia Bleotu, Carmen Steluta Ciobanu, Mikael Motelica-Heino, Stanislas Sizaret and Daniela Predoi. 2013. Magnetic properties and biological activity evaluation of iron oxide nanoparticles. Journal of Nanomaterials. Article ID 893970, 7.

Puigdollers, A.R., F. Illas and G. Pacchioni. 2016. Structure and properties of zirconia nanoparticles from density functional theory calculations. J. Phys. Chem. C 120: 4392–4402.

Qi, L.F., Z.R. Xu, X. Jiang, C.H. Hu and X.F. Zou. 2004. Preparation and antibacterial activity of chitosan nanoparticles. Carbohydrate Research 339(16): 2693–2700.

Ratner, B.D., A.S. Hoffman, F.J. Schoen and J.E. Lemons. 2004. Biomaterials Science—An Introduction to Materials in Medicine. Elsevier Academic Press, San Diego, 854.

Roeder, R.K., G.L. Converse, R.J. Kane and W. Yue. 2008. Hydroxyapatite-reinforced polymer biocomposites for synthetic bone substitutes. JOM 60: 38–45.

Roy, I., S. Mitra, A. Maitra and S. Mozumdar. 2003. Calcium phosphate nanoparticles as novel non-viral vectors for targeted gene delivery. Int. J. Pharm. 250: 25–33.

Sanoop, P. K., K.V. Mahesh, K.M. Nampoothiri, R.V. Mangalaraja and S. Ananthakumar. 2012. Multifunctional ZnO-biopolymer nanocomposite coatings for health-care polymer foams and fabrics. Journal of Applied Polymer Science 000: 000–000. VC 2012 Wiley Periodicals, Inc. DOI 10.1002/app.36831.

Santosh K. Jha, Hare R. Singh and Pragya Prakash. 2017. Dietary Fiber and Human Health: An Introduction. Fiber's Interaction Between Gut Micoflora, Sugar Metabolism, Weight Control and Cardiovascular Health, Academic Press. Rodney A. Samaan (ed.). Progessive MD, Los Angeles, CA, United States.

Savage, N., T.A. Thomas and J.S. Duncan. 2007. Nanotechnology applications and implications research supported by the US environmental protection agency star grants program. J. Environ. Monit. 9: 1046–1054.

Schulz-Siegmund, M., R. Hötzel, H. Peter-Georg and M.C. Hacker. 2011. Synthesis, Processing, and Characterization of Ceramic Nanobiomaterials for Biomedical Applications. Nanobiomaterials Handbook CRC Publication, 3–1.

Seil, J.T. and T.J. Webster. 2012. Antimicrobial applications of nanotechnology: Methods and literature. Int. J. Nanomedicine 7: 2767–2781.

Shen, Y., K.J. Xiao, J.Y. Xi and X.P. Qiu. 2015. Comparison study of few-layered graphene supported platinum and platinum alloys for methanol and ethanol electro-oxidation. J. Power Sources 278: 235–244.

Shojai, M.S., M.A. Nodehi and L.N. Khanlar. 2010. Hydroxyapatite nanorods as novel fillers for improving the properties of dental adhesives: Synthesis and application. Dental Materials 26: 471–482.

Shruti R. Saptarshi, Albert Duschl and Andreas L. Lopata. 2013. Interaction of nanoparticles with proteins: Relation to bio-reactivity of the nanoparticle. Journal of Nanobiotechnology 11: 26.

Shukla, S.K., Ajay K. Mishra, Omotayo A. Arotiba and Bhekie B. Mamba. 2013. Chitosan-based nanomaterials: A state-of-the-art review. Int. J. Biol. Macromol. 59: 46–58.

Siró, I. and D. Plackett. 2010. Microfibrillated cellulose and new nanocomposite materials: A review. Cellulose 17: 459–94.

Sónia, A. and C. Carabineiro. 2010. Gold highlights at the 11th. Trends in Nanotechnology. International Conference (TNT 2010) in Braga, Portugal, 6–10. Gold Bull; 2011. Doi: 10.1007/s13404-011-0008-7.

Sun, Y. and Y. Xia. 2004. Mechanistic study on the replacement reaction between silver nanostructures and chloroauric acid in aqueous medium. Journal of the American Chemical Society 126: 3892–3901.

Sunita R. Boddu, Veera R. Gutti, Tushar K. Ghosh, Robert V. Tompson and Sudarshan K. Loyalka. 2011. Gold, silver, and palladium nanoparticle/nano-agglomerate generation, collection, and characterization. Journal of Nanoparticle Research. DOI: 10.1007/s11051-011-0566.

Suzuki, S., T. Watanabe, T. Mikami and M. Suzuki. 1990. Proceedings of the 5th international conference on chitin and chitosan. USA, 96–105.

Thian, E.S., J. Huang, S.M. Best, Z.H. Barber and W. Bonfield. 2007. Silicon-substituted hydroxyapatite: The next generation of bioactive coatings. Materials Science and Engineering: C 27: 251–256.

Tripkovic, A.V., K.D. Popovic, B.N. Grgur, B. Blizanac, P.N. Ross and N.M. Markovic. 2002. Methanol electrooxidation on supported Pt and PtRu catalysts in acid and alkaline solutions. Electrochim. Acta 47: 3707–3714.

Upton, B.M. and A.M. Kasko. 2016. Strategies for the conversion of lignin to high-value polymeric materials: Review and perspective. Chem. Rev. 116: 2275–2306.

Uskokovic, V. and T.A. Desai. 2014. *In vitro* analysis of nanoparticulate hydroxyapatite/chitosan composites as potential drug delivery platforms for the sustained release of antibiotics in the treatment of osteomyelitis. J. Pharm. Sci. 103: 567–579.

Vallet-Regi, M. 2014. Bioceramics with Clinical Applications. John Wiley and Sons Ltd., 457.

Vallet-Regí, M. and F. Balas. 2008. Silica materials for medical applications. Open Biomed. Eng. J. 2: 1–9.

Veigas, B., A.R. Fernandes and P.V. Baptista. 2014. Aunps for identification of molecular signatures of resistance. Front. Microbiol. 5: 455.

Vigneshwaran, N., S. Kumar, A.A. Kathe, P.V. Varadarajan and V. Prasad. 2006. Functional finishing of cotton fabrics using zinc oxide–soluble starch nanocomposites. Nanotechnology 17: 5087.

Vo-Dinh, T. 2007. Nanotechnology in Biology and Medicine: Methods, Devices, and Applications. CRC Press, Taylor and Francis Group, 762.

von Maltzahn, G., J.-H. Park, A. Agrawal, Nanda Kishor Bandaru, Sarit K. Das, Michael J. Sailor and Sangeeta N. Bhatia. 2009. Computationally guided photothermal tumor therapy using long-circulating gold nanorod antennas. Cancer Research 69: 3892–3900.

Wang, G., F. Meng, C. Ding, P.K. Chu and X. Liu. 2010. Microstructure, bioactivity and osteoblast behaviour of monoclinic zirconia coating with nanostructured surface. Acta Biomater. 6: 990–1000.

Wang, R.H., J.H. Xin, X.M. Tao and W.A. Daoud. 2004. ZnO nanorods grown on cotton fabrics at low temperature. Chem. Phys. Lett. 398: 250–255.

Wang, Z.Y., E.H. Han and W. Ke. 2006. Effect of acrylic polymer and nanocomposite with Nano-SiO$_2$ on thermal degradation and fire resistance of APP-DPER-MEL coating. Polym. Degrad. Stab. 91: 1937–1947.

Waseem S. Khan, M. Ceylan and R. Asmatulu. 2012. Effects of Nanotechnology on Global Warming, ASEE Midwest Section Conference, Rollo, MO., 19–21.

Wen, Y. and J.K. Oh. 2014. Recent strategies to develop polysaccharide-based nanomaterials for biomedical applications. Macromol. Rapid Commun. 35: 1819–32.

Windeisen, E. and G. Wegener. 2012. 10.15—Lignin as building unit for polymers. pp. 255–265. *In*: Krzysztof Matyjaszewski and Martin Möller (eds.). Polymer Science: A Comprehensive Reference, Elsevier. ISBN 9780080878621.

Wu, C.W., Kevin, Y. Ya-Huei, L. Yung-He, C. Hui-Yuan, Y. Eric, Y. Yusuke and L.F. Huei. 2011. Facile synthesis of hollow mesoporous hydroxyapatite nanoparticles for intracellular bio-imaging. Current Nanoscience 7: 926–931.

Xiao, M.L., L.G. Feng, J.B. Zhu, C.P. Liu and W. Xing. 2015. Rapid synthesis of a PtRu nano-sponge with different surface compositions and performance evaluation for methanol electrooxidation. Nanoscale 7: 9467–9471.

Xie, G., J. Sun, G. Zhong, C. Liu and J. Wei. 2010. Hydroxyapatite nanoparticles as a controlled-release carrier of BMP-2: Absorption and release kinetics *in vitro*. Journal of Materials Science: Materials in Medicine 21: 1875–1880.

Xing-Jie, L. 2013. Nanopharmaceutics—The potential Applications of Nanomaterials. World Scientific Publishers, 600.

Xu, Tao and C.S. Xie. 2003. Tetrapod–like nano-particle ZnO/acrylic resin composite and its multi-function property. Progress in Organic Coatings 46: 297–301. 10.1016/S0300-9440(03)00016-X.

Xu, Y., H. Wang, Q. Wei, H. Liu and B. Deng. 2010. Structures and properties of the polyester nonwovens coated with titanium dioxide by reactive sputtering. J. Coat Technol. Res. 7: 637.

Ya, O., V.V. Uryupina, V.V. Vysotskii, A.V. Matveev, V.I. Gusel'nikova and O. Roldughin. 2011. Production of gold nanoparticles in aqueous solutions of cellulose derivatives. Colloid Journal 73: 551–556.

Yang, Z.P., X.Y. Gong and C.J. Zhang. 2010. Recyclable Fe_3O_4/hydroxyapatite composite nanoparticles for photocatalytic applications. Chemical Engineering Journal 165: 117–121.

Yasukawa, T., A. Suzuki, H. Miyamura, K. Nishino and S. Kobayashi. 2015. Chiral metal nanoparticle systems as heterogeneous catalysts beyond homogeneous metal complex catalysts for asymmetric addition of arylboronic acids to α,β-unsaturated carbonyl compounds. J. Am. Chem. Soc. 137: 6616–6623.

Zhang, B.G.X., D.E. Myers, G.G. Wallace, M. Brandt and P.F.M. Choong. 2014. Bioactive coatings for orthopaedic implants-recent trends in development of implant coatings. Int. J. Mol. Sci. 15: 11878–11921.

Zhang, S., J. Li, W. Gao and Y.Q. Qu. 2015a. Insights into the effects of surface properties of oxides on the catalytic activity of Pd for C-C coupling reactions. Nanoscale 7: 3016–3021.

Zhang, S., X.T. Shen, Z.P. Zheng, Y.Y. Ma and Y.Q. Qu. 2015b. 3D graphene/nylon rope as a skeleton for noble metal nanocatalysts for highly efficient heterogeneous continuous-flow reactions. J. Mater. Chem. A 3: 10504–10511.

Zhou, W.J., Z.H. Zhou, S.Q. Song, W.Z. Li, G.Q. Sun, P. Tsiakaras and Q. Xin. 2003. Pt based anode catalysts for direct ethanol fuel cells. Appl. Catal. B Environ. 46: 273–285.

2

Principles and Applications of Nanobiotechnology

Se-Kwon Kim[1] and *R. Nithya*[2,*]

1. Introduction

1.1 Nanotechnology

The term "nanotechnology" refers to the structure, development and application of materials, the dimensions of which are approximately between 1 and 100 nanometers (Emerich and Thanos 2003, Sahoo and Labhasetwar 2003). The prefix 'nano' is derived from Greek meaning 'dwarf' or something very small and measured as 10^{-9} m (Bayda et al. 2020). On the other side, the term "biotechnology" deals with the development and other physiological processes of biological substances including microorganisms. The combination of these two technologies which constitutes "nanobiotechnology" plays a vital role in developing and implementing many useful tools in the study of biology (Fakruddin et al. 2012). The dual fields, viz., nanotechnology and biotechnology, have got some common grounds in formulating the applications which have revolutionized the biological investigations.

1.2 Nanobiotechnology

Nanobiotechnology, which is a combination of both nanotechnology and biotechnology, has emerged only recently (Fakruddin et al. 2012). In other words, nanobiotechnology can be taken as "nanotechnology applied to biological systems". Nanobiotechnology is characteristic of pervading into biology, physics, chemistry, technology, medical science and clinical sciences to some extent, and can improvise our approaches and understanding towards development of new therapeutic systems (Chhikara 2017). The study of nanobiotechnology comprises of the research on biomolecules, such as nucleic acids, peptides, proteins, carbohydrates, lipids, botanical extracts, biodegradable polymers and microorganisms which are the basic building blocks of biological systems. The study ultimately helps in modifying the structures, functions and properties of these biomolecules and can be used intensively in the field of nanotechnology (Nagamune 2017, Rahman et al. 2019).

[1] Department of Marine Science & Convergence Engineering College of Science and Technology, Hanyang University ERICA Ansan-si, Gyeonggi-do 11558, South Korea. Email: sknkim@pknu.ac.kr
[2] Department of Chemistry, Dr. M.G.R. Educational and Research Institute, Chennai-95.
* Corresponding author: nithyar.22@gmail.com

However, to fully explore the potential of nanobiotechnology, it is essential to know what bionanomaterials are, how and why they differ from other materials, how to analyze/synthesize, and finally how to organize them. Thus, we can come to an understanding of some already proven application areas. Nanobiomolecules/biomaterials are also known as natural nanoparticles, utilized in medicinal field for morphogenesis or advancement. Biomolecules incorporate huge macromolecules such as, proteins, sugars, lipids, nucleic acids and essential metabolites. Most of the organic materials, comprising of biomolecules which are mostly the natural mixes of the four components, namely oxygen, carbon, hydrogen, and nitrogen and which make up 96% of the human weights, also play a major role in the formation of nanomaterials. In this chapter, we envisage a broad perspective of the field of nanobiomaterials. The chapter also provides the reader a bird's-eye view of the various aspects of nanobiomaterials and finally concludes with an observation of future prospects.

1.3 Syntheses of Nanomaterials

Techniques available are abundant in number to synthesize different types of nanomaterials in the form of colloids, clusters, powders, tubes, rods, wires, thin films, etc. Various physical, chemical, biological and hybrid methods are available for synthesizing nanomaterials, since the subject Nanobiotechnology is a transitional one.

There are two modes of synthesis in physical and chemical methods, namely "bottom up" and "top-down" (Tiwari et al. 2008). In the "bottom-up" mode, larger nanostructures are formed with single atoms and molecules by way of assemblage. This method proves to be generating identical structures with atomic precision. However, the materials generated by human brain in this way are still much simpler than nature's complex structures. The techniques involved in bottom-up method are pyrolysis, inert gas condensation, solvo-thermal reactions, sol-gel and micellar structured media.

On the other hand, the "top-down" approach inversely breaks down large pieces of material to generate the required nanostructures. This is a most useful method for making interconnected and integrated structures like those used in electronic circuitry. The techniques followed in this method are attrition, milling, laser treatment and vaporization followed by cooling (Banerjee et al. 2017).

The biological method of synthesizing nanomaterials involves the use of microorganisms, plant extracts or enzymes, templates like DNA, membranes, diatoms and viruses, whereas the hybrid method involves electrochemical, chemical vapour deposition, polymerization and micro emulsion.

1.4 Properties of Nanomaterials

The principal norms of nanoparticles are determined by their shape, size, surface characteristics and inner structure. The properties of nanomaterials depend upon these parameters (Rösslein et al. 2017). Nanoparticles have relatively large surface area (Akbari et al. 2011) which in combination with the quantum effects can alter the properties including reactivity, strength and electrical characteristics (Shivaramakrishnan et al. 2017).

Nanoparticles differ from bulk materials due to their nano size (Guo et al. 2014). The properties of nanoparticles can also be affected by crystal structure, shape, porosity, chemical composition, aggregation, etc. When the size of a nanoparticle is smaller than the de Broglie wavelength of the charge carrier (electrons and holes) or the wavelength of light, the atomic density on the amorphous particle surface is changed, which alters its physical properties (Schmid 2004).

They have superior penetration ability, antimicrobial activity (Beyth et al. 2015, Pelgrift and Friedman 2013) and exhibit unique magnetic, optical or biological properties (Moritz 2012). They are biocompatible, non-toxic, non-immunogenic, and biodegradable (Singh et al. 2016).

The nanoparticle has an important property acting as a targeted delivery vehicle by its ability to bind itself non-covalently to the carriers and selective carrier uptake by cells or tissues in the field of medicine (Mishra 2016). Gold nanoparticles exhibit the photothermal property and hence the energy of incident light can be partially converted into heat which is utilized for hyperthermia therapy and photoacoustic imaging (Yang et al. 2009).

Carbon nanomaterials, also called carbon dots, possess low toxicity and good biocompatibility which helps bioimaging, biosensing and drug delivery (Bayda et al. 2017). Their optical and electronic properties (Yang et al. 2009) pave way to the C-dots for catalysis, energy conversion, photovoltaic devices (Li et al. 2011) and for being nanoprobes for sensitive ion detection (Liu et al. 2011). Due to these salient features of nanomaterials, they have several applications in various fields such as biomedicine, pharmaceuticals, cosmetics, environment, etc. (Loos 2015).

1.5 Classification of Nanomaterials

Nanomaterials can be composed of metals, ceramics, polymers, organic materials and composites. Nanomaterials are classified mainly based on their structure and chemical composition. Based on the structure, nanomaterials are classified into nanofibrous and porous scaffold (Boisseau and Loubaton 2011) nanoparticles, nanoclusters, nanocrystals, nanotubes, nanowires, nanorods, nanofilms, dendrimers, and micelle formations (Edelstein and Cammarata 1996).

Based on the chemical classification, nanomaterials are classified into organic and inorganic. Organic structures include polymer-based nanomaterials and nanocomposites (Cai and Xu 2011) or dendrimers (Zolnik and Sadrieh 2009) and green synthesis. Inorganic structures include metal based nanomaterials (Fadeel and Garcia-Bennett 2010), semiconductor nanomaterials, quantum dots (Chan et al. 2006, Geszke et al. 2011) and carbon-based nanomaterials (Gan and Hu 2011).

1.6 Principles of Nanotechnology

There are some principles in the field of nanotechnology which allow the elimination of hazardous potential of nanomaterial. They are based on its morphology, alternate usage of materials instead of hazardous ones, functionalization, encapsulation and usage of less quantity when more quantity is hazardous. Functionalization is the bonding of atoms or molecules of nanoparticles. Encapsulation is the process of covering the potentially hazardous nanomaterial within a material that is less hazardous. Instead of using hazardous nanomaterial, an alternate material can be used so that potential hazard of the nanomaterial is eliminated. In an unavoidable situation, if hazardous nanomaterial is used, the quantity of the nanomaterial can be minimized (Morose 2009).

2. Applications of Nanobiotechnology

The unique physico-chemical properties of nanomaterials like magnetic property, biocompatibility, non-toxicity, non-immunogenic characteristic, biodegradability, optical and biological properties make them play significant role in sectors of biomedicine and healthcare, cosmetics, paints and coatings, textiles, food and nutritional ingredients, food packaging, agrochemicals, adsorbents, etc. Here, we have broadly classified the applications of these nanomaterials under three major categories, namely Biomedical and Healthcare sector, Industrial sector and Environmental sector.

2.1 Biomedical Applications

Nanobiomaterials are used in diagnosis, monitoring and treatment of diseases like cardiovascular diseases, cancer, diabetes, etc. (Bag 2016); as biocatalysts (Wang 2006); for gene and drug delivery

(Petkar et al. 2010), targeted drug delivery; as antimicrobial agent (Hajipour et al. 2013), dental implants, bioinks (Luo et al. 2020) and so on.

The application of nanobiomaterials in the treatment of cancer is unique and amazing. Carcinoma therapy for human gastric carcinoma cells (MGC803 cells) was achieved by chitosan-hyaluronic acid hybrid nanoparticles by co-loading irinotecan (IRN) and 5-fluorouracil (5-FU) (Gao et al. 2017). Nanoformulation of curcumin with alginate (ALG), chitosan (CS), and pluronic F127 were studied as a drug delivery for targeting cancer cells (Das et al. 2010). Curcumin diglutaric acid-loaded polyethylene glycol-chitosan oligosaccharide-coated superparamagnetic iron oxide nanoparticles (CG-PEG-CSO-SPIONs) were used as superior drug carrier for targeting towards colon cancer cells (Sorasitthiyanukarn et al. 2020).

Miltefosine-loaded alginate nanoparticles drug delivery system acts as non-toxic delivery system for the treatment of candidiasis and cryptococcosis (Spadari et al. 2019). The side effects of anti HIV drug Ritonavir was improved by altering it by loading alginate nanoparticles which showed better encapsulation and release of the drug (Biswas et al. 2009). Chitosan nanoparticles (CS NPs) fabricated with dextran sulphate (DS) were prepared as drug delivery system for the protection of protein (Katas et al. 2013). The effective entrapment of a protein called bovine serum albumin measuring up to 0.24 mg per 1 mg of polysaccharide was achieved by alginate and chitosan nanoparticles (Masalova et al. 2013).

Chitosan-alginate (CS/ALG) nanoparticles were studied for the release and delivery behaviour of nifedipine by Li et al. (2008). An enhanced printable bioink was prepared by mixing cellulose nanofiber (CNF) with gelatin-alginate bioinks which was used potentially in meniscus bioprinting (Luo et al. 2020). Hydrophilic nanocomposite of carboxymethyl chitosan (CMCS) and gold nanoparticles was prepared for enzyme immobilization by *in situ* process. Horseradish peroxidase (HRP) was immobilized on the silica sol-gel matrix containing the nanocomposite of carboxymethyl chitosan (CMCS) and gold nanoparticles to form a novel H(2)O(2) biosensor (Xu et al. 2006).

2.2 Industrial Applications

The ever increasing demand for high food quality and safety, and worries over the environment sustainable development warrant researchers in the food industry to explore the strong and green biodegradable nanocomposites, which provide new opportunities and challenges for the development of nanomaterials in the food industry (Huang et al. 2018). The society calls for the production of healthy and safe food with minimum use of synthetic inputs (including synthetic preservatives). In the field of food science, the development of nanotechnology plays a unique role in almost every aspect, namely, processing, packaging, storage, transportation, functionality, and other safety aspects of food.

The biodegradable cassava starch based films incorporated with cinnamon essential oil and sodium bentonite clay nanoparticles were used as a packaging material for meatballs to lengthen the shelf life (Iamareerat et al. 2018). Food grade nanoemulsions containing polyphenols (curcumin, gallic acid and quercetin), gelatin plus carrageenan were used as packaging material which increased the shelf-life of fresh broiler meat up to 17 days and showed excellent antioxidant activity (Khan et al. 2020). Calcium-alginate coating loaded with silver-montmorillonite nanoparticles was used to prolong the shelf-life of fresh-cut carrots for more than two months (Costa et al. 2012).

Different concentrations of chitosan and chitosan nanoparticles as edible coating affect postharvest quality extended shelf life and maintained the quality of banana fruits (*Musa acuminata* AAA group); in addition to that, the sensory quality also remained the same (Lustriane et al. 2018). Fresh Mazafati dates when packed on a nanocomposite of low-density polyethylene films with carbon nanotube to check the viability of the nanocomposite revealed that although the nanocomposite increased shelf life of Mazafati dates and decreased the fungal growth, some of the sensory characteristics were lost (Asgari et al. 2014).

The vogue of the world in recent decades is the widespread usage of nanostructured materials. The processing industries tend to use coating materials with antibacterial properties. Developing antibacterial coatings using simple organic as well as chemical methods can find way to eco-friendly applications.

Silver-nanoparticle-embedded antimicrobial paints in vegetable oil in a single step showed excellent antimicrobial properties by killing both Gram-positive human pathogens (*Staphylococcus aureus*) and Gram-negative bacteria (*Escherichia coli*) (Kumar et al. 2008). Silver nanoparticles (AgNPs) with polyurethane coatings in paints showed resistance against bacteria but not satisfactory resistance against fungi (Bechtold et al. 2020).

The effects of using nano gold and nano silver have been positively confirmed and they are popularly used for cosmetic material ingredients as they are non toxic against human cells. The study under microbiological tests on the safety of the use of silver and gold nanoparticle in cosmetics showed satisfactory fungicidal properties. However, the permeability of the metallic nanoparticles through skin ranges from 110–200 mg/kg, which showed the toxic properties of metallic nanoparticles for living organisms, which is a matter of concern and needs to be weeded out (Jolanta Pulit et al. 2019).

2.3 Environmental Applications

In addition to the means of application of nanotechnology as so far seen, namely, biomedical and industrial applications, it can also be efficiently used in detecting and cleaning up the environmental contaminants. Nanotechnology is the most useful application in the field of removing impurities in water. The abundant surface area, quantum and the capacity to affect molecular modification, etc., render nanoparticles ideal for metal removal in water treatment process. It can cater to the needs of the society by affording clean drinking water at minimal cost.

Nanoparticles were prepared from silver, iron, and chitosan by chemical processes, thus producing silver nanoparticle (Ag-NPs) from silver nitrate, nano iron (nZVI) by reduction of ferric chloride and chitosan nanoparticle (CS NPs) by degradation and by treatment with $AgNO_3$, respectively. Ag-NPs is capable of destroying *E. coli* bacteria, thus proving an antibacterial agent. CS NPs and nZVI were used to remove Cu(II) ions from aqueous solution (Naim et al. 2017). The removal of As(V), one of the most poisonous groundwater pollutants, by synthetic nanoscale zerovalent iron (NZVI) was studied (Kanel et al. 2006).

The nanoparticles prepared from Euphorbia macroclada act as bioaccumulator plant for removing heavy metals like Pb, Zn, Cu and Ni (Mohsenzadeh and Chehregani 2011). Multiwalled, single-walled, and hybrid carbon nanotubes removed ethylbenzene from aqueous solution at neutral pH, where it was analyzed that single-walled carbon nanotubes (SWCNTs) could remove ethylbenzene more efficiently than the other two (Bina et al. 2012).

Nanochitosan had higher removal efficiency for zinc metal ions due to their nano size, larger surface area and more functional groups (Seyedmohammadi et al. 2016). Chitosan nanorods were prepared and used as efficient biosorbent for removal of chromium (Sivakami et al. 2013). Nanocrystalline cellulose (NCC) was prepared by synthetic modifications on to cellulose for the removal 94.84% of Cr(III) and 98.33% of Cr(VI) (Singh et al. 2014).

A new type of nanocomposite microgel based on nanocellulose and amphoteric polyvinylamine (PVAm) was prepared which is usable in the removal of anionic dyes (Jin et al. 2015). Nano-cellulose hybrids containing polyhedral oligomeric silsesquioxane with multi-N-methylol (R-POSS) was potentially used as novel biosorbents for reactive dyes (Xie et al. 2011). Cellulose nanocrystal-alginate (CNC-ALG) hydrogel beads were found to be efficient in adsorption of methylene blue in consonance with Langmuir adsorption isotherm (Mohammed et al. 2016).

A novel polymer–inorganic nanocomposite membrane was generated from a cellulose acquired from recycled newspaper. This, in combination with TiO_2, nanorods acted as photocatalyst in wastewater treatment for removal of toxins (Mohamed et al. 2016).

3. Future Prospects and Conclusion

Nanotechnology manifests in the gigantic fields of biomedicine, industry and environment. It is looked as upcoming field with potential to yield health benefits and advantages for mankind. The present society is vested with the burden of overcoming the toxicities prevalent in our food, poison from our industries and pollution widespread in the environment. Nanotechnology is advantageous in overcoming these toxicities efficiently. However the results are not upto our expectations or we can say not upto the mark. Certain persistant nanoparticles may adversely affect human health and need to be used with careful examination.

In biomedical field, the biocompatibility and toxicity of the nanobiomaterials are of great concern. The safety of the used nanobiomaterial is baneful which determines the success rate of drug delivery. Only on confirmation that the interacted cells and the surrounding tissues are not affected by the nanobiomaterials, they become viable. Notwithstanding all these drawbacks, a suitable and result yielding phenomenon is quite essential for the society at large and the way out presently is the adherence to nanobiotechnology. Further and future studies may bring more prolific results.

References

Akbari, B., M.P. Tavandashti and M. Zandrahimi. 2011. Particle size characterization of nanoparticles—A practical approach. Iran. J. Mater. Sci. Eng. 8: 48–56.

Alan S. Edelstein and Robert C. Cammarata. 1996. Nanomaterials: Synthesis, Properties, and Applications. Taylor and Francis, Great Britain.

Asgari Parinaz, Omid Moradi and Behjat Tajeddin. 2014. The effect of nanocomposite packaging carbon nanotube base on organoleptic and fungal growth of Mazafati brand dates. Int. Nano Lett. 4: 98.

Ashavani Kumar, Praveen Kumar Vemula, Pulickel M. Ajayan and George John. 2008. Silver-nanoparticle-embedded antimicrobial paints based on vegetable oil. Nature Material. 7: 236–241.

Bayda, S., M. Hadla, S. Palazzolo, V. Kumar, I. Caligiuri, E. Ambrosi, E. Pontoglio, M. Agostini, T. Tuccinardi, A. Benedetti, P. Riello, V. Canzonieri, G. Corona, G. Toffoli and F. Rizzolio. 2017. Bottom-up synthesis of carbon nanoparticles with higher doxorubicin efficacy. J. Control. Release 248: 144–152.

Bayda Samer, Muhammad Adeel, Tiziano Tuccinardi, Marco Cordani and Flavio Rizzolio. 2020. The history of nanoscience and nanotechnology: From chemical–physical applications to nanomedicine. Molecules 25: 112.

Bechtold, M., A. Valério, A.A. Ulson de Souza, D. de Oliveira, C.V. Franco, R. Serafim, M.A. Guelli and U. Souza. 2020. Synthesis and application of silver nanoparticles as biocidal agent in polyurethane coating. J. Coat. Technol. Res. 17(3): 613–620.

Beyth, N., Y. Houri-Haddad, A. Domb, W. Khan and R. Hazan. 2015. Alternative antimi crobial approach: Nano-antimicrobial materials. Evid. Based Complement Alternat. Med. 2015: 246012.

Bina Bijan, Hamidreza Pourzamani, Alimorad Rashidi and Mohammad Mehdi Amin. 2012. Ethylbenzene removal by carbon nanotubes from aqueous solution. Journal of Environmental and Public Health 8.

Biswas Angshuman, Mahanta Rita, Bandhyopadhaya Sandip Kumar and S.K. Bhattacharjee. 2009. Development of alginate-based nanoparticulate drug delivery system for anti HIV drug Ritonavir. Journal of Pharmaceutical Research 8(2): 108–111.

Cai, X.J. and Y.Y. Xu. 2011. Nanomaterials in controlled drug release. Cytotechnology 63: 319–323.

Chan, W.H., N.H Shio and P.Z. Lu. 2006. CdSe quantum dots induce apoptosis in human neuroblastoma cells via mitochondrial-dependent pathways and inhibition of survival signals. Toxicology Letters 167: 191–200.

Chhikara, B.S. 2017. Current trends in nanomedicine and nanobiotechnology research. J. Mat. NanoSci. 4: 19–24.

Costa, C., A. Conte, G.G. Buonocore, M. Lavorgna and M.A. Del Nobile. 2012. Calcium-alginate coating loaded with silver-montmorillonite nanoparticles to prolong the shelf-life of fresh-cut carrots. Food Research International 48: 164–169.

Emerich, D.F. and C.G. Thanos. 2003. Nanotechnology and medicine. Expert Opin. Biol. Ther. 3: 655–663.

Fadeel, B. and A.E. Garcia-Bennett. 2010. Better safe than sorry: Understanding the toxicological properties of inorganic nanoparticles manufactured for biomedical application. Advanced Drug Delivery Reviews 62: 362–374.

Fakruddin, Md., Zakir Hossain and Hafsa Afroz. 2012. Prospects and applications of nanobiotechnology: A medical perspective. Journal of Nanobiotechnology 10: 31.

Gan, T. and S. Hu. 2011. Electrochemical sensors based on graphene materials. Microchimica Acta 175: 1–19.

Geszke, M., M. Murias, L. Balan, G. Medjahdi, J. Korczyński, M. Moritz J. Lulek and R. Schneider. 2011. Folic acid-conjugated core/shell ZnS:Mn/ZnS quantum dots as targeted probes for two photon fluorescence imaging of cancer cells. Acta Biomaterialia 7: 1327–1338.

Guo Dan, Guoxin Xie and Jianbin Luo. 2014. Mechanical properties of nanoparticles: Basics and applications. J. Phys. D: Appl. Phys. 47.

Hajipour, M.J., K.M. Fromm, A.A. Ashkarran, D. Jimenez de Aberasturi, I.R. de Larramendi, T. Rojo, V. Serpooshan, W.J. Parak and M. Mahmoudi. 2013. Antibacterial properties of nanoparticles. Trends Biotechnol. 31(1): 61–2.

Huang Yukun, Lei Mei, Xianggui Chen and Qin Wang. 2018. Recent developments in food packaging based on nanomaterials. Nanomaterials 8: 830.

Iamareerat Butsadee, Manisha Singh, Muhammad Bilal Sadiq and Anil Kumar Anal. 2018. Reinforced cassava starch based edible film incorporated with essential oil and sodium bentonite nanoclay as food packaging material. J. Food Sci. Technol. 55(5): 1953–1959.

Jin Liqiang, Qiucun Sun, Qinghua Xu and Yongjian Xu. 2015. Adsorptive removal of anionic dyes from aqueous solutions using microgel based on nanocellulose and polyvinylamine. Bioresour. Technol. 197: 348–55.

JolantaPulit-Prociak, Aleksandra Grabowska, Jarosław Chwastowski, Tomasz M.Majka and Marcin Banach. 2019. Safety of the application of nanosilver and nanogold in topical cosmetic preparations. Colloids and Surfaces B: Biointerfaces 183.

Kailash C. Petkar, Sandip S. Chavhan, Snezana Agatonovik-Kustrin and Krutika Sawant. 2010. Nanostructured materials in drug and gene delivery: A review of the state of the art. Crit. Rev. Ther. Drug Carrier Syst. 28: 101–164.

Katas Haliza, Maria Abdul Ghafoor Raja and Kai Leong Lam. 2013. Development of chitosan nanoparticles as a stable drug delivery system for protein/siRNA. International Journal of Biomaterials, 1–9.

Kiran Singh, Jyoti Kumar Arora, T. Jai Mangal Sinha and Shalini Srivastava. 2014. Functionalization of nanocrystalline cellulose for decontamination of Cr(III) and Cr(VI) from aqueous system: Computational modeling approach. Clean Techn. Environ. Policy 16: 1179–1191.

Li Ping, Ya-Ni Dai, Jun-Ping Zhang, Ai-Qin Wang and Qin Wei. 2008. Chitosan-alginate nanoparticles as a novel drug delivery system for nifedipine. Int. J. Biomed. Sci. 4(3): 221–228.

Li, Y., Y. Hu, Y. Zhao, G. Shi, L. Deng, Y. Hou and L. Qu. 2011. An electrochemical avenue to green-luminescent graphene quantum dots as potential electron-acceptors for photovoltaics. Adv. Mater. 23: 776–780.

Liu, L., Y. Li, L. Zhan, Y. Liu and C. Huang. 2011. One-step synthesis of fluorescent hydroxyls-coated carbon dots with hydrothermal reaction and its application to optical sensing of metal ions. Sci. China Chem. 54: 1342–1347.

Loos, M. 2015. Carbon nanotube reinforced composites. pp. 1–36. *In*: Nanoscience and Nanotechnology. William Andrew Publishing, Oxford.

Luo Wenbin, Zhengyi Song, Zhonghan Wang, Zhenguo Wang, Zuhao Li, Chenyu Wang, He Liu, Qingping Liu and Jincheng Wang. 2020. Printability optimization of gelatin-alginate bioinks by cellulose nanofiber modification for potential meniscus bioprinting. Journal of Nanomaterials 1–13.

Lustriane Cita, Fenny M. Dwivany, Veinardi Suendo and Muhammad Reza. 2018. Effect of chitosan and chitosan-nanoparticles on post harvest quality of banana fruits. J. Plant Biotechnol. 45: 36–44.

Masalova, O., V. Kulikouskaya, T. Shutava and V. Agabekov. 2013. Alginate and chitosan gel nanoparticles for efficient protein entrapment. Physics Procedia 40: 69–75.

Michał Moritz. 2012. Application of nanomaterials in medical sciences. CHEMIK 66(3): 219–226.

Mohamed, Mohamad Azuwa, W.N.W. Salleh, Juhana Jaafar, A.F. Ismail, Muhazri Abd Mutalib, N.A.A. Sani, S.E.A.M. Asri and C.S. Ong. 2016. Physicochemical characteristic of regenerated cellulose/N-doped TiO2 nanocomposite membrane fabricated from recycled newspaper with photocatalytic activity under UV and visible light irradiation. Chemical Engineering Journal 284: 202–215.

Mohammed Nishil, Nathan Grishkewich, Herman Ambrose Waeijen, Richard M. Berry and Kam Chiu Tam. 2016. Continuous flow adsorption of methylene blue by cellulose nanocrystal-alginate hydrogel beads in fixed bed columns. Carbohydr. Polym. 136: 1194–202.

Mohsenzadeh Fariba and Abdolkarim Chehregani Rad. 2011. Application of nano-particles of euphorbia macroclada for bioremediation of heavy metal polluted environments. International Conference on Nanotechnology and Biosensors. Singapore 25.

Morose, G. 2009. The 5 principles of "Design for Safer Nanotechnology". J. Clean Prod. 1: 5.

Muhammad Rehan Khan, Muhammad Bilal Sadiq and Zaffar Mehmood. 2020. Development of edible gelatin composite films enriched with polyphenol loaded nanoemulsions as chicken meat packaging material. CyTA-Journal of Food 18(1): 137–146.

Nagamune Teruyuki. 2017. Biomolecular engineering for nanobio/bionanotechnology. Nano Convergence 4: 9.

Naim, M.M., A.A. El-Shafei, M.M. Elew and A.A. Moneer. 2017. Application of silver-, iron-, and chitosan-nanoparticles in wastewater treatment. Desalination and Water Treatment 73: 268–280.

Patrick Boisseau and Bertrand Loubaton. 2011. Nanoscience and nanotechnologies: Hopes and concerns, Nanomedicine, nanotechnologies in medicine. Comptes Rendus Physique 12: 620–636.

Pelgrift, R.Y and A.J. Friedman. 2013. Nanotechnology as a therapeutic tool to combat microbial resistance. Adv. Drug Deliv. Rev. 65: 1803–15.

Rahman, K., S.U. Khan, S. Fahad, M.X. Chang, A. Abbas, W.U. Khan, L. Rahman, Z.U. Haq, G. Nabi and D. Khan. 2019. Nano-biotechnology: A new approach to treat and prevent malaria. Int. J. Nanomedicine 14: 1401–1410.

Ratul Kumar Das, Naresh Kasoju and Utpal Bora. 2010. Encapsulation of curcumin in alginate-chitosan-pluronic composite nanoparticles for delivery to cancer cells. Nanomedicine: Nanotechnology, Biology and Medicine 6(1): 153–160.

Rösslein, M., N.J. Liptrott, A. Owen, P. Boisseau, P. Wick and I.K. Herrmann. 2017. Sound understanding of environmental, health and safety, clinical, and market aspects is imperative to clinical translation of nanomedicines. Nanotoxicology 11: 147–49.

Sahoo, K.S. and V. Labhasetwar. 2003. Nanotech approaches to drug delivery and imaging. DDT 8(24): 1112–1120.

Sandip Bag. 2016. Application of nano-biomaterials in healthcare. J. Nanomed. Nanotechnol. 7: 4.

Schmid, G. 2004. Nanoparticles: From Theory to Application. Wiley-VCH. Weinheim. Germany.

Seyedmohammadi Jalal, Mohsen Motavassel, Mohammad Hossein Maddahi and Soudabeh Nikmanesh. 2016. Application of nanochitosan and chitosan particles for adsorption of Zn(II) ions pollutant from aqueous solution to protect environment Model. Earth Syst. Environ. 2: 165.

Shivaramakrishna, B., B. Gurumurthy and A. Balasubramanian. 2017. Potential biomedical applications of metallic nanobiomaterials: A review. IJPSR 8(3): 985–1000.

Sivakami, M.S., Thandapani Gomathi, Jayachandran Venkatesan, Hee-Seok Jeong, Se-Kwon Kim and P.N. Sudha. 2013. Preparation and characterization of nano chitosan for treatment wastewaters. Int. J. Biol. Macromol. 57: 204–12.

Sorasitthiyanukarn, F.N., C. Muangnoi, W. Thaweesest, P.R.N. Bhuket, P. Jantaratana, P. Rojsitthisak and P. Rojsitthisak. 2020. Polyethylene glycol-chitosan oligosaccharide-coated superparamagnetic iron oxide nanoparticles: A novel drug delivery system for curcumin diglutaric acid. Biomolecules 10(1): 73.

Spadari, C.C., F.W.M.S. de Bastiani, L.B. Lopes and K. Ishida. 2019. Alginate nanoparticles as non-toxic delivery system for miltefosine in the treatment of candidiasis and cryptococcosis. Int. J. Nanomedicine 14: 5187–5199.

Sulabha K. Kulkarni. 2014. Nanotechnology: Principles and Practices, Third Edition, Springer International Publishing.

Sundeep Mishra. 2016. Nanotechnology in medicine. Indian Heart Journal 68: 437–439.

Sushil Raj Kanel, Jean-Mark Grenèche and Heechul Choi. 2006. Arsenic(V) removal from groundwater using nano scale zero-valent iron as a colloidal reactive barrier material. Environ. Sci. Technol. 40: 2045–2050.

Sushmita Banerjee, Ravindra Kumar Gautam, Pavan Kumar Gautam, Amita Jaiswal and Mahesh Chandra Chattopadhyaya. 2016. Recent trends and advancement in nanotechnology for water and wastewater treatment: Nanotechnological approach for water purification. pp. 208–252. In: Anwar Khitab and Waqas Anwar (eds.). Advanced Research on Nanotechnology for Civil Engineering Applications. Engineering Science Reference.

Thakur Gurjeet Singh, Sunita Dhiman, Manish Jindal, Inderjeet Singh Sandhu and Mansi Chitkara. 2016. Nanobiomaterials: Applications in biomedicine and biotechnology. pp. 401–429. In: Fabrication and Self Assembly of Nanobiomaterials (ed.). Nanobiomaterials: Applications in Biomedicine and Biotechnology, Elsevier.

Tiwari, D.K., J. Behari and P. Sen. 2008. Application of nanoparticles in waste water treatment. World Applied Sciences Journal 3(3): 417–433.

Wang Ping. 2006. Nanoscale biocatalyst systems. Current Opinion in Biotechnology 17(6): 574–579.

Xie, K., W. Zhao and X. He. 2011. Adsorption properties of nano-cellulose hybrid containing polyhedral oligomeric silsesquioxane and removal of reactive dyes from aqueous solution. Carbohydrate Polymers 83(4): 1516–1520.

Xu, Q., C. Mao, N.N. Liu, J.J. Zhu and J. Sheng. 2006. Direct electrochemistry of horseradish peroxidase based on biocompatible carboxymethyl chitosan-gold nanoparticle nanocomposite. Biosensors & Bioelectronics 22(5): 768–773.

Yang, S.T., X. Wang, H. Wang, F. Lu, P.G. Luo, L. Cao, M.J. Meziani, J.H. Liu, Y. Liu, M. Chen, Y. Huang and Y.P. Sun. 2009. Carbon dots as nontoxic and high-performance fluorescence imaging agents. J. Phys. Chem. C 113: 18110–18114.

Yang, X., E.W. Stein, S. Ashkenazi and L.V. Wang. 2009. Nanoparticles for photoacoustic imaging. Wiley Interdiscip. Rev. Nanomed. Nanobiotechnol. 1: 360–368.

Zhuanglei Gao, Zhaoxia Li, Jieke Yan and Peilin Wang. 2017. Irinotecan and 5-fluorouracil-co-loaded, hyaluronic acid-modified layer-by-layer nanoparticles for targeted gastric carcinoma therapy. Drug Design, Development and Therapy 11: 2595–2604.

Zolnik, B.S and N. Sadrieh. 2009. Regulatory perspective on the importance of ADME assessment of nanoscale material containing drugs. Advanced Drug Delivery Reviews 61: 422–427.

3

Nanobiomaterials
A Scientometric Review of the Research

Ozcan KONUR

1. Introduction

Nanobiomaterials have been developed and increasingly used in biomedicine to diagnose and treat a number of diseases such as cancer (Daniel and Astruc 2004, Bruchez et al. 2013, Chan and Nie 1998, Michalet et al. 2005). Thus, these advanced materials have the utmost public importance.

However, for the efficient development and use of these nanobiomaterials, it is necessary to develop efficient incentive structures for the primary stakeholders and to inform these stakeholders about the research on the nanobiomaterials in the biomedical context (North 1990, 1991, Konur 2000, 2002a, b, c, 2006a, b, 2007a, b).

Scientometric analysis of the research offers ways to evaluate the research in a respective field (Garfield 1955, 1972). This method has been used to evaluate research in a number of research fields (Konur 2011, 2012a–n, 2017f, 2018a, b, 2019a).

Although there have been a number of studies on the nanotechnology in general (Porter et al. 2008, Schummer 2004), there have been limited number of pioneering scientometric studies on the various aspects of the nanobiomaterials in recent years (Konur 2016a–f, 2017a–e, 2019b).

This book chapter presents a study on the scientometric evaluation of the research on the nanobiomaterials within the biomedical context using two datasets. The first data set includes 100 most-cited papers whilst the second set includes over 249,000 papers published between 1980 and 2019.

The data on the indices, document types, authors, institutions, funding bodies, source titles, 'Web of Science' subject categories, key words, research fronts, and citation impact are presented and discussed.

2. Materials and Methodology

The search for the literature was carried out in the 'Web of Science' database in January 2020. It contains 'Science Citation Index Expanded' (SCI-E), Social Sciences Citation Index' (SSCI), 'Book

Formerly, Ankara Yildirim Beyazit University, Turkey. Email: okonur@hotmail.com

Citation Index–Science' (BCI-S), 'Conference Proceedings Citation Index-Science' (CPCI-S), 'Emerging Sources Citation Index' (ESCI), 'Book Citation Index-Social Sciences and Humanities' (BCI-SSH), 'Conference Proceedings Citation Index-Social Sciences and Humanities' (CPCI-SSH), and 'Arts and Humanities Citation Index' (A&HCI).

The keywords for the search of the literature are developed from the screening of abstract pages for the first 1,000 highly cited papers. Two sets of keywords are developed. The first and second keyword sets are related to biomedicine and nanotechnology, respectively. These keyword sets are provided in the appendix.

Two datasets are used for this study. The highly cited 100 papers comprise the first data set (sample data set, n = 100 papers) whilst all the papers form the second data set (population data set, n = over 249,000 papers).

The data on the indices, document types, publication years, institutions, funding bodies, source titles, countries, 'Web of Science' subject categories, citation impact, keywords, and research fronts are collected from these data sets. The key findings are provided in the relevant tables and one figure, supplemented with explanatory notes in the text. The findings are discussed and a number of conclusions are drawn and a number of recommendations for further study are made.

3. Results

3.1 Indices and Documents

There are over 309,000 papers related to nanobiomaterials in the biomedical context in the 'Web of Science' as of January 2020. This original population data set is refined for the document type (article, review, book chapter, book, editorial material, note, and letter) and language (English), resulting in over 249,000 papers comprising over 80.6% of the original population data set.

The primary index is SCI-E with 94.2% of this refined population data set. The papers on the social and humanitarian aspects of the nanobiomaterials are relatively negligible with over 2,000 and less than 100 papers for the SSCI and A&HCI, respectively. The BCI-S, CPCI-S, and ESCI form 3.2, 2.8, and 2.3% of this data set, respectively. On the other hand, the sole index for the sample data set is the SCI-E.

The brief information on the document types for both data set is provided in Table 1. The key finding is that article and review types of documents are the primary documents and they are under-represented and over-represented in the sample papers, respectively.

Table 1: Document types.

	Document Type	Sample Data Set %	Population Data Set %	Difference %
1	Article	55	90.6	−35.6
2	Review	42	7.1	+34.9
3	Book chapter	0	3.1	−3.1
4	Proc. paper	1	2.9	−1.9
5	Editorial mat.	2	1.6	+0.4
6	Letter	0	0.4	−0.4
7	Book	0	0.2	−0.2
8	Note	0	0.1	−0.1

3.2 Authors

A brief information about the most-prolific 16 authors is provided in Table 2.

The most-prolific author is 'Chad A. Mirkin' of University of the US working primarily on nanobiosensing. Eight and seven authors published 4 and 3 papers, respectively.

The most-prolific institutions for these top authors is 'Northwestern University' of the US with 4 authors. The other prolific institutions are 'Georgia Institute of Technology' of the US, 'University of California Berkeley' of the US, and 'University of Paris II' of France with 2 authors each. In total, 10 institutions house these top authors.

It is notable that 8 of these top researchers are listed in the 'Highly Cited Researchers' (HCR) in 2019 (Docampo and Cram 2019).

The most-prolific country for these top authors is the US with 13 authors. Other countries are France and Canada with two and one papers, respectively.

Table 2: Authors.

	Authors	No. Papers	Inst.	Country	Research Front	HCR Field*
1	Mirkin, Chad A.	7	Northwestern Univ.	US	Biosensing	Chem.
2	Chan, Warren C.W.	4	Univ. Toronto	Canada	Cytotoxicity, cell biology	Cross-field
3	Couvreur, Patrick	4	Univ. Paris 11	France	Biodrug delivery, cancer biotherapy	Cross-field
4	Dai, Hongjie	4	Stanford Univ.	US	Biodrug delivery	Phys., Chem., Mats. Sci.
5	El-Sayed, Mostafa A.	4	Georgia Inst. Technol.	US	Bioimaging, cancer biotherapy	
6	Langer, Robert S.	4	Mass. Inst. Technol.	US	Biodrug delivery, cancer biotherapy	Biol. Biochem., Mats. Sc., Pharm. Toxic.
7	Letsinger, Robert L.	4	Northwestern Univ.	US	Biosensing	
8	Nie, Shuming	4	Indiana Univ.	US	Biosensing, cancer biotherapy	Cross-field
9	Bruchez, Marcel P.	4	Univ. Calif. Berkeley	US	Bioimaging cancer biotherapy	
10	Alivisatos, A. Paul	3	Univ. Calif. Berkeley	US	Biosensing	Chem.
11	El-Sayed, Ivan H.	3	Georgia Inst. Technol.	US	Bioimaging, cancer biotherapy	
12	Farokhzad, Omid C.	3	Harvard Univ.	US	Biodrug delivery, cancer biotherapy	Pharm. Toxic.
13	Gref, Ruxandra	3	Univ. Paris 11	France	Biodrug delivery	
14	Mucic, Robert C.	3	Northwestern Univ.	US	Biosensing	
15	Shin, Dong M.	3	Emory Univ.	US	Bioimaging, cancer biotherapy	
16	Storhoff, James J.	3	Northwestern Univ.	US	Biosensing	

*HCR: 'Highly cited researchers'.

The most-prolific research front for these top authors is 'cancer biotherapy' with 8 authors. The other prolific research fronts are 'biosensing', 'biodrug delivery', and 'bioimaging' with 6, 5, and 4 authors, respectively. There is also one author each for 'cytotoxicity' and 'cell biology'.

It is further notable that there is a significant gender deficit among these top authors as only one researcher is female.

3.3 Publication Years

The information about the publication years for both data sets is provided in Figure 1.

This figure shows that whilst 77% of the sample papers are published in the 2000s, over 70% of the population papers are published in the 2010s. Additionally, 10% and over 2% of the sample and population papers are published in the 1990s, respectively. The research output in the 1980s is negligible for both data sets.

Similarly, the most-prolific publication years for the sample data set are 2008 and 2005 with 14 and 11 sample papers, respectively. On the other hand, the most-prolific publication years for the population data set are 2019, 2018, and 2017 with over 11, 10, and 9% of the population papers, respectively.

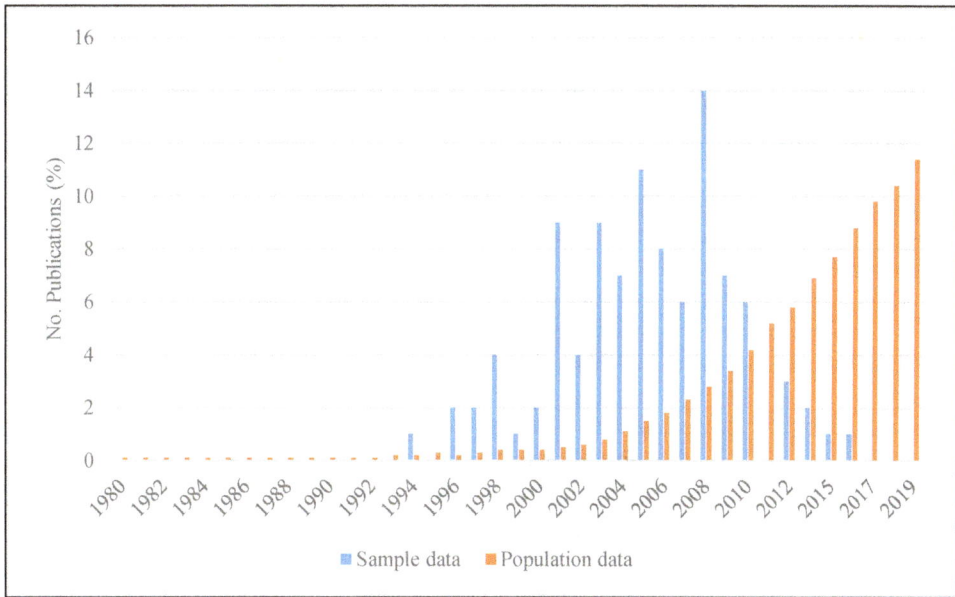

Figure 1: The research output between 1980 and 2019.

3.4 Institutions

A brief information on the top 16 institutions is provided in Table 3. In total, over 1,100 and 39,600 institutions contribute to the sample and population papers, respectively.

These top institutions publish 97.0 and 10.8% of the sample and population papers. The top institutions are 'Harvard University' and 'Northwestern University' of the US with 10 sample papers each. These institutions are significantly over-represented in the sample papers. The most-prolific countries for these top instructions are the US and France with 13 and 3 sample papers, respectively.

Table 3: Institutions.

	Institutions	Country	Sample No. Papers %	Population No. Papers %	Difference %
1	Harvard Univ.	US	10.0	1.4	8.6
2	Northwestern Univ.	US	10.0	0.5	9.5
3	Mass. Univ. Technol.	US	9.0	0.7	8.3
4	CNRS	France	7.0	2.3	4.7
5	Georgia Inst. Technol.	US	7.0	0.4	6.6
6	Stanford Univ.	US	7.0	0.5	6.5
7	Brigham Women S. Hosp.	US	6.0	0.4	5.6
8	Univ. Calif. L.A.	US	5.0	0.5	4.5
9	Emory Univ.	US	5.0	0.4	4.6
10	Univ. Paris Saclay	France	5.0	0.6	4.4
11	Univ. Paris 11	France	5.0	0.4	4.6
12	Univ. Texas System	US	5.0	1.0	4.0
13	Howard Hughes Med. Inst.	US	4.0	0.4	3.6
14	Natl. Inst. Hlth.	US	4.0	0.6	3.4
15	Univ. Calif. Berkeley	US	4.0	0.3	3.7
16	Univ. Calif. S.F.	US	4.0	0.4	3.6

3.5 *Funding Bodies*

The brief info about top 9 funding bodies is provided in Table 4. It is significant that only 49% of the sample papers declare any funding. This funding rate is less than the funding rate, 67%, reported in a study (Wang and Shapira 2011).

The top funding body is the 'National Cancer Institute' of the US, funding 15% and 1.3% of the sample and population papers, respectively. The 'National Institute of General Medical Sciences', 'National Science Foundation', and 'National Institute of Biomedical Imaging Bioengineering' of the US further fund 8, 5, and 5% of the sample papers, respectively.

It is notable that 'National Natural Science Foundation of China', 'National Basic Research Program of China', and 'Fundamental Research Funds for the Central Universities' of China fund

Table 4: Funding bodies.

	Institutions	Country	Sample No. Papers %	Population No. Papers %	Difference %
1	Natl. Cancer Inst.	US	15.0	1.3	13.7
2	Natl. Inst. Gen. Med. Serv.	US	8.0	1.8	6.2
3	Natl. Sci. Found.	US	5.0	3.7	1.3
4	Natl. Inst. Biomed. Imag. Bioeng.	US	5.0	0.4	4.6
5	CNRS	France	3.0	0.3	2.7
6	Eur. Union	Europe	3.0	1.8	1.2
7	Natl. Heart Lung Blood Inst.	US	3.0	0.4	2.6
8	Natl. Inst. Env. Hlth. Sci.	US	3.0	0.3	2.7
9	US Dept. Def.	US	3.0	0.9	2.1

18.5, 3.4 and 2.4% of the population papers, respectively. However, they are significantly under-represented in funding the sample papers. For example, the first Chinese funding body funds only 2% of the sample papers.

The most-prolific country for these top funding bodies is the US with 7 funding bodies. The other 2 funding bodies are from France and Europe.

3.6 Source Titles

A brief information about the top source titles is provided in Table 5. In total, 39 and over 9,200 source titles publish the sample and population papers, respectively. On the other hand, 11 top journals publish 67 and 3.7% of the sample and population papers, respectively.

The top journal is 'Science' publishing 16 sample papers with 15.8% publication surplus. This top journal is closely followed by 'Nature Biotechnology' with 12 sample papers and 11.8% publication surplus.

It is significant that only 2 and one journals are related to nanotechnology and biomedicine, respectively. The subject categories of 'Multidisciplinary Sciences', 'Chemistry Multidisciplinary', and 'Materials Science Multidisciplinary' dominate this top journal table.

Table 5: Source titles.

	Source Titles	WOS Subject Cat.	Sample No. Papers %	Population No. Papers %	Difference %
1	Science	Mult. Sci.	16	0.2	15.8
2	Nature Biotechnology	Biot. Appl. Microb.	12	0.2	11.8
3	Nature Materials	Chem. Phys. Mats. Sci. Mult., Phys. Cond. Matt., Phys. Appl.	6	0.2	5.8
4	Angewandte Chemie Int. Ed.	Chem. Mult.	5	0.5	4.5
5	Chemical Reviews	Chem. Mult.	5	0.2	4.8
6	Nature	Mult. Sci.	5	0.2	4.8
7	Advanced Drug Delivery Reviews	Pharm. Pharm.	4	0.2	3.8
8	Journal of the American Chemical Society	Chem. Mult.	4	0.7	3.3
9	Proceedings of the National Academy of Sciences of the United States of America	Mult. Sci.	4	0.3	3.7
10	Nano Letters	Chem. Mult., Chem. Phys., Nanosci. Nanotechnol., Mats. Sci. Mult., Phys. Appl., Phys. Cond. Matt.	3	0.8	2.2
11	Nature Nanotechnology	Nnaosci. Nanotechnol., Mats. Sci. Mult.	3	0.2	2.8
			67	3.7	63.3

3.7 Countries

A brief information about the top countries is provided in Table 6.

The top country is the US, publishing 72 and 23.1% of the sample and population papers, respectively. Other prolific countries are the United Kingdom, France, and Germany publishing 11, 9, and 6 sample papers, respectively.

Table 6: Countries.

	Countries	Sample No. Papers %	Population No. Papers %	Difference %
1	US	72	23.1	48.9
2	United Kingdom	11	4.8	6.2
3	France	9	3.6	5.4
4	Germany	6	5.4	0.6
5	India	4	8.3	−4.3
6	South Korea	4	5.4	−1.4
7	Canada	3	2.5	0.5
8	Ireland	3	0.5	2.5
9	Israel	3	0.9	2.1
10	China	3	26.7	−23.7
11	Australia	2	2.2	−0.2
12	Belgium	2	0.9	1.1
13	Mexico	2	0.6	1.4
14	Singapore	2	1.6	0.4
15	Spain	2	3.0	−1.0
16	Sweden	2	1.1	0.9
	Europe-7	35	19.3	15.7
	Asia-5	15	44.2	−29.2

The European and Asian counties represented in this table published altogether 35 and 15% of the sample papers whilst they published 19.3 and 44.2% of the population papers, respectively.

It is notable that the publication surplus for the US, these European and Asian countries is 48.9, 15.7, and −29.2%, respectively. It is further notable that the bulk of the Asian publication defect originates from China with −23.7% publication deficit.

3.8 Web of Science Subject Categories

A brief information about the top 'Web of Science' subject categories is provided in Table 7. The sample and population papers are indexed by 21 and 228 subject categories, respectively.

For the sample papers, the top subjects are 'Chemistry Multidisciplinary' and 'Multidisciplinary Sciences' with 26 and 25 papers, respectively. The other prolific subjects are 'Chemistry Physical', 'Materials Science Multidisciplinary', and 'Biotechnology Applied Microbiology' with 19, 19, and 16 papers, respectively.

It is notable that the publication surplus is most significant for 'Multidisciplinary Sciences' with 22.6% surplus. The other high-impact subjects are 'Biotechnology Applied Microbiology', 'Chemistry Multidisciplinary', 'Chemistry Physical', 'Physics Condensed Matter', and 'Physics Applied' with 11, 6.7, 6.5, 6.4, and 4.0% publication surplus, respectively.

On the other hand, the subjects with least impact are 'Engineering Biomedical', 'Chemistry Analytical', 'Materials Science Biomaterials', and 'Nanoscience Nanotechnology' with the publication deficit of −3,3, −5,2, −7,2, and −10,3%, respectively.

Table 7: Web of science subject categories.

	Subjects	Sample No. Papers %	Population No. Papers %	Difference %
1	Chemistry Multidisciplinary	26	19.3	6.7
2	Multidisciplinary Sciences	25	2.8	22.2
3	Chemistry Physical	19	12.5	6.5
4	Materials Science Multidisciplinary	19	20.2	−1.2
5	Biotechnology Applied Microbiology	16	5.0	11
6	Physics Applied	15	11.0	4
7	Nanoscience Nanotechnology	14	24.3	−10.3
8	Physics Condensed Matter	12	5.6	6.4
9	Pharmacology Pharmacy	11	11.2	−0.2
10	Materials Science Biomaterials	3	10.2	−7.2
11	Chemistry Analytical	2	7.2	−5.2
12	Electrochemistry	2	3.3	−1.3
13	Engineering Biomedical	2	5.3	−3.3
14	Medicine Research Experimental	2	3.1	−1.1
15	Oncology	2	1.4	0.6

3.9 Citation Impact

These sample papers receive over 269,000 citations as of January 2020. Thus, the average number of citations per paper is over 2,690.

3.10 Keywords

Although a number of keywords are listed in the appendix for the datasets of both nanotechnology and biomedicine, some of them are more significant.

For the data set of nanotechnology, the most-prolific keyword is '*nano*'. The other prolific keywords are 'quantum dot*', 'graphene*', 'mesoporous', and 'sers' at the first level.

The prolific keywords at the second level are 'plasmonic*', 'dendrimer*', 'langmuir-blodgett', '*fulleren*', '*c60*', 'cnt', 'mwcnt', 'mos2', 'phosphorene', 'atomic layer*', 'organic-inorganic hybrid*', 'metal-organic framework*', 'mof*', 'microporous', 'coordination polymer*', '*molecular sieve*', 'atomic force microscop*', 'afm', 'scanning tunneling microscop*', 'fluorescence microscop*', 'quantum well*', 'supramolecul*', 'single molecul*', 'atomic scale', 'single layer*', and 'nems'.

The most-prolific keywords for the biomedicine data set are 'bio*' and '*cell*'. The other prolific keywords are '*medic*', 'biosens*', 'photothermal', '*therap*', 'protein*', 'nanomed*', 'peptide*', '*nucleotide*', 'pharm*', '*cancer', '*drug*', 'immuno*', '*in vivo*', '*in vitro*', 'gene', 'neur*', 'tumor*', 'nanocapsule*', 'human*', 'glucose', '*bacter*', '*microb*', 'tissue*', 'bone*', '*dna', 'mice', '*toxic*', 'lipid*', and 'osteo*'.

3.11 Research Fronts

A brief information about the key research fronts is provided in Table 8.

The most-prolific research fronts are 'bioimaging and biodiagnostics using nanomaterials' and 'production and properties of nanobiomaterials' with 21 and 20 papers, respectively.

Table 8: Research fronts.

	Research Fronts	Sample No. Papers %
1	Production and properties	20
2	General applications	5
3	Bioimaging and biodiagnostics	21
4	Toxicity in general	2
5	Cytotoxicity	6
6	Biodrug delivery	16
7	Biosensing	16
8	Antimicrobial properties	6
9	Cancer nanobiotechnology	8
10	Tissue bioengineering	3
11	Cell interactions	8

The other prolific research fronts are 'biodrug delivery' and 'biosensing' with 16 papers each. The other research fronts are 'cancer nanobiotechnology', 'cytotoxicity', 'toxicity in general', 'antimicrobial properties', 'tissue engineering', and 'nanobiomaterials-cell interactions'.

4. Discussion

The size of the research on the nanobiomaterials within the biomedical context has increased to over 249,000 papers, forming over 15% of the research papers on the nanotechnology. The research has developed more in the technological aspects of nanobiomaterials, rather than the social and humanitarian pathways as evidenced by the indexing of the population papers by the indices of the 'Web of Science'.

The articles and reviews are under-represented and over-represented by about 35% each in the sample papers, respectively (Table 1). Thus, the contribution of reviews by 42% of the sample papers in this field is highly exceptional (c.f. Konur 2016a–f, 2017a–e, 2019b).

Sixteen authors from 10 institutions have 3 or more sample papers (Table 2). Thirteen of these authors are from the US and the remaining ones from France and Canada. These authors focus on 'cancer biotherapy', 'biosensing', 'biodrug delivery', and 'bioimaging'. It is significant that there is ample 'gender deficit' among these top authors as only one researcher is female.

Whilst 77% of the sample papers are published in the 2000s, over 70% of the population papers are published in the 2010s (Figure 1). This finding suggests that the population papers have built up on the sample papers, primarily published in the 2000s, in the 2010s.

The engagement of the institutions on the nanobiomaterials research within the biomedical context at the global scale is significant as over 1,100 and 39,600 institutions contribute to the sample and population papers, respectively.

Sixteen top institutions publish 97.0 and 10.8% of the sample and population papers, respectively (Table 3). The top institutions are 'Harvard University' and 'Northwestern University' of the US with 10 sample papers each as they are significantly over-represented in sample papers. As in the case of the top authors, most-prolific countries for these top instructions are the US and France with 13 and 3 sample papers, respectively.

It is significant that only 49% of the sample papers declare any funding. The most-prolific country for these top funding bodies is the US with 7 funding bodies (Table 4). The other 2 funding bodies are from France and Europe. The 'National Cancer Institute' and 4 other institutes of the

'National Health Institute' of the US dominate this top funding table. On the contrary, there is no Chinese funding body in this table although they are over-represented in the population papers. For example, 'National Natural Science Foundation of China' (NSFC) funds 18.5% of the population papers. These findings are in line with the studies showing the heavy research funding in China and the NSFC is the primary funding agency in China (Wang et al. 2012).

The sample and population papers are published by 39 and over 9,200 journals, respectively. It is significant that 11 top journals publish 67 and 3.7% of the sample and population papers, respectively (Table 5).

The top journals, 'Science' and 'Nature Biotechnology', publish together 28 sample papers with 27.6% publication surplus. It is significant that only 2 and one journals are related to nanotechnology and biomedicine, respectively. The subject categories of 'Multidisciplinary Sciences', 'Chemistry Multidisciplinary', and 'Materials Science Multidisciplinary' dominate this top journal table.

In total, 56 and 176 countries contribute to the sample and population papers, respectively. The top county is the US publishing 72 and 23.1% of the sample and population papers, respectively, with 48.9% publication surplus (Table 6). This finding is in line with the studies arguing that the US is not losing ground in science and technology (Leydesdorff and Wagner 2009).

The other prolific countries are the United Kingdom, France, and Germany publishing 11, 9, and 6 sample papers, respectively. These findings are in line with the studies showing that European countries have superior publication performance in nanotechnology (Youtie et al. 2008).

The European and Asian countries represented in this table publish altogether 35 and 15% of the sample papers whilst they publish 19.3 and 44.2% of the population papers, respectively.

It is notable that the publication surplus for the US, these European and Asian countries is 48.9, 15.7, and −29.2%, respectively. It is further notable that the bulk of the Asian publication deficit originates from China with −23.7% publication deficit. This finding is in contrast with China's efforts to be a leading nation in science and technology (Zhou and Leydesdorff 2006), but it is in line with the findings of Guan and Ma (2007) and Youtie et al. (2008).

The sample and population papers are indexed by 21 and 228 subject categories, respectively. For the sample papers, the top subject categories are 'Chemistry Multidisciplinary' and 'Multidisciplinary Sciences' with 26 and 25 papers, respectively (Table 7). The other prolific subjects are 'Chemistry Physical', 'Materials Science Multidisciplinary', and 'Biotechnology Applied Microbiology'.

It is notable that the publication surplus is most significant for 'Multidisciplinary Sciences' with 22.6% surplus. The other high-impact subjects are 'Biotechnology Applied Microbiology', 'Chemistry Multidisciplinary', 'Chemistry Physical', 'Physics Condensed Matter', and 'Physics Applied'.

On the other hand, the subjects with least impact are 'Engineering Biomedical', 'Chemistry Analytical', 'Materials Science Biomaterials', and 'Nanoscience Nanotechnology'.

These sample papers received over 269,000 citations as of January 2020. Thus, the average number of citations per paper is over 2,690.

Although a number of keywords are listed in the appendix for the datasets of both nanotechnology and biomedicine, some of them are more significant. For the data set of nanotechnology, the most-prolific keyword is '*nano*'. The other prolific keywords are 'quantum dot*', 'graphene*', 'mesoporous', 'sers' at the first level. The prolific keywords at the second level are 'plasmonic*', *'dendrimer*',* 'langmuir-blodgett', '*fulleren*', '*c60*', 'cnt', 'mwcnt', 'mos2', 'phosphorene', 'atomic layer*', 'organic-inorganic hybrid*', 'metal-organic framework*', 'mof*', 'microporous', 'coordination polymer*', '*molecular sieve*', 'atomic force microscop*', 'afm', 'scanning tunneling microscop*', 'fluorescence microscop*', 'quantum well*', 'supramolecul*', 'single molecul*', 'atomic scale', 'single layer*', and 'nems'.

However, the number of these keywords are reduced in the sample papers. 'Nanoparticle*' and 'quantum dot*' are the most-prolific keywords in these sample papers with 43 and 15 citations, respectively. The other prolific keywords are 'nanotube*', 'nanoscale', 'nanotechnology', 'nanocrystal', 'nanosensor*', and '*graphene'.

The most-prolific keywords for the biomedicine data set are 'bio*' and '*cell*'. The other prolific keywords are '*medic*', 'biosens*', 'photothermal', '*therap*', 'protein*', 'nanomed*', 'peptide*', '*nucleotide*', 'pharm*', '*cancer', '*drug*', 'immuno*', '*in vivo*', '*in vitro*', 'gene', 'neur*', 'tumor*', 'nanocapsule*', 'human*', 'glucose', '*bacter*', '*microb*', 'tissue*', 'bone*', '*dna', 'mice', '*toxic*', 'lipid*', and 'osteo*'.

However, the number of these keywords are reduced in the sample papers as in the case of nanotechnology keywords. The most prolific keywords are '*bio', 'cancer', 'cell*', and '*imaging' with 32, 13, 12, and 14 citations, respectively. The other prolific keywords are *toxic*', 'dna', 'detection', 'drug*', '*imaging', '*in vivo*', '*medic', and 'therap'.

As expected, these keywords provide valuable information about the pathways of the research on the nanobiomaterials within the biomedical context.

Pragmatically, 11 research fronts emerge from the examination of the sample papers (Table 8). The most-prolific research fronts are 'bioimaging and biosensing using nanomaterials' and 'production and properties of nanobiomaterials' with 21 and 20 papers, respectively. The other prolific research fronts are 'biodrug delivery' and 'biosensing' with 16 papers each. The other research fronts are 'cancer nanobiotechnology', 'cytotoxicity', 'toxicity in general', 'antimicrobial properties', 'tissue engineering', and 'nanobiomaterials-cell interactions'.

The key emphasis in these research front is the exploration of the structure-processing-property relationships for these nanobiomaterials (Konur and Matthews 1989, Gupta and Gupta 2005, Laurent et al. 2008).

5. Conclusion

This book chapter maps the research on the nanobiomaterials within the biomedical context using a scientometric method.

The sheer size of over 249,000 population papers shows the public importance of this interdisciplinary research field. However, it is significant that the research has developed more in the technological aspects of nanobiomaterials, rather than the social and humanitarian pathways. Articles and reviews dominate the sample papers, primarily published in the 2000s.

The data presented in the tables and in one figure show that a small number of authors, institutions, funding bodies, journals, keywords, research fronts, subject categories, and countries have shaped the research in this field.

It is notable that the authors, institutions, and funding bodies from the US dominate the research in this field. Furthermore, China and other Asian countries represented in the top country table are under-represented significantly in the sample papers.

These findings show the importance of the development of efficient incentive structures for the development of the research in this field as in other fields. It seems that the US and, to a lesser extent, European countries, have efficient incentive structures for the development of the research in this field, contrary to the Asian countries. However, it seems there is more to do to reduce significant gender deficit in this field as in other field of the biomedical research (Xie and Shauman 1998).

It further seems that the research funding is a significant element of these incentive structures, and it might not be the sole solution for increasing the incentives for the research in this field as in the case of Asian countries.

The data on the research fronts, keywords, source titles, and subject categories provides valuable evidence for the interdisciplinary nature of the research in this field. These findings are in line with studies arguing for the interdisciplinarity of the research in nanomaterials (Schummer 2004, Meyer and Persson 1998).

As only 24.3% of the population papers are indexed by the subject category of the 'Nanoscience and Nanotechnology', there is ample justification for the broad search strategy employed in this study. The search strategy employed in this study is in line with the search strategies employed for nanotechnology research in the relevant studies (Porter et al. 2008, Mogoutov and Kahane 2007, Arora et al. 2013).

Pragmatically, 11 research fronts emerge from the examination of the sample papers. The most-prolific research fronts are 'bioimaging and biosensing using nanomaterials' and 'production and properties of nanobiomaterials'. The other prolific research fronts are 'biodrug delivery' and 'biosensing'. The other additional research fronts are 'cancer nanobiotechnology', 'cytotoxicity', 'toxicity in general', 'antimicrobial properties', 'tissue engineering', and 'nanobiomaterials-cell interactions'.

It is recommended that further scientometric studies are carried out for each of these research fronts building on the pioneering studies in these fields. Such studies might also be carried out for special types of nanomaterials such as nanoparticles, graphene and quantum dots.

Acknowledgements

The contribution of the highly cited researchers in the field of nanobiomaterials within the medical context has been greatly acknowledged.

Appendix

Syntax: (I and II) or III.

I. Biomedicine-related keywords

(TI=(*medic* or biosens* or bioimag* or bio* or photothermal or *therap* or protein* or nanomed* or peptide* or *nucleotide* or corona* or theranost* or nacre or pharm* or *cell* or *cancer or heart or *drug* or immuno* or "*in vivo*" or "*in vitro*" or lymph* or oncol* or metastas* or "contrast agent*" or pathogen* or pulmon* or pegy* or gene or neur* or tumor* or tumour* or aptamer* or nanocapsule* or albumin or clinic* or cosmet* or dermat* or human* or glucose or *bacter* or *microb* or health or *infect* or tissue* or bone* or *dna or *rna or *rnai or clearance or her2 or synapse or mice or doxil or *toxic* or "nucleic acid*" or *diagnos* or micell* or albumin or p53 or liposome* or lipid* or "polyethylene glycol" or animal* or autophagy or mammal* or "photo-thermal" or rat or endocyto* or osteo*) or WC=(pharm* or oncol* or "materials science biomaterials" or anatom* or andro* or anest* or audio* or biophys* cardiac* or cell* or clin* or dent* or dermat* or emergency* or endoc* or "engineering biomed*" or gastro* or genet* or geriat* or health* or hemat* or imaging* or immun* or infect* or integrative* or med* or microbial* or mycol* or neur* or nurs* or obstet* or ophth* or ortho* or otorhin* or physiol* or psych* or "public env*" or radiol* or reprod* or respir* or rheum* or substance* or surgery or transplant* or tropical* or urol* or toxic*)) NOT TI=("fuel cell*" or photovoltaic* or "solar cell*" or *cellulos* or hydrogen or lithium or zeolite or energy).

II. Nanotechnology-related keywords

(TI=(*nano* or "quantum dot*" or "quantum-dot*" or plasmonic* or triboelectric* or "carbon dot*" or skyrmion* *or* dendrimer* or "carbon onion*" or "exfoliated graphite*" or filomicelle* or "atomic crystal*" or "langmuir-blodgett" or "c-dot*" or eletrospinning or ((graphit* or carbon) and *tubule*) or semimetal* or "polymer dot") OR TI=(*fulleren* or pcbm or *c60* or *c61* or *c70* or *c120* or "*butyric acid methyl ester*" or pc70bm or cnt or swnt or mwnt or swcnt or mwcnt or "c-60" or "c-61" or "c-70" or "c-120" or fulleri* or fullero* or buckyball*) OR TI=(mos2 or *graphene* or "transition metal dichalcogenide*" or mose2 or ws2 or wse2 or "black phosphorus" or phosphorene or "two-dimensional material*" or "low-dimensional material*" or "molybdenum disulfide" or "boron nitride layer*" or "g-c3n4" or mxene* or "2d *material*" or "atom thick" or "atomically thin" or graphane* or "graphitic carbon nitride*" or "atomic layer*" or "two-dimensional carbon*" or tmdc* or "waals heterostructure*" or "two-dimensional titan*" or "two-dimensional carbide*" or "2d crystal*" or "two-dimensional crystal*") OR TI=("inorganic-organic hybrid*" or "organic-inorganic hybrid*" or "metal-organic framework*" or mof* or mesoporous or microporous or "hybrid inorganic-organic" or "hybrid organic-inorganic" or "coordination polymer*" or "zeolitic imidazolate framework*" or "*molecular sieve*" or "metal-organic material*" or "covalent organic framework*" or cofs or mesopore* or "organometal halide perovskite*" or "hybrid halide perovskite*" or mesocrystal*) OR TI=("atomic force microscop*" or afm or "scanning tunneling microscop*" or "fluorescence microscop*" or sted or "stimulated emission depletion microscop*" or "localized surface plasmon resonance" or lspr or "surface-enhanced raman spectroscop*" or sers or "scanning probe microscop*" or "optical reconstruction microscop*" or "dynamic force spectroscop*") OR TI=("*molecular machin*" or "*molecular motor*" or "*molecular self-assembl*" or superlens or "quantum wire*" or "quantum well*" or "molecular photov*" or "*molecular assembl*" or "*molecular wire*" or "*molecular device*" or supramolecul* or "*molecular engineering" or "single atom*" or "single molecul*" or "dna machine*" or "molecular rotor*" or "molecular electronic device*" or "molecular biomimetic*" or "*molecular junction*" or "atomic scale" or "sub nm" or "single layer*" or "atomic sized" or nems) *OR* AU=("Kroto HW" or "Novoselov K*" or "Hell SW" or "Geim AK" or "Smalley RE" or "Curl RF" or "Betzig Eric" or "Binnig Gerd" or "Rohrer Hei*" or "Stoddart JF" or "Sauvage JP" or "Feringa BL" or "Yaghi OM" or "Castro Neto AH" or "Mirkin CA" "Nguyen ST" or "Alivisatos AP" or "Xia YN" or "Kitagawa Sus*" or "Yang Pei*" or "Colombo Lui*" or "Tour JM" or "Dai HJ" or "Lieber CM" or "Dekker Cees" or "Coleman JN" or "Meyer JC" or "Chan WCW" or "Ajayan PM" or "Dresselhaus MS" or "van Duyne RP" or "Astruc D" or "Iijima Sum*" or "Murphy CJ" or "Kotov NA" or "Faroqhzad OC" or "Stupp SI" or "McEuen PL" or "Liz-Marzan LM" or "Bawendi MG" or "Long JR" or "Pinnavaia TJ" or "Koppens FHL" or "Kang ZH" or "Kaner RB" or "Weissleder R" or "Wiesner U" or "Nowack B") OR SO=("ACS Nano" or "Advances in Nano*" or "Applied Nanoscience" or "Beilstein Journal of Nano*" or "Cancer Nano*" or Chem Nanomat or "Current Nano*" or "Environmental Science-Nano" or "Frontiers of Nano*" or Fullerenes* or "IEEE Transactions on Nano*" or "IET Nano*" or "International Journal of Nano*" or "Journal of Biomedical Nano*" or "Journal of Computational and Theoretical Nano*" or "Journal of Experimental Nano*" or "Journal of Nano*" or "Microporous and Mesoporous*" or Nano* or "Nature Nano*" or "NPJ 2d" or "Recent Patents on Nano*" or "RSC Nano*" or Small* or "Wiley Interdisciplinary Reviews-Nano*") or SO=("ACS Applied Nano*" or "Advances in Natural Sciences-Nano*" or "European Journal of Nano*" or "IEEE Nano*" or "International Journal of Nano*" or "International Nano*" or "Journal of Nano*" or Nano* or "Proceedings of the Institution of Mechanical Engineers Part N-Journal of Nano*")) NOT TI=(nanog or nanogram* or "nano-gram*" or nanosecond* or "nano-second*" or nanomol* or nanoliter* or nanolitre* or nanokelvin or nanoamper* or "nano-amper*" or nano2 or nano3 or nanoplan* or nanoflag* or nanoprot* or nanobacter* or nanoflare* or nanor or nanobody or nanostring or nanosatel*).

III. Cross-subject keywords

SO=("Internal journal of nanomedicine" or nanomedicine*).

References

Arora, S.K., A.L. Porter, J. Youtie and P. Shapira. 2013. Capturing new developments in an emerging technology: An updated search strategy for identifying nanotechnology research outputs. Scientometrics 95: 351–70.

Bruchez, M., M. Moronne, P. Gin, S. Weiss and A.P. Alivisatos. 1998. Semiconductor nanocrystals as fluorescent biological labels. Science 281: 2013–6.

Chan, W.C.W. and S.M. Nie. 1998. Quantum dot bioconjugates for ultrasensitive nonisotopic detection. Science 281: 2016–8.

Daniel, M.C. and D. Astruc. 2004. Gold nanoparticles: Assembly, supramolecular chemistry, quantum-size-related properties, and applications toward biology, catalysis, and nanotechnology. Chemical Reviews 104: 293–346.

Docampo, D. and L. Cram. 2019. Highly cited researchers: A moving target. Scientometrics 118: 1011–25.

Garfield, E. 1955. Citation indexes for science. Science 122: 108–11.

Garfield, E. 1972. Citation analysis as a tool in journal evaluation. Science 178: 471–9.

Guan, J.C. and N. Ma. 2007. China's emerging presence in nanoscience and nanotechnology—A comparative bibliometric study of several nanoscience 'giants'. Research Policy 36: 880–6.

Gupta, A.K. and M. Gupta. 2005. Synthesis and surface engineering of iron oxide nanoparticles for biomedical applications. Biomaterials 26: 3995–4021.

Konur, O. 2000. Creating enforceable civil rights for disabled students in higher education: An institutional theory perspective. Disability & Society 15: 1041–63.

Konur, O. 2002a. Access to nursing education by disabled students: Rights and duties of nursing programs. Nurse Education Today 22: 364–74.

Konur, O. 2002b. Assessment of disabled students in higher education: Current public policy issues. Assessment and Evaluation in Higher Education 27: 131–52.

Konur, O. 2002c. Access to employment by disabled people in the UK: Is the Disability Discrimination Act working? International Journal of Discrimination and the Law 5: 247–79.

Konur, O. 2006a. Participation of children with dyslexia in compulsory education: Current public policy issues. Dyslexia 12: 51–67.

Konur, O. 2006b. Teaching disabled students in Higher Education. Teaching in Higher Education 11: 351–63.

Konur, O. 2007a. A judicial outcome analysis of the Disability Discrimination Act: A windfall for the employers? Disability & Society 22: 187–204.

Konur, O. 2007b. Computer-assisted teaching and assessment of disabled students in Higher Education: The interface between academic standards and disability rights. Journal of Computer Assisted Learning 23: 207–19.

Konur, O. 2011. The scientometric evaluation of the research on the algae and bio-energy. Applied Energy 88: 3532–40.

Konur, O. 2012a. Evaluation of the research on the social sciences in Turkey: A scientometric approach. Energy Education Science and Technology Part B: Social and Educational Studies 4: 1893–908.

Konur, O. 2012b. Prof. Dr. Ayhan Demirbas' scientometric biography. Energy Education Science and Technology Part A: Energy Science and Research 28: 727–38.

Konur, O. 2012c. The evaluation of the biogas research: A scientometric approach. Energy Education Science and Technology Part A: Energy Science and Research 29: 1277–92.

Konur, O. 2012d. The evaluation of the educational research: A scientometric approach. Energy Education Science and Technology Part B: Social and Educational Studies 4: 1935–1948.

Konur, O. 2012e. The evaluation of the global energy and fuels research: A scientometric approach. Energy Education Science and Technology Part A: Energy Science and Research 30: 613–28.

Konur, O. 2012f. The evaluation of the research on the Arts and Humanities in Turkey: A scientometric approach. Energy Education Science and Technology Part B: Social and Educational Studies 4: 1603–18.

Konur, O. 2012g. The evaluation of the research on the biodiesel: A scientometric approach. Energy Education Science and Technology Part A: Energy Science and Research 28: 1003–14.

Konur, O. 2012h. The evaluation of the research on the bioethanol: A scientometric approach. Energy Education Science and Technology Part A: Energy Science and Research 28: 1051–64.

Konur, O. 2012i. The evaluation of the research on the biofuels: A scientometric approach. Energy Education Science and Technology Part A: Energy Science and Research 28: 903–16.

Konur, O. 2012j. The evaluation of the research on the biohydrogen: A scientometric approach. Energy Education Science and Technology Part A: Energy Science and Research 29: 323–38.

Konur, O. 2012k. The evaluation of the research on the microbial fuel cells: A scientometric approach. Energy Education Science and Technology Part A: Energy Science and Research 29: 309–22.

Konur, O. 2012l. The scientometric evaluation of the research on the production of bioenergy from biomass. Biomass and Bioenergy 47: 504–15. SCI.

Konur, O. 2012m. The scientometric evaluation of the research on the deaf students in higher education. Energy Education Science and Technology Part B: Social and Educational Studies 4: 1573–88.

Konur, O. 2012n. The scientometric evaluation of the research on the students with ADHD in higher education. Energy Education Science and Technology Part B: Social and Educational Studies 4: 1547–62.

Konur, O. 2016a. Scientometric overview in nanobiodrugs. pp. 405–428. *In*: Holban, A.M. and A.M. Grumezescu (eds.). Nanoarchitectonics for Smart Delivery and Drug Targeting. Amsterdam: Elsevier.

Konur, O. 2016b. Scientometric overview regarding nanoemulsions used in the food industry. pp. 689–711. *In*: Grumezescu, A.M. (ed.). Emulsions: Nanotechnology in the Agri-Food Industry. Amsterdam: Elsevier.

Konur, O. 2016c. Scientometric overview regarding the nanobiomaterials in antimicrobial therapy. pp. 511–535. *In*: Grumezescu, A.M. (ed.). Nanobiomaterials in Antimicrobial Therapy. Amsterdam: Elsevier.

Konur, O. 2016d. Scientometric overview regarding the nanobiomaterials in dentistry. pp. 425–453. *In*: Grumezescu, A.M. (ed.). Nanobiomaterials in Dentistry. Amsterdam: Elsevier.

Konur, O. 2016e. Scientometric overview regarding the surface chemistry of nanobiomaterials. pp. 463–486. *In*: Grumezescu, A.M. (ed.). Surface Chemistry of Nanobiomaterials. Amsterdam: Elsevier.

Konur, O. 2016f. The scientometric overview in cancer targeting. pp. 871–895. *In*: Holban, A.M. and A. Grumezescu (eds.). Nanoarchitectonics for Smart Delivery and Drug Targeting. Amsterdam; Elsevier.

Konur, O. 2017a. Recent citation classics in antimicrobial nanobiomaterials. pp. 669–685. *In*: Ficai, A. and A.M. Grumezescu (eds.). Nanostructures for Antimicrobial Therapy. Amsterdam: Elsevier.

Konur, O. 2017b. Scientometric overview in nanopesticides. pp. 719–744. *In*: Grumezescu, A.M. (ed.). New Pesticides and Soil Sensors. Amsterdam: Elsevier.

Konur, O. 2017c. Scientometric overview regarding oral cancer nanomedicine. pp. 939–962. *In*: Andronescu, E. and A.M. Grumezescu (eds.). Nanostructures for Oral Medicine. Amsterdam: Elsevier.

Konur, O. 2017d. Scientometric overview regarding water nanopurification. pp. 693–716. *In*: Grumezescu, A.M. (ed.). Water Purification. Amsterdam: Elsevier.

Konur, O. 2017e. Scientometric overview in food nanopreservation. pp. 703–729. *In*: Grumezescu, A.M. (ed.). Food Preservation. Amsterdam: Elsevier.

Konur, O. 2017f. The top citation classics in alginates for biomedicine. pp. 223–249. *In*: Venkatesan, J., S. Anil, S.K. Kim (eds.). Seaweed Polysaccharides: Isolation, Biological and Biomedical Applications. Amsterdam: Elsevier.

Konur, O. 2018a. Scientometric evaluation of the global research in spine: An update on the pioneering study by Wei et al. European Spine Journal 27: 525–9.

Konur, O. 2018b. Bioenergy and biofuels science and technology: Scientometric overview and citation classics. pp. 3–63. *In*: Konur, O. (ed.). Bienergy and Biofuels. Boca Raton: CRC Press.

Konur, O. and F.L. Matthews. 1989. Effect of the properties of the constituents on the fatigue performance of composites: A review. Composites 20: 317–28.

Konur, O. 2019a. Cyanobacterial bioenergy and biofuels science and technology: A scientometric overview. pp. 419–442. *In*: Mishra, A.K., D.N. Tiwari and A.N. Rai (eds.). Cyanobacteria: From Basic Science to Applications. Amsterdam: Elsevier.

Konur, O. 2019b. Nanotechnology applications in food: A scientometric overview. pp. 683–711. *In*: Pudake, R.N., N. Chauhan and C. Kole (eds.). Nanoscience for Sustainable Agriculture. Cham: Springer.

Laurent, S., D. Forge, M. Port, A. Roch, C. Robic, L.V. Elst and R.N. Muller. 2008. Magnetic iron oxide nanoparticles: Synthesis, stabilization, vectorization, physicochemical characterizations, and biological applications. Chemical Reviews 108: 2064–110.

Leydesdorff, L. and C. Wagner. 2009. Is the United States losing ground in science? A global perspective on the world science system. Scientometrics 78: 23–36.

Meyer, M. and O. Persson. 1998. Nanotechnology—Interdisciplinarity, patterns of collaboration and differences in application. Scientometrics 42: 195–205.

Michalet, X., F. Pinaudl, A. Bentolilaj, M. Tsays Doose, J.J. Li, G. Sundaresana, M. Wus, S. Gambhirand and S. Weiss. 2005. Quantum dots for live cells, *in vivo* imaging, and diagnostics. Science 307: 538–544.

Mogoutov, A. and B. Kahane. 2007. Data search strategy for science and technology emergence: A scalable and evolutionary query for nanotechnology tracking. Research Policy 36: 893–903.

North, D.C. 1991. Institutions, Institutional Change and Economic Performance. Cambridge, Mass: Cambridge University Press.

North, D.C. 1991. Institutions. Journal of Economic Perspectives 5: 97–112.

Porter, A.L., J. Youtie, P. Shapira and D.J. Schoeneck. 2008. Refining search terms for nanotechnology. Journal of Nanoparticle Research 10: 715–28.

Schummer, J. 2004. Multidisciplinarity, interdisciplinarity, and patterns of research collaboration in nanoscience and nanotechnology. Scientometrics 59: 425–65.

Wang, J. and P. Shapira. 2011. Funding acknowledgement analysis: An enhanced tool to investigate research sponsorship impacts: the case of nanotechnology. Scientometrics 87: 56–86.

Wang, X., D. Liu, K. Ding and X. Wang. 2012. Science funding and research output: A study on 10 countries. Scientometrics 91: 591–9.

Xie, Y. and K.A. Shauman. 1998. Sex differences in research productivity: New evidence about an old puzzle. American Sociological Review, 847–870.

Youtie, J., P. Shapira and A.L. Porter. 2008. Nanotechnology publications and citations by leading countries and blocs. Journal of Nanoparticle Research 10: 981–6.

Zhou, P. and L. Leydesdorff. 2006. The emergence of China as a leading nation in science. Research Policy 35: 83–104.

4

Electrospinning—Material, Techniques and Biomedical Applications

Sekar Vijayakumar,[1] *K. Vijayalakshmi,*[2,*] *V. Sangeetha*[3] *and E. Radha*[4]

1. Introduction

Due to their higher surface area and abundance of active surface locations, nanoscaled materials have seen an incredible growth in attention across numerous sectors (Lee and Im 2010). When compared to other nanoscaled materials, the ease of production procedures for nanofibers has led to their widespread use in industry. Since it has substantially increased the potential for addressing the issues of the modern world, the electro spinning technique has recently been the focus of intense research, particularly in generating and developing nanofibers with diameters ranging from 2 nm to several micrometres. Nanofibres are fibres with a diameter between ten and hundreds of nanometers that have a large surface area per unit mass. Electrospinning is a dry spinning process in which the fibers were drawn from the liquid polymer solution or melt using electrostatic forces (Tucker et al. 2012). Both electro spraying and traditional solution dry spinning of fibres have similarities to electro spinning (Ziabicki 1976). Complex chemical or biological materials can be treated safely - thanks to the electro spinning technique' benign processing conditions, such as room temperature and the lack of coagulation chemistry. This useful electro spinning technique is straightforward and reliable, making it simple to use in a lab setting and scale up to an industrial process (Ji et al. 2011).

1.1 History of Electro Spinning

The history of the physics underlying electro spinning dates back to the very beginning of scientific inquiry. The term "electrostatic spinning," from which the name "electrospinning" truly derives, was first used in 1897 (Bhardwaj and Kundu 2010). Until the mid-1990s, the electrospinning technique remained an obscure method of making fibers and this was due to competition from the mechanical

[1] Marine College, Shandong University, Weihai, P.R. China 264209.
[2] Department of Chemistry, Madras Christian College, Chennai, Tamil Nadu, India.
[3] Department of Chemistry, Sathyabama University, Chennai, Tamil Nadu, India.
[4] Biomaterials research lab, D.K.M. College for Women, Vellore, Tamil Nadu, India.
* Corresponding author: kumarhandins@rediffmail.com

drawing approach for creating polymeric fibres, it was not commercially adopted. William Gilbert made the first observation of the behaviour of magnetic and electrostatic phenomena, i.e., the electrostatic attraction of a liquid, in the year 1600. (Tucker 2013). Highly nitrated cellulose was created by Christian Friedrich Schönbein in 1846, and Charles Vernon Boys described the procedure in a paper on the production of nanofibers in 1887. The first patent for electro spinning was submitted by John Francis Cooley in 1900, which were followed by the production and manufacturing procedures. Zeleny discovered in 1914 that charged liquid droplets could release small fiber-like liquid jets when exposed to an electrical potential. This discovery is credited as the inspiration for current needle electro spinning.

Between 1931 and 1944, Anton Formhals obtained at least 22 patents for electro spinning. The first electro spun fibres were created by N.D. Rozenblum and I.V. Petryanov-Sokolov in 1938, and between 1964 and 1969, Sir Geoffrey Ingram Taylor produced the first theoretical foundation for electro spinning by mathematically simulating the shape of the (Taylor) cone formed by the fluid droplet under the influence of an electric field. Later studies examined the interaction between process variables and fibre shape while concentrating on the characterization of electrospun nanofibers. In 1971, Baumgarten published a paper on the electrospinning of acrylic fibres with a diameter range of 500 to 1100 nm (Baumgarten 1971). Due to the sub-micron diameter of the fibres produced, Reneker and his colleagues demonstrated the technique's potential for use in nanotechnology research. Early in the 1990s, a number of research teams-notably Reneker's, which is credited with popularising the term electrospinning-demonstrated electrospun nano-fibers. Since 1995, there have been exponential annual increases in the number of publications about electrospinning.

1.2 Basic Aspects of Electro Spinning

When using the electro spinning method, nanofibers were created from a polymer solution by applying a high voltage that causes a narrow liquid jet to develop (Ramakrishna et al. 2005). An example of a standard electro spinning device is shown in (Figure 1). The electrospinning apparatus's three main parts were a high voltage source, a capillary tube with a small-diameter pipette or needle, and a metal-collecting screen.

Standard electro spinning systems are straight forward in terms of equipment, but the physics driving the operation are rather intricate. In the conventional method, the polymer is dissolved in the appropriate solvent and placed into a syringe with a metal needle. High voltage is then used to create an electrically charged jet of polymer solution that emerges from the pipette or syringe at a steady flow rate. One electrode is inserted into the rotating solution, and the second electrode is fastened to the collector. The electric field that was applied to the end of the capillary tube, which contains the solution fluid held in place by its surface tension, was what caused the charge on the surface of the polymeric liquid.

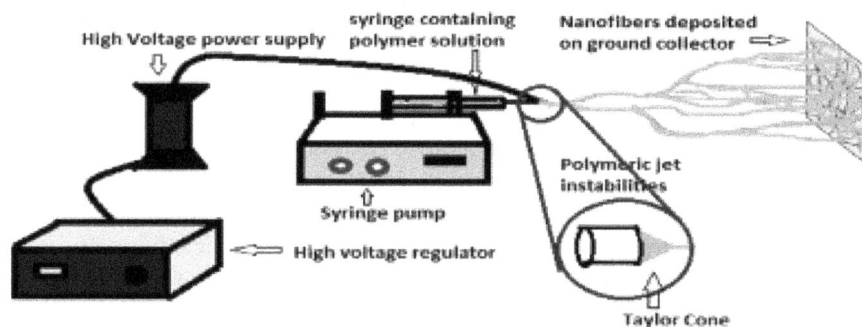

Figure 1: Electrospinning apparatus.

The mutual charge repulsion and contraction of the surface charges to the counter electrode produce a force that is directly opposed to surface tension. Surface tension is overcome by these repelling electrostatic forces, which cause the charged jet of fluid to be ejected from the tip where it causes the polymeric solution jet to solidify and evaporate. Finally, the solvent from the polymeric solution evaporates, resulting in the deposition of dry nanofibers as a web of interconnected tiny fibres on the collector. The mutual charge repulsion and contraction of the surface charges to the counter electrode produce a force that is directly opposed to surface tension. Surface tension is overcome by these repelling electrostatic forces, which cause the charged jet of fluid to be ejected from the tip where it causes the polymeric solution jet to solidify and evaporate. Finally, the solvent from the polymeric solution evaporates, resulting in the deposition of dry nanofibers as a web of interconnected tiny fibres on the collector.

Continuous stretching and whipping causes the polymeric jet's diameter to drop from a few hundred micrometres to as little as tens of nanometers. Low voltage causes the polymer solution to drip off the needle, but high enough voltage results in the development of a cone (the so-called Taylor cone). This cone's tip releases a charged fluid jet, which accelerates a polymeric fluid in the direction of a grounded collector. A charged polymer fibre is left behind by solvent evaporation after the discharged polymer solution jet goes through an instability and elongation phase that makes it extraordinarily long and thin. Smooth nanofibers were created under ideal circumstances, but bead-on-string morphology was created due to improper parameters used to create the nanofibers. This outcome was attained as a result of the polymer chains' incomplete stretching brought on by an inappropriate amount of electrostatic forces.

1.3 Parameters of Electrospinning

The electrospinning device can be used to create fibres, droplets, or a beaded structure depending on the different processing settings. A broad variety of processing parameters had a significant impact on the fibre production and structure (Sill et al. 2008). The obtained fibre diameter was used to calculate the electrospun fibre mats' mechanical, electrical, and optical characteristics (Yordem et al. 2008). The three main categories of processing parameters are: (i) properties of the solution used as the feedstock (solution parameters) which include the solvent, polymer concentration, viscosity and solution conductivity (ii) design, geometry, and operation-related parameters of the electrospinning apparatus (processing parameters), which include the applied electric field, distance between the needle and collector, flow rate, needle diameter; and (iii) other parameters (Stanger et al. 2005).

1.3.1 Effect of Solution Concentration

The concentrations of polymer solution used during the electrospinning process have a significant impact on how the fibres are formed. The solution concentration often causes an increase in fibre diameter, and by varying the solution concentration, the viscosity of the solution can also be modified. The process of electro spinning is where the phenomena of uniaxial stretching of a charged jet primarily occurs. Changing the concentration of the polymeric solution has a substantial impact on how much the charged jet stretches. For instance, before the polymeric solution reaches the collector, when its concentration is low, the surface tension and applied electric field cause the entangled polymer chains to fragment (Haider et al. 2013, Pillay et al. 2013), and these fragments result in the formation of beads or beaded nanofibers. The electro spray process rather than electrospinning results in the production of polymeric micro (nano) particles.

At low polymer concentrations due to the low viscosity and high surface tensions of the solution. A mixture of beads and fibres will be created as the polymer concentration is slightly raised. This might be as a result of the fact that as the polymeric solution's concentration rises, the viscosity of the solution rises as well, increasing the chain entanglement between the polymer chains, which subsequently overcomes the surface tension. The end product will be electrospun nanofibers and uniform beads. The polymer solution dries at the tip of the metallic needle and inhibits the flow of the solution, which ultimately leads to the formation of faulty or beaded nanofibers, when the concentration is raised above a critical threshold (the concentration at which beadless uniform nanofibers are created). Smooth nanofibers can be produced at the right concentration, however at very high concentrations, helix-shaped micro ribbons are seen instead of nanoscaled fibres.

1.3.2 *Effect of Solution Viscosity*

There must be a suitable viscosity for electrospinning because solution viscosity is the crucial factor in determining the fibre morphology. It has been established that continuous and smooth fibres cannot be produced in very low viscosities, whereas very high viscosities lead to the hard ejection of jets from solutions. To create beadless nanofibers, the critical concentration/viscosity value must be identified. A polyacrylonitrile (PAN) polymer solution experiment demonstrated that smooth electrospun nanofibers could be produced when the solution's viscosity was maintained between 1.7 and 215 cp. According to Shamin and his collaborators, the morphologies of the beads exhibit an intriguing round droplet-like shape with low viscosity polymer solutions, however with adequate viscosity, the round shape transforms into a stretched droplet or elliptical to smooth fibres (Shamim et al. 2012). Doshi and his colleagues discussed the impact of concentration and viscosity on the morphology of the nanofibers in their paper. They discovered via their work with PEO that 800-4000 cp is the ideal viscosity for the production of electrospun nanofibers (Reneker and Doshi 1995). Additionally, Zong et al. found that viscosity affects the shape of the beads when analysing poly (D,L-lactic acid) (PDLA) and poly(L-lactic acid) (PLLA) (Zong et al. 2002).

1.3.3 *Effect of Solution Conductivity*

The production of the Taylor cone by the conductive polymer solution, which has enough free charges to flow onto the fluid's surface and is required for the formation of Taylor cone, served as the catalyst for the start of the electrospinning process. The polymer, solvent, and salts are the key factors affecting the conductivity of a solution. The Taylor cone creation process is influenced by solution conductivity, which has a significant impact on determining the diameter of nanofibers. The electrospinning process is primarily governed by the Coulomb force between the charges on the fluid's surface and the force due to the applied external electric field, while the formation of the Taylor cone is largely controlled by the electrostatic force of the surface charges created by the applied external electric field. By raising the conductivity of the solution to a crucial value, the charge on the droplet's surface that forms the Taylor cone will be boosted, which will lead to a decrease in the diameter of the fibres (Sun et al. 2014), favouring the growth of thinner ones.

The Taylor cone formation as well as the electrospinning process was hindered by the increase of conductivity beyond a certain critical value. Poor fibre formation occurs with natural polymers as opposed to their synthetic counterparts, which may be related to their polyelectrolytic nature, in which the ions subjected to higher tension under the electric field increase the polymer jet's capacity to carry charge.

Organic acid can occasionally be used as the solvent to achieve high solution conductivity. According to a paper by Hou and his collaborators, the dissolution of nylon 6 using formic acid produces ultrathin (3 nm) electrospun nylon-beaded fibres (Hou et al. 2002). In their experiment, a small amount of pyridine was also added to the solution to make it more conductive, which helped to remove the beads. Ionic salts can be used to create nanofibers with small diameters. Ionic ions like KH2PO4 and NaCl, in particular, can improve the solution's electrical conductivity. A polymer solution's conductivity can be regulated by adding a suitable salt to the mixture.

1.3.4 Effect of Applied Voltage

According to reports, the diameter of the nanofibers increased as the applied voltage increased, and this increase in diameter was linked to a rise in the jet length with the applied voltage. When current flowed from a high-voltage power supply into a solution via a metallic needle, through the deformation of spherical droplet, to a Taylor cone, ultrafine nanofibers were created at a critical voltage (Laudenslager and Sigmund 2012). Each polymer has a different critical value of applied voltage. With an increase in the applied voltage, the smaller-diameter nanofibers were generated as a result of the stretching of the polymer solution in conjunction with the charge repulsion within the polymer jet (Sill and von Recum 2008). When the applied voltage was higher than the critical value, beads or beaded nanofibers were generated. Dietzel and his colleagues reported that beads formed as the applied voltage increased.

1.3.5 Effect of Humidity and Temperature

According to recent reports, several environmental (ambient) elements such the relative humidity and temperature have an impact on the diameter and morphology of the nanofibers in addition to the various electrospinning and solution parameters (Huan et al. 2015, Pelipenko et al. 2013). The humidity modifies the solidification of the charged jet, which alters the diameter of the nanofibers depending on the chemical makeup of the polymer. Pelipenko and his colleagues investigated the variations in nanofiber diameter with changes in humidity using PVA, PEO, and their mix solutions of PVA/hyaluronic acid (HA), and PEO/(chitosan (CS)) (Pilipenko et al. 2013). The observed results show that the diameter of the nanofibers dropped from 667 nm to 161 nm (PVA) and 252 nm to 75 nm with the rise in humidity from 4% to 60% (PEO).

When compared to the single materials, the nanofibers' diameters were found to be even furthermore decreased for the blend. For instance, the diameter of the nanofibers for PVA/HA and PEO/CS fell from 231 nm to 46 nm and 4% to 50%, respectively, as did the humidity. Bead fibre for individual polymers and essentially little electrospinning for blends were the results of increased humidity. Park and Lee also noted that increasing humidity was associated with a comparable reduction in the diameter of PEO nanofibers (Park and Lee 2010). The two conflicting impacts of temperature on nanofibers altered their average diameter. The viscosity of the solution is reduced and the rate of solvent evaporation is accelerated by temperature variations. The increase in solvent evaporation and the decrease in solution viscosity by two opposing mechanisms were responsible for the decrease in mean fibre diameter.

1.3.6 Effect of Flow Rate

The beginning droplet's shape was dramatically changed by the difference in applied flow rate, and some variations in fibre morphology also resulted. The variation in polymer solution flow rate had a significant impact on the distribution of fibre diameters, droplet size and shape at the capillary tip, jet trajectory, maintenance of Taylor cone, areal density, and nanofiber morphology. The lower flow rate is typically preferred since the polymer solution will have ample time to polarise. Due to the quick drying period before reaching the collector and the low stretching forces, if the flow rate of the solution is very high, bead fibres with thicker diameter were generated instead of smooth fibres with thin diameter.

Numerous flaws in the morphology of the fibres were revealed by raising the flow rate, primarily as a result of the higher droplet volume ratio and more unspun droplets. It was also feasible to see additional flaws like branched or splitting fibres and blobs due to the shorter flight duration and inadequate time for complete evaporation of the leftover solvents in the deposited web. At a flow rate of 0.66 ml/h, Yuan and his coworkers found that the effect of flow rate on the morphologies of PSF fibres from a 20% PSF/DMAC solution at 10 kV results in bead fibres with thicker diameters (Yuan et al. 2004).

2. Types of Electrospinning

2.1 Needle Based Electrospinning

A droplet will form at the tip of the needle when a polymer solution is continuously pushed via a syringe in needle-based electrospinning technique. Commonly, a high voltage between 5 and 30 kV is applied between the polymer solution feed and a collecting electrode. As a result of this electric field, which is applied to the end of a capillary tube that contains the polymer fluid held by its surface tension, the charge on the liquid's surface is induced. The mutual charge repulsion produced a force exactly opposite to the surface tension. The Taylor cone, a hemispherical surface of the solution at the tip of the capillary tube that elongates to form a conical shape, was also created as a result of the electric field's intensification.

In other words, we can say that a droplet of polymer solution directed toward the counter electrode forms into a cone under high voltage. Only when the electric field reaches a critical level where the repelling electric force outweighs the surface tension force does a charged jet of the solution shoot out of the cone's tip. The continuous solid fibre was created as the solvent evaporated by the fluid jet's real acceleration and the fluid jet's stretching caused by the coulomb force of the external electric field. The fluid jet rapidly thins as it moves in the direction of the collector, and as it gets smaller, the density of the surface charge rises. This causes the fluid jet to break up into multiple smaller jets due to the strong repulsive forces. The jet is significantly prolonged by bending and whipping processes brought on by electrostatic repulsion that start at tiny bends in the fibre and continue until the jet is eventually deposited on the collector.

2.1.1 Multineedle Electrospinning

The solution flow rate during the single-needle electrospinning procedure typically ranges from 0.1 to 0.3 mL/h. The production of nanofibers is very low as a result of this reduced flow rate, and it also severely restricts the large-scale manufacturing and marketing of electrospun nanofibrous products. The increase in nanofiber yield is the main objective of electrospinning research. Numerous researchers have created multineedle electrospinning devices, which are essentially better single-needle electrospinning devices, in response to existing difficulties.

Multineedle electrospinning process represents the simplest approach used to produce high yield of nanofibers. For this multineedle continuous electrospinning method, a high voltage is required due to the enormous mass of spinning solution delivered and the use of a greater number of needles in which the polymer solution is driven through multiple needles. In the same multiple spinneret arrangements, separate spinning solutions can be independently pumped to two different sets. This method has a few limitations, including clocking at the needle tips, cleaning many needles, an unstable electric field intensity, and a variable fibre size distribution. Even if multi-needle systems achieve high flow rates (1–18 mL/h), the repulsion from nearby jets is still a problem (Khalf et al. 2016).

2.1.2 Electro Blowing / Gas-assisted / Gas Jet Electrospinning

Industrial production of ultrafine thin fibers from both solution and melts was done using the newly emerged gas assisted electrospinning (electro blowing) technology; this technique mainly combines the spinning with velocity air flow. Ultrathin nanofibers were created using an electro blowing/gas-assisted/gas jet electrospinning technology by ejecting electrified polymer solution or melt when a high-speed gas stream is connected to a spinneret. Through this synergy between the air flow and electrostatic forces, the nanofibers were produced from high viscosity polymer melts and high surface tension polymer solutions which are otherwise non electrospinnable. The large scale production of nanofibers using an inexpensive gas assisted electrospinning method was proposed by Varabhas et al. (2008). A new mechanism for launching numerous jets was built with the help of low air pressure flow through the tube, and this is allowing the simultaneous ejection of many electrospinning jets from the surface of the porous tube. The nanofiber production rate can be monitored by altering the number of holes and tube length.

2.1.3 Centrifugal Electrospinning

An increasingly popular approach for producing high-output nanoscale fibres is centrifugal electrospinning (CES), which has accelerated its rapid growth. In the case of centrifugal electrospinning, it was discovered that a combination of centrifugal, electrostatic, gravitational, and auxiliary forces acts on the initial jet. If this combination is greater than the sum of the jet surface tension and viscous resistance, the jet is fully stretched out and ultrafine fibres are produced. Centrifugal electrospinning may successfully produce anisotropic structural features in polymer nanofibers because it combines a number of forces (Peng et al. 2016). In order to create nanofibers with a high degree of alignment and homogeneity on a wide scale, this CES (Centrifugal electrospinning) process merges the concepts of parallel-electrode electrospinning with centrifugal dispersion.

2.2 Needleless Electrospinning

A versatile and alternative method for creating an electrospun nanofiber web, needleless electrospinning technique has a number of advantages over syringe-based electrospinning in terms of productivity and processability. The drawbacks and limitations of the previous needle-based electrospinning systems are overcome by this needleless electrospinning technique. Several start-up businesses have recently completed the commercialization of nanofibers-based products made from needleless electrospinning nanofiber products (Ramakrishnan et al. 2016).

2.2.1 Edge Electrospinning

Exploiting unconfined polymer fluids (such as those from solution or melt), scaled-up manufacture of high-quality nanofibers was made possible using the edge electrospinning method. According to Thoppey and his coworkers, an edge-plate electrospinning configuration was used to increase fibre productivity (Thoppey et al. 2010). Polyethylene oxide (PEO) nanofibers were electrospun using a plate with a 40° inclination to the horizontal for holding fluid. In this procedure, the solution was delivered to the charged plate by one or more plastic pipettes attached to an electrically insulated reservoir, and each pipette served as the place where the jet was initiated. The results shown that even with a single spinning site (one pipette), the production rate may be raised by more than 5 times without resulting in coarser fibres or a broader diameter distribution.

2.2.2 Bubble Electrospinning

A new electrospinning technique called "bubble-electrospinning" uses compressed gas and an aerated solution in an electric field (Liu and He 2007). Essentially, the bubble electrospinning process relies on the production of gas bubbles on the surface of the electrospun polymeric solution to launch electrospun fibre jets. The invention of bubble electrospinning was made possible by the discharge of liquid jets from the soap bubbles apex. The liquid's surface tension and bubble radius were used in conjunction with the applied voltage to determine the production of a liquid jet.

Recently, the electrospinning polymer nanofibers from bubbles have been demonstrated using a single-nozzle apparatus (Liu et al. 2010). Several empirical relationships have been reported linking the fibres to particular process parameters, such as electric field strength (Liu et al. 2011) and solution concentration (Yang et al. 2010, Ren et al. 2011). The bubbles were started on the surface of the polymer solution (flat open surface) by blowing an inert gas into the polymer solution, and this will cause the bubble formation. It is possible for fibres to start from surfaces other than the bubble surface as a result of the electrospinning from a single bubble in a polymer bath, such as the container edge and solution surface.

2.2.3 Rotating Roller Electrospinning/Nanospider Technology

In industrial procedures, a special technique known as rotating electrospinning was used to continually generate nanofibers. With this technique, a rotating roller known as a nanospider is used to spin fibres directly out of a polymer solution. The rotating roller in this technology is connected to a high applied voltage source, which will cause the creation of nanofibers. Regarding the qualities of the nanofibers, the roller speed was quite important. The simultaneous formation of several Taylor cones on the rotating, spinning electrode makes the method extremely productive (Yener and Jirsak 2011, 2012).

This roller-based electrospinning technique assessed new quantitative measurement parameters, including the Taylor cone, density per surface area, spinning performance for one Taylor cone, overall spinning performance, Fiber Diameter Uniformity Coefficient (FDUC), and Non Fibrous Area Coefficient (NFA)17 (Cengiz et al. 2013). Mohammed H. El Newey and his colleagues described how nylon-6 nanofibers carrying 5,5-dimethyl hydantoin (DMH) as an antibacterial medication, with diameters of 15–328 nm, were produced using nanospider technology for use in biomedical applications. The acquired outcomes unmistakably show that the produced electrospun nylon-6 nanofiber is promising for applications in wound-healing (El-Newehy et al. 2011).

3. Electrospinning Technologies

3.1 Solution Spinning

Dry spinning and wet spinning are both parts of the solution spinning process. The polymer powder is first dissolved in a suitable solvent in the case of wet spinning, and the polymer solution is then extruded through a spinneret into a solvent-non solvent combination (coagulant). Due to the mutual diffusion of the solvent and non-solvent, this polymer solution coagulates to form fibres. The dry spinning technique is one that primarily entails the controlled evaporation of the fibre in the spinline, which transforms a high vapour pressure polymer solution into a solid fibre.

Since many polymer solutions have very low viscosities, solution electrospinning has the benefit of making very thin fibres (average fibre diameter ranges from 75 nm to 500 nm) relatively quickly. The dry-jet wet spinning of cellulose and cellulose/MWCNTs composite fibres employing a room temperature ionic liquid, ethyl methyl imidazolium acetate as a solvent for cellulose was reported by Sameer S. Rahatekar and his collaborators. According to the findings, solution spinning of cellulose/MWCNTs composite fibres with improved tensile strength, decreased thermal shrinkage, and high electrical conductivity may be successfully accomplished using the ethyl methyl imidazolium acetate solvent (EMIAc) (Rahatekar et al. 2009).

3.2 Melt Spinning

Melt-electrospinning, also known as electrospinning of molten polymer, is a processing method used to create fibrous structures from polymer melts for uses such as tissue engineering, textiles and filtration. Thin ribbons of metal or alloys with a particular atomic structure was also formed from the melt spinning process. The melting of polymers that are not decomposed or degraded by the temperatures to a suitable viscosity for extrusion process was achieved using melt spinning process. According to Rudolf Hufenus, a specially designed pilot melt-spinning factory can produce prototype mono-, bi-, and tri-component fibres with a throughput of up to several kg/hr and a variety of cross-sections and material combinations. Enhancing mechanical properties (Fornes et al. 2006, Sandler et al.

2004), implementing electric or magnetic functions (Kim et al. 2004, Liang et al. 2007), introducing biologically active species like drugs or silver composites (Yeo et al. 2003, Datsjerdi et al. 2008), and varying fibre morphology by bicomponent spinning are recent research activities in polymer melt spinning process (Shi et al. 2006, Huang et al. 2001).

3.3 Emulsion Spinning

Based on the solution, the electrospun fibers of different composition and structure were created from the concept of emulsion electrospinning process. A polar and non-polar solution typically make up an emulsion, which is made up of tiny droplets of one solution that is insoluble in the other. Emulsion electrospinning is intended to make it possible to load these medications into polymers that have organic solvent dissolved in them for spinning into fibres. It has also been used to create membranes with volatile aroma. The emulsion could be electrospun to create uniform nanofibers with a diameter of about 65 nm. The electrospinning of emulsions produced hollow fibres within the fibre, which may be related to the elongation and merging of droplets in the solution during electrospinning.

Zhang and his colleagues (Zhang et al. 2012) created axially aligned TiO_2 nanofibers by first electrospinning tetrabutyl titanate (TBT) with paraffin oil, followed by calcinations. For the purpose of encapsulating paraffin oil, butoxyl groups in TBT operate as an extra surfactant in addition to cetyltrimethylammonium bromide. Without paraffin oil, solid fibres were produced, and in rare cases, core-shell fibres may be created by placing immiscible droplets in a straight line along the fiber's core (Wang et al. 2014).

4. Electrospun Fibers: Morphologies and Unique Properties

In order to meet their various application needs, the fibers can be produced in different structural organization and morphologies by utilizing electrospinning technology. To correctly engineer the sample and achieve the necessary qualities, a variety of materials, including natural and synthetic polymers, and different types of solvents, including water and organic solvents, were used. Shamim Zargham and her colleagues looked at the impact of flow rate on the shape and deposition area of electrospun Nylon 6 nanofiber (Zargham et al. 2012). The droplet size and form at the capillary tip, the trajectory of the jet, the maintenance of the Taylor cone, the distribution of fibre diameter, and the morphology of the created nanofibers are all affected by the flow rate of the polymer solution, according to the results. Without enough solvent evaporation, fibres were gathered at a higher flow rate, which resulted in the production of several defects such blobs, split, branching, and flattened web-like structures.

Electrospun fibres have key characteristics that can be adjusted for a variety of applications, including fibre diameter, surface morphology, and mechanical strength (Huang et al. 2011). Although electrospun nanofiber mats are exceedingly porous, they typically have limited mechanical strength. The strength of the individual fibres as well as the bonding between fibres are both factors that affect the mechanical properties, which are frequently quantified as tensile strength and Young's modulus.

5. Emerging Applications of Electrospun Nanofibers

5.1 Drug Delivery and Release Control

Recently, it has been possible to use the nanofibers and microfibers created by the adaptable electrospinning technology as a medication delivery mechanism (DDS). Due to the large variety of materials that may be used with this electrospinning method, it has developed into a potent instrument for the creation of micro- or nanostructures ideal for applications in the domains of tissue engineering and drug delivery (Szentivanyi et al. 2011, Zamani et al. 2013, Zernetsch et al. 2013, Pfeiffer et al. 2014).

The introduction of innovative drug delivery approaches focusing on nanoscale delivery over the past ten years has improved drug release control, levels of dosing, local delivery, and also reduced undesirable side effects from systemic administration (Blanco et al. 2009, Vijayaraghavalu et al. 2007). The medications can be supplied via release systems through the electrospun fibres, which can be either in free form or by carrying electrospun fibres inside of other nanostructures or by making these agents available on the surface of the fibre. The electrospun nanofiber even affects cell development while acting as a mechanical support for the cell as a substrate, scaffold, or matrix.

5.2 Tissue Engineering Scaffolds

Langer and Vacanti defined tissue engineering as an interdisciplinary field that combines engineering and life science ideas to the creation of biological substitutes that restore, maintain, or improve tissue function in the year 1993. (Langer and Vacanti 1993). In order to modulate cellular behaviour, polymeric scaffolds have been designed with specific mechanical and biological properties similar to native extracellular matrix. Electrospun nanofibers have been used as promising tissue engineering scaffolds because they mimic the nanoscale properties of native extracellular matrix (ECM).

5.3 Wound Healing

Electrospinning, a facilitating nanotechnology, has drawn a lot of interest in the field of wound healing since it can create biomimetic nanofibrous materials from a wide range of natural and synthetic polymers with biologically relevant properties. For the purpose of promoting wound healing, several therapeutic substances (such as antimicrobials, antioxidants, anti-inflammatory medicines, anaesthetics, enzymes, and growth factors) have been added into electrospun nanofibers. When compared to conventional nanofiber drug-delivery systems, environment-responsive electrospun nanofibers may provide quicker response times and more accurate control over the release rate of therapeutic drugs for wound healing.

5.4 Neural Tissues, Blood Vessels, Bones and Cartilages

It remains a risky endeavour for many researchers to prepare small diameter blood arteries because of the occurrence of thrombus development, intimal hyperplasia, and aneurysmal dilation. The extracellular matrix (ECM) of native blood vessels must meet certain requirements for construction, and among the various tissue engineering approaches, the electrospinning technique was found to be the most promising one. It can produce seamless fibrous tubes with fibre diameter controllable in either nanoscale or microscale (polymer fibrous scaffolds), which promote cell adhesion, proliferation, and growth (Awad et al. 2018).

Good biomechanical qualities can be found in small diameter blood arteries made from a variety of synthetic polymers, natural polymers, and hybrid polymer-based scaffolds. Poly—caprolactone (PCL), poly-lactic acid (PLA), polyurethane (PU), and poly (lactide-co-caprolactone) are a few examples of synthetic polymers (PLCL). The thickness of the fibres, the treatment utilised before and after manufacturing, including sterilisation of α- or γ-radiations, all affect the mechanical properties as well as the cell response of fibrous scaffolds made of these polymers.

Beom Su Kim and his colleagues discussed how the nanofiber content of silk fibroin/poly(-caprolactone) nano/microfibrous composite scaffolds affected bone regeneration. The SF/PCL nano/microfibrous composite scaffold appears to offer a favourable environment for hMSC proliferation, adhesion, and differentiation into osteoblasts *in vitro*, and it was also clear that the scaffolds induced new bone formation in a rabbit calvarial defect model *in vivo* (Kim et al. 2015).

Febe Carolina Vazquez-Vazquez and his team examined the biocompatibility response of a 3D-printed tubular scaffold covered in a layer of 7% PLA nanofibers. Results revealed that the nanofiber coating on the surface of the 3D tubular scaffold improved cell adhesion, proliferation, and the shape of osteoblast cells when compared to a noncoated scaffold. Finally, a nanofiber layer could be applied to the surface of the 3D-printed tubular scaffold to create a more mimetic and active topography with great cellular biocompatibility for bone tissue applications (Vazquez-Vazquez et al. 2019).

The electrospinning process was discovered to be a useful tool for cartilage tissue engineering because it can create fibrous scaffolds that replicate the nanoscale and alignment of collagen fibres found in the superficial zone of articular cartilage (CTE). The application of kartogenin-loaded coaxial PGS/PCL aligned nanofibers for cartilage tissue engineering was discussed in a study by Joo C. Silva and colleagues. When compared to monoaxial PCL aligned scaffolds, coaxial PGS/PCL aligned electrospun scaffolds have the ability to produce a significantly more prolonged release of KGN. This is demonstrated by the successful manufacture and characterization of these scaffolds. Overall findings show that KGN-loaded coaxial aligned nanofibers were used for the creation of novel biomimetic MSC-based techniques to regenerate articular cartilage, specifically for the repair of lesions in its superficial zone (Silva et al. 2020).

6. Future Prospects for Research

Nowadays, science and technology continue to pay considerable attention to the recent new technologies. Numerous issues, including the repetitive manufacture of homogenous nanofibers at high volume levels with specified morphological, mechanical, and chemical properties for end user applications, have been addressed in the near future. To achieve desired results in the area of electrospun nanofiber processing parameter optimization, numerous investigations must be conducted. A variety of innovative hybrid polymer systems were enhanced to fit a variety of applications. These systems are electrospinnable and based on synthetic and natural polymers. Great efforts must be made to lower fibre widths while simultaneously generating a high degree of orientation in the structure in order to achieve the theoretically projected strengths. Numerous research projects on electrospinning techniques are currently being conducted for a variety of applications. In the near future, a sizable area of systematic research will need to be done in electrospinning techniques to develop new and more selectively responsive materials and ultimately couple them with other biomedical applications.

7. Conclusions

The manufacturing of nanofibers using electrospinning techniques has advanced significantly, thanks to its comparatively amazing properties in terms of fibre size, production economy, scale-up capability, and ease of adjustments while operating. Due to their unique properties, electrospun nanofibers are appealing for a variety of applications in the life sciences, including tissue engineering, drug delivery, and wound healing, protective clothing, sensors, filtration, reinforcement of composite materials, microelectronics, space applications, and microwave absorption. We anticipate that this chapter gave researchers attempting to identify novel materials with novel properties for the valuable biomedical uses of these electrospun nanofiber materials in cutting-edge domains fresh insights on the usage of electrospinning technologies.

Acknowledgement

The authors are grateful to authorities of Madras Christian College, East Tambaram, Chennai for the support. They are also thankful to the editors for the given opportunity to review such an innovative field.

References

Andrzej Ziabicki. 1976. Fundamentals of Fibre Formation: The Science of Fibre Spinning and Drawing. Wiley, 488 pp.

Awad Nasser K., Haitao Niu, Usman Ali, Yosry S. Morsi and Tong Lin. 2018. Electrospun fibrous scaffolds for small-diameter blood vessels: A review. Membranes (Basel). 8(1): 15.

Baumgarten, P.K. 1971. Electrostatic spinning of acrylic fibers. Journal of Colloid and Interface Science 36(1): 71–79.

Bhardwaj, N. and S.C. Kundu. 2010. Electrospinning: A fascinating fiber fabrication technique. Biotechnology Advances 28: 325–347.

Blanco, E., Chase W. Kessinger, Baran D. Sumer and Jinming Gao. 2009. Multifunctional micellar nanomedicine for cancer therapy. Exp. Biol. Med. Maywood. 234(2): 123–131.

Cengiz, C.F., O. Jirsak and M. Dayik. 2013. The influence of nonsolvent addition on the independent and dependent parameters in roller electrospinning of polyurethane. Journal of Nano Science and Nanotechnology 13(7): 4727–35.

Dae Hyun Yoo, Pawel Hrycaj, Pedro Miranda, Edgar Ramiterre, Mariusz Piotrowski, Sergii Shevchuk, Volodymyr Kovalenko, Nenad Prodanovic, Mauricio Abello-Banfi, Sergio Gutierrez-Ureña, Luis Morales-Olazabal, Michael Tee, Renato Jimenez, Omid Zamani, Sang Joon Lee, HoUng Kim, Won Park and Ulf Müller-Ladner. 2013. A randomised, double-blind, parallel-group study to demonstrate equivalence in efficacy and safety of CT-P13 compared with innovator infliximab when coadministered with methotrexate in patients with active rheumatoid arthritis: The PLANETRA study. Ann. Rheum. Dis. 72(10): 1613–20.

Dastjerdi, R., M.R.M. Mojtahedi and A.M. Shoshtari. 2008. Investigating the effect of various blend ratios of prepared masterbatch containing Ag/TiO2 nanocomposite on the properties of bioactive continuous filament yarns. Fibers and Polymers 9(6): 727–734.

Deitzel, J.M., J. Kleinmeyer, J.K. Hirvonen and N.C.B. Tan. 2001. Controlled deposition of electrospun poly (ethylene oxide) fibers. Polymer 42: 8163–8170.

Doshi, J. and D.H. Reneker. 1995. Electrospinning process and applications of electrospun fibers. J. Electrostat. 35(2-3): 151–160.

El-Newehy, Mohamed H., Salem S. Al-Deyab, El-Refaie Kenawy and Ahmed Abdel-Megeed. 2011. Nanospider technology for the production of nylon-6 nanofibers for biomedical applications. Journal of Nanomaterials Volume 2011, Article ID 626589, 8 pages.

Fatma YENER and Oldrich Jirsak. 2011. Improving performance of polyvinyl butyral electrospinning. 3rd International conference on NANOCON, 21. – 23. 9. 2011, 8000-0655, Brno, Czech Republic, EU.

Fatma Yener and Oldrich Jirsak. 2012. Fabrication and optimization of polyvinyl butyral nanofibres produced by roller electrospinning. 12th World Textile Conference Autex June 13th To 15th 2012, Zadar, Croatia.

Fornes, T.D., J.W. Baur, Y. Sabba and Edwin L. Thomas. 2006. Morphology and properties of melt-spun polycarbonate fibers containing single- and multi-wall carbon nanotubes. Polymer 47(5): 1704–1714.

Haider, S., Y. Al-Zeghayer, F. AhmedAli, A. Haider, A. Mahmood, W. Al-Masry, M. Imran and M. Aijaz. 2013. Highly aligned narrow diameter chitosan electrospun nanofibers. J. Polym. Res. 20(4): 1–11.

Hou, H., J. Zeng, A. Reuning, A. Schaper, J.H. Wendorff and A. Greiner. 2002. Poly(p-xylylene) nanotubes by coating and removal of ultrathin polymer template fibers. Macromolecules 35(7): 2429–2431.

Huan, S., G. Liu, G. Han, W. Cheng, Z. Fu, Q. Wu and Q. Wang. 2015. Effect of experimental parameters on morphological, mechanical and hydrophobic properties of electrospun polystyrene fibers. Materials 8(5): 2718.

Huang, J., Donald G. Baird, Alfred C. Loos, Priya Rangarajan and Aaron Powell. 2001. Filament winding of bicomponent fibers consisting of polypropylene and a liquid crystalline polymer. Composites Part A: Applied Science and Manufacturing 32(8): 1013–1020.

Huang Liwei, Nhu-Ngoc Bui, Seetha S. Manickam and Jeffrey R. McCutcheon. 2011. Controlling electrospun nanofiber morphology and mechanical properties using humidity. Journal of Polymer Science Part B: Polymer Physics 49: 1734–1744.

Ji, W., Y. Sun, F. Yang, J.J.P. Van den Beucken, M. Fan, Z. Chen and J.A. Jansen. 2011. Bioactive electrospun scaffolds delivering growth factors and genes for tissue engineering applications. Pharm. Res. 28: 1259–1272.

Khalf, A. and S.V. Madihally. 2016. Recent advances in multiaxial electrospinning for drug delivery. Eur. J. Pharm. Biopharm. http://dx.doi.org/10.1016/j.ejpb.2016.11.010.

Kim, B., Vladan Koncar, Eric Devaux, Dufour Claude and Pierre Viallier. 2004. Electrical and morphological properties of PP and PET conductive polymer fibers. Synthetic Metals 146(2): 167–174.

Kim Beom Su, Ko Eun Park, Min Hee Kim, Hyung Keun You, Jun Lee and Won Ho Park. 2015. Effect of nanofiber content on bone regeneration of silk fibroin/poly(ε-caprolactone) nano/microfibrous composite scaffolds. International Journal of Nanomedicine 10: 485–502.

Langer, R. and J.P. Vacanti. 1993. Tissue engineering. Science 260: 920.

Laudenslager, M.J. and W.M. Sigmund. 2012. Electrospinning Encyclopedia of Nanotechnology. Springer Publishers, 769–775.

Liang, Y., Xiaohong Xia, Yongsong Luo and Zhijie Jia. 2007. Synthesis and performances of Fe2O3/PA-6 nanocomposite fiber. Materials Letters 61(14-15): 3269–3272.

Liu, Y. and J.-H. He. 2007. Bubble electrospinning for mass production of nanofibers. Int. J. Nonlinear. Sci. 8: 393.

Liu, Y., L. Dong, J. Fan, R. Wang and J. Yu. 2011. Effect of applied voltage on diameter and morphology of ultrafine fibers in bubble electrospinning. J. Appl. Polym. Sci. 120: 592–598.

Liu, Y., Z.-F. Ren and J.-H. He. 2010. Bubble electrospinning method for preparation of aligned nanofibre mat. Mater. Sci. Technol. 26: 1309–1312.

Nick Tucker, Jonathan J. Stanger, Mark P. Staiger, Hussam Razzaq and Kathleen Hofman. 2012. The history of the science and technology of electrospinning from 1600 to 1995. J. Eng. Fib. Fabr, Special Issue-July 2012–Fibers, 63–73.

Park, J.-Y. and I.-H. Lee. 2010. Relative humidity effect on the preparation of porous electrospun polystyrene fibers. J. Nanosci. Nanotechnol. 10(5): 3473–3477.

Pelipenko, J., J. Kristl, B. Janković, S. Baumgartner and P. Kocbek. 2013. The impact of relative humidity during electrospinning on the morphology and mechanical properties of nanofibers. International Journal of Pharmaceutics 456: 125–134.

Peng Hao, Yong Liu and Seeram Ramakrishna. 2017. Recent development of centrifugal electrospinning. J. Appl. Polym. Sci. DOI: 10.1002/APP.44578.

Pfeiffer, D., C. Stefanitsch, K. Wankhammer, M. Muller, L. Dreyer, B. Krolitzki, H. Zernetsch, B. Glasmacher, C. Lindner, A. Lass, M. Schwarz, W. Muckenauer and I. Lang. 2014. Endothelialization of electrospun polycaprolactone (PCL) small caliber vascular grafts spun from different polymer blends. J. Biomed. Mater. Res. A 102: 4500–4509.

Pillay, V., C. Dott, Y.E. Choonara, C. Tyagi, L. Tomar, P. Kumar, L.C. du Toit and V.M.K. Ndesendo. 2013. A review of the effect of processing variables on the fabrication of electrospun nanofibers for drug delivery applications J. Nanomater. 2013: 22.

Ramakrishna, S., K. Fujihara, W.E. Teo, T.C. Lim and Z. Ma. 2005. An introduction to electrospinning and nanofibers. World Scientific, New Yersey.

Ramakrishnan Ramprasath, Jolius Gimbun, Fahmi Samsuri, Vigneswaran Narayanamurthy, Natarajan Gajendran, Yamini Sudha Lakshmi, Denisa Stranska and Balu Ranganathan. 2016. Needleless electrospinning technology— An entrepreneurial perspective. Indian Journal of Science and Technology 9(15). DOI: 10.17485/ijst/2016/ v9i15/91538.

Ren, Z. and J. He. 2011. Single polymeric bubble for the preparation of multiple micro/nano fibers. J. Appl. Polym. Sci. 119: 1161–1165.

Reneker, D.H., A.L. Yarin, E. Zussman and H. Xu. 2007. Electrospinning of nanofibers from polymer solutions and melts. Adv. in App. Mechanics 41: 43–195.

Rudolf Hufenus. Fiber Development by Multicomponent Melt-Spinning. https://www.researchgate.net/publication/ 255825756.

Sameer S. Rahatekar, Asif Rasheed, Rahul Jain, Mauro Zammarano, Krzysztof K. Koziol, Alan H. Windle, Jeffrey W. Gilman and Satish Kuma. 2009. Solution spinning of cellulose carbon nanotube composites using room temperature ionic liquids. Polymer 50: 4577–4583.

Sandler, J.K.W., S. Pegela, M. Cadek, F. Gojny, M. van Es, J. Lohmar, W.J. Blau, K. Schulte, A.H. Windle and M.S.P. Shaffer. 2004. A comparative study of melt spun polyamide-12 fibres reinforced with carbon nanotubes and nanofibers. Polymer 45(6): 2001–2015.

Shamim Zargham, Saeed Bazgir, Amir Tavakoli, Abo Saied Rashidi and Rogheih Damerchely. 2012. The effect of flow rate on morphology and deposition area of electrospun nylon 6 nanofiber. Journal of Engineered Fibers and Fabrics 7(4): 42–49.

Shi, X.Q., H. Ito and T. Kikutani. 2006. Structure development and properties of high-speed melt spun poly(butylene terephthalate)/poly(butylene adipate-coterephthalate) bicomponent fibers. Polymer 47(2): 611–616.

Silva João C., Ranodhi N. Udangawa, Jianle Chen, Chiara D. Mancinelli, Fábio F.F. Garrudo, Paiyz E. Mikael, Joaquim M.S. Cabrala, Frederico Castelo Ferreira and Robert J. Linhardt. 2020. Kartogenin-loaded coaxial PGS/PCL aligned nanofibers for cartilage tissue engineering. Materials Science & Engineering C 107: 110291.

Sill, T.J. and H.A. Von Recum. 2008. Electrospinning: Applications in drug delivery and tissue engineering. Biomaterials 29: 1989–2006.

Stanger, J., N. Tucker and M. Staiger. 2005. Electrospinning. Rapra Review Report, Report 190(16): 10.

Sun, B., Y.Z. Long, H.D. Zhang, M.M. Li, J.L. Duvail, X.Y. Jiang and H.L. Yin. 2014. Advances in three-dimensional nanofibrous macrostructures via electrospinning. Prog. Polym. Sci. 39(5): 862–890.

Szentivanyi, A., T. Chakradeo, H. Zernetsch and B. Glasmacher. 2011. Electrospun cellular microenvironments: Understanding controlled release and scaffold structure. Adv. Drug Deliv. Rev. 30: 209–220.

Thoppey, N.M., J.R. Bochinski, L.I. Clarke and R.E. Gorga. 2010. Unconfined fluid electrospun into high quality nanofibers from a plate edge. Polymer 51(21): 4928–4936.

Varabhas, J.S., G.G. Chase and D.H. Reneker. 2008. Electrospun nanofibers from a porous hollow tube. Polymer 49(19): 4226–4229.

Vazquez-Vazquez Febe Carolina, Osmar Alejandro Chanes-Cuevas, David Masuoka, Jesús Arenas Alatorre, Daniel Chavarria-Bolaños, José Roberto Vega-Baudrit, Janeth Serrano-Bello and Marco Antonio Alvarez-Perez. 2019. Biocompatibility of developing 3d-printed tubular scaffold coated with nanofibers for bone applications. Journal of Nanomaterials Article ID 6105818, 13 pages.

Vijayaraghavalu, S., D. Raghavan and V. Labhasetwar. 2007. Nanoparticles for delivery of chemotherapeutic agents to tumors. Curr. Opin. Investig. Drugs 8: 477–484.

Wang, W., L. Wang and M. Wang. 2014. Evolution of core-shell structure: From emulsions to ultrafine emulsion electrospun fibers. Materials Letters, Article in Press.

Wendy Zhang, Nadia O'Brien, Jamie I. Forrest, Kate A. Salters, Thomas L. Patterson, Julio S.G. Montaner, Robert S. Hogg and Viviane D. Lima. 2012. Validating a Shortened Depression Scale (10 Item CES-D) among HIV-Positive People in British Columbia, Canada.

Xiaoyan Yuan, Yuanyuan Zhang, Cunhai Dong and Jing Sheng. 2004. Morphology of ultrafine polysulfone fibers prepared by electrospinning 53(11): 1704–1710.

Yang, Y., N. Zhang, M. Xue and S. Tao. 2010. Environmental pollution impact of soil organic matter on the distribution of polycyclic aromatic hydrocarbons (PAHs) in soils, pp. 2170–2174.

Young-Seak Lee and Ji Sun Im. 2010. Preparation of functionalized nanofibers and their applications. DOI: 10.5772/8150.

Zeleny, J. 1914. The electrical discharge from liquid points and a hydrostatic method of measuring the electric intensity at their surface. Phys Rev. 3: 69–91.

Zong, X., K. Kim, D. Fang, S. Ran, B.S. Hsiao and B. Chu. 2002. Structure and process relationship of electrospun bioabsorbable nanofiber membranes. Polymer 43(16): 4403–4412.

5

Toxicity Towards Nanostructured Biomaterials in Human Health

M.I. Niyas Ahamed[1,]* and *V. Ragul*[2]

1. Introduction

In the emerging field of science and technology, nanotechnology plays a great impact by its structure, devices and functional materials at nanodimension ranges between 0.1 to 100 nm. In the 21st century, its most efficient and commercial methodologies has great impact on day to day life of each and every individual to develop and protect human health (Burgos 2005). In addition, all the living entities are exposed to a number of nanosized materials. In general, it inspires the attention of the scientific community, industries and general public in the world. The Asian, American and European government agencies interestingly (Nal et al. 2006).

The USA government offers 3.67 dollars to carry out the nanotechnology based research work through National Nanotechnology Research and Development focused to enhance the science, medicine & engineering field. In the year of 2015, the worldwide market had around one trillion dollars by this technology (Englert 2004). The Nobel Prize winner Richard Feynman acknowledged that 'there's plenty of room at the bottom', to the American Physical Society. In 1985, H.W. Kroto and his team achieved the milestone by the discovery of C_{60} fullerene in the field of nanotechnology. For this extraordinary discovery, in the year of 1996 Sir Harold W. Kroto, Robert F. Curl and Richard E. Smallley won the Nobel Prize in Chemistry (Oberdorster 2002).

Two decades ago, man-made nanomaterials were initially synthesized on the laboratory level like nanotubes, nanowires, quantum dots, dendrimers, nanoclusters, nanocoatings, which are worldwide emerging materials. A wide number of nanostructured materials possess unique size with the range of 0.1 to 100 nm in a significant manner (Brown et al. 2006).

The molecular alignment, size, diameter and surface properties of the nanostructured materials were determined by scanning electron microscopy (SEM), transmission electron microscopy (TEM), scanning tunneling microscopy (STM), high resolution transmission electron microscopy (HRTEM) and atomic force microscopy (AFM). It has effective range of applications in electronics, optoelectronics, biomedical engineering, photonics, biotechnology, skin care, consumer products, pharmaceuticals, environment, composites, coatings, agriculture and engineering technologies. Figure 1 represents the application of nanoparticles (Ahamed 2014).

[1] Assistant Professor, Department of Biochemistry, Sacred Heart College (Autonomous), Tirupattur, Tamil Nadu, India.
[2] Research Scholar, Department of Biochemistry, Sacred Heart College (Autonomous), Tirupattur, Tamil Nadu, India.
* Corresponding author: driniyasahamed@shctpt.edu

APPLICATIONS OF NANOPARTICLES

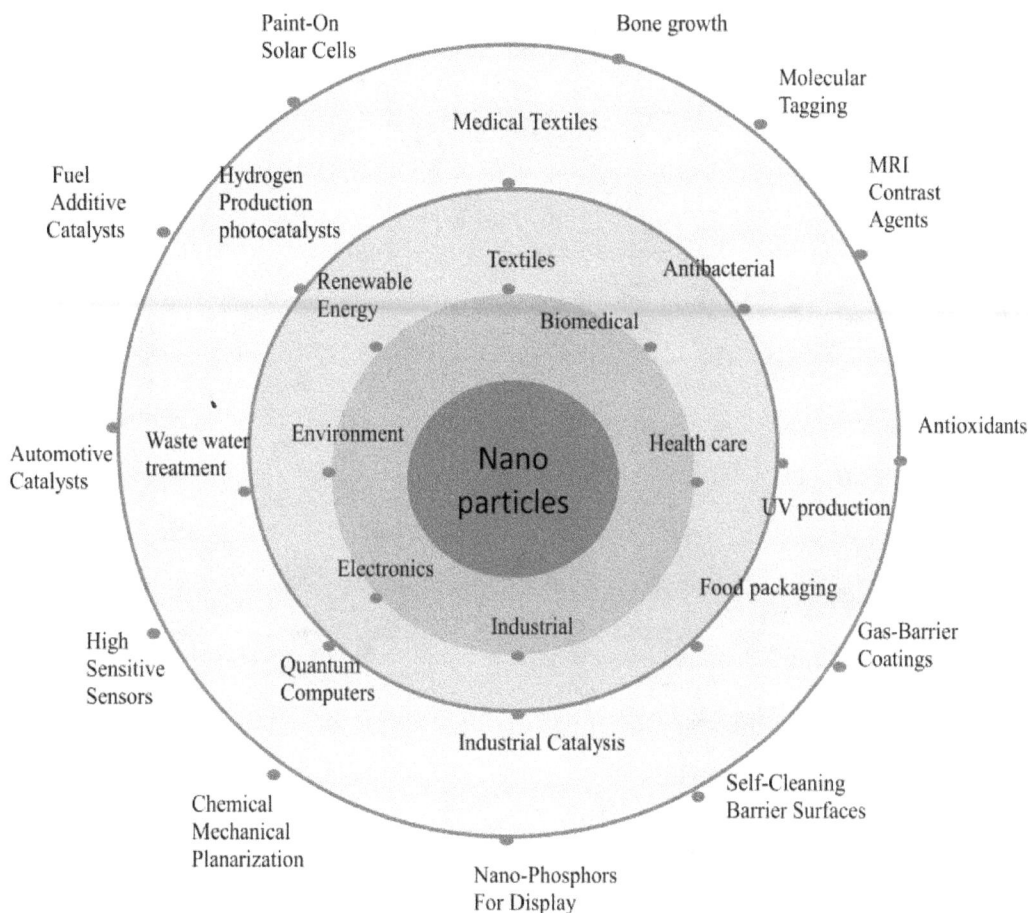

Figure 1: Application of nanoparticles (Ahamed 2014).

The toxicity effect of nanosized materials on the environment and human health care demands great attention. The emerging scientific communities, police and common people are very much eager to learn about adverse effects and biocompatibility for the upcoming generations (Singh and Nalwa 2007). The inhaled nanomaterials directly penetrate into the brain through the lungs. Another great impact is industrial exposure. This is a very new topic in research called nanotoxicology, a science dealing with the toxic effect of nanosized materials on the biological system including plants and laboratory animals (Qiu et al. 2018).

The interaction of nanomaterials with the biological system has produced safety and curiosity concerns from eminent researchers working in this area in industries, public and government sectors (Carisp et al. 2016). The Environmental Protection Agency (EPA) of the United States provides extreme knowledge about nanotoxicity towards the biological system. This initiative enhances the researchers' focus in the field of nanotoxicology were shown in Figure 2. This was effectively undertaken by Europe, USA and Asia to promote nanotoxicity studies (Hermans et al. 2016).

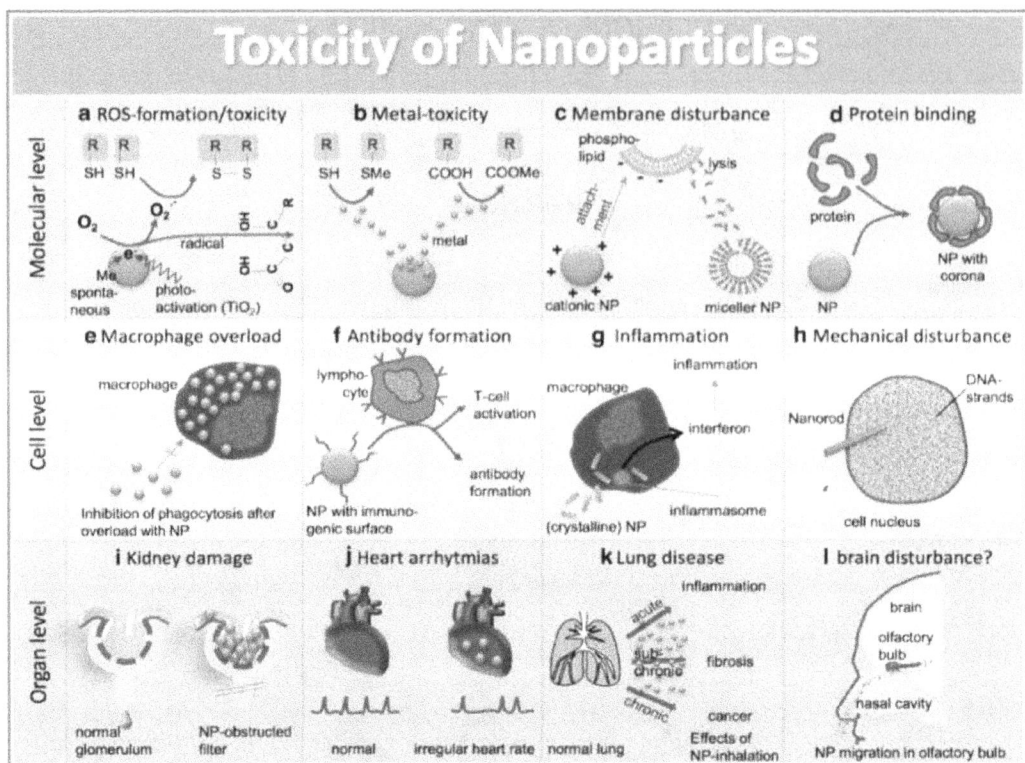

Figure 2: Toxicity of nanoparticles (Mensch et al. 2017).

2. Nanostructured Materials

The term nanoscale (1 meter to 10^{-9} m) was simply expressed, the nanostructure fiber is 1000 times thinner than that of human hair and its characteristic range is 1–100 nm (Figures 3 and 4). The nanoscale materials offer great and improved unique properties to all other bulk counterparts. Here the thickness of 6 nanometer copper grains is improved fivefold when compared with that of normal copper bulk materials (Mensch et al. 2017).

The obtained fibers from carbon nanotube based materials expressed greater hardness than steel, spider silk or Kevlar. The term nano offers 10^{-9} mete. It is expressed as 1 nm = 1/1000000000 m. The one billionth of meter (10^{-9} m) refers to a nanometer (Hussian et al. 2014). It means that if the diameter of the fiber were decreased from a micrometer to nanometer scale, it would be 1000 times thinner. In layman's language, we can have 1000s of nanometer scale fibers within the specified area of micrometer scale fiber (Von Moos and Slaveykova 2014).

As the diameter of the particle size went down from micrometer scale to nanometer scale, there was an great impact on the particles number in the specific chemical reactivity and as a consequence there will be much interaction among the nanoparticles and all the biological systems (Feng et al. 2015).

From the toxicological approach, nanoparticles enlighten the biological interactions between significant developments in toxicological eliminations in the system. The nanostructured particles more effectively affect the different organs absorbed by the biological system compared to micrometer scale and big particles (Jacobson et al. 2015). In the positive impact, these nanostructured particles provide us with guidance to reach the target and expose the medicine to different parts of the human body and all the biological systems. For example, the nanostructured materials effectively and easily reach out the brain very easily when compared to other organs (Mensch et al. 2018).

NANOMETER SCALE

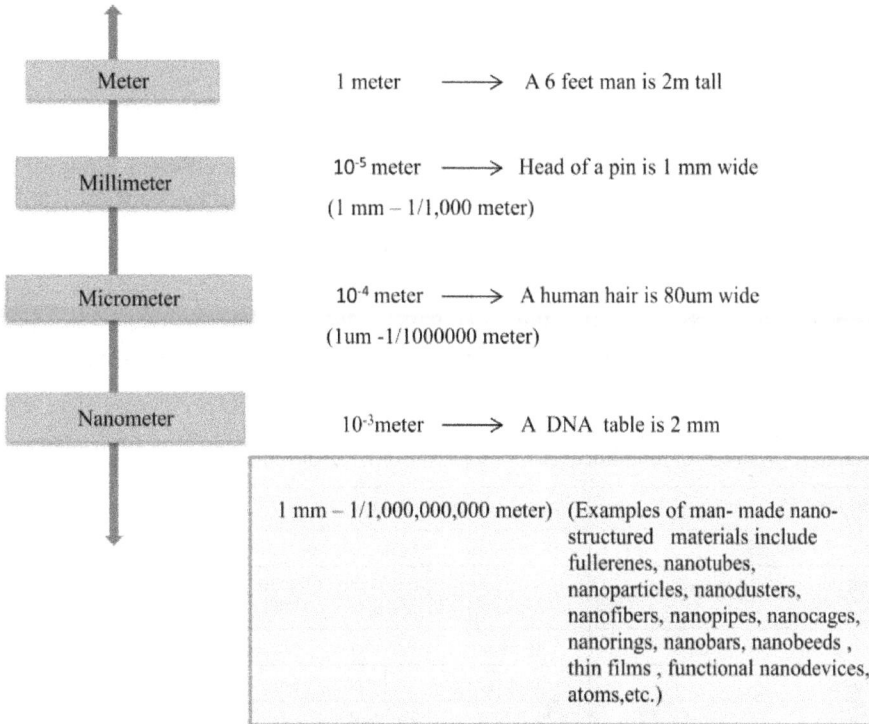

Meter

1 meter ⟶ A 6 feet man is 2m tall

Millimeter

10^{-5} meter ⟶ Head of a pin is 1 mm wide
(1 mm – 1/1,000 meter)

Micrometer

10^{-4} meter ⟶ A human hair is 80um wide
(1um -1/1000000 meter)

Nanometer

10^{-3}meter ⟶ A DNA table is 2 mm

1 mm – 1/1,000,000,000 meter) (Examples of man- made nanostructured   materials include
fullerenes, nanotubes,
nanoparticles, nanodusters,
nanofibers, nanopipes, nanocages,
nanorings, nanobars, nanobeeds ,
thin films , functional nanodevices,
atoms,etc.)

Figure 3: Nanometer scale for nanoparticles (Mensch et al. 2018).

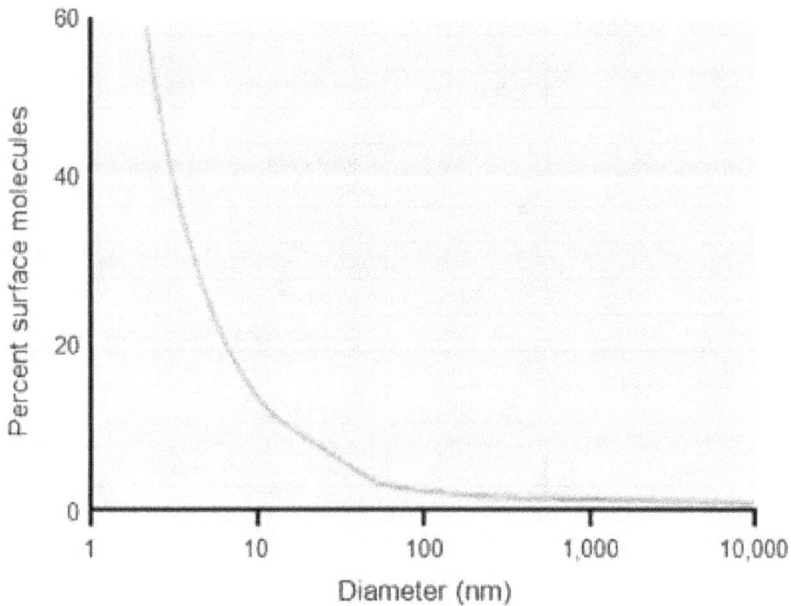

Figure 4: Diameter and percentage of surface molecule (Mensch et al. 2018).

This chapter discusses the toxicological impact of man-made nanoparticles on health hazards from pharmaceuticals, product manufacture, smoking, and other applications. It provides a brief account on the toxicological impacts, risk assessments and safety evaluation for nanomaterials on the biological system (Lai et al. 2017).

3. List of Nanostructured Materials and its Human Exposure

In day to day life, each and every living organism is exposed to different kinds of airborne nanomaterials like drugs, paints, soaps, detergents, tennis rackets, coatings, concrete, car bumpers, photocopier toners, fabrics, diesel fuel, catalysts, tubes, forest fires, industrial plants, batteries, volcanoes, bacteria, smoke, cosmetics, shampoos, sunscreens, video screens, tires, boot polis, electronics, cloths, gasoline, plastics, wire, coal mines, manufacturing factories, combustion processes, fuel cells, environmental pollution and manmade nanoscale materials. There are a wide range of nanomaterials; their synthesis process and toxicity were first expressed in the book titled <u>Nanotoxicology</u> written by Nalwa and Zhao (Table 1) (Williams et al. 2018).

Table 1: Synthesis of nanostructured materials with different dimensions (Williams et al. 2018).

Nanostructured Materials	Dimension/size	Method
Ceria (CeO_2) nanowires	50–100 nm	Sonochemical method
Aluminium nitride (AlN) nanowires	40–150 nm	Nitridation of a mixture of Al and ammonium chloride powders
Selenium (Se) nanowire bundles	30–50 nm	Polymer-controlled chemical method
ZnTe nanowires	30–80 nm	Vapor phase transport method
ZnO nanowires	20–120 nm	Thermal evaporation under high vacuum
Magnesium oxide (MgO) nanowires	20–30 nm	Thermal evaporation of Mg powder
Nanorods		
Molybdenum disulfide (MoS_2) nanorods	10–20 nm	Hydrothermal method
ZnO nanorods	100–400 nm	Self-assembly method (transformation of ZnO nanorods into ZnO nanotubes at ambient temperature)
ZnO nanotubes	20–40 mm	
Nanofibers		
Mesoporous V-Mg-O nanofibers	0.6 nm	Catalytic synthesis using vanadium (V) oxide and magnesium chloride
Quantum dots		
Mn quantum dots	1–4 nm	Bioreduction with water hyacinth
InAs/GaAs quantum dots	37 nm	Molecular beam epitaxy (MBE)
GaN/AlGaN quantum dots	6–40 nm	Self organization
Si/SiO_2 quantum dots	2–7 nm	Etching
CdSSe quantum dot thin films	7.5 nm	Self assembly
Bimetallic Au–Pt, Au–Pd, and Pt–Pd dendrimer nanocomposites	2–4 nm	Reduction with sodium borohydride in the presence of poly(amidoamine) dendrimers
Nanoparticles		
Silver (Ag) nanoparticles	12 nm	Synthesis by a core–shell method
Metal selenide (ZnSe, CdSe, HgSe) nanoparticles	15 nm	Thermolysis of M $(SeCH_2CH_2CH_2NMe_2)_2$
Silver nanoparticles	8–47 nm	Solution irradiation and hydrothermal treatment
Gold nanoparticles	9–6 ± 1–8 nm	Seed mediated approach in organic media (toluene)

The fabrication and synthesis of nanoscale materials like nanoparticles, nanotubes, fullerenes, nanopowders, nanowires, quantum dots, nanocrystals and nanomaterials can be synthesized by physical, chemical and engineering means where the scientists are exposed in laboratories and pilot level industrial plants. Here are some of the examples for methods of nanoparticles' synthesis and their dimension (Anjem et al. 2012).

4. Nanostructured Materials' Uptake by the Human Body and Nanotoxicity

There are three most common effective ways of nanomaterial uptake by the human body. Some of the nanoparticles' structures are given in Figure 5 (Xia et al. 2019).

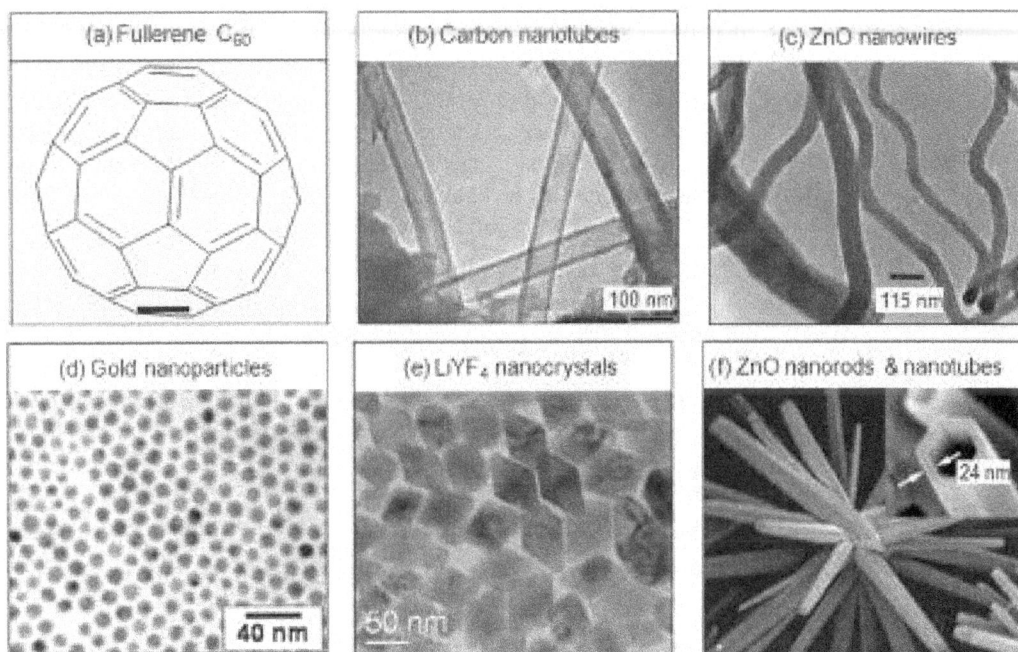

Figure 5: SEM image for different nanomaterials (Naatz et al. 2017).

5. Exposure through the Respiratory System

In general, the inhalation of nanoparticles may lead to the accumulation of nanomaterials in the respiratory tract and lungs which leads to asthma, bronchitis, etc. The intake and translocation of nanomaterials effectively lead to increased health complications. There is epidemiological evidence that inhaled nanoparticles widely affect human health (Naatz et al. 2017).

6. Exposure through the Skin

Skin is the primary vital part in order to encounter the particles inside the body. Skin exposure to sunscreens, dusts and cosmetics will effectively result in the accumulation of nanoparticles. Metallic nanoparticles with less size than 10 nm could penetrate the stratum corneum and hair follicle through the viable epidermis. However, metallic nanoparticles cannot penetrate the transversal patch (Duran et al. 2016, Niyas Ahamed et al. 2020).

7. Exposure through Ingestion

Intake and translocation of different nanomaterials into the alimentary canal can occur after intake of regular food, medicines and drinks. Nanomaterials and nanoparticles may get absorbed into all the biological system by means of cytotoxicity effects were represent in Table 2 (Gunsolus et al. 2015). Apoptosis refers to a process of deliberated cell self-destruction in an organism. The toxicological effects of ultrafine copper particles were determined to be mediated depending on the size. The increasing toxicity observed linearly as the size of copper particles decreased (Hudson-smith et al. 2016, Niyas Ahamed et al. 2018).

In general, nanoparticles show promise for cancer related research in target analysis, drug delivery and diagnosis properties (Hang et al. 2016). Quantum dots express surface manipulation as well as fictionalization; therefore, they also express effective advantage for biomedical research nanoscale contest agents that express effective application in magnetic resonance molecular imaging for dynamic clinical diagnosis. The cadmium telluride nanoparticles show wide and strong fluorescence that may be used in solid state lighting and biological probing (Hang et al. 2018). Carbon nanotubes offer specific and targeted drug delivery in the tumor cells. These nanotubes encapsulated within magnetic nanoparticles could also help in transporting medicine to a target location. It seems to be an effective strategy, particularly in drug delivery systems. The biocompatibility of these nanoparticles, carbon nanotubes and quantum dots remains unknown due to their adverse effects on the biological system (Gallagher et al. 2018).

Table 2: Median lethal dose (LD50, 95% confidence interval, and toxicity class for micro scale, nanoscale, and ion–copper particles) (Hang et al. 2016).

Copper Particle	Specific Surface Area (cm²/g)	Particle Number (per μg)	LD50 (mg/kg)	95% PL(mg/kg)[a] 95% FL(mg/kg)[b]	Toxicity Class (Hodge and Sterner Scale)
Microscale copper (17 μm)	3.99×10^2	44	$> 5000^a$ 5610^b	N/A^a 5075–6202^b	Non-toxic Class5
Nanoscale copper (23.5 nm)	2.95×10^5	1.7×10^{10}	413^a 413^b	305–560^a 328–522^b	Moderately toxic Class3
Ion–copper (0.072 nm)	6.1×10^5	9.4×10^{15}	110^a 119^b	93–145^a 102–139^b	Moderately toxic Class3

8. Incidental Nanomaterials

Volcanic eruptions, forest fires and photochemical reactions are some of the naturally existing processes which help to prevent or protect from natural nanoparticles as mentioned. In general, hair shedding and skin damage in animals and plants were observed in frequent manner. The naturally existing processes like forest fires, volcanic eruptions and photochemical reactions offer great quantities of nanomaterials in the air with huge quantities (Jiang et al. 2014, Kaweeteerawat et al. 2015). The transportation, charcoal burning and industrial operations showed effective attention towards the nanoparticles in the human health by synthetic nanoparticles. In the atmosphere, the existing nanoparticles in the atmospheric aerosols by nature are around 90% and by man are around 10% only (Dominguez et al. 2018).

9. Cosmic Dust and Dust Storms

6500 light years away from the Earth, Eagle Nebula stars are born with disk-like clouds and the ability to create solar systems by the combination of dust and gases, i.e., mostly hydrogen gas.

In general, observing astronomy IR spectroscopy and direct stardust experiments in meteorite collections and space missions identify oxide, nitride, carbon, carbide, and organic nanomaterials as stardust's main components (Joshi et al. 2015). Diamonds of a few nanometers in diameter have been observed in Murchison meteorite, which is an exact and unique example of the nanoparticles' origin in the planetary system other than stars. Electromagnetic radiation, dramatic temperature, pressure gradients, shock waves and physical collisions in forcing nanoparticles in space may lead to the greatest range of nanoscale materials with distinct phase and isomerization and mixing along the chemical spectrum (Gunsolus et al. 2017).

The investigation supported by satellite figures expressed that dust storms in one region can migrate the nano and micro sized minerals and anthropogenic pollutants to 1000s of kilometers away from their origin. In atmospheric aerosol, around 50% of particles that arise from dust storms in deserts are in the range of 100–200 (Arakha et al. 2015). The effect of aerosol particles on the climate and environment was impressively reviewed by Posfai and Buseck. The widespread transport of aerosols across oceans has had a major effect on life in the food chain. In addition, Prashant Kumar and Al-Dabbous found that airborne nanoparticles in the range of 5–1000 nm were present in dust episodes and in the summertime on the terrestrial sides of Kuwait (Pramanik et al. 2018).

Emphysema and asthma are two effective health problems in humans that are led by terrestrial airborne dust particles. Dust is a nanoparticle that effectively damages the lung tissues by forming reactive oxygen species (ROS). More recently, dust storms help to form iron oxide nanoparticles in clouds which form pH fluctuations, and mineralogical, physical and chemical properties were observed in the Saharan desert region. Figure 2 is an extensive aggregated nanoparticle existing in the dust storm region during and after dust storms (Zhi et al. 2018).

10. Engineered Nanomaterials

Very simple combustion in vehicles, fuel oil, cooking and coal for power generation, chemical manufacturing, welding, smelting, airplane engines and welding are great anthropogenic activities that lead to nanoparticles' formation. Nanomaterials such as carbon, titanium and hydroxyapatites exist in commercial sporting goods, cosmetics, toothpaste and sunscreens. These effective synthesis nanoparticles stimulate adverse environmental and human health effects (Manthiram 2017, Nitta et al. 2015).

11. Nanoparticles from Diesel and Engine Exhaust

From the diesel exhaust, the carcinogens may be liberated in the environment and may lead to tremendous effects on the human health system. Quantity of nanoparticles discovered in the urban atmosphere, and the elimination of nanoparticles was determined to account for more than 36–44% of the total nanoparticles concentration. Both petroleum and diesel fueled vehicles produced different nanoparticles sized around < 10 nm (Zhu et al. 2015). According to the findings of the tests, there were significant quantities of nanoparticles present in the atmosphere. These particles ranged in size from 0.056 to 0.1 micrometres and had an average concentration of between 0.55 and 1.16 micrograms per cubic metre. The chemical evaluation of ultrafine particles expressed almost 50% organic compounds, an increased amount of metal oxides, sulfate, sodium, carbon, nitrate and chloride. The emission ratio of this particle's range is more than 10 metric tons per day, which was observed in the south coast air Basin and Los Angeles from the vehicles, automobiles and fuel combustion sources (Table 3). Many reports show that particles exhaled from the diesel exhaust and carbon black nanoparticles stimulate adverse carcinogenic effects in lung tumors and health defects in the rat model (Zhi et al. 2018).

Table 3: Different types of nanomaterials and their cytotoxicity (Ahamed et al. 2016).

Nanomaterials	Size (nm)	Cytotoxicity
Titanium dioxide (TiO_2)	< 100	ROS generation/apoptosis
Fullerene C_{60}	≈ 60	ROS generation Plasma membrane damage
Cadmium selenide (CdSe, CdSe/ZnS)	2–8	ROS generation Apoptosis and necrosis
Cadmium telluride (CdTe) Single-walled carbon nanotubes (SWCNTs)	≈ 2 or ≈ 5	ROS generation, apoptosis ROS generation
SWCNTs Silver (Ag)	15 and 100	Necrosis and apoptosis ROS generation
Aluminum (Al)	30	Apoptosis
Molybdenum (MoO_3)	30	Apoptosis
Gold (Au) cationic	6–10	Necrosis
Silica (SiO_2) unmodified	40–70	Cellular senescence

Almost all types of automobiles exhaust and diesel engines eliminate 20–130 nm sized particles and in gasoline engines release 20–60 nm sized particles. In these fuels, around 90% of carbon nanoparticles were existing in the atmosphere exhausted by fuels. The researchers found that the nanoparticles affect the broncho alveolar lavage fluids from asthmatic Parisian children (Xie et al. 2018). This expresses the presence of carbon nanotube in the cells which cause granulomatous reactions, inflammation and oxidative stress which lead to fibroplasias and neoplasia in lungs. Also, benzo pyrene is a polynuclear aromatic hydrocarbon and a carcinogen present in the diesel exhaust that leads to cardiopulmonary mortality, myocardial infarction, proinflammatory, and hemolytic responses, which are some of the health problems observed in humans. Particles from diesel exhaust consist of polycyclic aromatic hydrocarbons, carbons and transition metals. These fuel nanoparticles and other manufactured nanomaterials damage animals' DNA and increase the risk of cancer (Ahamed et al. 2016).

12. Cigarette Smoke and Building Demolition

Building demolitions and cigarette smoking are highly anthropogenic activities that lead to the spread of nanoparticles into the environment. Cigarette smoke has a very defined composition of about 100000 chemical compounds in the form of nanoparticles ranging from 10–700 nm. Likewise, nano and micro particulates quite smaller than 10 micrometer are released into the environment. The building debris, glass, lead, respirable asbestos fibers and other heavy toxic substances from household materials are released as defined nanosized particles in and around the site of building demolition. Mainly cigarette smoke can cause chronic respiratory illness, pancreatic cancer, cardiovascular disease, exacerbated asthma and middle ear disease. The most dangerous effects of demolition substances and their long term effects towards humans are still unknown. These key indications were extensively studied and carried out amongst workers of demolition sites to identify the ill effect of substances that are dissipated (Mehoney et al. 2016).

13. Nanoparticles in Biomedical and Healthcare Products

Nanomaterials are combined in the sunscreens and cosmetics as antioxidants and anti reflections. It is used for commercial applications as synthesised nanoparticles that are produced using chemical, biological and physical methods. As these prepared nanoparticles are embedded to a firm surface, the effect of unbound and causing health issues is lessened. Nanomaterials are extensively prepared and used for the commercial purpose and personal care productions in paints. Titanium oxide nanoparticles larger than 100 nm are widely utilized as a white pigment in sunscreens and cosmetics. In addition, silver nanoparticles have been used in wide applications, which include air sanitizer sprays, food storage containers, wet wipes, shampoos and toothpastes (Wu et al. 2017).

13.1 According to Toxicological Data, The Toxicity of NMs Depends on Various Factors

- Dose and exposure time effect. The numbers of NMs that penetrate the cells directly depend on the molar concentration of NPs in the adjacent medium multiplied by the exposure time.

- Aggregation and concentration effect. There are many contradictory reports on the toxicity of NPs at different concentrations. Increasing the NP concentration promotes aggregation. Most NP aggregates are micrometer in size, so that a significant quantity of aggregated NPs may not penetrate cells thereby losing their toxicity.

- Particle size effect. NPs show a size-dependent toxicity. Ag NPs with $\approx$ 10 nm diameter show a higher capacity to penetrate and disturb cellular systems of many organisms than Ag+ ions and Ag NPs of larger diameters (20–100 nm).

- Particle shape effect. NPs exhibit shape-dependent toxicity, that is, different toxicity levels at different aspect ratios. For example, asbestos fibers of 10 μm length can cause lung cancer, shorter asbestos fibers (5–10 μm) can cause mesothelioma and 2 μm length fibers can cause asbestosis.

- Surface area effect. Typically, the toxicological effect of NPs increases with decreasing particle size and increasing surface area. It can also be noted that nano and microparticles with the same mass dose react differently with the human cells.

- Crystal structure effect. Based on the crystal structure, NPs may exhibit different cellular uptake, oxidative mechanisms and subcellular localization. For example, the two crystalline polymorphs of TiO_2 (rutile and anatase) show different toxicity. In the dark, rutile NPs (200 nm) lead to DNA damage via oxidation, while anatase NPs (200 nm) do not induce DNA damage in dark conditions.

- Surface fictionalization effect. The surface properties of NPs have shown drastic effects relating to translocation and subsequent oxidation processes.

- Pre-exposure effect. The cellular phagocytic activity can be stimulated by shorter exposure time or the pre-exposure of lower NP concentrations. This pre-exposure results in the adaptability of the human body against NPs to some degree (Ahamed et al. 2018).

14. Conclusion

Although few studies have been completed, several research groups are currently in the process of analyzing data pertaining specifically to the toxicity of nanomaterials. Current literatures are focusing on the toxicity to living communities. As per the available literature, not sufficient work was carried out on the discussed topic. So, the present work will give the detailed outcome for the various research problems associated with the nanotoxicity as well as the ecotoxicity studies of nanomaterials with special reference to human health.

References

Ahamed, M., M.A.M. Khan, M.J. Akhtar, H.A. Alhadlaq and A. Alshamsan. 2016. Role of Zn doping in oxidative stress mediated cytotoxicity of TiO2 nanoparticles in human breast cancer MCF-7 cells. Sci. Rep. 6: 30196.

Anjem, A. and J.A. Imlay. 2012. Mononuclear iron enzymes are primary targets of hydrogen peroxide stress. J. Biol. Chem. 287: 15544–15556.

Arakha, M., M. Saleem, B.C. Mallick and S. Jha. 2015. The effects of interfacial potential on antimicrobial propensity of ZnO nanoparticle. Sci. Rep. 5: 9578.

Burgos, J.S. 2005. Involvement of the Epstein-Barr virus in the nasopharyngeal carcinoma pathogenesis. Med. Oncol. 22: 113–121.

Caruso, G., M. Azzaro, C. Caroppo, F. Decembrini, L.S. Monticelli, M. Leonardi, G. Maimone, R. Zaccone and R. La Ferla. 2016. Microbial community and its potential as descriptor of environmental status. ICES J. Mar. Sci. 73: 2174–2177.

Domínguez, G.A., M.D. Torelli, J.T. Buchman, C.L. Haynes, R.J. Hamers and R.D. Klaper. 2018. Size dependent oxidative stress response of the gut of daphnia magna to functionalized nanodiamond particles. Environ. Res. 167: 267–275.

Durán, N., M. Durán, M.B. de Jesus, A.B. Seabra, W.J. Fávaro and G. Nakazato. 2016. Silver nanoparticles: A new view on mMechanistic aspects on antimicrobial activity. Nanomedicine 12: 789–799.

Englert, N. 2004. Fine particles and human health—A review of epidemiological studies. Toxicol. Lett. 149: 235–242.

Feng, Z.V., I.L. Gunsolus, T.A. Qiu, K.R. Hurley, L.H. Nyberg, H. Frew, K.P. Johnson, A.M. Vartanian, L.M. Jacob, S.E. Lohse, M.D. Torelli, R.J. Hamers, C.J. Murphy and C.L. Haynes. 2015. Impacts of gold nanoparticle charge and ligand type on surface binding and toxicity to gram-negative and gram-positive bacteria. Chem. Sci. 6: 5186–5196.

Gallagher, M.J., J.T. Buchman, T.A. Qiu, B. Zhi, T.Y. Lyons, K.M. Landy, Z. Rosenzweig, C.L. Haynes and D.H. Fairbrother. 2018. Release, detection and toxicity of fragments generated during artificial accelerated weathering of CdSe/ZnS and CdSe quantum dot polymer composites. Environ. Sci. Nano 5: 1694–1710.

Gunsolus, I.L., M.N. Hang, N.V. Hudson-Smith, J.T. Buchman, J.W. Bennett, D. Conroy, S.E. Mason, R.J. Hamers and C.L. Haynes. 2017. Influence of nickel manganese cobalt oxide nanoparticle composition on toxicity toward Shewanella oneidensis MR-1: Redesigning for reduced biological impact. Environ. Sci.: Nano 4: 636–646.

Gunsolus, I.L., M.P.S. Mousavi, K. Hussein, P. Bühlmann and C.L. Haynes. 2015. Effects of humic and fulvic acids on silver nanoparticle stability, dissolution, and toxicity. Environ. Sci. Technol 49: 8078–8086.

Hang, M.N., I.L. Gunsolus, H. Wayland, E.S. Melby, A.C. Mensch, K.R. Hurley, J.A. Pedersen, C.L. Haynes and R.J. Hamers. 2016. Impact of nanoscale lithium Nickel Manganese Cobalt Oxide (NMC) on the bacterium Shewanella oneidensis MR-1. Chem. Mater. 28: 1092–1100.

Hang, M.N., N.V. Hudson-Smith, P.L. Clement, Y. Zhang, C. Wang, C.L. Haynes and R.J. Hamers. 2018. Influence of nanoparticle morphology on ion release and biological impact of Nickel Manganese Cobalt Oxide (NMC) complex oxide nanomaterials. ACS Appl. Nano Mater. 1: 1721–1730.

Hermans, S.M., H.L. Buckley, B.S. Case, F. Curran-Cournane, M. Taylor and G. Lear. 2016. Bacteria as emerging indicators of soil condition. Appl. Environ. Microbiol. 83.

Hudson-Smith, N.V., P.L. Clement, R.P. Brown, M.O.P. Krause, J.A. Pedersen and C.L. Haynes. 2016. Research highlights: Speciation and transformations of silver released from Ag NPs in three species. Environ. Sci. Nano 3: 1236–1240.

Hussain, S., S. Garantziotis, F. Rodrigues-Lima, J.-M. Dupret, A. Baeza-Squiban and S. Boland. 2014. Intracellular signal modulation by nanomaterials. Adv. Exp. Med. Biol. 811: 111–134.

Jacobson, K.H., I.L. Gunsolus, T.R. Kuech, J.M. Troiano, E.S. Melby, S.E. Lohse, D. Hu, W.B. Chrisler, C.J. Murphy, G. Orr, F.M. Geiger, C.L. Haynes and J.A. Pedersen. 2015. Lipopolysaccharide density and structure govern the extent and distance of nanoparticle interaction with actual and model bacterial outer membranes. Environ. Sci. Technol. 49: 10642–10650.

Jiang, Y., Y. Dong, Q. Luo, N. Li, G. Wu and H. Gao. 2014. Protection from oxidative stress relies mainly on derepression of OxyRDependent KatB and Dps in Shewanella oneidensis. J. Bacteriol. 196: 445–458.

Joshi, N., B.T. Ngwenya, I.B. Butler and C.E. French. 2015. Use of bioreporters and deletion mutants reveals ionic silver and ROS to be equally important in silver nanotoxicity. J. Hazard. Mater. 287: 51–58.

Kaweeteerawat, C., A. Ivask, R. Liu, H. Zhang, C.H. Chang, C. Low-Kam, H. Fischer, Z. Ji, S. Pokhrel, Y. Cohen, D. Telesca, J. Zink, L. Mädler, P.A. Holden, A. Nel and H. Godwin. 2015. Toxicity of metal oxide nanoparticles in Escherichia coli correlates with conduction band and hydration energies. Environ. Sci. Technol. 49: 1105–1112.

Lai, L., S.-J. Li, J. Feng, P. Mei, Z.H. Ren, Y.-L. Chang and Y. Liu. 2017. Effects of surface charges on the bactericide activity of CdTe/ZnS quantum dots: A cell membrane disruption perspective. Langmuir 33: 2378–2386.

Mahoney, S., M. Najera, Q. Bai, E.A. Burton and G. Veser. 2016. The developmental toxicity of complex silica-embedded nickel nanoparticles is determined by their physicochemical properties. PLoS One 11: e0152010.

Manthiram, A. 2017. An outlook on lithium ion battery technology. ACS Cent. Sci. 3: 1063–1069.

Mensch, A.C., J.T. Buchman, C.L. Haynes, J.A. Pedersen and R.J. Hamers. 2018. Quaternary amine-terminated quantum dots induce structural changes to supported lipid bilayers. Langmuir 34: 12369–12378.

Mensch, A.C., R.T. Hernandez, J.E. Kuether, M.D. Torelli, Z.V. Feng, R.J. Hamers and J.A. Pedersen. 2017. Natural organic matter concentration impacts the interaction of functionalized diamond nanoparticles with model and actual bacterial membranes. Environ. Sci. Technol. 51: 11075–11084.

Mousavi, M.P.S., I.L. Gunsolus, C.E. Pérez De Jesús, M. Lancaster, K. Hussein, C.L. Haynes and P. Bühlmann. 2015. Dynamic silver speciation as studied with fluorous-phase ion-selective electrodes: effect of natural organic matter on the toxicity and speciation of silver. Sci. Total Environ. 537: 453–461.

Naatz, H., S. Lin, R. Li, W. Jiang, Z. Ji, C.H. Chang, J. Köser, J. Thöming, T. Xia, A.E. Nel, L. Mädler and S. Pokhrel. 2017. Safe-by-design of CuO nanoparticles via Fe-Doping, Cu-O bond lengths variation, and biological assessment in cells and Zebrafish embryos. ACS Nano 11: 501–515.

Nel, A., T. Xia, L. Madler and N. Li. 2006. Toxic potential of materials at the nanolevel. Science 311: 622–627.

Nitta, N., F. Wu, J.T. Lee and G. Yushin. 2015. Li-ion battery materials: Present and future. Mater. Today 18: 252–264.

Niyas Ahamed, M.I. 2014. Ecotoxicity concert of nano zero-valent iron particles—A review. Journal of Critical Review 1: 36–39.

Niyas Ahamed, M.I., S. Rajeshkumar, V. Ragul, S. Anand and K. Kaviyarasu. 2018. Chromium remediation and toxicity assessment of nano zerovalent iron against contaminated lake water sample (Puliyanthangal Lake, Tamil Nadu, India). South African Journal of Chemical Engineering (Elsevier) 25: 128–132.

Niyas Ahamed, M.I., S. Sathya and V. Ragul. 2020. An in vitro study on Hexavalent Chromium [Cr(VI)] remediation using iron oxide nanoparticles based beads. Environmental Nanotechnology, Monitoring & Management 14: 1–5.

Niyas Ahamed, M.I., V. Ragul, S. Anand, K. Kaviyarasu, V. Chandru and B. Prabhavathi. 2018. Green synthesis and toxicity assessment of nanozerovalent iron against chromium contaminated surface water. International Journal of Nanoparticles 10: 312–325.

Oberdörster, G. 2002. Toxicokinetics and effects of fibrous and nonfibrous particles. Inhalation Toxicol. 14: 29–56.

Powers, K.W., S.C. Brown, V.B. Krishna, S.C. Wasdo, B.M. Moudgil and S.M. Roberts. 2006. Research strategies for safety evaluation of nanomaterials. Part VI. Characterization of nanoscale particles for toxicological evaluation. Toxicol. Sci. 90: 296–303.

Pramanik, S., S.K.E. Hill, B. Zhi, N.V. Hudson-Smith, J.J. Wu, J.N. White, E.A. McIntire, V.S.S.K. Kondeti, A.L. Lee, P.J. Bruggeman, U.R. Kortshagen and C.L. Haynes. 2018. Comparative toxicity assessment of novel SI quantum dots and their traditional Cd-based counterparts using bacteria models Shewanella oneidensis and Bacillus subtilis. Environ. Sci.: Nano 5: 1890–1901.

Qiu, T.A., P.L. Clement and C.L. Haynes. 2018. Linking nanomaterial properties to biological outcomes: Analytical chemistry challenges in nanotoxicology for the next decade. Chem. Commun. 54: 12787–12803.

Surya Singh and Hari Singh Nalwa. 2007. Nanotechnology and health safety—Toxicity and risk assessments of nanostructured materials on human health. Journal of Nanoscience and Nanotechnology 7(9): 3048–3070.

Von Moos, N. and V.I. Slaveykova. 2014. Oxidative stress induced by inorganic nanoparticles in bacteria and aquatic microalgae—State of the art and knowledge gaps. Nanotoxicology 8: 605–630.

Williams, D.N., S. Pramanik, R.P. Brown, B. Zhi, E. McIntire, N.V. Hudson-Smith, C.L. Haynes and Z. Rosenzweig. 2018. Adverse interactions of luminescent semiconductor quantum dots with liposomes and Shewanella oneidensis. ACS Appl. Nano Mater. 1: 4788–4800.

Wu, F., B.J. Harper and S.L. Harper. 2017. Differential dissolution and toxicity of surface functionalized silver nanoparticles in small-scale microcosms: Impacts of community complexity. Environ. Sci.: Nano 4: 359–372.

Xie, C., J. Zhang, Y. Ma, Y. Ding, P. Zhang, L. Zheng, Z. Chai, Y. Zhao, Z. Zhang and X. He. 2019. Bacillus subtilis causes dissolution of ceria nanoparticles at the nano-bio interface. Environ. Sci. Nano 6: 216–223.

Xie, X., C. Mao, X. Liu, L. Tan, Z. Cui, X. Yang, S. Zhu, Z. Li, X. Yuan, Y. Zheng, K.W.K. Yeung, P.K. Chu and S. Wu. 2018. Tuning the bandgap of photo-sensitive polydopamine/Ag3PO4/Graphene oxide coating for rapid, noninvasive disinfection of implants. ACS Cent. Sci. 4: 724–738.

Zhi, B., M.J. Gallagher, B.P. Frank, T.Y. Lyons, T.A. Qiu, J. Da, A.C. Mensch, R.J. Hamers, Z. Rosenzweig, D.H. Fairbrother and C.L. Haynes. 2018. Investigation of phosphorous doping effects on polymeric carbon dots: fluorescence, photostability, and environmental impact. Carbon 129: 438–449.

Zhi, B., S. Mishra, N.V. Hudson-Smith, U.R. Kortshagen and C.L. Haynes. 2018. Toxicity evaluation of boron- and phosphorus-doped silicon nanocrystals towards Shewanella oneidensis MR-1. ACS Appl. Nano Mater. 1: 4884–4893.

Zhu, S., Y. Song, X. Zhao, J. Shao, J. Zhang and B. Yang. 2015. The photoluminescence mechanism in carbon dots (graphene quantum dots, carbon nanodots, and polymer dots): Current state and future perspective. Nano Res. 8: 355–381.

6

Fabrication of Micro and Nano Devices for Tissue Engineering Applications

Thandapani Gomathi,[1,*] *R. Sastrekha,*[1] *J. John Joseph*[2]
and *Govindasamy Rajakumar*[3,4,*]

1. Introduction

Tissue Engineering (TE) is a branch of multidisciplinary science, including fundamental principles from materials engineering and molecular biology in efforts to develop biological substitutes for failing tissues and organs. In general, tissue engineering is used to fabricate living replacement parts for the body. Langer and Vacanti (1993) reported that the most common approach for engineering biological substitutes is based on living cells. Tissue engineering (TE) is a rapidly growing scientific area that aims to create, repair, and/or replace tissues and organs by using combinations of cells, biomaterials, and/or biologically active molecules.

The main target of TE is to combine biocompatible materials, cells, drugs, and active molecules to restore or improve biological functions using biodegradable materials. This field is expanding on a daily basis, through integrating knowledge, such as cell biology, chemistry, materials science, nanotechnology, and micro- and nanofabrication (Hwang et al. 2015, Webber et al. 2015, Luu et al. 2015). In this way, TE intends to help the body to produce a material that resembles as much as possible the body's own native tissue. By doing so, TE strategies promise to revolutionize current therapies and significantly improve the quality of life of millions of patients.

The classical TE strategy consists of isolating specific cells through a biopsy from a patient, growing them on a biomimetic scaffold under controlled culture conditions, delivering the resulting construct to the desired site in the patient's body and directing the new tissue formation into the scaffold that can be degraded over time. Most of the presently existing TE techniques rely on the use of macro structured porous scaffolds, which act as supports for the initial cell attachment and sub-sequent tissue formation, both *in vitro* and *in vivo*.

Nanomaterials have a growing role to play in the development of novel TE instruments due to the combination of their bulk and surface properties with their overall behaviour. The creation of useful nanostructured materials meets the needs of ailing or damaged bone, cartilage, muscle, and nervous system tissues during the healing process (Mauro et al. 2012, Zhu et al. 2014). Tissue

[1] Department of Chemistry, D.K.M. College for Women, Vellore, Tamil Nadu, India.
[2] School of Advanced Sciences, Vellore Institute of Technology, Vellore, Tamil Nadu, India.
[3] College of Chemistry and Chemical Engineering, Hubei University, Wuhan 430062, China.
[4] Department of Orthodontics, Saveetha Dental College and Hospitals, Saveetha Institute of Medical and Technical Sciences, Saveetha University, Chennai, Tamil Nadu, India.
* Corresponding authors: drgoms1@gmail.com; microlabsraj@gmail.com

function and physiopathology must be outlined *in vitro* experiments and integrated *in vivo* into host tissue. Materials for TE must be chosen so that the extracellular space results as close as possible to the original environment.

Surfaces characterized by submicron scale features have been used to study cells' responses to nanometer-scale topographical cues that can influence a wide range of cellular functions such as morphology, adhesion, and migration (Gattazzo et al. 2014, Flemming et al. 1999, Gentile et al. 2012).

Nowadays, nanotechnology, together with the most advanced micro fabrication and post-processing modification techniques, supports the realization of a wide range of two- and three-dimensional (2D and 3D) bioengineered substitutes for *in vitro* (Gentile et al. 2010) and implantation tests (Ainslie and Desai 2008, Park et al. 2010, Curtis et al. 2005). During the last 15 years, the scientific publication rate of paper treating arguments regarding the application of nanostructure materials in the TE field has increased year by year.

With the development of nanotechnology coupled with advance microfabrication techniques, biocompatible-nanostructured materials closely fulfill the requirement in the recovery of native tissues for TE applications. The goal of this chapter is to discuss the fabrication and application of novel micro structured, nano structured natural and synthetic polymers and scaffolds for TE research. Microfabrication, nanolithography, and miscellaneous nanolithography techniques are discussed in detail along with different applications in many branches of the modern medicine such as neuroscience, cardiology, orthopedics, skin, and dermatology.

2. Biomaterials

Biomaterials are ubiquitous in biomedical applications which include drug delivery systems, engineered tissues, and biomedical devices. Biomaterials are defined as synthetic or natural materials which are in contact with the biological environment in order to treat any malfunctions of the body (Ratner et al. 2004). Biomaterials have been widely investigated and implanted in the human body for applications such as skeletal repair, organ replacement, and improvement of senses, among many others with remarkable success. With advances in sciences, polymeric materials can now be synthesized from a combinatorial array of monomers, oligomers and polymers with tunable chemical, mechanical and geometrical properties to create new, biocompatible substances (Slaughter et al. 2009). Cellulose and chitin as biopolymers are the most abundant organic compounds in Nature and estimated to be at levels approaching 1011 tons annually. Chitosan is typically obtained by deacetylation of chitin under alkaline conditions, which is one of the most abundant organic materials, being second only to cellulose in the amount produced annually by biosynthesis. In addition, chitosan is expected to be useful in the development of composite materials such as blends or alloys with other polymers, since chitosan has many functional properties.

3. Chitosan

Chitosan is a linear polysaccharide, composed of glucosamine and N-acetyl glucosamine units linked by β (1–4) glycosidic bonds. Generally, chitosan has three types of reactive functional groups, an amino group as well as both primary and secondary hydroxyl groups at the C(2), C(3), and C(6) positions, respectively. These groups allow modification of chitosan like graft copolymerization for specific applications, which can produce various useful scaffolds for tissue engineering applications. It exhibits anticancer activity via immunomodulatory, antiangiogenic, and anti-inflammatory actions. The modification of chitosan by introducing various functional groups and recently, their applications were seen in various artificial organs such as skin, bone, cartilage, liver, nerve and blood vessel. Chitosan has been combined with alginate, hydroxyapatite, hyaluronic acid, calcium phosphate, PMMA, poly-L-lactic acid (PLLA), and growth factors for potential application in orthopedics. Overall, chitosan offers broad possibilities for cell-based tissue engineering (Hu et al. 2006). Additionally, chitosan is claimed to be a dietary fibre that can process losing weight. The aim of this review is to discuss the recent developments on the biopolymer like chitosan fabricated for the tissue engineering applications.

3.1 Chitosan as a Tissue Supporting Material

The scaffold should have the following requirements:

- Biocompatibility with the tissues,
- Biodegradability at the ideal rate corresponding to the rate of new tissue formation,
- Nontoxicity and non-immunogenicity,
- Optimal mechanical property, and
- Adequate porosity and morphology for transporting of cells, gases, metabolites, nutrients and signal molecules both within the scaffold and between the scaffolds.

Being a promising material, chitosan-based scaffolds meet the requirement for the scaffolds suitable for tissue engineering application and also it possess some special properties such as interconnected-porous structures. The porous structure of chitosan is a promising characteristic for the development and optimization of a variety of tissue scaffolds and regeneration aids (Madihally and Matthew 1999). Chitosan based scaffolds possess some special properties for use in tissue engineering. When calcium carbonate gels are added to the chitosan solution, it forms "internal bubbling process" (IBP), and also generate specific shapes by using suitable molds (Chow and Khor 2000).

Niamsa et al. successfully prepared the nanocomposite blend films containing methoxy poly (ethylene glycol)-b-poly (D, L-lactide) nanoparticles with different chitosan/silk fibroin ratios. These biodegradable nanocomposite blend films may have potential for use in drug delivery, wound dressing and tissue engineering applications (Niamsa et al. 2009). Moreover, Nishikawa et al. (2000) reported that chitosan structurally resembles GAG consisting of long chain, unbranched, and repeating disaccharide units, which is regarded to play a key role in modulating cell morphology, differentiation and function. These mechanical properties of chitosan-based scaffolds are dependent on the pore sizes and pore orientations.

Chitosan has lately undergone alteration to increase its solubility, add desired qualities, and broaden the range of potential applications. Ding et al. (2004) reported that chitosan graft-polymerized onto poly (L-lactide) (PLA) surface by plasma coupling reaction can be used to control the morphology and function of cells and potential applications in tissue engineering. Generally, poly (α-hydroxyacid)s, homopolymers and copolymers based on glycolide and lactide, have been widely used as biomaterial in sutures, drug release systems and tissue engineering owing to their biocompatibility and biodegradability (Morita and Ikada 2002). Poly (glycolide) (PGA) and its copolymers such as lactide–glycolide copolymer (PLGA) degrade too quickly when used as a scaffold, because their tensile strength reduces to the half within two weeks.

In contrast, PLA degrades too slowly, requiring 3–6 years for complete resorption. Owing to this inadequate resorption property of PGA and PLA, naturally derived-polymers such as alginate, collagen, hyaluronic acid and chitosan have been preferably employed in recent studies on tissue engineering. Moreover, poly (α-hydroxyacid)s degrades through non-enzymatic hydrolysis, whereas naturally derived polymers undergo enzymatic hydrolysis (Betancourt and Brannon-Peppas 2006, Voldman et al. 1999, Li et al. 2003, Ratner et al. 2004, Slaughter et al. 2009).

Naturally derived-polymers are hydrophilic and yield products with low mechanical strength, which leads to limited applications (Morita and Ikada 2002, Ikada 2006). But modification of biopolymers leads to the suitable biopolymer derivative that can overcome the drawbacks. Tissue engineering approach with more expanded understanding of articular cartilage and associated pathologies may provide the chitosan-based material that supports chondrogenesis, which can improve the quality of neocartilage produced and the integration with the host tissue as well as the long-term outcomes of cartilage repair in clinical settings (Kim et al. 2007).

Therefore, chitosan derivatives can be modified for tissue engineering applications and also it can be introduced in new techniques for using them as a scaffold in different types of organs. Fabrication of chitosan and its derivatives in micro and nano forms enhances its applications.

4. Fabrication Technologies

In this chapter, the advantages of microscale and nanoscale materials in tissue engineering applications are reviewed. Henceforth, the usefulness of micro- and nanotechnologies in fabricating biomimetic scaffolds with increased complexity and vascularization will be detailed. Furthermore, these technologies can be used to control the cellular microenvironment (i.e., cell–cell, cell–matrix and cell–soluble factor interactions) in a reproducible manner and with high temporal and spatial resolution (Chung et al. 2007). Micro- and nanotechnologies can be used to fabricate materials with specified structures and functional properties to address these limitations.

4.1 Microscale Technologies

Biocompatible scaffolds and surfaces should successfully mimic the macro-, micro-, and nanostructure of systems and organs (macroscale range), cells (microscale range) and biomolecules (nanoscale range). Microscale technologies are the currently developing, useful assets for tissue fabrication and examinations. These technologies are potentially powerful in addressing some of the challenges in tissue engineering (Andersson and van den Berg 2004). A wide array of microfabrication techniques has been optimized for the purpose of creating efficient biomimetic devices. Microtechnologies that have been adapted from the microelectronic industry typically involve top-down fabrication approaches, such as photolithography, microcontact printing and micromolding (Ainslie et al. 2009, Lima et al. 2014).

4.1.1 Photolithography

Photolithography is a widely used technique for microscale material fabrication. The photolithographic technique has been reviewed thoroughly previously (Voldman et al. 1999, Li et al. 2003). Photolithographic process consists of a number of steps to arrive at a desired pattern of the surface. In this technique, the pattern will be transferred to a photosensitive or photoresistive polymer by exposure to a light source through an optical mask. An optical mask usually consists of opaque patterns (usually chrome or iron oxide) on a transparent support (usually quartz) placed on the top of the substrate and photoresist. Then, it will be irradiated with UV light (Figure 1).

Depending on the type of photoresist utilized, the photoresist polymer will undergo one of the two possible transformations, i.e., it will become more soluble (positive photoresist) or crosslinked (negative photoresist) after UV light exposure, thus generating the appropriate pattern upon developing. The resulting photoresist patterns are then used to protect the covered substrate from etching, or from the deposition of compounds or biomolecules on its surface. After the desired process is completed, the photoresist can be removed, leaving the pattern design on the substrate.

Chitosan's bioscaffolding property has garnered attention for microscale biodevices (Koev et al. 2007, Park et al. 2006, McDermott et al. 2006, Yi et al. 2005, Co et al. 2005). With the numerous enzyme–substrate combinations which have already been identified to be compatible with chitosan, multiple forms of these devices are possible with the ability to microfabricate chitosan structures. Khademhosseini and coworkers used a photolithographic technique to fabricate

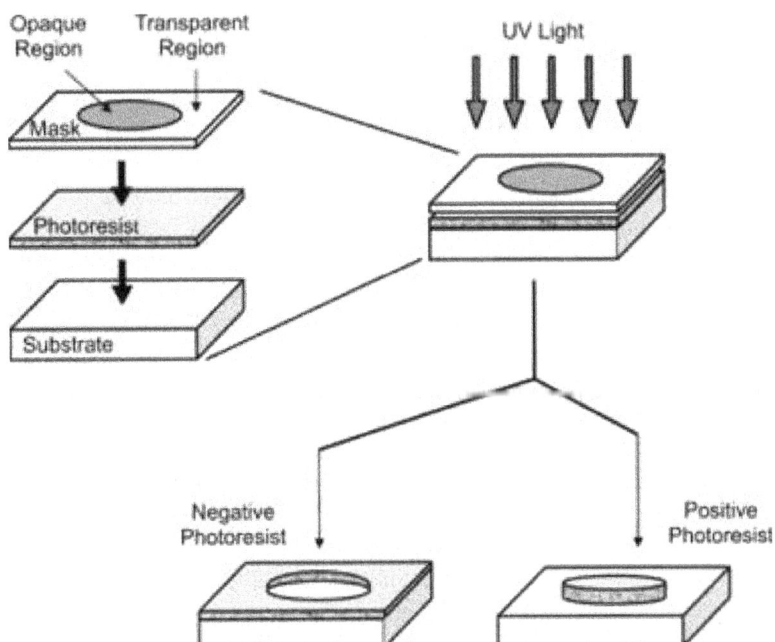

Figure 1: Process of photolithography (Betancourt and Brannon-Peppas 2006).

cell-laden gelatin methacrylate hydrogels with varying length scales (Nichol et al. 2010). The ability of photolithography is to create three-dimensional (3D) materials efficiently using layer-by-layer method; multilayered 3D materials can be easily developed by simply repeating the fabrication process to create a layer on top of the previous layer (Liu Tsang and Bhatia 2004).

4.1.2 Soft Lithography

Soft lithography, similar to photolithography, is a method used to transfer a pattern onto a surface. Soft lithography is a set of microfabrication techniques that utilizes a soft, flexible material, often an elastomer, to generate micron- and submicron-scale structures or molecules on a surface (Xia and Whitesides 1998). The most commonly used soft lithography techniques include replica-moulding, nano- and microcontact printing (mCP) (Li et al. 2003) and microfluidics. Soft lithography refers to a collection of techniques that enable the fabrication of structures or the transfer of materials using elastomeric "soft" masters (e.g., stamps, molds, and photomasks).

Soft lithography is broadly used in bio-imprinting (Vozzi et al. 2010). The typical imprinting stamp is produced by self-assembly of the template biomolecules on top of a smooth support under appropriate conditions and the self-assembled template stamp is then softly pressed over a prepolymer film for a certain period of time (Mujahid et al. 2013). The template biomolecules should be strictly packed in order to attain high imprinting density (Dickert et al. 2004, Mujahid et al. 2013). Overall, this procedure allows the production of geometrical features of a wide range of biomolecules and translates the chemical information involved in noncovalent interactions between the polymeric functional groups and the receptor surface (Mujahid et al. 2013).

Soft lithography brings to microfabrication low cost, simple procedures, rapid prototyping of custom-designed devices, three-dimensional capability, easy integration with existing instruments such as optical microscopy, molecular level control of surfaces, and biocompatibility (Jiang and Whitesides 2003, Whitesides et al. 2001). These techniques allow patterning of cells and their environments with convenience and flexibility at dimensions smaller than micrometers.

4.1.2.1 Microfluidic Technologies

Microfluidics is defined as science and technology that control the flow of fluids using channels with micrometer dimensions (Whitesides 2006). Microfluidic devices allow for the manipulation of multiple fluids in highly complex channel configurations, and are ideal for performing cell-based high throughput experiments because they are cheap, reduce expensive reagent volumes and reduce cell numbers (Sia and Whitesides 2003). Microfluidic arrays can be used to perform screening experiments, to test drug toxicity and to optimize culture conditions for inducing specific cell fates. To create such systems, it is important to be able to culture cells inside microfluidic devices in a reliable and long-term manner and to be able to combine these devices in such a way as to test many conditions simultaneously. In TE, the ability to control the direction and quantity of fluid flows using micrometer-scale channels is utilized in several applications, including cellular patterning, perfusion bioreactors, and droplet-based biomaterials.

Current techniques include the use of complicated valves and pumps to fabricate large arrays of micro channels that can be individually addressed (Thorsen et al. 2002). The ability to test various combination sand doses of drugs can be useful for performing factorial experiments, which can optimize culture conditions for inducing various biological fates (Prudhomme et al. 2004). Microfluidic devices are increasingly used as perfusion bioreactors to provide cell culture media to tissue constructs (Leclerc et al. 2003, Gómez-Sjöberg et al. 2007). It is also possible to create 3D cell culture conditions by stacking multiple device layers connected via channels or membranes.

4.1.2.2 Replica-moulding

Replica moulding is to replicate the shape, form, structure and other information of the master, and can accept formative information of materials in a wider range than photolithography. Furthermore, it allows the replication of three-dimensional morphology through only one processing step, which is impossible in photolithography. Replica molding has been used for mass production of objects that have stable surface structures such as diffraction grating (Ramos and Choquette 1996), holograms (Nakano 1979), CD (Haverkorn et al. 1982), and microtools (Kiewit 1973). Replica molding that uses appropriate materials can replicate reliably down to nanometer unit even materials with very complex structures in a simple, cheap method. The excellent replication property of replica molding is determined by Van der waals interaction, wet method, and dynamic factors used for filling the mold. Due to this physical interaction, replica molding enables more accurate replication in smaller sizes than 100 nm which cannot be done with photolithography because of its limitation by diffraction. In replica moulding, the original master is not required to be a negative of the final piece. After producing a PDMS stamp of the original master, the stamp then is used as a secondary master and another stamp is made from it. In this manner, the original master cannot be damaged or degraded as multiple copies are made.

4.1.2.3 Microcontact Printing

Microcontact printing is perhaps the simplest method for patterning surfaces. It also provides the highest resolution in patterns with the greatest flexibility in the shape and size of the patterns generated. It provides the best control when one needs to pattern only two types of ligands or proteins. Microtopographies can be useful for certain experiments where micropatterning alone is not sufficient. In microcontact printing the stamp is 'inked' with selected chemicals, normally alkanethiols. The stamp is then pressed onto the substrate and removed leaving a 1 molecule thick layer with features down to 300 nm.

4.1.2.4 Inject Printing

Three dimension micro-fabrication technique is one of the essential technologies in manufacturing complicated 3D biological tissues by "inkjet 3D biofabrication approach" as a technology to build up designed 3D biological tissues with micro- to macro-structures by handling different multi-types of living cells. In this technique, using inkjet printing technique with gel precursor and gel reactant, 3D hydrogel structures can be fabricated. Inkjet printing, present in households and most offices, is a common method for transferring digital data to paper or transparencies. The aromatic dipeptides, FF, can form both spherical and tubular structures. These structures were used as bio-ink to be efficiently patterned on surfaces via a desktop inkjet printer (Adler-Abramovich and Gazit 2008). A variety of cells can be encapsulated into the peptide hydrogel. Therefore, in principle, cell-containing peptide hydrogels can be patterned on the surface either by soft lithography or inject printing.

4.1.2.5 Advantages and Applications of Soft Lithography

Soft lithography is a low cost production method that allows for the creation of three-dimensional patterns at room temperatures and pressures. Additionally, because the stamp is flexible, the substrate material need not be perfectly flat. It can be flat curved, spherically curved or in some instances, contain surface features or roughness. Soft lithography can be used for the production of lab-on-chip systems, biosurfaces, biochips, microfluidics, microreactors, sensors, microelectromechanical systems (MEMS), and microoptics.

4.1.3 Micro Imprinting Lithography

Micro imprinting lithography, also known as "hot embossing" or "compression molding," is one of the most widely used processes to fabricate microstructures for data storage, wavegratings, or microfluidic applications (Miller et al. 2001, Chou et al. 1995, Lehmann et al. 1983, Gottschalch et al. 1999). Hot embossing has many advantages such as less cost required for manufacturing, flexible fabrication method, and a single mold can be used for mass production, which has demonstrated polymer high aspect ratio microstructures as well as nanoimprinting patterns. Hot embossing (HE) is an easy operation and high accuracy can be obtained with hot embossing technique. It uses polymer or glass substrates to imprint structures created on a master stamp. This allows the stamp to produce many fully patterned substrates using a wide range of materials. Hot embossing is therefore suited for applications from rapid prototyping to high volume production. Hot embossing can be applied in a wide variety of fields: µTAS, microfluidics (micromixers, microreactors), microoptics (wave guides, switches), etc.

Using hot embossing method, a master with a microscale relief structure on the surface can be fabricated by standard MEMS techniques. A thin layer of chromium was coated on a quartz wafer by evaporation coating. A layer of electron beam resist (a polymer that is sensitive to electron beams; PMMA is often used) was then spun-cast on the chromium layer. The electron beam resist was selectively exposed to a focused electron beam in an electron lithographic machine. Then the exposed resist was dissolved in the developer to keep the resist layer in the desired patterns. The wafer was exposed to chromium etchant and the pattern was transferred to the chromium layer. The quartz substrate was etched using reactive ion etching through the chromium mask, leaving areas covered by chrome. The quartz substrate was silanized after removing the chrome. The silanization was done by exposure to the vapor of $CF_3(CF_2)_6(CH_2)_2SiCl_3$ for about 30 min. The biodegradable biopolymers can be used generally as micropatterns.

Hot embossing process relies on raising the temperature of a sheet of polymer up to its melting range and on pressing a heated master plate into the polymer for triggering a local flow of the material to fill the cavities to be replicated (Figure 2). This technique has attracted increased attention in recent years in particular due to the relatively simple set-up and low cost associated with its implementation in comparison to other replication techniques (Heckele and Schomburg 2004, He et al. 2011, Becker and Heim 2000).

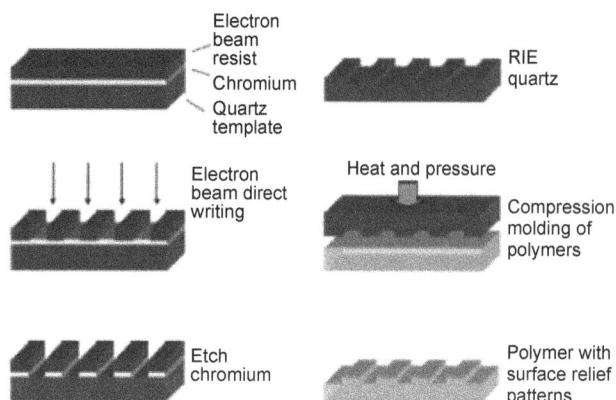

Figure 2: Hot embossing procedures (Omar 2013).

4.1.4 Film Deposition

Film deposition is a surface engineering or surface modification technique, which encompasses an overlay referred to as hard facing, a protective material with outstanding physico-chemical properties. This is a form of surface coating which requires the deposition of thin film layers on the surface to alter the properties of the material surface. The coating techniques can be vapor phase processes, solution state and fusion process. A thin film is a layer of material ranging from fractions of a nanometer (monolayer) to several micrometers in thickness. Like all materials, the structure of thin films is divided into amorphous and polycrystalline structure depending on the preparation conditions as well as the material nature. Thin films comprise two parts: the layer and the substrate where the films are deposited on it.

The term thin film is usually applied to the surface of the thickness range below 1 micron. To produce a thin film with good quality, physical and chemical deposition methods can be used. The physical deposition method comprises evaporation techniques, vacuum thermal evaporation, electron beam evaporation, molecular beam epitaxy, ion plating evaporation, laser beam evaporation, arc evaporation and sputtering techniques. The chemical deposition method consists of sol-gel method, chemical bath deposition, spray pyrolysis technique, plating, and chemical vapor deposition (Jilani et al. 2017).

4.2 Nanoscale Fabrication Techniques

Top-down and bottom-up are the two fundamentally different approaches to the fabrication of nanoscale devices. In the nanofabrication context, the bottom-up approaches seek to use smaller (usually molecular) components to construct larger and more complex, but still nanoscale, devices, while top-down approaches aim to create nanoscale devices by using larger, externally controlled equipment to direct their assembly.

4.2.1 Nanolithography

In Greek, words "nanos", mean dwarf; "lithos", mean rock; and "grapho" mean to write; therefore, the term nanolithography means a "tiny writing on rocks". Nanolithography is the science of etching, writing or printing to modify a material surface with structures under 100 nm. Conventional photolithography is limited by the wavelength of incident light. Therefore, to overcome this limit, several nanolithography techniques such as electron beam lithography (EBL), focused ion beam (FIB), nanoimprint lithography (NIL), interferometric lithography (IL), and sphere lithography (SL) have been developed.

4.2.1.1 Electron Beam Lithography (EBL)

It is a fabrication process using the beam of electrons emitted in a pre-defined path on the surface which is covered with a thin film of electron sensitive resist. Electron beam (e-beam) lithography is used for primary patterning directly from a computer-designed pattern (Sze 1985, McCord and Rooks 1997). Similar to photolithography, it will create small structures in the resist which can transfer to the substrate by reactive ion etching (RIE) or other etching processes. Though, unlike conventional lithography, EBL patterns the resist without masks but in a direct writing mode, it resist (Chen and Pepin 2001). EBL can produce features as small as 20 nm but is very expensive and time consuming. For example, a lithographic process that would take 5 minutes using photolithography would take approximately 5 hours using EBL.

4.2.1.2 Focused Ion Beam (FIB)

Focused ion beam lithography is analogous to e-beam lithography (Melngailis et al. 1989, Gamo 1996), but here magnetic lenses are replaced by electrostatic lenses because of the much heavier ion masses. It operates in the range of 10–200 keV and is an essential tool for highly localized implantation doping, mixing, micromachining, controlled damage as well as ion-induced deposition (Tseng 2005). It is often known as a powerful defect-repair tool in the semiconductor industry. Recently, FIB, especially FIB milling, has been attracting great attention in the nanofabrication because it is capable of generating specific nanoscale patterns directly on hard substrates without requiring masks or photoresists. Another application of FIB in nanofabrication of nanofluidic devices is to generate nanochannels parallel to the surface using a scanning mode.

4.2.1.3 Nanoimprint Lithography (NIL)

Nanoimprint is a nonconventional and low-cost technique for high resolution pattern replication. In addition to the above two direct writing based techniques, NIL is an important nanolithography method with high-throughput capability. Unlike conventional lithography, NIL replicates nanoscale features by mechanically pressing predefined molds into imprint resist and thus overcomes the diffraction limit. It has been widely used in recent decades to fabricate 1-D and 2-D nanochannels in varied nanofluidic systems. NIL can generate nanoscale features (10 nm) over a large area with relatively low cost compared with the aforementioned EBL and FIB as its molds can be reused, which makes this technique promising in high-throughput nanofluidic device fabrication.

4.2.1.4 Interferometric Lithography (IL)

Like NIL, IL technique is also capable of fabricating large-area, nanometer-sized, periodically patterned structures (Xia et al. 2011). This technique has been variously referred to as holographic lithography, or interference lithography. A coherent laser source was used in this technique, which split into two different beams and then projected onto the photoresist. Typical sinusoidal interference pattern with certain pitch is formed on the photoresist based on interferometric exposure of these two coherent beams. The resulting patterning line width is determined by the incident light wavelength, angle, and developing time. This process is maskless and only relies on light and material properties. Nanoscale pattern of the photoresist can be transferred to the substrate via etching process. Up to now, the width of nanochannel immediately after etching can reach 200 nm and further thermal oxidation may narrow the width down to 100 nm or smaller.

4.2.1.5 Sphere Lithography (SL)

SL, also named as colloidal lithography, is another low-cost technique to pattern largescale two dimensional ordered nanostructure arrays (Zhang and Wang 2009, Li et al. 2008, Vogel et al. 2012), especially nanopore arrays. The key step in the SL based nanofabrication approach is the preparation of a high quality monolayer of nanoparticles. Most monolayers are formed by classic assembly-based methods, including evaporation induced self-assembly (Denkov et al. 1992), Langmuir-Blodgett deposition (Huang et al. 2005, Hsu et al. 2008) or roll-to-roll process (Jeong et al. 2010). Even some of the above methods have demonstrated their ability in wafer-scale fabrication. The process of compatibility of the assembly-based nanoparticle monolayer preparation with traditional microfabrication still remains a challenge due to requirements of special instruments and tricky manual operations, which limit the applications of sphere lithography in nanofluidics. Recently, monolayer of nanoparticles has been achieved by spin-coating a photoresist or hybrid sol with particles doped (Lu et al. 2008). These new methods, in principle, are compatible with traditional microfabrication processes and thereby hold great potential in developing another useful tool for the fabrication of micro/nanoscale devices.

4.2.2 Molecular Self Assembly

Self-assembly has become a prominent strategy in chemistry and materials science as a whole. It has become significant in highly diverse applications, in fields ranging across electronic materials, synthetic biology, structural materials, chemical biology, and biomaterials. Molecular self-assembly is a type of bottom–up approach by which molecules adopt a defined arrangement without guidance from an outside source. It is the spontaneous organization of molecules due to their mutual interactions into ordered aggregates (Whitesides and Grzybowski 2002, Whitesides et al. 2005, 1991). There are two types of self-assembly. These are: intramolecular self-assembly and intermolecular self-assembly. Commonly, the term molecular self-assembly refers to intermolecular self-assembly, while the intramolecular analog is more commonly called folding.

The key to self-assembly in a variety of systems such as colloidal suspensions, micro- and nanoemulsions, and biological systems lies in achieving force balance between various attractive and repulsive interactions, also including hydrogen bonding, hydrophobic interactions, Coulombic interactions, π-stacking, and van der Waals forces (Lee 2007). It is imperative to gain a thorough understanding of the nature and complexity behind these processes to manipulate the arrangement of nanoparticles in a desired manner. The desired structure is programmed in the shape and functional groups of the molecules using molecular self-assembly. According to Whitesides, 6 of the components of a molecular self assembling system consist of a group of molecules or segments of a macromolecule that interact with one another. These molecules, or molecular segments, may be the same or different. It has a number of advantages.

- It carries out many of the most difficult steps in nanofabrication (those involving atomic-level modifications) using very highly developed techniques of synthetic chemistry.
- It draws from the enormous wealth of examples in biology for inspiration, since self-assembly is one of the most important strategies used in biology for the development of complex, functional structures.
- It can incorporate biological structures directly as components in the final systems.
- It requires that the target structures be the most stable thermodynamically, and tends to produce structures that are relatively defect-free and self-healing (Whitesides 1995).

Much of biology is a result of self-assembly which, taken to an extreme, produces the high molecular weight networks that constitute the cytoskeleton and the extracellular matrix. Self-assembling strategies in biomaterials can likewise produce highly structured, compositionally defined, multicomponent, multifunctional materials from a discrete set of molecular building blocks (Mendes et al. 2013).

4.2.3 Soft Lithography for Nanoscale

In soft lithography, an elastomeric stamp with patterned relief structures on its surface is used to generate patterns and structures with feature sizes ranging from 30–100 m. This is a convenient, effective, and low-cost method, which has been widely used in nanofabrication, particularly infabrication of various biomedical nanodevices (Patel et al. 1998, Chen et al. 1997). Xia et al. have provided a detailed review of soft lithographic techniques (Xia and Whitesides 1998).

One currently developed soft-lithographic method uses PDMS elements, not as stamps or molds, but as optical components for patterning constructions in photosensitive materials (Jeon et al. 2004a, 2004b). The PDMS optical element, which features as a segment mask, forms an ideal, conformal contact with a photopolymer film. General floor forces (i.e., van der Waals kind interactions) drive this contact; externally utilized strain is no longer required. Figures 3a,b,c illustrates the method. This passive process yields optical alignment of the masks to the polymer with nanometer precision within the everyday direction. Ultraviolet (UV) light passing through this element, which has elements of relief similar in dimension to the wavelength, generates a 3-dimensional depth distribution that exposes a photopolymer movie at some point of its thickness. Creating a method for the polymer's non-UV light-crosslinked constituents to result in 3D nanostructures with function sizes as small as 50 nm within the geometry of the depth distribution.

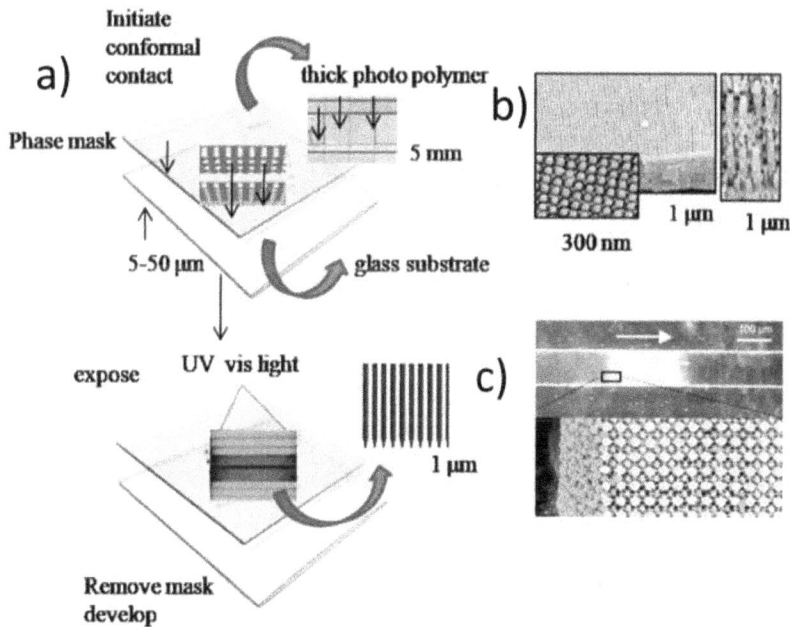

Figure 3: (a) Schematic illustration of the use of a PDMS phase mask for three-dimensional nanofabrication of structures in a photosensitive polymer. (b) A typical structure and optical modeling of the intensity distribution that defined it. (c) Use of this type of structure as a filtering element in a microfluidic channel. (Part (b) reprinted with permission from Jeon, S. et al., Adv. Mater. (2004) 16(15): 1369. © 2004 Wiley-VCH. Part (c) reprinted with permission from Jeon, S. et al., Proc. Natl. Acad. Sci. USA (2004) 101(34): 12428. © 2004 National Academy of Sciences, USA).

Because liability of the polymer occurs in the proximity field region of the mask, refer to the technique as proximity field nanopatterning (PnP) (Jeon et al. 2004a, 2004b). The proximity geometry places requirements on the spatial and temporal coherence of the light source that can be easily met even with low-cost setups (e.g., a handheld lamp with an interference filter is sufficient). Only the spot size of the light source and the size of the phase mask limit the size of the patterned areas. Nanostructures with thicknesses up to 100 µm can be achieved; only the structural integrity and optical absorption of the photopolymer limit this thickness. This method appears to hold exceptional promise as a general platform technology for large-area, three-dimensional nanopatterning—a capacity that conventional photolithography currently lacks. Figure 3b shows an SEM of a typical three-dimensional structure formed with this PnP technique. Rigorous coupled wave analysis can model accurately the optics associated with this method (Jeon et al. 2004a). Simulations that quantitatively describe some of the fabricated structures appear in Figures 3a and b. A wide range of periodic and aperiodic structures can be generated by adjusting the geometry of the mask. There are many potential applications for this technique. Figure 3c shows a simple example in microfluidics. Here, PnP forms a three-dimensional polymer nanostructure integrated directly into a microfluidic channel for filtration, separation, and mixing purposes. The colorized SEM shows the separation of a suspension of 500 nm poly (styrene) beads (red) from a flow that moves from left to right (Rogers and Nuzzo 2005).

4.2.4 Rapid Prototyping

The method which is used for the production of models, patterns or simple prototypes by using generative production methods is known as rapid prototyping. In 1984, Chuck Hull found the basis for the rapid prototyping with the invention of the stereolithography process and also other manufacturing processes can be developed. The utilization of rapid prototyping techniques was given by a collaborating group from the University of Mexico and Sandia National Laboratories (Fan et al. 2000).

The following are the manufacturing techniques of the prototyping:

- Colorjet
- Fused Deposition Modeling (FDM)
- Stereolithography (SLG)
- PolyJet
- Selective Laser Sintering (SLS)
- HP Multi Jet Fusion
- Direct metal laser sintering (DMLS)
- Binder jetting

Advantages of rapid prototyping: Rapid prototying is used to

- generate micro- and nano-structures in a layer by layer approach
- study for the fabrications of scaffolds for tissue engineering (Yeong et al. 2004)
- reduce design and development time
- reduce overall product development cost
- eliminate or reduce risk
- allow functionality testing.

4.2.5 3D Nanoprinting

Through the use of various engineering technologies, tissue engineering can restore a tissue's original function by replacing the damaged area with a new tissue or organ that has been grown from scratch. This technology supports the development of three-dimensional (3D) tissues using a scaffold. The architecture, pore shape, porosity, and interconnectivity of conventional scaffolds are not controlled, which limits their capacity to promote cell growth and the formation of new tissue. These drawbacks of traditional fabrication techniques may be overcome by 3D printing technologies. Computers are used in these technologies to aid in design and fabrication, allowing for the standardisation and production of 3D scaffolds. These techniques have been examined in several fields of tissue engineering, in part because nanofabrication technology based on two photon absorption (2PA) and on controlled electrospinning may produce structures with submicron resolution. Improvements in tissue regeneration are now possible because to recent integrations of 3D nanoprinting technology with cell dynamics and molecular biology techniques. A key tool in tissue engineering will be 3D nanoprinting if the interactions of cells, scaffold systems, and biomolecules can be understood, controlled, and an ideal 3D environment for tissue regeneration can be achieved (Lee 2015).

Various 3D printing technologies (Zein et al. 2002) including stereolithography, deposition modeling, inkjet printing, selective laser sintering, and electrospinning technology have been developed. Those technologies have been widely used in studies of regeneration of tissues such as bone, cartilage, ligament, muscle, skin, and neurons and of organs such as trachea, liver, kidney, and heart. 2PP (Photon Polymerization) and electrospinning are used to fabricate and construct the submicron precision.

4.2.6 *Electrically Induced Nanopatterning*

Electrically induced nanopatterning techniques utilize electrostatic interactions between a thin dielectric materials' liquid film and an electric field gradient to produce lateral patterns and structures at the nanometer scale (Schaffer et al. 2000, Betancourt et al. 2006). Electric field-induced micro-/nanopatterns in thin polymer films is also known as electrohydrodynamic patterning; it is a technique to fabricate micro-/nanostructures, and its advantages are in microstructure level (easy demolding) and low cost. The systematic study of the effect of different parameters on electrohydrodynamic patterning with a numerical phase field model can be done.

In the past, researchers used lubrication approximation (i.e., long-wave approximation) to simplify the numerical model. This approximation is valid if the structural height is equal to or greater than the wavelength, which happens in the majority of cases. Thus, they deserted the lubrication approximation and solved the full governing equations for fluid flow and electric field. The nanofabrication system consists of two parallel capacitor electrodes separated by an air gap of less than a micron. One of the electrode is coated by spin coating with polymer such as polystyrene (PS), and at high temperature, there is a different in the voltage across the electrodes, which causes destabilization of the polymeric film and forms "wave" like structure (Schaffer et al. 2000, Liu et al. 2003, Betancourt et al. 2006). In this model, the deformation of polymer film is described by the phase field model. As to the electric field, the leaky dielectric model is adopted in which both electrical permittivity and conductivity are considered. The fluid flow together with electric field is coupled together in the framework of phase field. By this model, the effect of physical parameters, such as external voltage, template structure height, and polymer conductivity, is studied in detail. After that, the governing equations are nondimesionalized to analyze the relationship between different parameters. A dimensionless parameter, Electrical Reynolds number ER, is defined, for which, a large value would simplify the electric field to perfect dielectric model and a small value leads it to steady leaky model. These findings and results may enhance the understanding of electrohydrodynamic patterning and may be a meaningful guide for experiments (Ju et al. 2018).

4.3 Replication Techniques

Replication technologies have proven useful for biodegradable polymer microfabrication because the principles behind these processes are direct method and well known in the macro world. The underlying principle is the replication of a microfabricated mold tool, which represents the inverse geometry of the desired polymer structure. The expensive microfabrication step is only necessary for the initial fabrication of this master structure, which can be replicated many times into the polymer substrate. In addition to the cost advantage, replication techniques also offer the benefit of the freedom of design—the master can be fabricated with a large number of different microfabrication technologies, which allow various geometries to be realized. Replication based fabrication techniques are advantageous because of their simplicity, low cost and speed, compared to direct patterning processes. When used lithographically (e.g., NIL), they can replace costly techniques like EBL and address the resolution limits of photolithographic patterning too.

Polymer replication techniques like hot embossing and injection moulding specialise in the production of monolithic parts that can be of use in optical and microfluidic applications and the fact that they can be applied to many different materials makes them highly applicable to patterning of materials for tissue engineering applications, both clinically and in continuing biological research (Gadegaard et al. 2003).

5. Nanotechnologies for Tissue Engineering

Nanotechnology can be used to create nanofibers, nanopatterns and controlled-release nanoparticles with applications in tissue engineering. These techniques are particularly useful for mimicking native tissues because many biological structures, such as ECM fibers, are in the range of tens of nanometers (Pham et al. 2006).

For example, polymeric nanofibers can be fabricated by electrospinning (Pham et al. 2006) and self-assembly (Zhang 2003). In general, the synthesis of nano-structured materials can be generated by any one of two approaches. In one approach, nanomaterials are synthesized by miniaturizing existing materials with nanoscale resolution. These techniques include nanopatterning and electrospinning. In the other approach, molecular build-up, such as self-assembly (Zhang 2003) and layer-by-layer deposition (Thierry et al. 2003) can be used to generate nano materials (Decher et al. 2003).

Chitosan is a biodegradable polysaccharide that is obtained from deacetylated chitin (66% to 95% deacetylation). Chitosan may undergo protonation leading to its solubilization in water because it contains several amino groups in acidic pH. It also establishes electrostatic interactions with the negatively charged DNA to form complexes (polyplexes). Chitosan is a non-toxic material widely used in regenerative medicine.

Zhang and Chang (2009) suggested a static method to fabricate 3D fibrous tubes composed of ultrafine electrospun fibers. They can be used as a 3D collecting template based on manipulation of electric fields and forces to fabricate 3D architecture. This technique can fabricate micro- and macrotubes with multiple micropatterns, and multiple interconnected tubes with the same or different sizes, structure, shapes, and patterns. It is helpful to control the patterened architecture, which has many biomedical and industrial applications (Zhang and Chang 2009). The use of chitosan as a biomaterial is approved by the Food and Drug Administration (FDA) for application in biomedical devices, in particular, in drug delivery and in tissue engineering, with the final goal to restore the functionality of defective or lost tissues (Masil et al. 2019). Chitosan has been used alone or in combination with other materials, in order to enhance the mechanical properties and degradation time for scaffolds. Chitosan has been mixed with hydroxyapatite to create an ideal matrix for osteoblast proliferation and mineral deposition (Zo et al. 2012).

Yadav et al. (2014) suggested that graphene-based chitosan nanocomposites have been used in tissue engineering application for wound healing purpose. Fonseca-Santos et al. (2017) reported faster cell attachment on graphene/CS nanocomposite films displaying good biocompatibility and non-cytotoxicity. Similarly, Wan et al. (2014) and Cheung et al. (2015) prepared CS/GO nanocomposites containing nano-hydroxyapatite composites with improved mechanical properties. These materials can show high cell proliferation rate (Sultana et al. 2015, Venkatesan and Kim 2014).

6. Conclusion

Tissue engineering, termed as 'Regenerative Medicine', is also regarded as a fundamentally ideal medical treatment for diseases that have been too difficult to be cured by existing methods. For regeneration of failed tissues, this biomedical engineering utilizes three fundamental tools: living cell, signal molecules, and scaffold. These are developed in miniature- micro and nano levels.

Chitosan is one of the most promising biomaterials in tissue engineering because it offers a distinct set of advantageous physico-chemical and biological properties that qualify them for a variety of tissue regeneration. The micro and nanoscale engineering is a powerful tool for tissue engineering and biological applications. This chapter gave an overview of microfabrication and nanofabrication techniques that are being used for tissue engineering. It will be a good initiative to know about the technologies used for fabricating the suitable material.

References

Adler-Abramovich, L. and E. Gazit. 2008. Controlled patterning of peptide nanotubes and nanospheres using inkjet printing technology. J. Pept. Sci. 14(2008): 217–223.

Ainslie, K.M. and T.M. Desai. 2008. Microfabricated implants for applications in therapeutic delivery, tissue engineering, and biosensing. Lab Chip 8(11): 1864–1878. Doi: 10.1039/b806446f.

Ainslie, K.M. and T.M. Desai. 2009. *In vitro* inflammatory response of nanostructured titania silicon oxide, and polycaprolactone. J. Biomed. Mater. Res. A 91(3): 647–655.

Ainslie, K., R. Lowe, T. Beaudette, L. Petty, E. Bachelder and T. Desai. 2009. Microfabricated devices for enhanced bioadhesive drug delivery: Attachment to and small-molecule release through a cell monolayer under flow. Small (Weinheim an der Bergstrasse, Germany) 5: 2857–63. 10.1002/smll.200901254.

Anderson and vanden Berg. 2004. Microfabrication and microfluidics for tissue engineering: State of the art and future opportunities. Lab on a Chip 4: 98–103.

Andrea Masotti and Giancarlo Ortaggi. 2009. Chitosan micro- and nanospheres: Fabrication and applications for drug and DNA delivery mini-reviews in medicinal chemistry. Mini Rev. Med. Chem. 9(4): 1389–5575/09.

Becker, H. and C. Gärtner. 2000. Polymer microfabrication methods for microfluidic analytical applications. Electrophoresis 21: 12–26.

Becker, H. and Ulf Heim. 2000. Hot embossing as a method for the fabrication of polymer high aspect ratio structures. Sensors and Actuators A: Physical 83(1-3): 130–135.

Betancourt, T. and L. Brannon-Peppas. 2006. Micro- and nanofabrication methods in nanotechnological medical and pharmaceutical devices. Int. J. Nanomedicine 1(4): 483–95. Doi: 10.2147/nano.2006.1.4.483. PMID: 17722281; PMCID: PMC2676643.

Chen, C.S., S. Mrksich, Huang, G.M. Whitesides and D.E. Ingber. 1997. Geometric control of cell life and death. Science 176: 1425.

Chen, Y. and A. Pepin. 2001. Nanofabrication: Concentional and nonconventional methods. Electrophoresis 22: 187–207.

Cheung, R.C., T.B. Ng, J.H. Wong and W.Y. Chan. 2015. Chitosan: An update on potential biomedical and pharmaceutical applications. Mar. Drugs. 13: 5156–5186.

Chou, S.Y., P.R. Krauss and P.J. Renstrom. 1995. Imprint of sub-25 nm vias and trenches in polymers. Appl. Phys. Lett. 67: 3114.

Chow, K.S. and E. Khor. 2000. Novel fabrication of open-pore chitin matrixes. Biomacromolecules 1: 61–67.

Chung Bong Genn, Lifeng Kang and Ali Khademhosseini. 2007. Micro-nanoscale technologies for tissue engineering and drug discovery applications. December 2007 2(12): 1653–1668. Doi: 10.1517/17460441.2.12.1653.

Co, C., Y.C. Wang and C.C. Ho. 2005. Biocompatible micropatterning of two different cell types. J. Amer. Chem. Soc. 127(6): 1598–1599.

Curtis, A.S., C.D. Wilkinson, J. Crossan, C. Broadley, H. Darmani, K.K. Johal, H. Jorgensen and W. Monaghan. 2005. An *in vivo* micro fabricated scaffold for tendon repair. Eur. Cell Mater. 9: 50–57.

Dai, H., X.K.W. Jiang and H.Q. Mao. 2006. Chitosan-DNA nanoparticles delivered by intrabiliary infusion enhance liver-targeted gene delivery. Int. J. Nanomedicine 1(4): 507–22.

Decher. 2003. Multilayer thin films: Sequential assembly of nanocomposite materials. *In*: Decher, G. and J. Schlenoff (eds.). Wiley Interscience, Weinheim.

Denkov, N., O. Velev, P. Kralchevski, I. Ivanov, H. Yoshimura and Y. Nagayama. 1992. Mechanism of formation of two-dimensional crystals from latex particles on substrates. Langmuir 8(12): 3183–3190.

Dickert, F.L., O. Hayden, R. Bindeus, K.J. Mann, D. Blass and E. Waigmann. 2004. Bioimprinted QCM sensors for virus detection-screening of plant sap. Analytical & Bioanalytical Chemistry 378: 1929–1934.

Fan, H., Y. Lu and A. Stump. 2000. Rapid prototyping of patterned functional structures. Nature 405: 56–60.

Flemming, R.G., C.J. Murphy, G.A. Abrams, S.L. Goodman and P.F. Nealey. 1999. Effects of synthetic micro- and nano-structured surfaces on cell behavior. Biomaterials 20(6): 573–588.

Fonseca-Santos, B. and M. Chorilli. 2017. An overview of carboxymethyl derivatives of chitosan: Their use as biomaterials and drug delivery systems. Mater. Sci. Eng. C Mater. Biol. Appl. 77: 1349–1362.

Gadegaard, N., S. Thoms, D.S. Macintyre, K. McGhee, J. Gallagher, B. Casey and C.D.W. Wilkinson. 2003. Arrays of nano-dots for cellular engineering. Microelectron. Eng. 67-68: 162–168.

Gamo, K. 1996. Microelectron. Engineer 32: 159–171.

Gattazzo, F., A. Urciuolo and P. Bonaldo. 2014. Extracellular matrix: A dynamic microenvironment for stem cell niche. Biochim. Biophys. Acta 1840(8): 2506–2519. Doi: 10.1016/j.bbagen.2014.01.010.

Gentile, F., L. Tirinato, E. Battista, F. Causa, C. Liberale, E.M. di Fabrizio and P. Decuzzi. 2010. Cells preferentially grow on rough substrates. Biomaterials 31(28): 7205–7212. Doi: 10.1016/j.biomaterials.2010.06.016.

Gentile, F., R. La Rocca, G. Marinaro, A. Nicastri and A. Toma. 2012. Differential cell adhesion on mesoporous silicon substrates. ACS Appl. Mater. Interfaces 4(6): 2903–2911. Doi: 10.1021/am300519a.

Gómez-Sjöberg, R., A.A. Leyrat, D.M. Pirone, C.S. Chen and S.R. Quake. 2007. Versatile, fully automated, microfluidic cell culture system. Anal. Chem. 79(22): 8557–8563.

Gottschalch, F., T. Hoffmann, C.M.S. Torres, H. Schulzband and H.C. Scheer. 1999. Polymer issues in nanoimprinting technique. Solid-State Electronics 43: 1079–1083.

Grzybowski, B.A., C.E. Wilmer, J. Kim, K.P. Browne and K.J.M. Bishop. 2009. Self-assembly: From crystals to cells. Soft Matter 5: 1110–1128.

Haverkorn Van Rijsewijik, H.C., P.E.J. Legierse and G.E. Thomas. 1982. Manufacture of LaserVision video discs by a photo—Polymerization process. Philips Tech. Rev. 40: 287.

He, L., C. Mu, J. Shi, Q. Zhang, B. Shi and W. Lin. 2011. Modification of collagen with a natural cross-linker, procyanidin. Int. J. Biol. Macromol. 48(2): 354–359.

Heckele, M. and W.K. Schomburg. 2004. Review on micro molding of thermoplastic polymers. J. Micromech. Microeng. 14: R1.

Hsu, C.M., S.T. Connor, M.X. Tang and Y. Cui. 2008. Wafer-scale silicon nanopillars and nanocones by Langmuir–Blodgett assembly and etching. Appl. Phys. Lett. 93(13): 133109.

Hu, C. Jerry, Athanasiou and A. Kyriacos. 2006. A self-assembling process in articular cartilage tissue engineering. Tissue Engineering 12(4): 969–979. Doi: 10.1089/ten.2006.12.969. ISSN 1076-3279. PMID 16674308.

Huang, J., F. Kim, A.R. Tao, S. Connor and P. Yang. 2005. Spontaneous formation of nanoparticle stripe patterns through dewetting. Nature Mater. 4(12): 896–900.

Hwang, J., Y. Jeong, J.M. Park, K.H. Lee, J.W. Hong and J. Choi. 2015. Biomimetics: Forecasting the future of science, engineering, and medicine. Int. J. Nanomed. 8(10): 5701–5713. Doi: 10.2147/IJN.S83642.

Ikada. 2006. Challenges in tissue engineering. Yoshito Ikada, Published: 18 April 2006 https://doi.org/10.1098/rsif.2006.0124.

In-Yong Kim, Seog-Jin Seo, Hyun-Seuk Moon, Mi-Kyong Yoo, In-Young Park, Bom-Chol Kim and Chong-Su Cho. 2007. Research review paper Chitosan and its derivatives for tissue engineering applications. Doi: 10.1016/j.biotechadv.2007.07.009.

Jameela, S.R. and A. Jayakrishnan. 1995. Glutaraldehyde cross-linked chitosan microspheres as a long acting biodegradable drug delivery vehicle: studies on the *in vitro* release of mitoxantrone and *in vivo* degradation of microspheres in rat muscle. Biomaterials 16: 769–775.

Jeon, S., J.U. Park, R. Cirelli, S. Yang, C.E. Heitzman, P.V. Braun, P.J.A. Kenis and J.A. Rogers. 2004. Fabricating complex three-dimensional nanostructures with high-resolution conformable phase masks (Article) (Open Access) Proceedings of the National Academy of Sciences of the United States of America Volume 101, Issue 34, 24 August 2004, Pages 12428–12433. Doi: 10.1073/pnas.0403048101.

Jeon, S., E. Menard, J.U. Park, J. Maria, M. Meitl, J. Zaumseil and J.A. Rogers. 2004. Three-dimensional nanofabrication with rubber stamps and conformable photomasks (Review). Advanced Materials Volume 16, Issue 15 SPEC. ISS., 3 August 2004, Pages 1369–1373. Doi: 10.1002/adma.200400593.

Jeong, S., L. Hu, H.R. Lee, E. Garnett, J.W. Choi and Y. Cui. 2010. Fast and scalable printing of large area monolayer nanoparticles for nanotexturing applications. Nano Lett. 10(8): 2989–2994.

Jiang, X., R. Ferrigno, M. Mrksich and G.M. Whitesides. 2003. Electrochemical desorption of self-assembled monolayers non-invasively releases patterned cells from geometrical confinements. J. Am. Chem. Soc. 125: 2366–2367.

Jilani Asim, Mohamed Shaaban Abdel-wahab and Ahmed Hosny Hammad. 2017. Advance Deposition Techniques for Thin Film and Coating, Edited book—Modern Technologies for Creating the Thin-film Systems and Coatings, Intech Publisher.

JinWoo Lee. 2015. 3D Nanoprinting technologies for tissue engineering applications, nanomaterials for medical and dental applications, Volume 2015 |Article ID 213521 | https://doi.org/10.1155/2015/213521.

John A. Rogers and Ralph G. Nuzzo. 2005. Recent progress in soft lithography by John A. Rogers and Ralph G. Nuzzo February 2005 ISSN: 1369 7021 © Elsevier Ltd 2005.

Ju Fenhong Song Dapeng, Fangwei Gu, Yan Liu, Yuan Ji, Yulin Ren, Xiaocong He, Baoyong Sha, Ben Q. Li and P.M. Qingzhen Yang. 2018. Parametric study on electric field-induced micro-/nanopatterns in thin polymer films ID: 29542932 Doi: 10.1021/acs.langmuir.8b00007 Langmuir 2018 Apr 10; 34(14): 4188–4198.

Kang, M.L., H.L. Jiang, S.G. Kang, D.D. Guo, D.Y. Lee, C.S. Cho and H.S. Yoo. 2007. Pluronic F127 enhances the effect as an adjuvant ofchitosan microspheres in the intranasal delivery of Bordetella bronchiseptica antigens containing dermonecrotoxin. Vaccine 25(23): 4602–10.

Kiewit, D.A. 1973. Microtool fabrication by etch pit replication. Rev. Sci. Instrum. 44: 1741.

Koev, S.T., M.A. Powers, H. Yi, L.Q. Wu, W.E. Bentley, G.W. Rubloff, G.F. Payne and R. Ghodssi. 2007. Mechano-transduction of DNA hybridization and dopamine oxidation through electrodeposited chitosan network. Lab Chip 7(1): 103–111, Jan. 2007.

Leclerc, E., Y. Sakai and T. Fujii. 2003. Cell culture in 3-dimensional microfluidic structure of PDMS (polydimethylsiloxane). Biomed. Micro Devices 5(2): 109–114.

Lee, Y.S. 2007. Self-Assembly and Nanotechnology: A Force Balance Approach John Wiley & Sons, Inc.

Lehmann, H.W., R. Widmer, M. Ebnoether, A. Wokaun, M. Meier and S.K. Miller. 1983. Fabrication of submicron crossed square wave gratings by dry etching and thermoplastic replication techniques. J. Vac. Sci. Technol. B 1: 1207.

Li, N., A. Tourovskaia and A. Folch. 2003. Biology on a chip: Microfabrication for studying the behavior of cultured cells. Crit. Rev. Biomed. Eng. 31(5-6): 423–88.

Li, Y., W. Cai and G. Duan. 2008. Ordered micro/nanostructured arrays based on the monolayer colloidal crystals. Chem. Mater. 20(3): 615–624.

Lima, M.J., V.M. Correlo and R.L. Reis. 2014. Micro/nano replication and 3D assembling techniques for scaffold fabrication. Mater. Sci. Eng. Part C 42: 615–621. Doi: 10.1016/j.msec.2014.05.064.

Liu Tsang, V. and S.N. Bhatia. 2004. Three-dimensional tissue fabrication. Adv. Drug Delivery Rev. 56(11): 1635–1647.

Lu, Z.X., A. Namboodiri and M.M. Collinson. 2008. Self-supporting nanopore membranes with controlled pore size and shape. ACS Nano 2(5): 993–999.

Luu, T.U., S.C. Gott, B.W. Woo, M.P. Rao and W.F. Liu. 2015. Micro- and nanopatterned topographical cues for regulating macrophage cell shape and phenotype. ACS Appl. Mater. Interfaces 7(51): 28665–28672. Doi: 10.1021/acsami.5b10589.

Madihally, S.V. and H.W. Matthew. 1999. Porous chitosan scaffolds for tissue engineering. Biomaterials 20: 1133–1142.

Masil Alessia De, Ilaria Tonazzini, Cecilia Masciullo, Roberta Mezzena, Federica Chiellini, Dario Puppi, Marco Cecchini. 2019. Chitosan films for regenerative medicine: Fabrication methods and mechanical characterization of nanostructured chitosan films. Received: 23 July 2019 /Accepted: 2 September 2019 /Published online: 16 September 2019 International Union for Pure and Applied Biophysics (IUPAB) and Springer-Verlag GmbH Germany, part of Springer Nature 2019.

Mauro, N., A. Manfredi, E. Ranucci, P. Procacci, M. Laus, D. Antonioli, C. Mantovani, V. Magnaghi and P. Ferruti. 2013. Degradable poly(amidoamine) hydrogels as scaffolds for *in vitro* culturing of peripheral nervous system cells. Macromol. Biosci. 13(3): 332–347. Doi: 10.1002/mabi.201200354.

McCord, M.A. and M.J. Rooks. 1997. Handbook of microlithography. pp. 139–250. *In*: Rai-Choudhury, P. (ed.). Microfabrication and Micro-systems, SPEI Press, Washington.

McDermott, M.K., C. Zhu, R. Ghodssi, G.F. Payne and L.Q. Wu. 2006. Mimicking biological phenol reaction cascades to confer mechanical function. Adv. Funct. Mater. 16(15): 1967–1974.

Melngailis, J., A.A. Mondelli, I.L. Berry III and R. Mohondro. 1989. A review of ion projection lithography. J. Vac. Sci. Technol. B 16(3): 927–957.

Mendes, Ana C., Erkan T. Baran, Rui L. Reis and Helena S. Azevedo. 2013. Self-assembly in nature: Using the principles of nature to create complex nanobiomaterials. WIREs Nanomed. Nanobiotechnol. 2013. Doi: 10.1002/wnan.1238.

Miller, C., H. Shanks, A. Witt, G. Rutkowski and S. Mallapragada. 2001. Oriented Schwann cell growth on micropatterned biodegradable polymer substrates. Biomaterials 22: 1263.

Morita, S.I. and Y. Ikada. 2002. Lactide copolymers for scaffolds in tissue engineering. pp. 111–122. *In*: Lewandrowski, K.O., D.L. Wise, D.J. Trantolo, J.D. Gresser, M.J. Yaszemski and D.E. Attobelli (eds.). Tissue Engineering and Biodegradable Equivalents Scientific and Clinical Applications. Marcel Dekker, New York, Basel.

Mujahid, M., M. Khalida, S. Aminb, R.S. Rawatc, A. Nusaird and G.R. Deenc. 2013. Effect of surfactant and heat treatment on morphology, surface area and crystallinity in hydroxyapatite nanocrystals. Ceramics International 39(1): 39–50. January 2013. https://doi.org/10.1016/j.ceramint.2012.05.090.

Nakano. 1979. Dislocation in Solid, North-Holland Publishing Company, Amsterdam.

Niamsa, N., Y. Srisuwan, Y. Baimark, P. Phinyocheep and S. Kittipooms. 2009. Preparation of nanocomposite chitosan/silk fibroin blend films containing nanopore structures. Carbohydrate Polymer 78: 60–65.

Nichol, J.W., S.T. Koshy, H. Bae, C.M. Hwang, S. Yamanlar and A. Khademhosseini. 2010. Cell-laden microengineered gelatin methacrylate hydrogels. Biomaterials 31(21): 5536–5544.

Omar, Fuad. 2013. Hot embossing process parameters: Simulation and experimental studies. Materials Science, 2013.

Park, C.W., Y.S. Rhee, S.H. Park, S.D. Danh, S.H. Ahn, S.C. Chi and E.S. Park. 2010. *In vitro/in vivo* evaluation of NCDS-micro-fabricated biodegradable implant. Arch. Pharm. Res. 33(3): 427–432. Doi: 10.1007/s12272-010-0312-4.

Park, J.J., X. Luo, H. Yi, T.M. Valentine, G.F. Payne, W.E. Bentley, R. Ghodssi and G.W. Rubloff. 2006. Chitosan-mediated *in situ* biomolecule assembly in completely packaged microfluidic devices. Lab Chip 6(10): 1315–1321, Oct. 2006.

Patel, N., R. Padera, G.H.W. Sanders, S.M. Cannizzaro, M.C. Davies, R. Langer, C.J. Roberts, S.J.B. Tendler, P.M. Williams and K.M. Shakessheff. 1998. Spatially controlled cell engineering on biodegradable polymer surfaces. FASEB 12: 1447.

Pham, Q.P., U. Sharma and A.G. Mikos. 2006. Electrospun poly (ε-caprolactone) microfiber and multilayer nanofiber/microfiber scaffolds: Characterization of scaffolds and measurement of cellular infiltration. Biomacromolecules 7(10): 2796–2805.

Prudhomme, W., G.Q. Daley, P. Zandstra and D.A. Lauffenburger. 2004. Proc. Natl. Acad. Sci. USA 101: 2900–2905.

Ramos, B.L. and S.J. Choquette. 1996. Embossable grating couplers for planar waveguide optical sensors. Analytical Chemistry 68: 1245–1249. DOI: 10.1021/ac950579x.

Ratner, B.D., A.S. Hoffman, F.J. Schoen and J.E. Lemons. 2004. Biomaterials Science: An Introduction to Materials in Medicine. Academic Press.

Schaefer, D., I. Martin, P. Shastri, R.F. Padera, R. Langer, L.E. Freed and G. Vunjak-Novakovic. 2000. *In vitro* generation of osteochondral composites. Biomaterials 21: 2599.

Sia, S.K. and G.M. Whitesides. 2003. Microfluidic devices fabricated in poly(dimethylsiloxane) for biological studies. Electrophoresis 24: 3563–3576.

Slaughter, B.V., S.S. Khurshid and O.Z. Fisher. 2009. Hydrogels in regenerative medicine. Adv. Mater. 21: 3307–3329.

Sultana, N., M. Mokhtar, M.I. Hassan, R.M. Jin, F. Roozbahani and T.H. Khan. 2015. Chitosan-based nanocomposite scaffolds for tissue engineering applications. Mater. Manuf. Process 30: 273–278.

Sze, S.M. 1985. Semiconductor Devices: Physics and Technology, John Wiley, New York.

Tahara, K., H. Yamamoto, H. Takeuchi and Y. Kawashima. 2007. Development of gene delivery system using PLGA nanospheres. YakugakuvZasshi 127(10): 1541–48.

Tania Betancourt and Lisa Brannon-Peppas. 2006. Micro- and nanofabrication methods in nanotechnological medical and pharmaceutical devices. Int. J. Nanomedicine 2006 Dec; 1(4): 483–495.

Thierry, B., F.M. Winnik, Y. Merhi and M. Tabrizian. 2003. Nanocoatings onto arteries via layer-by-layer deposition: Toward the *in vivo* repair of damaged blood vessels. J. Am. Chem. Soc. 125(25): 7494–7495.

Thorsen, T., S.J. Maerkl and S.R. Quake. 2002. Microfluidic large-scale integration. Science 298: 580–584.

Tseng, A. 2005. Recent developments in nanofabrication using focused ion beams. Small 1(10): 924–939.

Venkatesan, J. and S.K. Kim. 2014. Nano-hydroxyapatite composite biomaterials for bone tissue engineering—A review. J. Biomed. Nanotechnol. 10: 3124–3140.

Vogel, N., C.K. Weiss and K. Landfester. 2012. From soft to hard: The generation of functional and complex colloidal monolayers for nanolithography. Soft Matter 8(15): 4044.

Voldman, J., M.L. Gray and M.A. Schmidt. 1999. Microfabrication in biology and medicine. Annu. Rev. Biomed. Eng. 1: 401–25.

Vozzi Giovanni Vozzi, Ilaria Morelli, Federico Vozzi, Chiara Andreoni, Elisabetta Salsedo, Annagiulia Morachioli, Paolo Giustiand and Gianluca Ciardelli. June 2010. A novel microfabrication technique integrating soft-lithography and molecular imprinting for tissue engineering applications. https://doi.org/10.1002/bit.22740.

Wan, Y., X. Chen, G. Xiong, R. Guo and H. Luo. 2014. Synthesis and characterization of three-dimensional porous graphene oxide/sodium alginate scaffolds with enhanced mechanical properties. Mater. Express 4: 429–434.

Webber, M.J., E.A. Appel, E.W. Meijer and R. Langer. 2015. Supramolecular biomaterials. Nat. Mater. 15(1): 13–26. Doi: 10.1038/nmat4474.

Whitesides, G.M. 1995. Molecular self-assembly and nanochemistry—A chemical strategy for the synthesis of nanostructures. Science 254: 1312–1319. DOI: 10.1126/science.1962191.

Whitesides, G.M. 2006. The origins and the future of microfluidics. Nature 442(7101): 368–373.

Whitesides, G.M. and B. Grzybowski. 2002. Self-assembly at all scales. Science 295: 2418–2421.

Whitesides, G.M., J.K. Kriebel and B.T. Mayers. 2005. Self-assembly and nanostructured materials. pp. 217–239. *In*: Huck, W.T.S. (ed.). Nanoscale Assembly: Chemical Techniques. New York: Springer.

Whitesides, G.M., J.P. Mathias and C.T. Seto. 1991. Molecular self-assembly and nanochemistry—a chemical strategy for the synthesis of nanostructures. Science 254: 1312–1319.

Whitesides, G.M., S.K.W. Dertinger, D.T. Chiu and N.L. Jeon. 2001. Generation of gradients having complex shapes using microfluidic networks. Anal. Chem. 73: 1240–1246.

Xia, D., Z. Ku, S.C. Lee and S.R. Brueck. 2011. Nanostructures and functional materials fabricated by interferometric lithography. Adv. Mater. 23(2): 147–179.

Xia, Y.N. and G.M. Whitesides. 1998. Soft lithography. Annu. Rev. Mater. Sci. 28: 153.

Yadav, M., K.Y. Rhee and S.J. Park. 2014. Synthesis and characterization of grapheneoxide/carboxymethylcellulose/alginate composite blend films. Carbohydr. Polym. 110: 18–25.

Yeong, W.Y., C.K. Chua, K.F. Leong and M. Chandrasekaran. 2004. Rapid prototyping in tissue engineering: Challenges and potential. Trends Biotechnol. 22(12): 643–652.

Yi, H., L. Wu, W.E. Bentley, R. Ghodssi, G.W. Rubloff, J.N. Culver and G.F. Payne. 2005. Biofabrication with chitosan. Biomacromolecules 6(6): 2881–2894, Nov./Dec. 2005.

Zein Iwan, Zeina Dietmar, W. Hutmacher, Kim Cheng, Tanc Swee and Hin Teo ha. 2002. Fused deposition modeling of novel scaffold architectures for tissue engineering applications. Biomaterials 23(4): 1169–1185. 15 February 2002. https://doi.org/10.1016/S0142-9612(01)00232-0.

Zhang, S. 2003. Fabrication of novel biomaterials through molecular self-assembly. Nat. Biotechnol. 21(10): 1171–1178.

Zhang, G. and D. Wang. 2009. Colloidal lithography—The art of nanochemical patterning. Chem. Asian J. 4(2): 236–245.

Zhu, W., C.O. Brien, J.R.O. Brien and L.G. Zhang. 2014. 3D nano/microfabrication techniques and nanobiomaterials for neural tissue regeneration. Nanomedicine 9(6): 859–875. Doi: 10.2217/nnm.14.36.

Zo, S.M., D. Singh and A. Kumar. 2012. Chitosan-hydroxyapatite macroporous matrix for bone tissue engineering. 103: 1438–46.

7

Characterization of Nanostructures

A. Dinesh Karthik,[1,]* *K. Geetha*[2] *and D. Shakila*[3]

1. Introduction

Nanomaterials behave in a different way as the dimensions changes with respect to the bulk materials. It is necessary to characterize physical, structural and optical properties of a material to qualify as nanomaterial. Various characterization techniques are used to know the characteristics of the nanomaterials. When materials' dimensions are reduced to nanoscale, they demonstrate unique properties which are different from those of their bulk counterparts. For example, their electronic and optical properties alter, their chemical activities can be increased or decreased and mechanical/ structural stabilities are changed. Such features make nanomaterials attractive for unique sensing applications, and also at the same time, cause complications in their characterization processes. Therefore, the challenge lies in finding the right characterization techniques that have the optimum capabilities for studying the characteristics of nanomaterials. A large number of techniques can be employed for nanomaterials characterization. Some of the most common characterization techniques in nanotechnology will be presented in this chapter. In addition, applicability of these techniques for investigating different types of nanomaterials and their relevance to sensor technology will also be described.

Nanotechnology is the technology of the small, the very small. It is the use and manipulation of being counted at a tiny scale. Nanoparticles are not necessarily produced through cutting-edge synthesis laboratories, but have obviously existed in nature for a long time, and therefore their use can be traced back to ancient times. The important properties of nanoparticles that could affect nanoparticles' behavior and toxicity includes particle size, shape, surface properties, aggregation state, solubility, shape and chemical make-up. Therefore, whilst analyzing nanoparticles in different matrices, It isn't first-class the composition and attention at the way to need to be determined but moreover the physical and chemical properties of the engineered nanoparticles in the sample and the chemical characteristics of any capping/functional layer on the particle surface. Methods are available that have been advanced for natural nanomaterials or engineered nanomaterials in easy matrices, which could be optimized to provide the necessary

---

[1] Unit of Nanotechnology and PG and Research Department of Chemistry, Shanmuga Industries arts and Science College, Tiruvannamalai, Tamil Nadu.

[2] PG and Research Department of Chemistry, Muthurangam Govt. Arts College (Autonomous), Vellore - 632 002, Tamil Nadu, India.

[3] PG and Research Department of Chemistry, D.K.M College for Women, (Autonomous), Vellore - 632 002, Tamil Nadu, India.

* Corresponding author: dineshkarthik2008@gmail.com

information. These include microscopy, chromatography, spectroscopy, centrifugation as well as filtration and related techniques. A combination of these is frequently required. This describes the characterization of nanoparticles. The verification of the particle characterization was carried out through the spectroscopic and microscopic techniques' transmission. A general discussion of the experimental findings, in view of the reports, concludes the chapter.

Nanoparticles are typically characterized through their size, morphology and surface charge, using such advanced microscopic techniques as scanning electron microscopy (SEM), transmission electron microscopy (TEM) and atomic force microscopy (AFM). The average particle diameter, their size distribution and charge affect the physical stability and the *in vivo* distribution of the nanoparticles. Electron microscopy techniques are very useful in ascertaining the overall shape of polymeric nanoparticles, which may determine their toxicity.

The surface charge of the nanoparticles affects the physical stability and redispersibility of the polymer dispersion as well as their *in vivo* performance. Particle size distribution and morphology are the most important parameters of characterization of nanoparticles. Morphology and size are measured by electron microscopy.

There are several tools for determining nanoparticle size as discussed below.

1.1 Dynamic Light Scattering (DLS)

Currently, the fastest and most popular method of determining particle size is photon-correlation spectroscopy (PCS) or dynamic light scattering (DLS). DLS is widely used to determine the size of Brownian nanoparticles in colloidal suspensions in the nano and submicron ranges. Shining monochromatic light (laser) onto a solution of spherical particles in Brownian motion causes a Doppler shift when the light hits the moving particle, changing the wavelength of the incoming light. This change is related to the size of the particle. It is possible to extract the size distribution and give a description of the particle's motion in the medium, measuring the diffusion coefficient of the particle and using the autocorrelation function. The photon correlation spectroscopy (PCS) represents the most frequently used technique for accurate estimation of the particle size and size distribution based on DLS (Betancor and Luckarift 2008).

1.2 Scanning Electron Microscopy (SEM)

A scanning electron microscope (SEM) is a type of electron microscope that images a sample by scanning it with a high energy beam of electrons in a raster scan pattern (Sitterberg et al. 2009, Molpeceres et al. 2008). The electrons interact with the atoms that make up the sample producing signals that contain information about the sample's surface topography, composition, and other properties such as electrical conductivity. SEM can produce very high-resolution images of a sample surface, revealing details about less than 1 to 5 nm in size. Due to the very narrow electron beam, SEM micrographs have a large depth of field, yielding a characteristic three-dimensional appearance useful for understanding the surface structure of a sample (Figure 1). Under vacuum, electrons generated by a source are accelerated in a field gradient. The beam passes through electromagnetic lenses, focusing onto the specimen. As result of this bombardment, different types of electrons are emitted from the specimen. A detector catches the secondary electrons and an image of the sample surface is constructed by comparing the intensity of these secondary electrons to the scanning primary electron beam. Finally, the image is displayed on a monitor. In most of the applications, the data collected is over a pre-selected area of the sample surface and following this, a 2D image is generated that shows the various spatial variations. Conventional SEMs with a magnification range of 20X–30000X with a spatial resolution of 50–100 nm can scan areas which vary from 1 cm to 5 μm in width (Couvreur et al. 1995, Goldberg et al. 2007). SEMs also have the ability to analyze particular points as can be seen during EDX operations which help in determining the chemical composition of the sample concerned.

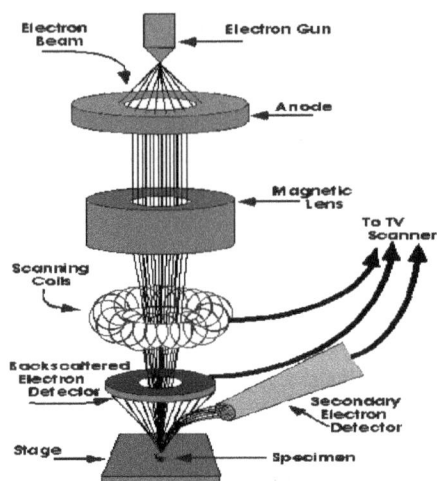

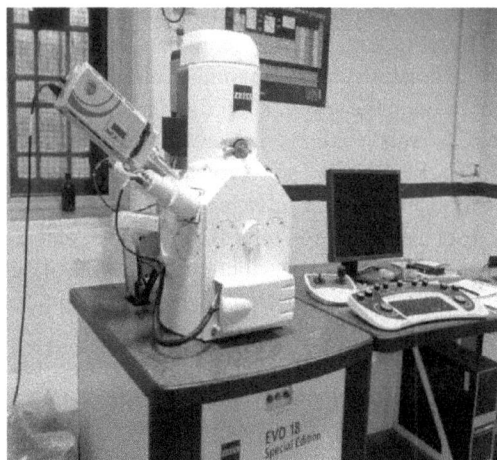

Figure 1: Schematic diagram of Scanning Electron Microscopy (SEM).

1.3 Transmission Electron Microscope (TEM)

TEM operates on different principle than SEM, yet it often brings same type of data. The sample preparation for TEM is complex and time consuming because of its requirement to be ultra-thin for the electron transmittance (Gurny et al. 1981, Koosha et al. 1989, Polakovic 1999). The nanoparticles' dispersion is deposited onto support grids or films. To make nanoparticles withstand the instrument vacuum and facilitate handling, they are fixed using either a negative staining material, such as phosphor tungstic acid or derivatives, uranyl acetate, etc., or by plastic embedding. Alternate method is to expose the sample to liquid nitrogen temperatures after embedding in vitreous ice.

1.3.1 High Resolution Transmission Electron Microscopy (HRTEM)

High Resolution Transmission Electron Microscopy (HRTEM) or Transmission Electron Microscope (TEM) allows the imaging of the crystallographic structure of a sample at an atomic scale (Quintanar-Guerrero et al. 1998). Because of its high resolution, it is an invaluable tool in the study of nanoscale properties of crystalline materials such as semiconductors and metals. At present, the highest resolution realized is 0.8 Å. In comparison to conventional microscopy, HRTEM does not use amplitudes, i.e., absorption by the sample, for image formation. Instead of this formation, contrast arises from the electron wave with itself. As the recording of the phase of these waves is difficult, the amplitude resulting from this interference is generally measured.

1.4 Atomic Force Microscopy (AFM)

Atomic force microscopy (AFM) offers ultra-high resolution in particle size measurement and is based on a physical scanning of samples at sub-micron level using a probe tip of atomic scale. Instrument provides a topographical map of sample based on forces between the tip and the sample surface. Samples are usually scanned in contact or noncontact mode depending on their properties. In contact mode, the topographical map is generated by tapping the probe on to the surface across the sample and probe hovers over the conducting surface in noncontact mode (Figure 2). The prime advantage of AFM is its ability to image non conducting samples without any specific treatment, thus allowing imaging of delicate biological and polymeric nano and microstructures. AFM provides the most accurate description of size and size distribution and requires no mathematical treatment.

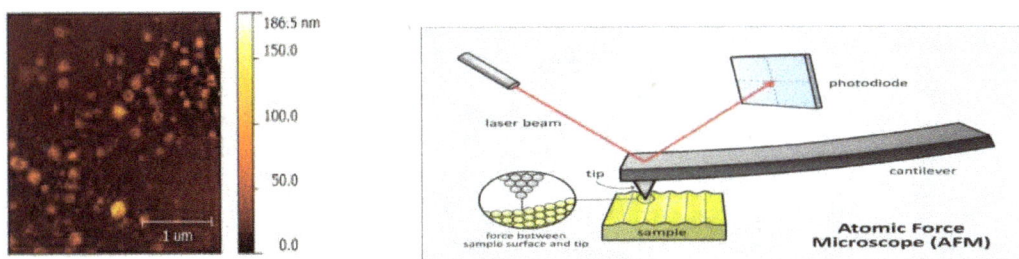

Figure 2: Schematic diagram of an Atomic Force Microscopy (AFM).

Moreover, particle size obtained by AFM technique provides real picture which helps understand the effect of various biological conditions (Quintanar-Guerrero et al. 1998).

1.5 Scanning Tunneling Microscopy (STM)

The atomic resolution of STM is attributed to the imaging mechanism based on the quantum tunneling phenomena. STM measures the tunneling current I that is generated by voltage V between the atomically sharp STM tip and surface. The tunneling current increases by an order of the current for every 1 Å reduction in distance. The distance in the xyz-directions is also controlled by a piezoelectric scanner, which provides angstrom-order changes in the distance, and a feedback loop, which controls the z-direction, and is installed to keep the tunneling current constant. By observing the tunneling current as the tip scans a surface, the morphology of the surface can be precisely detected (Figure 3). Measurements can be carried out at room temperature in solution as well as at low temperature under ultra-high vacuum (UHV) conditions. We can obtain various electrical properties of an object by analyzing the tunneling current, such as the local density of states (LDOS) from dI/dV and local barrier height from dI/Dv (Kaparissides et al. 2006, Lakshmana Prabu et al. 2009, Mukhopadhyay et al. 2016, Santanu Sasidharan et al. 2014).

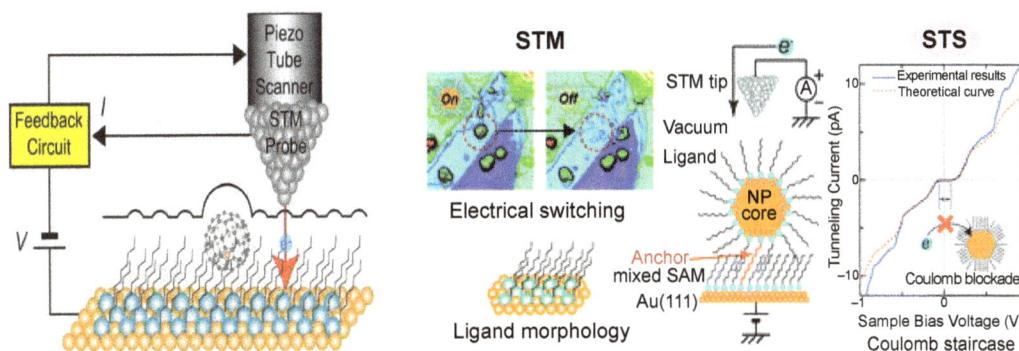

Figure 3: Schematic diagram of a Scanning Tunneling Microscopy (STM).

1.6 Small Angle X-Ray Scattering (SAXS)

Small angle X-ray scattering (SAXS) is a well-established technique to investigate structural properties of materials at the nanoscale (Figure 4). Due to the short wavelength that is typically well below 1 nm, X-rays are perfectly suited for the investigation of nanoparticles. As an ensemble method, SAXS provides information already averaged over a large number of NPs, and there is no need to analyze many images as in microscopy to obtain statistically relevant results. On the other hand, there is no directly visible real space image from SAXS: The scattering information has to

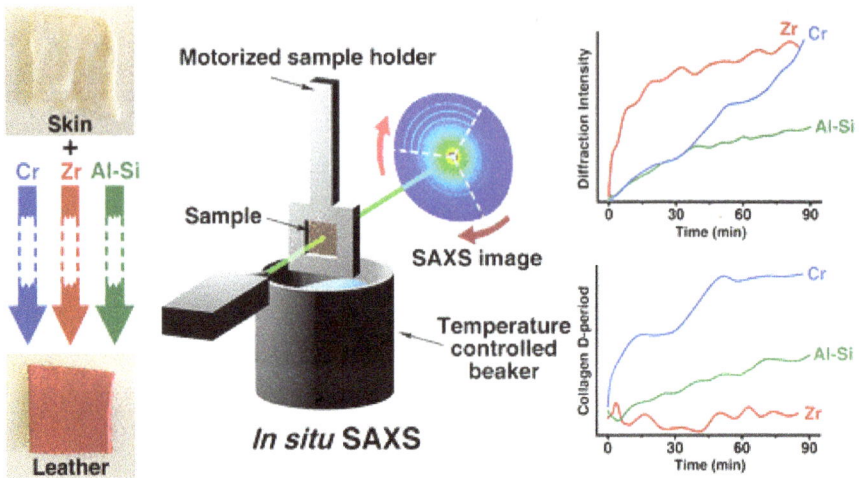

Figure 4: Schematic diagram of a Small Angle X-ray Scattering (SAXS).

be analyzed in reciprocal space, and modeling or fitting is required to extract the information. One of the great advantages of SAXS is that it can be applied directly to the NPs in suspension without extensive sample preparation. If an almost parallel and monochromatic X-ray beam impinges on material with density inhomogeneities on the nanoscale like NPs in a suspension, a small fraction of the radiation is scattered in forward direction around the transmitted direct beam. This effect is called small angle x-ray scattering (SAXS) as the scattering angles are typically below about 5 degrees (Brian 2013). For randomly distributed particles, the scattering pattern is independent of the azimuthal angle; thus, the scattering image can be circularly integrated to obtain the scattered intensity as function of the scattering angle. As the scattering depends also on the wavelength λ (or photon energy EPh) of the x-rays, the scattered intensity is usually analyzed as function of the momentum transfer q, which is given by

$$q = 4\pi/\lambda \sin \Theta = 4 \pi/(hc) E_{Ph} \sin \Theta$$

where Θ is half of the scattering angle, h the Planck constant, and c the speed of light (Hu et al. 2012). If the nanoparticles are sufficiently monodispersed, the scattering image consists of concentric rings.

1.7 X-Ray Diffractometer

The crystal structure and also the particle size were studied by X-ray diffraction. 1 ml of the nanoparticle solution was spread on a glass slide and dried at 40°C in an oven. The process was repeated 3–4 times to obtain a thin film. The spectra were recorded in a Phillips Xpert Pro Diffractometer (Cu Kα radiation, $\lambda 1$ = 1.54056; $\lambda 2$ = 1.54439) running at 40 kV and 30 mA. The diffracted intensities were recorded at 2θ from 2 35.01° to 79.99°. The crystalline size was calculated from the half-height width of the diffraction peak of XRD pattern using the Debye Scherrer equation:

Kλ

$$D = K\lambda/\beta \cos \theta$$

where, D = crystalline size, Å

K = crystalline-shape factor

λ = X-ray wavelength

θ = observed peak angle, degree

β = line broadening at half the maximum intensity (FWHM)

The crystallinity of the particles was evaluated through a comparison of crystallite size from XRD and TEM particle size determination by the following equation:

$$I_{CRY} = D_P \text{ (TEM, SEM)}/D_{CRY} \text{ (XRD) } (I_{CRY} \geq 1)$$

where, Icry is the crystallinity index

Dp is the particle size (obtained from either TEM or SEM morphological analysis)

Dcry is the particle size (calculated from the Scherrer equation)

XRD is based on the principle of constructive interference of X-rays and the sample concerned which should be crystalline. The X-rays which are generated by a CRT are filtered, collimated and then directed towards the sample. The interaction that follows produces constructive interference based on Bragg's law which relates wavelength of the incident radiations to the diffraction angle and lattice spacing (Yao et al. 2015, Pawar et al. 2014).

X-ray powder diffraction (XRD) is a rapid analytical technique primarily used for phase identification of the crystalline material and can provide information on unit cell dimension and atomic spacing. The X-ray is generated by cathode ray tube, filtered to produce monochromatic radiation, collimated to concentrate, and directed towards the sample. The interaction of the incident monochromatic rays with the sample produces constructive interference (and diffracted ray) when condition satisfies Bragg's Law $n\lambda = 2d \sin\theta$. This equation relates the wavelength (λ) of electromagnetic radiation to the diffraction angle (θ) and the lattice spacing (d) in a crystalline sample by scanning the sample through arrangement of 2θ angles. All the possible diffraction directions of the lattice are attained due to the random orientation of the powdered materials.

1.8 Energy Dispersive X-Ray (EDX)

Energy dispersive X-ray spectroscopy (EDS or EDX) is an analytical technique used for the elemental analysis or chemical characterization of a sample. It is one of the variants of X-ray fluorescence spectroscopy which relies on the investigation of a sample through interactions between electromagnetic radiation and matter, analyzing X-rays emitted by the matter in response to hitting the charged particles. Its characterization capabilities are due, in large part, to the fundamental principle that each element has a unique atomic structure, allowing X-rays that are characteristic of an element's atomic structure to be identified uniquely from one another. To stimulate the emission of characteristic X-rays from a specimen, a high-energy beam of charged particles such as electrons or protons or a beam of X-rays is focused onto the sample being studied. At rest, an atom within the sample contains ground state (or unexcited) electrons in discrete energy levels or electron shells bound to the nucleus. The incident beam may excite an electron in an inner shell, ejecting it from the shell while creating an electron hole where the electron was. An electron from an outer, higher-energy shell then fills the hole, and the difference in energy between the higher-energy shell and the lower energy shell may be released in the form of an X-ray. The number and energy of the X-rays emitted from a specimen can be measured by an energy dispersive spectrometer. The atomic structure of the element from which they were emitted and the energy of the X-rays are both characteristics of the energy difference between the two shells (Tan et al. 2013). This allows the elemental composition of the specimen to be measured.

A transmission electron microscope is constituted of: (1) two or three condenser lenses to focus the electron beam on the sample, (2) an objective lens to form the diffraction in the back focal plane and the image of the sample in the image plane, (3) some intermediate lenses to magnify the image or the diffraction pattern on the screen. If the sample is thin (< 200 nm) and constituted of light chemical elements, the image presents a very low contrast when it is focused. To obtain amplitude contrasted image, an objective diaphragm is inserted in the back focal plane to select the transmitted beam (and possibly few diffracted beam): the crystalline parts in Bragg orientation appear dark and the amorphous or not Bragg oriented parts appear bright. This imaging mode is called bright field mode (BF). In diffraction mode, other intermediate lens is inserted to image on the screen of the diffraction pattern of the back focal plane. If the diffraction is constituted by many diffracting phases, each of them can be differentiated by selecting one of its diffracted beams with the objective diaphragm. To do that, the incident beam must be tilted so that the diffracted beam is put on the objective lens axis to avoid off- axis aberrations. This mode is called dark field mode, or DF. The BF and DF modes are used for imaging materials to nanometer scale. SAED and micro diffraction patterns of a crystal permit to obtain the symmetry of its lattice and calculate its interplanar distances (with the Bragg law). This is useful to confirm the identification of a phase, after assumptions generally based on the literature of the studied system and on chemical analyses (Reverberi et al. 2016).

1.9 Ultraviolet / Visible (UV / VIS) Spectroscopy

Ultraviolet-visible spectroscopy or ultraviolet-visible spectrophotometer (UV-Vis) involves the spectroscopy of photons in the UV-Visible region. It uses light in the visible and adjacent near ultraviolet (UV) and near infrared (NIR) ranges. In this region of the electromagnetic spectrum, molecules undergo electronic transitions. UV-Vis spectrophotometers are mainly used to measure transmission or absorption in liquids and transparent or opaque solids. It does so by sending a beam of light through the sample and then monitoring the remaining light in a detector. In the case of a UV-Vis spectrophotometer, the light is in the wavelength of 800–200 nm, probing electronic transitions in the sample. It is hard to reach a lower wavelength than 200 nm as oxygen starts to absorb light below that wavelength. When the light passes through the sample, some of the molecules in the sample will absorb light at various wavelengths of this spectrum, depending on their chemical bonds and structure. As a rule, energetically favored electron promotion will be from the highest occupied molecular orbital (HOMO) to the lowest unoccupied molecular orbital (LUMO), and the resulting species is called an excited state. When sample molecules are exposed to light having an energy that matches a possible electronic transition within the molecule, some of the light energy will be absorbed as the electron is promoted to a higher energy orbital. A spectrophotometer records the wavelengths at which absorption occurs, together with the degree of absorption at each wavelength. The resulting spectrum is presented as a graph of absorbance versus wavelength (Shrikant et al. 2006). These measurements are compared at each wavelength to quantify the sample's wavelength dependent extinction spectrum. The data is typically plotted as extinction as a function of wavelength. Each spectrum is background corrected using a "blank"—a cuvette filled with only the dispersing medium to guarantee that spectral features from the solvent are not included in the sample extinction spectrum. Our standard UV-Vis analysis is performed with an Agilent 8453 single beam diode array spectrometer, which collects spectra from 200–1100 nm using a slit width of 1 nm. Deuterium and tungsten lamps are used to provide illumination across the ultraviolet, visible, and near infrared electromagnetic spectrum. Spectra are typically collected from 1 mL of a sample dispersion, but we can test volumes as small as 100 μL using a microcell with a path length of 1 cm. Additionally, we have assembled a variety of light source/spectrometer custom setups for measuring optical properties of materials from the ultraviolet to the deep infrared (200 nm to m), and can customize analytical systems to measure scattering or absorption from both μ 20 liquid and solid samples. We also have a highly instrumented chamber for aerosolizing nanoparticles and measuring the optical properties of the suspended particles.

1.10 Particle Size Analyser (PSA)

The technique of PSA is ideally suited for the determination of the size of particles in the nanometer size range. The Malvern Zetasizer Nano Series uses patented optics that provides exceptional levels of sensitivity and allows the determination of the size of samples that contain very small particles and/or particles that are present at very low concentrations (Varghese et al 2012). In addition, the backscatter optics allows for the measurement of samples at much higher concentrations than is possible using conventional DLS instruments using a 90° detection angle (Figure 5).

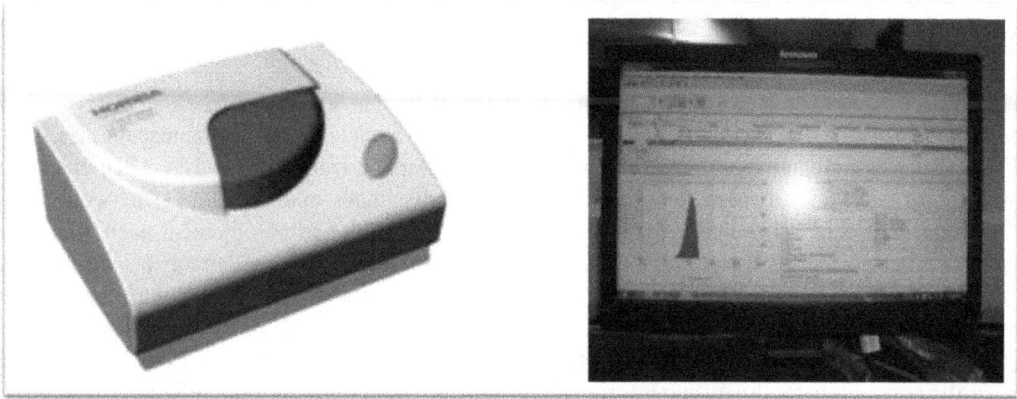

Figure 5: Schematic diagram of a 4-Circle diffractometer.

Particle size is an important parameter for the characterization of nanoparticles. The technique of dynamic light scattering is well suited to the measurement of the size of nanoparticles' dispersions. Conventionally, measurement of very small and/or poorly scattering particles or samples that are very dilute was difficult unless high powered lasers were used.

1.11 Static Light Scattering Technique (SLS)

In Static Light Scattering (SLS), time averaged scattering intensities are detected and observed at one specific scattering angle (Nelson et al. 2006). It has been determined that colloids, from which the incident laser beam is scattered, are no point scatter. Their scattering pattern depends on their shape and size. If primary light waves are scattered on several scattering centers, the resulting secondary waves differ in their path lengths. This difference of the path lengths results into a phase factor difference of the scattered light waves. Classical light scattering involves measurement of the total scattered intensity of light as a function of angle, concentration, or both. This is commonly summarized in a Zimm plot, which is described by the following equation:

$$\frac{H_c}{R_{(\theta,c)}} = \frac{1}{M_w}\left[1 + \frac{r_{g^2}K^2}{3}\right] + 2A_2C$$

where K is the magnitude of the scattering vector, c is the concentration, H is an optical constant, and $R(\theta,c)$ is the excess Raleigh ratio.

Scattered light intensities are measured at several angles for each solution concentration and the pure solvent. It is then possible to determine the molecular weight "Mw", the radius of gyration "rg", and the second virial coefficient "A2" for the species under investigation.

1.12 Electron Energy Loss Spectroscopy (EELS)

In Electron Energy Loss Spectroscopy (EELS), a material is exposed to a beam of kinetic energies. Some of the electrons will lose energy by inelastic scattering, which is primarily an interaction of the beam electron with an electron in the sample. This inelastic scattering results in both a loss of energy and a change in momentum. These interactions may be plasmon excitations or inner shell ionization, phonon excitations, and inter- and intra-band transitions. It is useful for detecting the elemental components of a material, as the energy transferred in such an interaction is related to the ionization potential of the atom, and therefore, the spectrum can be compared to that of known samples.

2. Conclusion

The impact of nanotechnology on the various fields of human life is far reaching. This review described the role of several different techniques for the characterization of nanoscale materials. Through this comprehensive summary of nanoparticles' characterization methods, we demonstrated the uses of each one of them, emphasizing on their advantages and limitations, as well as on explaining how they can be effectively combined and how they can complement each other. By presenting the role of each technique in a comparative way, our review will act as a robust guide, helping the scientific community to understand better the discussed topic. In this way, researchers will be helped for the choice of the most suitable techniques for their characterization, together with the ability to assess their use in a more precise manner. Of course, there are challenges in the scientific community for the further improvement of the accuracy and resolution of many techniques. Therefore, we finally hope that a careful reading of this review will help to identify which valuable techniques merit efforts for further technical improvements.

References

Aejaz, A., K. Azmail, S. Sanaullah and A. Mohsin. 2010. Formulation and *in vitro* evaluation of Aceclofenac solid dispersion incorporated gels. International Journal of Applied Sciences 2: 7–12.

Betancor, L. and H.R. Luckarift. 2008. Bioinspired enzyme encapsulation for biocatalysis. Trends Biotechnol. 26: 566–572.

Brian Richard Pauw. 2013. Everything SAXS: Small-angle scattering pattern collection and correction. J. Phys.: Condens. Matter. 25: 383201

Chorney, M., H. DAneuberg and G. Golomb. 2002. Lipophilic drug loaded nanospheres by nano precipitation effect of the formulation variables on size, drug recovery and release kinetics. J. Control Release 83: 389–400.

Couvreur, P., C. Dubernet and F. Puisieux. 1995. Controlled drug delivery with Nano particles: Current possibilities and future trends. Eur. J. Pharm. Biopharm. 41: 2–13.

Goldberg, M., R. Langer and X. Jia. 2007. Nanostructured materials for applications in drug delivery and tissue engineering. J. Biomater. Sci. Polym. 18: 241–68.

Gurny, R., N.A. Peppas, D.D. Harrington and G.S. Banker. 1981. Development of biodegradable and injectable lattice for control release of potent drugs. Drug Dev. Ind. Pharm. 7: 1–25.

Hu, C.H., Y.I. Li, L. Xiong, H.M. Zhnag, J. Song and M.S. Xia. 2012. Comparative effects of nano elemental selenium and sodium selenite on selenium retention in broiler chickens. Anim. Feed Sci. and Technol. 177: 204–210.

Johannes Sitterberg, Aybike Özcetin, Carsten Ehrhardt and Udo Bakowsky. 2009. Utilising atomic force microscopy for the characterisation of nanoscale drug delivery systems. European Journal of Pharmaceutics and Biopharmaceutics 6411(09): 00283–00285.

Kaparissides, C., S. Alexandridou, K. Kotti and S. Chaitidou. 2006. Recent advances in novel drug delivery systems 2026–2027.

Kargar Razi, M.R. Sarraf Maamoury and S. Banihashemi. 2011. Preparation of nano selenium particles by water solution phase method from industrial dust. Int. J. Nao. Dim. 1: 261–267.

Koosha, F., R.H. Muller, S.S. Davis and M.C. Davies. 1989. The surface chemical structure of poly (β-hydroxybutyrate) microparticles produced by solvent evaporation process. J. Control Release 9: 149–57.

Lakshmana Prabu, S., A.A. Shirwaikar, A. Shirwaikar and A. Kumar. 2009. Formulation and evaluation of sustained release microspheres of rosin containing Aceclofenac. Ars. Pharm. 50(2): 5162.

Molpeceres, J., M.R. Aberturas and M. Guzman. 2008. Biodegradable nanoparticles as a delivery system for cyclosporine: Preparation and characterization. J. Microencapsul. 29: 599–61.

Mukhopadhyay, S., N.V. Madhav Satheesh and K. Upadhyaya. 2016. Formulation and evaluation of bionanoparticulated drug delivery of Rivastigmine. World Journal of Pharmaceutical Sciences 4: 264–272.

Nelson, D.J., M. Strano and Richard Smalley. 2006. Saving the world with nanotechnology. Nat. Nanotechnol. 1(2): 96–97.

Pawar, K. and G. Kaul. 2014. Toxicity of titanium oxide nanoparticles causes functionality and DNA damage in buffalo (Bubalus bubalis) sperm *in vitro*. Toxicol. Ind. Health 30: 520–533.

Pawar, S.A., R.S. Devan, D.S. Patil, V.V. Burungale, T.S. Bhat, S.S. Mali, S.W. Shin, J.E. Ae, C.K. Hong, Y.R. Ma, J.H. Kim and P.S. Patil. 2014. Hydrothermal growth of photoelectrochemically active titanium dioxide cauliflower-like nanostructures. Electrochimica Acta 117: 470–479. https://doi.org/10.1016/j.electacta.2013.11.182.

Polakovic, M., T. Gorner, R. Gref and E. Dellacherie. 1999. Lidocaine loaded biodegradable nanospheres. II. Modelling of drug release. J. Control Release 60: 169–177.

Quintanar-Guerrero, D., E. Allemann, H. Fessi and E. Doelker. 1998. Preparation techniques and mechanism of formation of biodegradable nanoparticles from preformed polymers. Drug Dev. Ind. Pharm. 24: 1113–28.

Reverberi, A.P., N.T. Kuznetsov, V.P. Meshalkin, M. Salerno and B. Fabiano. 2016. Systematical analysis of chemical methods in metal nanoparticles synthesis. Theor. Found. Chem. Eng. 50: 59–66.

Rudakovskya, C. and H. Karnik. 2014. Synthesis and characterization of terpyridine type ligand protected gold-coated Fe3O3 nano particles. Mendeleev Communications 20(3): 150–160.

Santanu Sasidharan and R. Balakrishnaraja. 2014. Comparison studies on the synthesis of selenium nanoparticles by various micro-organisms. Int. J. Pure & Appl. Biosci. 2: 112–117.

Shrikant C. Watawe, Vishnu Mujumdar and Ramesh S. Chaughule. 2006. Nanoparticles: Synthesis characterization and applications. Edited by Chaughule, R.S. and R.V. Ramanujan, pp. 1–20.

Tan, K.S. and K.Y. Cheong. 2013. Advances of Ag, Cu, and Ag–Cu alloy nanoparticles synthesized via chemical reduction route. J. Nanopart. Res. 15: 1537.

Varghese, P.I. and T. Prdeep. 2012. Text Book of Nanoscience and Nanotechnology. Mcgraw Hill. ISBN: 9781259007323.949.

Yao, M., D.J. McClements and H. Xiao. 2015. Improving oral bioavailability of nutraceuticals by engineered nanoparticle-based delivery systems. Curr. Opin. Food Sci. 2: 14–19.

Zhang, X.F., Z.G. Liu, W. Shen and S. Gurunathan. 2016. Silver nanoparticles: Synthesis, characterization, properties, applications, and therapeutic approaches. Int. J. Mol. Sci. 17: 1534.

Zhou, B. and J.J. Zhu. 2006. A general route for the rapid synthesis of one-dimensional nanostructured single-crystal Te, Se powders. Nanotechnology 17: 1763–1769.

Zhou, H. and J. Lee. 2011. Nanoscale hydroxyapatite particles for bone tissue engineering. Acta Biomater. 7: 2769.

Part II

Biological and Biomedical Applications

8

Bioabsorbable Engineered Nanomaterials for Antimicrobial Chemotherapy

Guhapriya U.,[1] *Vidhya A.*[1,*] and *Muthu Thiruvengadam*[2]

1. Introduction

Nanotechnology has started leaving the confines of laboratories and conquering applications to change our lives. Nanotechnology is a field with its application in science and technology for manufacturing new material at the nano scale level (Albrecht et al. 2006). Nanotechnology is growing day-by-day, making an impact in all spheres of human life (Vaidyanathan et al. 2009). Nanotechnology is the hot buzzword of the new millennium. Nanotechnology research is interdisciplinary with physics, chemistry, biology, material science and medicine (Saha et al. 2010). The word "nano" is used to indicate one billionth of a metre or 10^{-9} m. The term nanotechnology was coined by Norio Taniguchi, a researcher at University of Tokyo, Japan (Taniguchi 1974). Nanoparticles possess increased structural integrity as well as unique chemical, optical, mechanical, electric and magnetic properties (Bhushan 2004). These unique properties are due to the variations in specific characteristics such as size, distribution and structure of particles. Due to these incredible properties, nanoparticles have become popular in recent years and nano products are coming to the market in large number.

The fast growing field of nanotechnology presents great potential to influence various sectors in the areas of environment, agriculture, health care and consumer goods. Nanomedicine is a new branch of nanotechnology applying the methods in medicine. It allows the use of nanotechnology and applying the techniques for the treatment of diseases. Therefore, it has built great expectations not only in the academic community but also among the investors, governments and industries. The worldwide nano product market is estimated to reach 1 trillion by the year 2015 (Roco 2005). To overcome this demand, the synthesis of nanomaterials of specific composition, shape and size is a burgeoning area of research in the field of nanotechnology. In recent years, in the field of nanotechnology, polymers are the major area of research due to their characteristics and applications (Munoz-Bonilla et al. 2014).

[1] PG and Research Department of Microbiology, DKM College for Women, Vellore, Tamil Nadu, India.
[2] Department of Applied Bioscience, Konkuk University, Seoul 05029, Korea.
* Corresponding author: vidhyasur76@gmail.com

2. Nanomaterials

Both the chemical and physical properties of nanomaterials have made it to be extensively used in various fields. The different nanomaterials including gold nanoparticles, iron oxide nanoparticles, cerium oxide nanoparticles, carbon based nanoparticles and polymeric nanoparticles have been developed. The nanomaterials exhibit various activities including medicinal and antibacterial activity. They have the ability to inhibit or lyse the pathogenic bacteria by their cellular toxicity (Mitchell et al. 2009). Recently, the polymer based nanomaterials such as chitosan, polylactic acid (PLA), polyglutamic acid (PGA), poly glycolic acid (PGLA), and polycaprolactone (PCL) are becoming popular as they are biologically safe and have good biodegradability attributes (Mitchell et al. 2007).

2.1 Properties of an Ideal Polymeric Nanomaterial for Biomedical Applications

- Easy to synthesize
- Economical
- Biodegradable
- Biocompatible
- Water soluble
- Non toxic
- Non immunogenic

2.2 Natural Polymer-Based Nanomaterials

Natural polymer is an environmentally friendly renewable resource that can be degraded into water, carbon dioxide and inorganic molecules. The commonly used natural polymers include chitosan, starch, alginate, cellulose, hyaluronic acid, chondroitin sulfate, collagen, etc. (Table 1).

2.2.1 Chitosan

Chitosan is a polysaccharide derived from chitin. Its molecular weight is between 300–1000 kDa depending on the source of chitin. The chemical structure of chitin is made up of 1-4 linked 2-acetamido-2-deoxy-β-D-glucopyranose. Chitosan is a copolymer of N-acetyl-D-glucose amine and D-glucose amine as shown in Figure 1. It has a heterogeneous chemical structure made up of both 1-4 linked 2-acetamido-2-deoxy-β-D-glucopyranose as well as 2-amino-2-deoxy-β-D-glucopyranose.

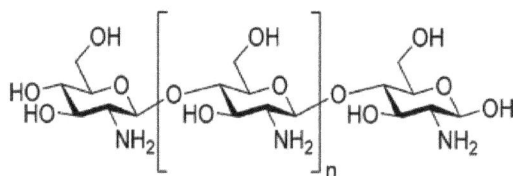

Figure 1: Chitosan.

Chitosan, a partially or fully deacetylated chitin, is a polymer with good biocompatibility and biodegradability. Chitosan has an antibacterial and antifungal property and has been investigated as a drug carrier because of its non toxic and biocompatible properties. It also has good adsorbability, permeability and moisture retention. The chemical modification through acylation, alkylation, sulfation, hydroxylation, quaterization and carboxymethylation of chitosan produces chitosan derivatives. These chitosan derivatives have significant biological properties and have great potential in the medical field (Bugnicourt 2016).

Table 1: Natural and synthetic polymeric nanomaterials.

Classification	Materials	Application Area	Advantages	Disadvantage
Natural polymeric material	Chitosan	Hydrogel, hemostasis material, drug delivery carrier, medical dressing, gene transfer (Ravi Kumar 2000)	Biocompatibility, biodegradable adsorbability, antimicrobial, innocuous, film formation (Badawy et al. 2005, Yang et al. 2005, Jayakumar Nwe et al. 2007, Kojima et al. 2004, Liu et al. 2001)	Poor spinnability, poor strength, low water-solubility (Badawy et al. 2005, Jayakumar et al. 2007)
	Starch	Hemostasis material, drug delivery carrier, tissue-engineered scaffold, bone repair material (Moran et al. 2003, Cicalleri et al. 1989)	Extensive sources, low price, non-toxic, non-antigenic and degradation products safe (Moran et al. 2003, Woggum et al. 2014)	Poor mechanical properties, poor blocking performance, resistance to water (Moran et al. 2003, Woggum et al. 2014)
	Alignate	Pharmaceutical excipient, pepcid complete, medical dressing (Parkes 1900, Augst et al. 2006, Lee et al. 2012, Rebecca 2016, Gou et al. 2014)	Biocompatibility, hypotoxicity, enhances immunity, suppresses tumor growth (Parkes 1900, Augst et al. 2006, Lee et al. 2012, Rebecca 2016, Gou et al. 2014)	Poor biodegradability and poor cell attachment (Parkes 1900, Augst et al. 2006, Lee et al. 2012, Rebecca 2016, Gou et al. 2014)
	Cellulose	Pharmaceutical adjuvant (Moran et al. 2003)	Extensive sources, economical (Moran et al. 2003)	Rare adverse reactions (Moran et al. 2003)
Biosynthesis material	Poly β-hydroxybutyrate (PHB)	Drug delivery carrier, tissue engineering material (Wang et al. 2013, Tao et al. 2016, Fidkowski et al. 2005)	Biodegradable, safe, nontoxic, good physical and chemical properties (Tao et al. 2016, Chaturvedi et al. 2015)	High crystallinity, bad thermal stability (Tao et al. 2016, Chaturvedi et al. 2015)
Chemosynthesis material	Polylactic (PLA)	Anti-adhesion materials, patch, drug-delivery carrier, bone-fixing device, suture, tissue-engineered scaffold (Zhao et al. 2016)	Biocompatibility, good mechanical properties, safe, non-toxic (Zhao et al. 2016, Kim et al. 2003, Sha et al. 2016)	Poor toughness, degradation speed slow, hydrophobicity, lack of reactive side chain groups (Zhao et al. 2016, Kim et al. 2003, Sha et al. 2016)
	Polyurethane	Excipients, medical bandage (Noreen et al. 2016, He et al. 2011, Ou et al. 2014)	Rich resource, economical, good mechanical properties (Noreen et al. 2016, He et al. 2011, Jiang et al. 2011, Shendi et al. 2017)	Degradation speed slow (Noreen et al. 2016, Jiang et al. 2011, Shendi et al. 2017)
	Poly (Lactic-glycolic acid) (PLGA)	Absorbable structure, drug delivery, tissue repair, bone screw fixation (Panyam 2003, Danhier et al. 2012, Makadia et al. 2011, Zhang et al. 2014)	Controllable biodegradability, biocompatibility (Panyam 2003, Danhier et al. 2012, Makadia et al. 2011)	Higher cost, drug-loading capacity and stability can be improved (Panyam 2003, Danhier et al. 2012, Makadia et al. 2011)

2.2.2 Starch

Starch is an abundant biopolymer widely available and stored in the form of granules in grains, beans and potato crops. It contains two types of homopolysaccharides–amylose and amylopectin. Amylose is an unbranched homopolysaccharide formed by glucose units, linked by α-(1→4) glycosidic bonds (Figure 2). Amylopectin is a branched molecule formed by thousands of glucose units, joined by α-(1→4) glycosidic bonds. Approximately, at every 25–30 glucose units, a branch point or lateral chain is present, formed by glucose units as well, joined to the main chain by a α-(1→6) glycosidic bonds.

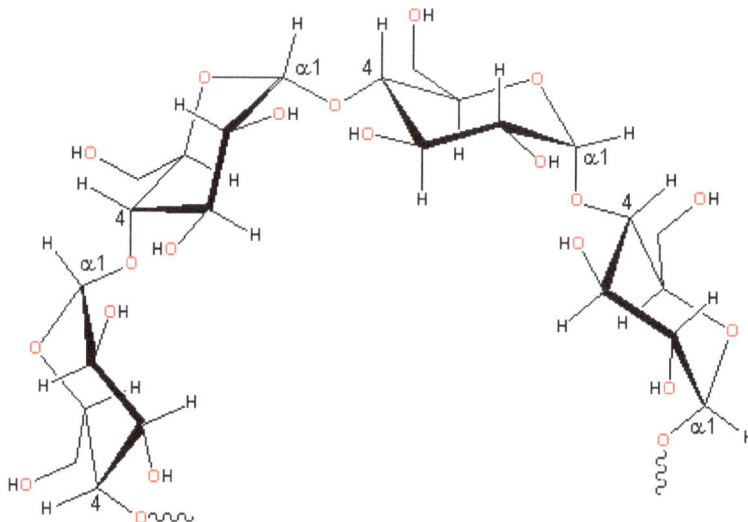

Figure 2: Starch.

By using different nanotechnological methods, these microsized starch granules can be broken down into nanosize particles. Starch has concentric semi-crystalline multistate structures that are involved in the production of starch nanocrystal by acid hydrolysis, while gelatinized starch will form nanoparticles. The methods such as high pressure miniemulsion cross linking, homogenization, emulsion, microemulsion and nanoprecipitation are known to produce starch nanoparticles (SNP) (Chin et al. 2014). These starch nanoparticles are non toxic and have biomedical application, including drug carriers, as an adsorbent in waste water treatment and as a packaging material in food and other industries. Starch nanoparticles are used to encapsulate antibiotics for antimicrobial chemotherapy as the starch in this particle can be digested easily and does not create health problems.

Due to its various applications, starch nanoparticles can be prepared by physical methods, regeneration and acid hydrolysis (Kawaljit Singh and Vikash 2017).

2.2.3 Alginate

Alginate is a natural anionic polysaccharide derived from marine brown algae cell walls. It consists of a chain of (1-4)-linked β-D-mannuronic acid and α-L-guluronic acid in different arrangements of residues (Figure 3).

It is biodegradable, non toxic, biocompatible, good moisture absorption and mucoadhesive polymer (Draget and Taylor 2011, Downs et al. 1992). The hydrophilic property of alginate nanoparticles (NPs) prolongs the antigen release and enhances the immunogenicity due to their adjuvant properties. Due to its mucoadhesive properties and enhancing permeability, they can reduce degradation in acidic environment when it is administered through nasal and oral routes. The reports have revealed the importance of NPs over microparticles (Fundueanu et al. 1998, Jani

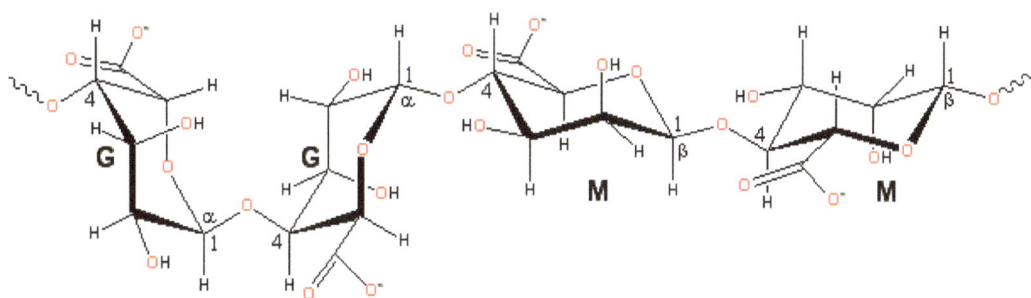

Figure 3: Alginate.

et al. 1989) due to its improved drug encapsulation, pharmacokinetics, bioavailability as well as therapeutic efficacy (Pandey and Khuller 2004). Alginate NPs can be prepared easily by inducing gelation with divalent cations such as calcium ions (Li and Xu 2008, Machado et al. 2012).

2.2.4 Cellulose

Cellulose is the most abundant, natural, renewable and sustainable biopolymer on Earth. It is available in plants, tunicates and some bacteria belonging to the genera *Acetobacter*, *Achromobacter*, *Agrobacterium*, *Rhizobium*, *Azotobacter*, *Pseudomonas*, *Escherichia*, *Sarcina* and *Alcaligenes* (Lin et al. 2013). It is found along with the lignin and hemicelluloses in the cell wall of wood. It is made up of many 1-4 linked D-glucose units in a linear fashion (Ulery et al. 2011, Kalia et al. 2014). The cellulose fibers are made up of microfibrils which in turn consists of elementary fibrils. These elementary fibrils or nanofibers are the basic structural units which measure about 2–20 nm in diameter and a few micrometers in length (Mishra et al. 2018).

2.2.5 Hyaluronic Acid

Hyaluronic acid (HA) is a basic component of the extracellular matrix of the skin, mucosal tissue, joints, eyes and many other organs and tissues. It plays a major role in tissue repair processes and in the resurfacing of the skin and in the prevention of the formation of scar. It contains repeating units of β-1,4-D-glucuronic acid and β-1,3-*N*-acetylglucosamine units (Figure 4).

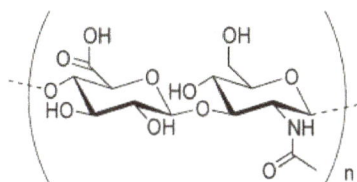

Figure 4: Hyaluronic acid.

It possesses antioxidant properties, including the ability to eliminate free radicals (Chen and Abarangelo 1999, Weindl et al. 2004, Rai et al. 2009), which are tissue damaging byproducts (Cruzat 2010). It restores tissue hydration during the inflammatory process due to its osmotic capability and helps to prevent the passage of bacteria and viruses into the pericellular area due to its viscous nature. It also stimulates the inflammatory process because it acts as a barrier to tissue degradation. It has been observed that the films of HA doped with metal nanoparticles improves skin recovery and prevents the bacterial infections of wounds.

2.2.6 Chondroitin Sulfate

Chondroitin sulfate is present in the bone, cartilage and cornea of animals. It consists of many repeated units of D-glucuronic acid and D-*N*-acetylgalactosamine units. The *N*-acetylgalactosamine is substituted with a sulfate at either its 4' or 6' position. Chondroitin sulphate has been widely used as nano sized carriers in drug/gene delivery systems (Lili Zhao et al. 2015). Chondroitin sulfate polymer coated gold nanoparticles overcome multidrug resistance in cancer cells and also suppress thrombo inflammation triggered by the chemotherapeutic drug (Deepanjali Gurav et al. 2016).

2.2.7 Collagen

Collagen is the most abundant protein in mammals and promotes cell proliferation (Lin et al. 2009, Wang et al. 2003, Nehrer et al. 1997). It influences the cell physiology and morphology, induces platelet aggregation, promotes blood clotting and promotes wound healing (Heimbach et al. 1988). In the field of nanotechnology, collagen scaffold has been widely used in biological experiments for introducing chemical and pharmaceutical substances. Metal nanoparticles, especially silver nanoparticles, have established a broad range of biomedical applications (Neto et al. 2008), due to their antibacterial ability and selective toxicity to microorganisms (Wong and Liu 2005). The silver nanoparticles stabilized with type I collagen (AgNPcols) have a very good antibacterial activity against Gram positive and Gram negative bacteria.

2.3 Biosynthesized Polymer based Nanomaterials

2.3.1 Poly β-hydroxybutyrate (PHB)

The PHB monomer consists of hydroxybutyric acid ($C_4 H_6 O_2$ units) and a methyl group as side chain (Figure 5).

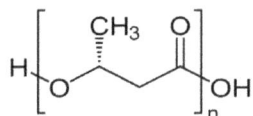

Figure 5: Poly β-hydroxybutyrate (PHB).

Biosynthesized polymers such as poly β-hydroxybutyrate (PHB) are obtained using enzymes by hydrolysis. PHB is a polymer produced by some microorganisms under unfavourable conditions (Wang et al. 2013). PHB has biodegradability, biocompatibility, optical activity and piezoelectricity properties (Fidkowski et al. 2005). These properties make it to be used as a drug delivery carrier, as a scaffold material in tissue engineering, as a bone repair material, wound support material and in surgical treatment.

2.4 Chemically Synthesized Polymer based Nanomaterials

The polymer nanomaterials include poly lactic acid, poly lacticglycolic acid, polyurethane, etc., which are produced through chemical methods.

2.4.1 Polylactic Acid (PLA)

Polylactic acid (PLA) or polylactide (Figure 6) is aliphatic semicrystalline polyester produced through the direct condensation reaction of its monomer, lactic acid, as the oligomer, and followed by a ring-opening polymerization of the cyclic lactide dimer.

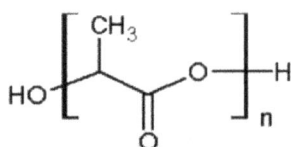

Figure 6: Polylactic acid.

Polylactic acid is biodegradable, biocompatible and non toxic in nature. This polymer has good mechanical strength and elasticity which makes it to be used in bone tissue engineering, cartilage repair and regeneration. It is also used as a carrier in the controlled release of drugs (Zhao et al. 2012), enhancing the drug concentration and minimize the side effects of drugs (Kim et al. 2003).

2.4.2 Polylactic Glycolic Acid

PLGA is a biodegradable polymer with good biocompatibility. Poly (D,L-lactic acid-co-glycolic acid) is a copolymer and a polyester macromolecule composed of 2-hydroxypropanoyl and 2-hydroxyacetyl units (Figure 7).

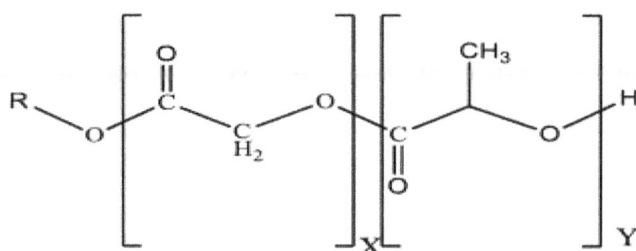

Figure 7: Polylactic glycolic acid.

PLGA nanoparticles are used as drug carriers for sustained release of drugs that increase the pharmacokinetic properties and reduce the adverse reaction (Ou et al. 2014, Panyam and Labhasetwar 2003, Danhier et al. 2012, Makadia et al. 2011).

2.4.3 Polyurethane (PU)

Polyurethane (PUR and PU) is a polymer composed of organic units joined by carbamate (urethane) links (Figure 8). Polyurethane materials are biological and tissue compatible, have high strength and are elastic in nature. Hence, they have many applications in the biomedical field, in the production of artificial organs and catheter interventions (Noreen et al. 2016). It is less toxic, non allergic and non carcinogenic but the only disadvantage is it cannot be degraded and lead to environmental pollution. Currently, the biodegradable polyurethane includes oligosaccharide derived PU, cellulose derivatives of PU and starch derivatives of PU (He et al. 2017, Shendi et al. 2011), which are superior in the biomedical field.

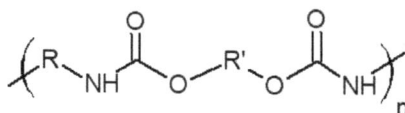

Figure 8: Polyurethane.

2.4.4 Polycaprolactone (PCL)

Polycaprolactone (PCL) (Figure 9) is a semi-crystallline, biodegrable thermoplastic polyester with a melting point of about 60°C. PCL is used as an additive for resins due to its toughness, flexibility and tear strength. Polycaprolactone nanoparticles can be prepared by a simple and reproducible solvent displacement method (Sinha et al. 2004). Polycaprolactone (PCL) film polymer with metal oxide nanoparticles exhibits food antibacterial effect. Polycaprolactone polymer-metal nanoparticles composites appear to extend the use of biocide metals. CuO nanoparticles work together with the PCL film as a wound dressing material to protect from bacterial contamination.

PCL

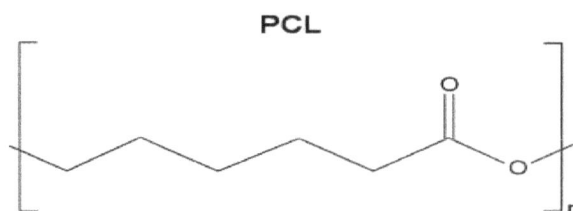

Figure 9: Polycaprolactone.

3. Antimicrobial Chemotherapy

Modern medicine has undergone revolutionary advances in the past 100 years, which can be attributed in part to the availability of effective antibiotics based on the initial discovery of penicillin by Nobel Prize laureate Dr. Alexander Fleming. Selman Waksman in 1941 first used the term antibiotics to describe the antimicrobial agents produced by certain microorganisms (Waksman and Tishlar 1942). Generally, these antimicrobial agents are required to kill the microorganisms without being toxic to the adjacent tissues.

Antibacterial agents are very important to control the pathogenic diseases. They may be either bacteriostatic or bactericidal in nature. The massive use of antibiotics over the last few decades in fields of medicine, farming and agriculture has led to the development of antibiotic resistance among the pathogens. However, the abuse of antibiotics has led to the emergence of drug resistance to most of the available antibiotics. The drug resistance is most often due to the evolutionary processes or by gene transfer by conjugation, transformation and transduction (Witte 2004). The World Health Organization has identified the antibiotic resistance as the major threat to the mankind.

Emergence of resistance to antibiotics led to an increasing range of infections, including urinary tract infection, pneumonia, and bloodstream infections caused by bacteria (Gwynn et al. 2010). The list of the most pathogenic drug-resistant bacteria includes sulfonamide-resistant, methicillin-resistant, vancomycin-resistant *S. aureus*, macrolide-resistant *Streptococcus pyogenes*, penicillin-resistant *Streptococcus pneumoniae*, vancomycin-resistant *Enterococcus*, multidrug-resistant *Mycobacterium tuberculosis*, methicillin-resistant *Neisseria gonorrhoeae*, *Enterobacter cloacae*, *Escherichia coli*, *Klebseilla pneumoniae*, *Pseudomonas aeruginosa*, *Vibrio cholera*, and beta-lactamase-expressing *Haemophilus influenzae* (Alanis 2005, Riley et al. 2012, Levy and Marshall 2004, Davies and Davies 2010). The emergence of novel infections and the growing resistance to the antibiotics available have led to the development of new antimicrobial drugs.

Currently available major groups of antibiotics generally affect the bacteria by inhibiting: cell wall synthesis, translational machinery, and DNA replication. Unfortunately, bacterial resistance may develop against each one of these modes of action. The different mechanisms of resistance developed by the bacteria include enzymes that modify or degrade the antibiotic such as β-lactamases and aminoglycosides, modification of cell components such as cell wall as seen in vancomycin resistance and ribosomes in tetracyclines resistance, and finally efflux pumps that provide multidrug resistance against numerous antibiotics (Magiorakos et al. 2012).

Nanomaterials may be advantageous as an active antibacterial group since their surface area is exceedingly large relative to their size. Nanosized particles may provide high activity although only a small dose of the particles is used. Hence, the nanomaterials could serve as an alternative to antibiotics to control bacterial infections (Magiorakos et al. 2012).

The mode of action of nanomaterials is mainly by direct contact with the bacterial cell wall, without the need to penetrate the cells; hence, most of the resistance mechanisms seen with antibiotics are irrelevant. This shows that nanomaterials would be less prone than antibiotics to promote resistant bacteria (Beyth et al. 2015). Nanomolecules can be used as an alternative to conventional antimicrobial agents and can also act as carriers for antibiotics and other drugs in human and veterinary medicine.

3.1 Polymer Nanocomposites

Polymers and nanoparticle blends, referred to as polymer nanocomposites, have found many applications in various fields (Fang et al. 2015, Giannelis 1996, Roco et al. 2000, Starr et al. 2004, Vaia and Giannelis 2001, Wypych 1999). Nanocomposite organo-inorganic materials are not simply physical blends but materials with attractive qualities of dissimilar oxide and polymer components that are intimately mixed. The scale of mixing would influence the properties of the nanocomposite materials when the component mixture reached the nanometer range. Polymer based nanosystems are categorized into polymeric nanoparticles with intrinsic antibacterial properties and polymers acting as antibiotic delivery systems (Parisi et al. 2017).

In the last decade, the biopolymer nanocomposites emerged as a modern area of research in nanotechnology. The challenges in this area are the effective separation route for extraction of nano reinforcements from renewable resources, achieving the compatibility between nano-reinforcement and the polymer matrix and finding techniques suitable for processing bionanocomposites (Uddin et al. 2012).

Bionanocomposites consist of an organic matrix in which inorganic materials such as nanoparticles, nanosheets, nanotubes, nanowires and nanoclay are dispersed (Othman 2014). The properties of these polymer matrixes have been improved in the aspects of optical, thermal, magnetic, mechanical and optoelectronic because of the synergy between the polymer matrix and inorganic materials (Tunc and Duman 2013). There is an urgent need to develop antimicrobial nanocomposites that can either control or prevent microbial colonization. This can be achieved by incorporating metal or metal oxide nanoparticles with known antibacterial activity into the polymer matrix (Venkatesan and Rajeswari 2016). Bionanomaterials as the nanomedicines provide targeted, durable and potent antimicrobial activity at lower doses (Nathan and Cunningham-Bussel 2013).

3.1.1 Antimicrobial Polymeric Nanocomposites

Polymeric nanomaterials with antimicrobial properties can be used in the situation of problems associated with antibiotic resistance. There are numerous antimicrobial polymers which are available, comprised of cationic polymeric systems with quaternary nitrogen groups, chitosan derivatives, silver and titania nanocomposites.

3.1.2 Mechanism of Antimicrobial Activity

Nano sized polymeric materials with high surface area and high reactivity have better antimicrobial properties, inhibiting the growth of bacteria and fungi. There are several mechanisms exhibited by the nanomaterials towards the microorganisms. The nanomaterials either interact with the microbial cells by interrupting transmembrane electron transfer, penetrating the cell envelope, producing secondary products or oxidizing cell components that leads to cell damage (Li et al. 2008). These nanomaterials can kill or inhibit the growth by penetrating the cell walls of bacteria, causing the condensation of DNA molecules and preventing the growth or reproduction (Feng et al. 2000). The mechanism of the antibacterial activity of nanomaterials is due to the harmful metal ions produced from the termination of metals and oxidative stress through creation of reactive oxygen species (ROS) on the nanomaterials' surface (Khezerlou et al. 2018, Zhang et al. 2008). The positively charged nanomaterials bind to the negatively charged bacteria, which enhances their bactericidal properties.

3.1.3 Cell Membrane Damage

The mechanism of action of nanoparticles on cell membranes is recognized as non-specific. The cell permeability is altered when the cell is in contact with the nanoparticles. Researches indicated the formation of a "hole" or "pore" in the living cell membranes as a possible mechanistic hypothesis (Leroueil et al. 2007). Although the meaning of the term "hole" or "pore" still requires clarification, images of cell damage have given clear evidence of this effect. In extreme cases, a literal hole in the bilayer membrane exists which leads to the complete loss of the plasma membrane (Niskanen et al. 2010).

3.1.4 Release of Toxic Ions

It has been demonstrated that the different metal ions of Cd^{2+}, Zn^{2+}, and Ag^+ can react with different groups of proteins in bacteria. The ability of Ag^+ in forming sparingly soluble salts is also considered as one of its major mechanisms for attacking bacterial cells. For instance, when the chloride ions precipitate as silver chloride in the cytoplasm of the cells, the cell respiration is inhibited. Hence, the silver nanoparticles releases silver ions and also penetrate the cells interfering in their metabolic systems. There are good evidences that Ag^+ ions can also inhibit the DNA replication. The concentrations required for bactericidal activity of silver nanoparticles are low, usually in the range of 10^{-9} mol/L. Although silver has received much attention as an antimicrobial agent, Cd^{2+} and Zn^{2+} ions can also bind to sulfur-containing proteins of the cell membrane and interfere in cell permeability (Niskanen et al. 2010).

3.1.5 Interruption of Electron Transport, Protein Oxidation and Membrane Collapse

Researches prove that the positive charge of nanoparticles is critical for antimicrobial activity since bacterial cell membrane is negatively charged. Although the mechanism is still not clear, it has been suggested that silver ions can affect membrane-bound respiratory enzymes as well as affect efflux bombs of ions that can result in cell death (Allaker 2010). In general, the chain of reaction occurs after nanoparticle contact with bacteria can start with possible oxidation of respiratory enzymes, and so help facilitate the production of Reactive Oxygen Species (ROS) and radical species that will eventually affect cell physiology and promote DNA degradation (Spacciapoli et al. 2001).

3.1.6 Generation of ROS (Reactive Oxygen Species)

Even though the oxygen is the best acceptor of electrons during respiration and it is a powerful oxidant agent that can be lethal for some bacterial. There are many theories which explain the mechanisms of action of nanoparticles involving the liberation of ROS and inhibition of the cellular adhesion. Triplet oxygen ($3O_2$) is the ground state of the oxygen molecule and can be extremely poisonous for cells, but singlet oxygen (O_2) can also be deadly for bacteria. It leads to the peroxidation of cellular constituents such as proteins and lipids. Singlet oxygen promotes spontaneous and undesirable oxidations inside the cell, during which H_2O_2 is formed by the respiratory burst which consumes O_2 with the production of free radicals. Hence, the free hydroxyl radicals lead to the oxidation of DNA, proteins and membrane lipids (Bronshteint et al. 2006). Bacteria lose the integrity of their membranes progressively, disabling them to adhere to the surfaces, to maintain an appropriate communication with other bacteria, or to express other functions with efficiency.

3.1.7 Photocatalytic Antimicrobial Properties of Nanocomposites

There is a significant interest in the development of antimicrobial biomaterials for application in the health and biomedical devices, food, and personal hygiene industries; hence, it prevents the growth and attachment of pathogenic microorganisms. Currently, titania (TiO_2) is identified as a potential nanomaterial for polymer modification with many advantages. The activity of TiO_2 is great under UV light excitation forming energy-rich electron-hole pairs. On the surface of the material, these charge carriers interact with microorganisms rendering microbicidal properties to the corresponding polymer-based nanocomposite films.

Metal doping has long been known to be one of the most effective ways to change the intrinsic band structure of TiO_2, and consequently, to improve its visible light sensitivity as well as increase its photocatalytic activity under UV irradiation. This task has been typically attempted by controlling the morphological-structural-defect characteristics of the oxide and/or by extending its absorption power into the visible region (Kubacka et al. 2009). TiO_2 is excited by near-UV irradiation and a metal such as Ag shows a very intense localized surface plasmon (LSP) absorption band in the near-UV-visible region. The LPS resonance can allow extending the absorption light into the visible region of the electromagnetic spectrum and, due to the enhancement of the electric near-field in the vicinity of the Ag, would allow boosting the excitation of electron-hole pairs. Through the plasmonic effect, there is an overall improvement of the oxide-polymer nanocomposite performance upon excitation on a region ranged from the near-UV (above ca. 280 nm) to the visible light (below ca. 500–525 nm). This would yield an improved performance with respect to TiO_2-alone nanocomposites, and have the potential of working under sunlight.

3.1.8 Biomedical Applications of Antimicrobial Nanocomposites

Nanosized polymers have been found to be a better antimicrobial agent than the macromolecular polymers (Gonzalez et al. 2015). It has been proved that the nanomaterials are very good absorbents, antibacterial agent and catalysts due to their large surface area. Bioabsorbable nanomaterial ranges from 1 to 1000 nm in particle size (Gref et al. 1994). Due to their particle size, they have been investigated in drug delivery and long circulation in blood (Allemenn 1993, Yih and Al-Fandi 2006).

Polymer based nanomaterials are a promising tool for the controlled release and delivery of antibiotics (Kalhapure et al. 2015a). These nanomaterials are taken up by the microorganisms which subsequently release the drug. Polylactic acid and its co-polymer polylactic co-glycolic acid can be used to encapsulate drugs by microencapsulation for the enhanced drug delivery. Other natural polymers such as hyaluronic acid, chitosan and dextran also have applications in the medical field as a drug carrier.

3.1.9 Antimicrobial Application

Chitosan was discovered by Rouget in 1859 and its antibacterial activity was first proposed by Allan and Hadwinger in 1979. The antimicrobial activity can be explained by two different mechanisms. In the first mechanism, chitosan binds with the DNA and inhibits the microbial RNA synthesis (Chung and Chen 2008). The antimicrobial activity depends on its molecular weight, concentration, hydrophilic/hydrophobic character, chelating capacity, pH, temperature and the type of microbes involved (Rabea et al. 2003).

It has been proved that chitosan possesses inherent antimicrobial effect against bacteria, fungi and yeast, due to the positive charge of its amino groups, which binds electrostatically to the surface of the microbial cell membrane and inactivates the microbial enzymes (Kurniasih et al. 2018). Chitosan exhibits a broad spectrum of activity against both the Gram-positive and Gram-negative bacteria, with lower toxicity in mammalian cells. Hence, it is extensively used in the biomedical field for antibacterial and wound healing applications (Hirano and Nagao 1989, Kendra and Hadwiser 1984, Uchida et al. 1989, Ueno et al. 1997). Chitosan has been used in different forms including hydrogels, membranes, nanofibers, beads, micro/nanoparticles, scaffolds, and sponges (Franklin and Snow 1981, Takemono et al. 1989, Nagahama et al. 2008, Jayakumar et al. 2006, 2008, 2010, Prabaharan and Mano 2005, Anitha et al. 2009, Madhumathi et al. 2010, Muramatsu et al. 2003).

A variety of metal and metal oxide nanoparticles such as silver, copper, zinc oxide, etc., can be encapsulated into chitosan. Chitosan nanoparticles have the ability to control the release of bioactive agents and avoid the use of any hazardous organic solvents while fabricating nanoparticles. The combination of chitosan with nanoparticles is an efficient approach to produce improved antibacterial materials with enhanced properties. It has been found that Chitosan-Ag-ZnO composite showed enhanced antimicrobial activity (Li et al. 2010).

The polycationic character of the chitosan confers the antimicrobial properties, which helps to interact with the negatively charged microbial cell walls and cytoplasmic membranes. This results in the decreased osmotic stability, membrane disruption and eventual leakage or intracellular elements. Chitosan may also bind to the microbial DNA and inhibit the mRNA and protein synthesis (Qi et al. 2005). Although gaining a polycationic charge in a weakly acidic environment, it is weakly soluble in physiological solvents that limit its clinical use (Chavez de Paz et al. 2011). Chitosan-alginate NPs could be prepared and delivered in physiological parameters, and show antimicrobial activity against cutaneous bacterium for acne therapy. Alginate form complexes with the polycationic material, which stabilizes and reduces the porosity of the gel. Chitosan has received considerable attention among the polycations; because of its cationic characterisitics, it can interact with an anionic polymer that has carboxylic groups such as alginate by ionic binding (Coppi et al. 2001).

3.1.10 Wound Healing Application

Nanotherapies represent an emerging field with a new class of treatment with improved standard and prognosis for wound healing. The different nanoscale strategies, the carrier, drug related and scaffold can target the different phases of wound repair. The current researches explain the mechanism of wound healing at the molecular and cellular level which helps to develop new therapeutic strategies. This also helps to overcome the barriers of the currently used other therapeutic methods for wounds, to reach the specific site and by exerting the therapeutic action of nanomaterial at the site of wound. In order to achieve the targeted site reaction, the nanoparticles are incorporated with the ligands like biomaterials (Wang and Uludag et al. 2008). This also helps in tissue regeneration that could supplement the wound repair process.

The features which support the application of bionanocomposites in wound healing are their low dimension (1–100 nm) and the variety of shapes. The advantage of these nanomaterials for the tissue repair is their suitability to make them into thin films and their light weight, and the mechanical strength prevents the compression of wounded tissue *in situ*. These properties provide an extensive advantage in the rapid healing process, particularly when a large surface area of the skin is to be repaired, for example, in burn cases (Rohiwal et al. 2015).

Chitosan nanomaterial has the wound healing property due to its inherent antimicrobial properties. Hence, this polymeric bioabsorable nanomaterial is used extensively to treat wound infections. It also has the ability to deliver extrinsic antimicrobial agents that have been blended with it (Dai et al. 2011, Jayakumar et al. 2011). TiO_2 nanoparticles have antimicrobial effect against both Gram positive and Gram negative bacteria; it can be blended with chitosan to prepare antimicrobial wound dressings (Archana et al. 2013).

Cellulose is considered to be the most abundant available natural polymer, with a history of use in fiber fabrication (Frey 2008). In recent days, electrospun nanofibers made from cellulose and its derivatives, such as cellulose acetate and cellulose hydroxyl propyl, are finding wide use in antimicrobial applications. Nanocellulose does not possess any antimicrobial property of its own. Therefore, it needs to be functionalized with antimicrobial agents. To achieve this, sorbic acid, benzalkonium chloride, copper nanoparticles and silver chloride nanoparticles (Son et al. 2006) have been successfully incorporated into cellulose nanofibers and by chemically grafting functional groups onto the surface of cellulose nanofiber network to obtain antimicrobial cellulose membranes. Quaternary ammonium groups could be immobilized onto cellulose nanofibers through reversible addition-fragmentation chain transfer (RAFT) (Roy et al. 2008). Cellulose nanofibers functionalized with amino groups and aminosilane groups (Saini et al. 2016) have been successfully prepared, which effectively inhibited the growth of *Staphylococcus aureus* and *Escherichia coli*.

The wound dressing made up of nanocellulose impregnated with silver nanoparticles possesses potent antimicrobial activity when demonstrated against *Staphylococcus aureus* and *Escherichia coli* (Maneerang et al. 2008).

The preparation of PCL-CuONPs was simple, fast, and low cost for practical application as wound dressings. Copper oxide nanoparticles (CuONPs) have been widely used as they possess very well-known antibacterial properties (Chaudhury et al. 2018). CuONPs are effective against a wide range of bacterial pathogens involved in nosocomial infections. They are also a powerful biocide toward fungi and they have antiviral capabilities (Ren et al. 2009). Their antimicrobial activity depends on the ability to release metallic ions and the extent of their surface area (Mohanraj et al. 2017). Cu^{2+} ions released from CuONPs are negatively charged that damage the proteins and lipids of the cell membrane. Additionally, Cu^{2+} ions possess high redox properties, which produce reactive oxygen species, producing alterations in subcellular components, including damage to DNA double-helix molecule (Taran et al. 2017).

3.1.11 Nanosystems as Antibacterial Drug Delivery Agents

Nano sized particles have better drug loading efficiency of both hydrophilic and lipophilic antibiotics with enhanced antibacterial effect (Lambardo et al. 2019). The antibiotics loaded nanomedicines are internalized by passing the reticuloendothelial system (Abed and Coaverear 2014). The nanosystems interact with proteins, and with the various components of tissues, thus affecting biodistribution. The macrophages with anionic nature attract the positively charged nanosystems (Dacoba et al. 2017). The hydrophobicity of nanosystems plays a major role in targeted drug delivery related to the interactions with the phospholipid layer of the bacterial membrane (Patra et al. 2018), whereas the hydrophilic nanosystems interact less with opsonins, and thus have longer circulation time in the blood.

The various mechanisms exhibited by the nanosystems include the ability to optimize the physiochemical characteristics of the entrapped antibacterial drugs, enhance the ability to accumulate near the cytoplasm, their electrostatic interaction with the bacterial membrane, producing ROS, high oxidizing power and prevention of drug degradation (Vallet- Regi et al. 2019). It also restrains the drug resistance by overcoming the resistance strategies that involve drug degradation by β-lacatamse, efflux pumps or thickening of bacterial cell walls (Baptisa et al. 2018).

Other natural polymers include dextran sulfate and chondroitin sulfate; nanoparticles of these polymers were formulated with high encapsulation efficiency of around 65%. The intracellular uptake of the antibiotic by suing antibiotic loaded dextran sulfate nanoparticles was 4 fold higher than that of antibiotic loaded chondroitin sulfate nanoparticles (Kiruthika et al. 2015).

Among the synthetic polymers, PLGA and PCL are commonly used for the antibacterial drug delivery. Although the proper mechanism was not clear, the PLGA-loaded nanopaticles could carry out either fusion or adsorption (Lotfipur et al. 2016). The antibiotic clarithromycin was successfully encapsulated within PLGA particles, which showed enhanced antibacterial efficacy against *Helicobacter pylori* than with the free antibiotic. A significant increase in the uptake of anti-tubercular rifampicin into macrophages was observed when it was encapsulated with PCL nanoparticles compared to free antibiotic, thus improving its efficiency towards the *Mycobacterium tuberculosis* (Trousil et al. 2017).

3.1.12 Nanomedicines for Biofilm Forming Bacteria

To understand the application of nanomaterials as antibacterial delivery agents, it is necessary to understand the mechanism by which bacteria form colonies that escape antibiotic chemotherapies. There exist two different forms of bacterial growth—the planktonic growth or a free swimming unicellular phase and the biofilm growth phase or a multicellular sessile state (Berlanga et al. 2017).

Biofilm is an evolved system where the bacteria live and form permanent colonies. The bacterial growth is dense, and the group of bacteria attach to each other and to the surface and form an external matrix composed mainly of exo polysaccharide, amino acids and extracellular DNA (Rizzato et al. 2019, Majamdar and Pal 2017). The biofilm bacteria are 1000 times more resistant to the conventional antibiotics and are associated with diseases of lung, colon, urethra, eye and ear, and also to infective endocarditis, gum related infections and wound related infections (Valappil 2018).

Microorganisms may adhere and proliferate on any surface in a moist environment when they come in contact with them. These populations rapidly increase on the surface and eventually build up a biofilm which consists of a complex polysaccharide matrix with many other components. This microbial cell in the biofilm is much less susceptible to most of the available antibiotics and biocides. The main approach to prevent the growth of microorganisms in the biofilm and to inhibit the spread of microbial infections is the use of creating antimicrobial surfaces (Shady et al. 2013). Such surfaces repel the microorganisms, prevent the attachment to the surface or kill the microbes. Surgical sutures and implants can be exposed to microorganisms in the environment leading to contamination, formation of biofilm and infections (Tummalapalli et al. 2016). These postoperative infections lead to a great part of morbidity and mortality. The antimicrobial nanomaterials are an alternative since they can be used in wound sutures, artificial tendons, bone cements or medical packaging.

Nanomedicine plays an important role in improving the effectiveness of the currently available therapeutics, by enhancing the stability of antibiotics, its biofilm internalization, prolonging antibiotic release, and offering targeted drug delivery at the site of infection with less side effects (Patra et al. 2018).

References

Alanis, A.J. 2005. Resistance to antibiotics: Are we in the post-antibiotic era? Arch. Med. Res. 36(6): 697–705, 1284.

Albrecht, M.A., C.W. Evan and C.R. Raston. 2006. Green chemistry and health implications of nanoparticles. Green Chem. 8: 417–432.

Allaker, R.P. 2010. The use of nanoparticles to control oral biofilm formation. J. Dent. Res. 89(11): 1175–1185.

Allan, C.R. and L.A. Hadwiger. 1979. The fungicidal effect of chitosan on fungi of varying cell wall composition. Exp. Mycol. 3: 285–287.

Allemann, E., R. Gurny and E. Doelker. 1993. Drug-loaded nanoparticles: Preparation methods and drug targeting issues. European Journal of pharmaceutical and Biopharmaceuticals 39: 173–191.

Alves, D. and M. Olivia Pereira. 2014. Mini-review: Antimicrobial peptides and enzymes as promising candidates to functionalize biomaterial surfaces. Biofouling 30: 483–499.

An, J., X. Zhang, Q. Guo, Y. Zhao, Z. Wu and C. Li. 2015. Glycopolymer modified magnetic mesoporous silica nanoparticles for MR imaging and targeted drug delivery. Colloids and Surfaces A. Physicochemical Eng. Aspects 482: 98–108.

Anitha, A., V.V. Divya Rani, R. Krishna, V. Sreeja, N. Selvamurugan and S.V. Nair. 2009. Synthesis, characterization, cytotoxicity and antibacterial studies of chitosan, O-carboxymethyl, N, O-carboxymethyl chitosan nanoparticles. Carbohydr. Polym. 78: 672–677.

Anitha, A., S. Sowmya, P.T.S. Kumar, S. Deepthi, K.P. Chennazhi, M. Ehrlich, M. Tsurkan and R. Jayakumar. 2014. Chitin and Chitosan in selected biomedical applications. Prog. Polym. Sci. 39: 1644–1667.

Anwunobi, A.P. and M.O. Emeje. 2011. Recent application of natural polymers in nanodrug delivery. J. Nanomedic. Nanotechnol. S4: 002.

Archana, D., J. Dutta and P.K. Dutta. 2013. Evaluation of chitosan nano dressing for wound healing characterization, *in vitro* and *in vivo* studies. Int. J. Biol. Macromol. 57: 193–203.

Augst, A., H. Kong and D.J. Mooney. 2006. Alginate hydro gels as biomaterials. Macromol. Biosci. 6: 623–633.

Badawy, M.E.I., E.I. Rabea, T. Rogge, C.V. Stevens, G. Smagghe, W. Steurbaut and M. Hofte. 2005. Synthesis and fungicidal activity of new N,O-acyl Chitosan derivatives. Biomacromolecules 5: 589–595.

Ball, P. 1999. Engineering shark skin and other solutions. Nature 400: 507–509.

Baptista, P.V., M.P. McCusker, A. Carvalho, D.A. Ferreira, N.M. Mohan, M. Martins and Alexandra R. Fernandes. 2018. Nano-strategies to fight multidrug resistant bacteria—A Battle of the Titans. Front. Microbiol. 9: 1441.

Bell, B.G., F. Schellevis, E. Stobberingh, H. Goossens and M. Pringle. 2014. A systematic review and meta-analysis of the effects of antibiotics consumption on antibiotic resistance. BMC Infect. Dis. 14: 13–19.

Berlanga, M. and R. Guerrero. 2017. Living together in biofilms: The microbial cell factory an its biotechnological implications. Microb. Cell Factories 15: 165.

Beyth, N., Y.H. Haddad, A. Domb, W. Khan and R. Hazan. 2015. Alternative antimicrobial approach. Nano Antimicrobial Materials 246012.

Bhushan, B. 2004. In Springer Handbook of Nanotechnology, Spinger-Verlag Berlin, Heidelberg.

Bronshteint, I., S. Aulova, A. Juzeniene, V. Iani and L.W. Ma. 2006. *In vitro* and *in vivo* photosensitization by protoporphyrins possessing different lipophilicities and vertical localization in the membrane. Photochem. Photobiol. 82(5): 1319–1325.

Bugnicourt, L. and C. Ladaviere. 2016. Interests of chitosan nanoparticles ionically cross-linked with tripolyphosphate for biomedical applications. Prog. Polym. Sci. 60: 1–17.

Carlmark, A., E. Larsson and E. Malmstrom. 2012. Grafting of cellulose by ring-opening polymerization—A review. Eur. Polym. J. 48: 1646–1659.

Chaturvedi, K., K. Ganguly, A.R. Kulkarni, W.E. Rudzinski, L. Krauss and M.N. Nadagouda. 2015. Oral insulin delivery using deoxycholic acid conjugated PEGylated polyhydroxybutyrate co-polymeric nanoparticles. Nanomedicine 10: 1569–1583.

Chavez de Paz, L.E., A. Resin, K.A. Howard, D.S. Sutherland and P.L. Weise. 2011. Antimicrobial effect of Chitosan nanoparticles on *Streptococcus mutans* biofilms. Appl. Environ. Microbiol. 77: 3892–5.

Chen, D. and S. Liu. 2015. Nanofibers used fir drug delivery of analgesics. Nanomedicine 10: 1785–1800.

Chen, W.Y.J. and G. Abatangelo. 1999. Functions of hyaluronan in wound repair. Wound repair and Regeneration 2: 79–89.

Chin, S.F., S.N.A. Mohd Yazid and S.C. Pang. 2014. Preparation and characterization of starch nanoparticles for controlled release of curcumin. Int. J. Polym. Sci. 1–8.

Chung, Y.C. and C.Y. Chen. 2008. Antibacterial characterisitics and activity of acid-soluble chitosan. Bioresour. Technol. 99: 2806–2814.

Civalleri, D., G. Scopinaro, N. Balletto, F. Claudiani, F. Decian, G. Camerini, M. Depaoli and U. Bonalumi. 1989. Changes in vascularity of liver tumours after hepatic arterial embolization with degradable starch microspheres. Br. J. Surg. 76: 699–703.

Clermont, O., D. Gordon and E. Denamur. 2015. A guide to the various phylogenetic classification schemes for *Escherichia coli* and the correspondence among schemes. Microbiology 161: 980–988.

Coppi, G., V. Iannuccelli, E. Leo, M.T. Bernabei and R. Cameroni. 2001. Chitosan-alginate micropaticles as a protein carrier. Drug Dev. Ind. Pharm. 27: 393–400.

Crandall, B.C. (ed.). 1996. Nanotechnology. MIT Press, Cambridge.

Dacoba, T.G., A. Olivera, D. Torres, J. Crecente-Campo and M.J. Alonso. 2017. Modulating the immune system through nanotechnology. In Seminars in Immunology; Elsevier: Amsterdam, the Netherlands.

Dai, T., M. Tanaka, M. Huang and Y.Y. Hamblin. 2011. Chitosan preparations for wounds and burns: Antimicrobial and wound-healing effects. Expert Rev. Anti-Infect. Ther. 9: 857–879.

Danhier, F., E. Ansorena, J.M. Silva, R. Coco, A.L. Breton and V. Preat. 2012. PLGA- based nanoparticles: An review of biomedical applications. J. Control Release 161: 505–522.

Daoud, W.A. and W.S. Tung. 2008. Self- cleaning fibers via nanotechnology—A virtual reality. In Proceedings of the 8th IEEE Conference on Nanotechnology, IEEE-NANO, Aelington, TX, USA, 18–21 August, 1–2.

Dastjerdi, R. 2010. A review on the application of inorganic nano-structured materials in the modification of textiles: Focus on anti-microbial properties. Colloids Surf. Biointerfaces 79: 5–18.

Dickson, M.N., E.I. Liang, L.A. Rodriguez, N. Vollereaux and A.F. Yee. 2015. Nanopatterned polymer surfaces with bactericidal properties. Biointerphases 10.

Downs, E.C., N.E. Robertson, T.L. Riss and M.L. Plunket. 1992. Calcium alginate beads as a slow-release system for delivering angiogenic molecules *in vivo* and *in vitro*. J. Cell Physiol. 152: 422–9.

Draget, K.I. and C. Tylor. 2011. Chemical, physical and biological properties of alginates and their biomedical implications. Food Hydrocoll. 25: 251–6.

Du, J. and Y.-L. Hsieh. 2007. PEGylation of Chitosan for improved solubility and fiber formation via electrospinning. Cellulose 14: 543–552.

Elsabee, M.Z., H.F. Naguib and R.E. Morsi. 2012. Chitosan bassed nanofibers, review. Mater. Sci. Eng. C 32: 1711–1726.

Fang, Y., G. Sun, T. Wang, Q. Congo and L. Ren. 2007. Hydrophobicity mechanism of non-smooth pattern on surface of butterfly wing. Chin. Sci. Bull. 52: 711–716.

Feng, QL., J. Wu, G.Q. Chen, F.Z. Cui, T.N. Kim and J.O. Kim. 2000. A mechanistic study of the antibacterial effect of silver ions on *Escherichia coli* and *Staphylococcus aureus*. J. Biomed. Mater. Res. 52(4): 662–668.

Fernandes, S.C.M., P. Sadocco, A. Alonso-Varona, T. Palomares, A. Eceiza, A.J.D. Silvestre, Inaki Mondragon and Carmen S.R. Freire. 2013. Bioinspired antimicrobial and biocompatible bacterial cellulose membranes obtained by surface functionalization with amino alkyl groups. ACS Appl. Mater. Interfaces 5: 3290–3297.

Fidkowski, C., M.R. Kaazempurmofrad, J.T. Borenstain, J.P. Vacanti, R. Langer and Y. Wang. 2005. Endothelialized microvasculature based on a biodegradable elastomer. Tissue Eng. 11: 302–309.

Franklin, T.J. and G.A. Snow. 1981. Biochemistry of Antimicrobial Action. Chapman and Hall, London, 175.

Frey, M.W. 2008. Electrospinning cellulose derivatives. Polym. Rev. 11: 1785–1800.

Fundueanu, G., C. Nastruzzi, A. Carpov, J. Desbrieres and M. Rinaudo. 1999. Physico-chemical characterization of Ca-alginate micro particles produced with different methods. Biomaterials 20: 1427–35.

Fundueanu, G., E. Esposito, D. Mihai, A. Carpov, J. Desbrieres, M. Rinaudo and C. Nastruzzi. 1998. Preparation and characterization of Ca-alginate microspheres by a new emulsification method. International Journal of Pharmaceutics 170: 11–2.

Geng, X., O.-H. Kwon and J. Jang. 2005. Electrospinning of Chitosan dissolved in concentrated acetic acid solution. Biomaterials 26: 5427–5432.

Giannelis, E.P. 1996. Polymer layered silicate nanocomposites. Adv. Mater. 8: 29.

Gonzalez, M.I., S. Perni and P. Prokopovich. 2015. Antimicrobial Polymer based on Nanostructures A New Generation of Materials with Medical Applications, CRC Press, Taylor and Francis Group.

Gonzales, P.R., M.W. Pesesky, R. Bouley, A. Ballard, B.A. Biddy, M.A. Suckow, W.R. Walter, V.A. Schroeder, C.A. Burnham, S. Mobashery, M. Chang and G. Dantas. 2015. Synergistic, collaterally sensitive-lactam combinations suppress resistance in MRSA. Nat. Chem. Biol. 11: 855–864.

Gou, S., P. Zhu, C.H. Dong, J. Liu, D. Guo and A.F. Amp. 2014. Selective oxidized modification of alginate fiber. Dyeing Finish 8: 482–490.

Gref, R., Y. Minamitake, M.T. Peracchia, V. Trubetskoy, V. Torchilin and R. Langer. 1994. Biodegradable long-circulating polymeric nanospheres. Science 263: 1600–1603.

Gullin Wang and Hasan uludag. 2008. Recent developments in nanoparticles-based drug delivery and targeting systems with emphasis on protein-based nanoparticles. Drug Delivery 5: 499–515.

Guo, Z., F. Zhou, J. Hao and W. Liu. 2005. Stable biomimetic super-hydrophobic engineering materials. J. Am. Chem. Soc. 127: 15670–15671.

Gurav, D., Oommen P. Varghese, Osama A. Hamad, B. Nilsson, J. Hilbom and Oommen P. Oommen. 2016. Chondroitin sulfate coated gold nanoparticles: A new strategy to resolve multidrug resistance and thromboinflammation. Chemical Communications 5: 1–4.

Gwynn, M.N., A. Portnoy, S.F. Rittenhouse and D.J. Payne. 2010. Challenges of antibacterial discovery revisited. Ann. N.Y. Acad. Sci. 1213: 519.

Hasan, J., H.K. Webb, V.K. Truong, S. Pogodin, V.A. Baulin, G.S. Watson, J.A. Watson, R.J. Crawford and E.P. Ivanova. 2013. Selective bactericidal activity of nanopatternes super hydrophobic cicada Psaltoda claripennis wing surfaces. Appl. Microbiol. Biotechnol. 97: 9257–9262.

He, C., M. Wang, X. Cai, X. Huang, L. Li, H. Zhu, J. Shen and J. Yuan. 2011. Chemically induced graft copolymerization of 2-hydroxyethyl methacrylate onto polyurethane surface for improving blood compatibility. Appl. Surf. Sci. 258: 755–760.

Heimbach, D., A. Luterman, J. Burke, A. Cram, D. Herndon, J. Hunt, M. Jordan, W. McManus, L. Solem and G. Warden. 1988. Artificial dermis for major burns. Ann. Surg. 208: 313–320.

Heunis, T., O. Bshena, B. Klumperman and L. Dicks. 2011. Release of bacteriocins from nanofibers prepared with combinations of poly (D,L-lactide) (PDLLA) and poly (ethylene oxide) (PEO). Int. J. Mol. Sci. 12: 2158–2173.

Heunis, T.D.J., C. Smith and L.M.T. Dicks. 2013. Evaluation of a nisin-eluting nanofiber scaffold to treat Staphylococcus aureus-induced skin infections in mice. Antimicrob. Agents Chemother. 57: 3928–3935.

Hirano, S. and N. Nagao. 1989. Effects of chitosan, pectic acid, lysozyme, and chitinase on the growth of several phytopathogens. Agri. Biol. Chem. 53: 3065–3066.

Ignatova, M., K. Starbova, N. Markova, N. Manolova and I. Rashkov. 2006. Electrospun nanofiber mats with antibacterial properties from quaternished Chitosan and poly (vinyl alcohol). Carbohydr. Res. 341: 2098–2107.

Imark, S., M. Kurtulus, A. Hasanoglu Hesenov and O. Eebatur. 2013. Gasification efficiencies of cellulose, hemicelluloses and lignin fractions of biomass in aqueous media by using Pt on activated carbon catalyst. Biomass Energy 49: 102–108.

Ivanova, E.P., J. Hasan, H.K. Webb, V.K. Truong, G.S. Watson and J.A. Watson. 2012. Natural bactericidal surfaces: Mechanical rupture of Pseudomonas aeruginosa cells by cicada wings. Small 8: 2489–2494.

Ivanovo, E.P., J. Hasan, H.K. Webb, G. Gervinskas, S. Juodkazis, V.K. Truong, A.H. Wu, R.N. Lamb, V.A. Baulin, G.S. Waston, J.A. Watson, D.E. Mainwaring and R.J. Crawford. 2013a. Bactericidal activity of black silicon. Nat. Commun. 4.

Ilvanova, E.P., S.H. Nguyen, H.K. Webb, J. Hasan, V.K. Truong, R.N. Lamb, X. Duan, M.J. Tobin, P.J. Mahon and R.J. Crawford. 2013b. Molecular organization of the nanoscale surface structures of the dragonfly *Hemianaxpappuensis* wing epicuticle. PLoS ONE 8: e67893.

Jani, P., G.W. Halbert, J. Langridge and A.T. Florence. 1989. The uptake and translocation of latex nanospheres and microspheres after oral administration to rats. J. Pharm. Pharmacol. 41: 809–12.

Jayakumar, R., N. Nwe, S. Tokura and H. Tamura. 2007. Sulfated chitin and Chitosan as novel biomaterials. Int. J. Biol. Macromol. 40: 175–181.

Jayakumar, R., N. Nwe, H. Nagahama and H. Tamura. 2008. Synthesis, characterization and biospecific degradation behavior of sulfated chitin. Macromol. Symp. 264: 163–167.

Jayakumar, R., M. Prabaharan, S.V. Nair and H. Tamura. 2010. Novel chitin and Chitosan nanofibers in biomedical applications. Biotechnol. Adv. 28: 142–150.

Jayakumar, R., M. Prabaharan, M. Sudhessh Kumar, P.T. Nair and S.V. Tamura. 2011. Biomaterials based on chitin and chitosan in wound dressing applications. Biotechnol. Adv. 29: 322–337.

Jayakumar, R., R.L. Reis and J.F. Mano. 2006. Phosphorous containing chitosan beads for controlled oral drug delivery. J. Bioact. Compat. Polym. 21: 327–340.

Jeong, E.H., J. Yang and J.H. Youk. 2007. Preparation of polyurethane cationomer nanofiber mats for use in antimicrobial nanofilter applications. Mater. Lett. 61: 3991–3994.

Jiang, X., K. Wang, M. Ding, J. Li, H. Tan, Z. Wang and Q. Fu. 2011. Quantitative grafting of peptide onto the nontoxic biodegradable waterborne polyurethanes to fabricate peptide modified scaffold for soft tissue engineering. J. Mater. Sci. Mater. Med. 22: 819–827.

Jin, L., W. Guo, P. Xue, H. Gao, M. Zhao, C. Zheng, Y. Zhang and D. Han. 2015. Quantitative assay for the colonization ability of heterogenous bacteria on controlled nanopillar structures. Nanotechnology 26.

Kalhapure, R.S., N. Suleman, C. Mocktar, N. Seedat and T. Govender. 2015. Nanoengineered drug delivery systems for enhancing antibiotic therapy. J. Pharm. Sci. 104: 872–905.

Kalhapure, R.S., S.J. Sonawane, D.R. Sikwal, M. Jadhav, S. Rambharose and C. Mocktar. 2015. Solid lipid nanoparticles of clotrimazole silver complex: An efficient nano antibacterial against Staphylococcus aureus and MRSA. Colloids Surf. B Biointerfaces 136: 651–658.

Kalia, S., S. Boufi, A. Celli and S. Kango. 2014. Nanofibrillated cellulose: Surface modification and potential applications. Colloid Polym. Sci. 292: 5–31.

Kampalanonwat, P., P. Supaphol and G.E. Morlock. 2013. Electrospun nanofiber layers with incorporated photoluminescence indicator for chromatography and detection of ultraviolet-active compounds. J. Chromatogr. A 1299: 110–117.

Kawaljit Singh, S. and N. Vikash. 2017. Starch nanoparticles: Their preparation and nanoparticles. Plant Biotechnology: Recent Advancements and Developments, 213–232.

Kendra, D.F. and L.A. Hadwiser. 1984. Characterization of the smallest chitosan oligomer that is maximally antifungal to *Fusarium solani* and *Elicitis pisatin* formation in *Pisum sativum*. Exp. Mycol. 8: 276–281.

Kim, K., M. Yu, X. Zong, J.B. Chiu, D. Fang, Y. Seo, B.S. Hsiao, B. Chu and M. Hadjiargyrou. 2003. Control of degradation rate and hydrophilicity in electrospun non-woven poly (D,L-lactide) nanofiber scaffolds for biomedical applications. Biomaterials 24: 4977–4985.

Kiruthika, V., S. Maya, M.K. Suresh, V.A. Kumar, R. Jayakumar and R. Biswas. 2015. Comparative efficacy of chloramphenicol loaded chondroitin sulfate and dextran sulfate nanoparticles to treat intracellular Salmonella infections. Colloids Surf. B Biointerfaces 127: 33–40.

Koch, K., B. Bhushan, Y.C. Jung and W. Barthlott. 2009. Fabrication of artificial Lotus leaves and significance of hierarchical structure for superhydrophobicity and low adhesion. Soft Matter 5: 1386–1393.

Kojima, K., Y. Okamoto, K. Kojima, K. Miyatake, H. Fujise, Y. Shigemasa and S. Minami. 2004. Effects of chitin and Chitosan on collagen synthesis in wound healing. J. Vet. Med. Sci. 66: 1595–1598.

Kong, H., J. Song and J. Jang. 2010. Photo catalytic antibacterial capabilities of TiO_2-biocidal polymer nanocomposites synthesized by a surface-initiated photopolymerization. Environ. Sci. 44: 5672–5676.

Kubacka, A., M. Ferrer, M.L. Cerrada, C. Serrano, M. Sanchez-Chaves, M. Fernandez-Garcia, M.D. Santos, R. Bargiela and D. Roja. 2009. Boosting TiO_2-anatase antimicrobial activity: Polymer-oxide thin films. Appl. Catal. B 89: 441–447.

Lee, K.Y. and D.J. Mooney. 2012. Alginate: Properties and biochemical applications. Prog. Polym. Sci. 37: 106–126.

Leroueil, P.R., S. Hong, A. Mecke, J.R. Baker, B.G. Orr and M.M. Banasazak. 2007. Nanoparticles interaction with biological membranes: Does nanotechnology present a Janus face? Acc. Chem. Res. 40(5): 335–342.

Levy, S.B. and B. Marshall. 2004. Antibacterial resistance worldwide causes, challenges and responses. Nat. Med. 10: 6S122–S129.

Li, M. and Z. Xu. 2008. Quercetin in a lotus leaves extract may be responsible for antibacterial activity. Arch. Pharm. Res. 31: 640–644.

Li, Z., R. Yang, M. Yu, F. Bai, C. Li and Z.L. Wang. 2010. Cellular level biocompatibility and biosafety of ZnO nanowires. The Journals of Physical Chemistry 112(51): 20114–20117.

Li, Y., G. Liu, X. Wang, J. Hu and S. Liu. 2016. Enzyme-responsive polymeric vesicles for bacterial—strain-selective delivery of antimicrobial agents. Angew. Chem. Int. Ed. Engl. 55: 1760–1764.

Lili Zhao, L., M. Liu, J. Wang and G. Zhai. 2015. Chondroitin sulfate-based nanocarriers for drug/gene delivery. Carbohydrate Polymers 133: 319–399.

Lin, Y.C., F.J. Tan, K.G. Marra, S.S. Jan and D.C. Liu. 2009. Synthesis and characterization of collagen/hyaluronan/chitosan composite sponges for potential biomedical applications. Acta Biomater. 5: 2591–2600.

Liu, K., X. Lin, L. Chen, L. Huang, S. Cao and H. Wang. 2013. Preparation of micro fibrillated cellulose/chitosan-benzalkonium chloride biocomposite for enhancing antibacterium and strength of sodium alginate films. J. Agric. Food Chem. 61: 6562–6567.

Liu, X.F., Y.L. Guan, D.Z. Yang, Z. Li and K. Yao. 2001. Antibacterial action of Chitosan and carboxymethylated Chitosan. J. Appl. Polym. Sci. 79: 1324–1335.

Liu, Y. and G. Li. 2012. A new method for producing "Lotus Effect" on a biomimetic shark skin. J. Colloid Interface Sci. 388: 235–242.

Liu, Y., J. Tang, R. Wang, H. Lu, L. Li, Y. Kong, K. Qi and J.H. Xin. 2007. Artificial lotus leaf structures from assembling carbon nanotubes and their applications in hydrophobic textiles. J. Mater. Chem. 17: 1071–1078.

Lombardo, D., M.A. Kiselev and M.T. Caccamo. 2019. Smart nanoparticles for drug delivery applications: Development of versatile nanocarrier platforms in biotechnology and nanomedicine. J. Nanomater. 2019: 3702518.

Lotfipour, F., H. Valizadeh, M. Milani, N. Bahrami and R. Ghotaslou. 2016. Study of antimicrobial effects of clarithromycin loaded PLGA nanoparticles against clinical strains of Helicobacter pylori. Drug Res. 66: 41–45.

Ma, B., Y. Huang, C. Zhu, C. Chen, X. Chen, M. Fan and D. Sun. 2016. Novel Cu@SiO$_2$/bacterial cellulose nanofibers: Preparation and excellent performance in antibacterial activity. Mater. Sci. Eng. C 62: 656–661.

Machado, A.H., D. Lundberg, A.J. Ribeiro, F.J. Veiga, B. Lindman, M.G. Miguel and U. Olsson. 2012. Preparation of calcium alginate nanoparticles using water-in-oil (W/O) nanoemulsions. Langmuir 28: 4131–41.

Madhumathi, K., P.T. Sudheesh Kumar, S. Abilash, V. Sreeja, H. Tamura and K. Manzoor. 2010. Development of novel chitin/nanosilver composite scaffolds for wound dressing applications. J. Mater. Sci. Mater. Med. 21: 807–813.

Magiorakos, A.P., A. Srinivasan, R.B. Carey, Y. Carmeli, M.E. Falagas, C.G. Giske, S. Harbarth, J.F. Hindler, G. Kahlmeter, B. Olsson-Liljequist, D.L. Paterson, L.B. Rice, J. Stelling, M.J. Struelens, A. Vatopoulos, J.T. Weber and D.L. Monnet. 2012. Multidrug resistant, extensively drug-resistant and pan drug resistant bacteria. An internation expert proposal for interim standard definitions for acquired resistance. Clinical Microbiology and Infection 18: 268–281.

Majumdar, S. and S. Pal. 2017. Bacterial intelligence: Imitation games, time-sharing, and long-range quantum coherence. J. Cell Commun. Signal 11: 281–284.

Makadia, H.K. and S.J. Siegel. 2011. Poly lactic-co-glycolic acid (PLGA) as biodegradable controlled drug delivery carrier. Polymers 3: 1377–1397.

McCrmick, K. and N. Kautto. 2013. The bioeconomy in Europe: An overview Sustainability 5: 2589–2608.

Mishra, R.K., A. Sabu and S.K. Tiwari. 2018. Materials chemistry and the futurist eco-friendly applications of nanocellulose: Status and prospect. J. Saudi Chem. Soc. 22(8): 949–978.

Mitchell, L.A., J. Goa, R.V. Wal, A. Gigliotti, S. Burchieal and J. Mcdonald. 2007. Pulmonary and systemic immune response to inhaled multiwalled carbon nanotubes. Toxicol. Sci. 100: 203–214.

Mitchell, L.A., F.T. Lauer, S.W. Burchiel and J.D. Mcdonald. 2009. Mechanisms for how inhaled multiwalled carbon nanotubes suppress systemic immune function in mice. Nat. Nanotechnol. 4: 451–456.

Mojtaba Taran, Maryam Rad and Mehran. 2017. Antibacterial activity of copper oxide (CuO) nanoparticles biosynthesize by Bacillus sp. FU4: Optimization of experiment design. Pharm. Sci. 23: 198–206.

Mokhothu, T.H. and M.J. John. 2015. Review on hygroscopic aging of cellulose fibres and their biocomposites, Carbohydr. Polym. 131: 337–354.

Molbak, K. 2004. Spread of resistant bacteria and resistance genes from animals to humans—The public health consequences. J. Vet. Med. Ser. B 51: 364–369.

Moran, J.I., A. Vazquez and V.P. Cyras. 2013. Bio-nanocomposites based on derivatized potato starch and cellulose, preparation and characterization. J. Mater. Sci. 48: 7196–7203.

Munoz-Bonilla, A., A. Cerrada and M. Fernandez-Garcia. 2014. Polymeric Materials with Antimicrobial activity: From synthesis to applications; RSC: London, UK.

Muramatsu, K., S. Masuda, A. Yoshihara and A. Fujisawa. 2003. In vitro degradation behavior of freeze-dried carboxymethyl-chitin sponges processed by vacuum heating and gamma irradiation. Polymer Degradation and Stability 81: 327–332.

Nagahama, H., N. New, R. Jayakumar, S. Koiwa, T. Furuike and H. Tamura. 2008. Novel biodegradable chitin membranes for tissue engineering applications. Carbohydrate Polymers 73: 295–302.

Nathan, C. and A. Cunningham-Bussel. 2013. Beyond oxidative stress: An immunologist's guide to reactive oxygen species. Nature Reviews Immunology 13(5): 349–361.

Neamnark, A., R. Rujiravanit and P. Supaphol. 2006. Electrospinning of hexanoyl Chitosan. Carbohydr. Res. 341: 2098–2107.

Nehrer, S., H.A. Breinan, A. Ramappa, G. Young, S. Shortkroff, L.K Louie et al. 1997. Matrix collagen type and pore size influence behavior of seeded canine chondrocytes. Biomaterials 18: 769–776.

Niskanen, J., J. Shan, H. Tenhu, H. Jiang, E. Kauppinen and V. Barranco. 2010. Synthesis of copolymer stabilized silver nanoparticles for coating materials. Colloid Polym. Sci. 288(5): 543–553.

Noreen, A., K.M. Zia, M. Zuber, S. Tabasum and A.F. Zahoor. 2016. Bio-based polyurethane: An efficient and environment friendly coating systems: A review. Prog. Org. Coat. 91: 25–32.

Nosonovsky, M. and B. Bhushan. 2007. Biomometic super hydrophobic surfaces: Multiscale approach. Nano Lett. 7: 2633–2637.

Ohkawa, K., K.I. Minato, G. Kumagai, S. Hayashi and H. Yamamoto. 2006. Chitosan nanofiber. Biomacromolecules 7: 3291–3294.

Othman, S.H. 2014. Bio-nanocomposites materials for food packaging applications: Types of biopolymer and nano-sized filter. Agric. Agric. Sci. Procedia 2: 296–303.

Ou, C., C. Su, U. Jeng and S. Hsu. 2014. Characterization of biodegradable polyurethane nanoparticles and thermally induced self-assembly in water dispersion. ACS Appl. Mater. Interfaces 6: 5685–5694.

Palza, H. 2015. Antimicrobial polymers with metals nanoparticles. Int. J. Mol. Sci. 16: 2099–2116.

Pandey, R. and G.K. Khuller. 2004. Polymer based drug delivery systems for mycobacterial infections. Curr. Drug Deliv. 1: 195–201.

Panyam, J. and V. Labhasetwar. 2003. Biodegradable nanoparticles for drug and gene delivery to cells and tissue. Adv. Drug Deliv. Rev. 55: 329–347.

Parisi, O.I., L. Scrirano, M.S. Sinicrupi and F. Puoci. 2017. Polymeric nanoparticle constructs as devices for antibiotic therapy. Curr. Opin. Pharmacol. 36: 72–77.

Parkes, J. 1900. A clinical in-market evaluation of an alignate fibre dressing. Br. J. Nurs. 24: 30–35.

Patra, J.K. and K-H. Baek. 2014. Green nanobiotechnology: Factors affecting synthesis and characterization techniques. J. Nanomater. 2014: 219.

Patra, J.K., G. Das, L.F. Fraceto, E.V.R. Campos, M. del Pilar Rodriguez-Torres and L.S. Acosta-Torres. 2018. Nano based drug delivery systems: Recent developments and future prospects. J. Nanobiotechnol. 16: 71.

Prabaharan, M. and J.F. Mano. 2005. Chitosan-based particles as controlled drug delivery systems. Drug Deliv. 12: 41–57.

Prasanta Sani, Swadhin Kr. Saha, Priya Roy, Pranesh Chowdhury and P. Sinha babu. 2016. Evidence of reactive oxygen species (ROS) mediated apoptosis in *Setaria cervi* induced by green silver nanoparticles from *Acacia auriculiformis* at a very low dose. Exp. Parasitol. 160: 39–48.

Qi, L., Z. Xu, X. Jiang, C. Hu and X. Zou. 2005. Preparation and antibacterial activity of Chitosan nanoparticles. Carbohydr. Res. 339: 2693–700.

Rabea, E.I., M.E. Badawy, C.V. Stevens, G. Smagghe and W. Steurbaut. 2003. Chitosan as antimicrobial agent: Applications and mode of action. Biomacromolecules 4: 1457–1465.

Rai, M., A. Yadav and A. Gade. 2009. Silver nanoparticles as a new generation of antimicrobials. Biotechnology Advances 1: 76–83.

Ravi Kumar, M.N. 2000. A review of chitin and Chitosan applications. React. Funct. Polym. 46: 1–27.

Rebecca, J. 2016. Alginate fiber from brown algae. Det. Pharm. Lett. 8: 68–71.

Riley, M.A., S.M. Robinson, C.M. Roy, M. Dennis, V. Liu and R.L. Dorit. 2012. Resistance is futile: The bacteriocin model for addressing the antibiotic resistance challenge. Soc. Trans. 40(6): 1438–1442.

Rizzzato, C., J. Torres, E. Kasamatsu, M. Camorlinga, M.M. Bravo, F. Canzian and I. Kato. 2019. Potential role of biofilm formation in the development of digestive tract cancer with special reference to Helicobacter pylori infection. Front. Microbiol. 10: 846.

Roco, M.C. 2005. International perspective on government nanotechnology funding in 2005. J. Nanopart. Res. 7: 707–712.

Roco, M.C., S. Williams and P. Alivisatos (eds.). 2000. Nanotechnology Research Directions: IWGN Workshop Report Vision for Nanotechnology in the Next Decade. Kluwer Academic Publishers, Dordrecht.

Roemhild, K., C. Wiegand, U.C. Hipler and T. Heinze. 2013. Novel bioactive amino-functionalized cellulose nanofibers. Macromol. Rapid Commun. 34: 1767–1771.

Rouget, C. 1859. Des substances amylacees dans les tissus des animaux, specialement des Articules (chitine), Comptes Rendus 48: 792–795.

Roy, D., J.S. Knapp, J.T. Guthrie and S. Perrier. 2008. Antibacterial cellulose fiber via RAFT surface graft polymerization. Biomacromolecules 9: 91–99.

Saha, S.J., J. Sarkar, D. Chattopadhyay, S. Patra, A. Chakraborty and K. Acharya. 2010. Production of silver nanoparticles by phytopathogenic fungus *Bipolaris nodulosa* and its antimicrobial activity. Digest Journal of Nanomaterials and Biostructures 5(4): 887–895.

Saini, S., M.N. Belgacem, M.C.B. Salon and J. Bras. 2016. Non leaching biomimetic antimicrobial surfaces via surface functionalisation of cellulose nanofibers with aminosilane. Cellulose 23: 1–16.

Salwiczek, M., Y. Qu, J. Gardiner, R.A. Strugnell, T. Lithow, K.M. McLean and H. Thissen. 2014. Emerging rules for effective antimicrobial coatings. Trends Biotechnol. 32: 82–90.

Schiffman, J.D. and C.L. Schauer. 2007. One-step electrospinning of cross-linked Chitosan fibers. Biomacromolecules 8: 2665–2667.

Schiffman, J.D. and C.L. Schauer. 2008. A review: Electrospinning of biopolymer nanofibers and their applications. Polym. Rev. 48: 317–352.

Schiffman, J.D., L.A. Stulga and C.L. Schauer. 2009. Chitin and Chitosan: Transformations due to the electrospinning process. Polym. Eng. Sci. 49: 1918–1928.

Seil, J.T. and T.J. Webster. 2012. Antimicrobial applications of nanotechnology: Methods and literature. Int. J. Nanomed. 7: 2767–2781.

Sethi, S., L. Ge, L. Ci, P.M. Ajayan and A. Dhinojwala. 2008. Gecko-inspired carbon nanotubes-based self-cleaning adhesives. Nano Lett. 8: 822–825.

Sha, L., Z. Chen, Z. Chen, A. Zhang and Z. Yang. 2016. Polylactic acid based nanocomposites: Promising safe and biodegradable materials in biomedical field. Int. J. Polym. Sci. 2016: 1–11.

Shady, S.F., P. Garhwal, R. Leahy, C. Ellis, E. Crawford, K. Schmidt and D.F. McCarthy. 2013. Synthesis and characterization of pullulan-polycaprolactone core-shell nanospheres encapsulated with ciprofloxacin. J. Biomed. Nanotechnol. 9: 1644–1655.

Shendi, H.K., I. Omrani, A. Ahmadi, A. Farhadian, N. Babanejad and M.R. Nabid. 2017. Synthesis and characterization of a novel internal emulsifier derived from sunflower oil for the preparation of waterborne polyurethane and their application in coatings. Prog. Org. Coat. 105: 303–309.

Simoncic, B. and B. Tomsic. 2010. Structures of novel antimicrobial agents for textiles—A review. Text. Res. J. 80: 1721–1737.

Son, W.K., J.H. Youk and W.H. Park. 2006. Antimicrobial cellulose acetate nanofibers containing silver nanoparticles. Carbohydr. Polym. 65: 430–434.

Spacciapoli, P., D. Buxton, D. Rothstein and P. Friden. 2001. Antimicrobial activity of silver nitrate against periodontal pathogens. J. Period Res. 36(2): 109–113.

Starr, F.W., S.C. Glotzer, J.R. Dutcher and A.G. Marangoni (eds.). 2004. Soft Materials, Structure and Dynamics. Marcel Dekker, New York.

Subbiah, T., G.S. Bhat, R.W. Tock, S. Parameswaran and S.S. Ramkumar. 2005. Electrospinning of nanofibers. J. Appl. Polym. Sci. 96: 557–569.

Sudarshan, N.R., D.G. Hoover and D. Knorr. 1992. Antibacterial action of chitosan. Food Botechnol. 6: 257–272.

Takemono, K., J. Sunamoto and M. Askasi. 1989. Polymers and Medical Zare. Mita, Tokyo (Chapter IV).

Taniguchi, N. 1974. On the basic concept of nanotechnology. Proc. Intl. Conf-Prod. Eng. (ICPE). Tokyo, Japan, 18–23.

Tao, F., S. Cazeneuve, Z. Wen, L. Wu and T. Wang. 2016. Effective recovery of poly-β-hydroxybutyrate (PHB) biopolymer from Cupriavidus necator using a novel and environmentally friendly solvent system. Biotechnol. Prog. 32: 678–685.

Teo, W.E. and S. Ramakrishna. 2006. A review on electrospinning design and nanofiber assemblies. Nanotechnology 17: R89–R106.

Thawatchai Maneerung, Seiichi Tokura and Ratna Rujiravanit. 2008. Impregnation of silver nanoparticles into bacterial cellulose for antimicrobial wound dressing. Carbohydrate Polymer 72: 43–51.

Tong, S.Y.C., J.S. Davis, E. Eichenberger, T.L. Holland and V.G. Fowler. 2015. Staphylococcus aureus infection: Epidemiology, pathopysiology, clinical manifestation and management. Clin. Microbiol. Rev. 28: 603–661.

Torres, N.I., K.S. Noll, S. Xu, J. Li, Q. Huang, P.J. Sinko, M.B. Wachsman and M.L and Chikindas. 2013. Safety, formulation, and *in vitro* antiviral activity of the antimicrobial peptide subtilosin against herpes simplex virus type 1. Probiotics Antimicrob. Proteins 5: 26–35.

Torres-Giner, S., M.J. Ocio and J.M. Lagaron. 2008. Development of active antimicrobial fiber based Chitosan polysaccharide nanostructures using electrospinning. Eng. Life Sci. 8: 303–314.

Tummalapalli, M., S. Anjum, S. Kumari and B. Gupta. 2016. Antimicrobial surgical sutures: Recent developments and strategies. Polymer Rev. 56: 607–630.

Tunc, S. and O. Duman. 2013. Preparation and characterization of biodegradable methyl cellulose/montrillonite nanocomposites films. Appl. Clay Sci. 48: 414–424.

Uchida, Y., M. Izume and A. Ohtakara. 1989. Preparation of Chitosan oligomers with purified chitosanase and its application. London Elsevier, 373–382.

Ueno, K., T. Yamaguchi, N. Sakairi, N. Nishi and S. Tokura. 1997. Advances in Chitin Science Jacques Andre, Lyon. 156.

Ulery, B.D., L.S. Nair and C.T. Laurencin. 2011. Biomedical applications of biodegradable polymers. J. Polym. Sci. Part B Polym. Phys. 49: 832–864.

Vaia, R.A. and E.P. Giannelis. 2001. Polymer nanocomposites status and opportunities. MRS Bull. 26(5): 394–401.

Vaidyanathan, R., K. Kalimuthu, G. Gopalram and S. Gurunathan. 2009. Nanosilver the burgeoning therapeutic molecule and its green synthesis. Biotechnol. Adv. 27: 924.

Venkatesan, R. and N. Rajeswari. 2016. Preparation, mechanical and antimicrobial properties of SiO2/poly(butylenes adipate-co-terephthalate) films for active food packaging. Silicon, 1–7.

Vermerris, W. 2011. Survey of genomics approaches to improves bioenergy traits in maize, sorghum and sugarcane, J. Integr. Plant Biol. 53: 105–119.

Viana, J.F.C., J. Carrijo, C.G. Freitas, A. Paul, J. Alcaraz, C.C. Lacorte, L. Migliolo, C.A. Andrade, R. Falcao, N.C. Santos, S. Goncalves, A.J. Otero-Gonzalez, A. Khademhosseini, S.C. Dias and O.L. Franco. 2015. Antifungal nanofibers made by controlled release of sea animal derived peptide. Nanoscale 7: 6238–6246.

Waksman, S.A. and M. Tishler. 1942. The chemical nature of actinomyccin, an anti microbial substance produced by actinomyces antibioticus. J. Biol. Chem. 142: 519–528.

Wan, Y., Q. Cong, X. Wang and Z. Yan. 2008. The wettability and mechanism of geometric non-smooth structure of dragonfly wing surface. J. Bionic Eng. 5: 40–45.

Wang, G. and H. Uludag. 2008. Recent developments in nanoparticle-based drug delivery and targeting systems with emphasis on protein-based nanoparticles. Expert Opinion on Drug Delivery 5: 499–515.

Wang, X.H., D.P. Li, W.J. Wang, Q.L. Feng, F.Z. Cui, Y.X. Xu, X.H. Song and M.V. Derwerf. 2003. Cross linked collagen/chitosan matrix for artificial livers. Biomaterials 24: 3213–3220.

Wang, S., J. Bai, C. Li and J. Zhang. 2012. Functionalization of electrospun-cyclodextrin/polyacrylonitrile (PAN) with silver nanoparticles: Broad spectrum antibacterial property. Appl. Surf. Sci. 261: 499–503.

Wang, Z., B. Sun, M. Zhang, L. Ou, Y. Che, J. Zhang and D. Kong. 2013. Functionalization of electrospun poly (-caprolactone) scaffold with heparin and vascular endothelial growth factors for potential application as vascular grafts. J. Bioact. Compat. Polym. 28: 154–166.

Webb, H.K., R.J. Crawford and E.P. Ivanova. 2014. Wettability of natural super hydrophobic surfaces. Adv. Colloid Interface Sci. 210: 58–64.

Weindl, G., M. Schaller, M. Schafer-korting and H.C. Korting. 2004. Hyaluronic acid in the treatment and prevention of skin diseases: Molecular biological, pharmaceutical and clinical aspects. Skin Pharmacology and Physiology 5: 207–213.

Witte, W. 2004. International discussion of anti resistant strains of bacterial pathogens. Infect. Genet. Evol. 4: 181–191.

Wogggum, T., P. Sirivongpaisal and T. Wittaya. 2014. Properties and characteristics of dual modified rice starch based biodegradable films. Int. J. Biol. Macromol. 67: 490–502.

Wong, K.K.Y. and X. Liu. 2005. Silver nanoparticles-the real "silver bullet" in clinical medicine? Med. Chem. Commun. 1: 125–131.

Wu, D., Q.D. Chen, H. Xia, J. Jiao, B.-B. Xu and X. Lin. 2010. A facile approach for artificial biomimetic with both super hydrophobicity and iridescence. Soft Matter 6: 263–267.

Wypych, G. 1999. Handbook of Fillers, 4th edn. Chem Tech Publishing, Toronto.

Yang, T., C. Chou and C. Li. 2005. Antibacterial activity of N-alkylated disaccharide Chitosan derivatives. Int. J. Food Microbiol. 97: 237–245.

Yih, T.C. and M. Al-Fandi. 2006. Engineered nanoparticles as precise drug delivery systems. J. Cell. Biochem. 97(6): 1184–90.

Zasloff, M. 2002. Antimicrobial peptides of multicellular organisms. Nature 415: 389–395.

Zhang, K., X. Tang, J. Zhang, W. Lu, X. Lin, Y. Zhang, B. Tian, H. Yang and H. He. 2014. PEG-PLGA copolymers: Their structure and structure-influenced drug delivery applications. J. Control 183: 77–86.

Zhang, L., D. Pornpattananangkul, C-M.J. Hu and C.-M. Huang. 2010. Development of nanoparticles for antimicrobial drug delivery. Current Medical Chemistry 17(6): 585–594.

Zhang, L., F.X. Gu, J.M. Chan, A.Z. Wang, R.S. Langer and O.C. Farokhzad. 2008. Nanoparticles in medicine: Therapeutic applications and developments. Clinical Pharmacology & Therapeutics 83(5): 761–769.

Zhao, C., A. Tan, G. Pastorin and H.K. Ho. 2013. Nanomaterial scaffolds for stem cell proliferation and differentiation in tissue engineering. Biotechnol. 31: 654–668.

Zhao, N., J. Xu, Q. Xie, L. Weng, X. Guo, X. Zhang and L. Shi. 2005. Fabrication of biomimetic superhydrophobic coating with a micro-nano-binary structure. Mcromol. Rapid Commun. 26: 1075–1080.

Zhao, S.S., L.I. Yuan, H.L. Cao and W. Peng. 2012. Preparation and characterization of PLA/MMT nanocomposites with microwave irradiation. J. Mater. Eng. 2: 4.

Zinoviadou. K.G., M. Gougouli and C.G. Biliaderis. 2016. Innovative biobased materials for packaging sustainability. In Innov. Strateg. Food Ind. Tools Implement, 167–189.

9

Hybrid Polymeric Nanobiomaterials for Applications in Drug Delivery Systems and Tissue Engineering

Sajjad Husain Mir

1. Introduction

Polymers are large molecules created by linking a series of different monomers. In addition to synthetically generated polymers, there are biopolymers that are the common components of proteins, nucleic acids, and sugars. These biopolymers are the key factor in controlling and regulating several biochemical and biophysical functions of living cells, thus, in turn, participating in all major natural processes leading to be an integral part of cell machinery. These biopolymers can undergo highly cooperative interactions, which give rise to their property of nonlinear response to external stimuli. This high degree of cooperativity is thought to be the cause behind the strongly coherent biochemical and biophysical functions in living cell machinery. The mechanism of the cooperative interaction of these biopolymers and their interaction with the external stimuli are a major field of study for generating synthetic polymers that can mimic the cooperative behavior of biopolymers. These polymers can then be utilized as biomaterials and can be employed to interface with biological systems for various functions of a living cell. Biocompatibility is an essential requirement for a polymer to be characterized as a biomaterial. Biocompatibility of a material is its ability to execute a response in a specific application with respect to a particular host. The transient existence of the polymeric biomaterials is suitable for *in vivo* applications since it can overcome biostability and biocompatibility issues in the long term. Among several applications, one of the major applications of the biodegradable polymers is in the field of drug delivery. In recent years, controlled and targeted delivery of drugs has become an extremely important field of research considering the fact that the delivery of the pharmaceutics in a very controlled manner and targeted only at the diseased region will increase the efficacy of the pharmaceutics by several folds and will also help to get rid of undesired side effects of the drugs.

The global market for biomaterials was worth nearly $75.1 billion until 2017. This market is expected to grow at a compound annual growth rate of 6.7% between 2018 and 2020, resulting in a $79.1 billion global market in 2018 and a $109.5 billion global market in 2019 (Piskin 1995).

Advanced Materials and BioEngineering Research Centre (AMBER) & CRANN, Trinity College Dublin, The University of Dublin, Dublin 2, Ireland.
Emails: sajjad.mir@tcd.ie, sajjad.h.mir@gmail.com

Although biomedical applications of natural enzymatically degradable polymers date back thousands of years, the application of synthetic biodegradable polymers began only in the second half of the 1960s (Taylor 2015, accessed March 2019). Considering their advantages over biostable materials in terms of long-term biocompatibility along with the technical and ethical issues accompanying revision surgeries, investigations into the application of biodegradable biomaterials rather than permanent prosthetic devices for assisting in tissue repair and regeneration has vigorously increased recently (Barbucci 2002, Langer and Vacanti 1993, Dvir et al. 2011). As a result, polymeric biomaterials are quickly replacing other material classes, such as metals, alloys, and ceramics, for use as biomaterials due to their versatility (Vacanti and Langer 1999). Within the global biomaterial market, the polymeric biomaterials sector is expected to show the highest growth, at a CAGR of 22.1%, because of its promising potential in a wide range of biomedical applications (Piskin 1995). This chapter is to highlight the most often-studied polymeric biomaterials and underscore their immense potential in the areas of drug design, development, and therapy.

Given the complexity of the human body and the scope of applications that polymeric biomaterials are currently utilized for, no single polymeric system can be considered the ideal biomaterial for all medical applications. Thus, recent advances in biodegradable biomaterial synthesis have been directed toward developing and synthesizing polymers with properties tailored for specific biomedical applications. Moreover, current developments incorporating multifunctional and combinatorial approaches in biomaterial design have accelerated the innovation of novel biodegradable biomaterials. Another hotspot in biomaterials research is the development of therapeutic devices, including temporary prostheses, three-dimensional (3D) porous scaffolds for tissue engineering, and delivery vehicles for pharmacological applications. Most recently, 3D bio-printing has also been acknowledged, and preliminary data collection for biomaterial use as potential bio-ink for printing of 3D scaffolds has begun. Moreover, by properly engineering the structure and degradation parameters, these biodegradable materials can be used to generate micro- or nanoscale drug-delivery vehicles for controlled drug delivery in an erosive or diffusive manner, or as a combination of both (Dvir et al. 2011, Kohane and Langer 2008).

Because of the interest aroused in the areas demanding biodegradable biomaterials, including controlled drug delivery, gene therapy, and nanotechnology, there has been a robust expansion of biomedical applications of synthetic biodegradable polymers and analogous natural polymers, and this chapter will focus on exploring the most current development of these polymers for drug delivery systems.

2. Naturally Abundant Polymeric Biomaterials

Biodegradable biomaterials can be roughly divided into two categories, natural and synthetic, based on their source and whether they are composed of naturally occurring extracellular matrix (ECM). Natural biodegradable polymeric biomaterials generally include proteins (collagen, fibrin, silk, etc.), and polysaccharides (starch, alginate, chitin/chitosan, hyaluronic acid derivatives, etc.) (Vert 2005, Gollwitzer et al. 2003, Freyman et al. 2001). Furthermore, a family of native polyesters-polyhydroxyalkanoates (PHA) has been recognized as natural biodegradable biomaterials and, more recently, sundew adhesives (natural polysaccharide-based hydrogels) and ivy nanoparticles (macromolecular compositions of nanospherical arabinogalactan proteins) have garnered more attention for their ability to create effective nanocomposite adhesives and for their potential use as nano-carriers in drug delivery, respectively (Marijinissen et al. 2002, Park et al. 2013).

2.1 Collagen

As the most prevalent protein in the human body, collagen offers physical support to tissues by inhabiting the intercellular space, acting not only as native structural support for organizing cells within connective tissues, but also as a mobile, dynamic, and flexible substance essential to cellular behaviors and tissue function (Lenaghan 2011) (Scheme 1). Generally, collagen is a rodtype polymer, approximately 300-nm long with a molecular weight of approximately 300 kDa. Free amino acids in the body are synthesized into subunit chains of collagen, which then undergo transcription, translation, and post-translational modification processes in suitable cells such as osteoblasts and fibroblasts (Marijinissen et al. 2002, Lenaghan 2011).

Collagen plays a critical role in preserving the biological and structural integrity of the ECM and is highly dynamic, undergoing continuous remodeling for proper physiological functions. For most soft and hard connective tissues (e.g., blood vessels, cornea, skin, tendon, cartilage, and bone), collagen fibrils and their networks function through their highly organized 3D structure. Tissue regeneration attempts to repair both the structural integrity and the intricate remodeling process of the native ECM, particularly restoring the delicate collagen networks under which normal physiologic regeneration occurs; thus, recent efforts have been focused on replacing native collagen-based ECM by developing novel biomaterials that imitate its intricate fibrillar architecture and function as cell scaffolding.

With regard to drug delivery, collagen has been notably studied for the delivery of low-molecular-weight drugs, proteins, genes, and plasmids. Currently, a few collagen-based gentamicin-delivery vehicles are available in the global market (e.g., Sulmycin-Implant and Collatamp-G, Innocoll Pharmaceuticals Ltd). These delivery systems permit a sustained local delivery of antibiotics with limited systemic exposure. Moreover, another product, Septocoll (Biomet), achieved prolonged collagen delivery by incorporating two gentamicin salts possessing different solubility, and has been approved for infection prevention (Huang et al. 2015). More recently, a new biodegradable, collagen-based chlorhexidine chip has been shown to provide a longer, more sustained release of chlorhexidine in the confines of the periodontal pocket as compared to simple subgingival irrigation of chlorhexidine.

Collagen as a biomaterial has already seen significant use in several specific applications in skin repair, hemostatic agents, and drug delivery, but its slight immunogenicity, high cost, and varying physicochemical and degradation properties prevent further expansion of collagen-based biomaterials. Although recombinant collagens have the potential to catapult collagen-based biomaterials for widespread use, significant limitations such as their lack of post-translational modification will need to be continually studied for them to have any noticeable impact for use as a biomaterial.

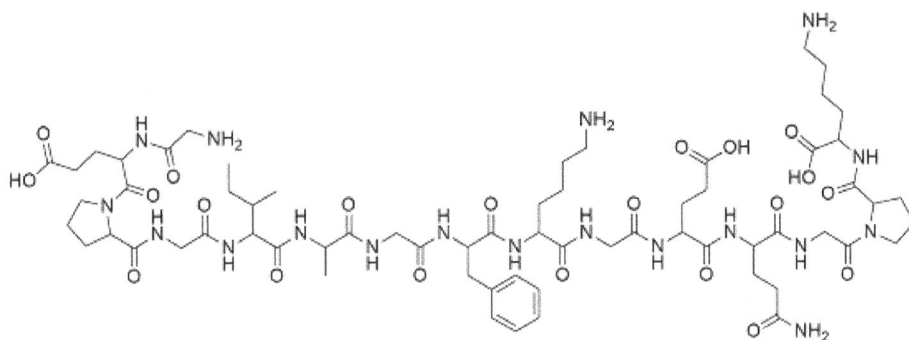

Scheme 1: Chemical structure of collagen.

2.2 Gelatin

Gelatin is a natural biopolymer derived from collagen via controlled alkaline, acid, or enzymatic hydrolysis (Gelse et al. 2003). As a result of its biological origin, it has excellent biodegradability and biocompatibility and because it is widely available, gelatin is a relatively low-cost polymer

(Burke and Hsu 1999). Gelatin has been used in medical and pharmaceutical fields as a matrix for implants, and as stabilizers in vaccines such as measles, mumps, and rubella (Gou et al. 2018). Moreover, gelatin is water permeable and soluble in water and has multifunctional properties as a drug-delivery carrier (Gou et al. 2018). Gelatin's mechanical properties, swelling behavior, thermal properties, and many other physiochemical properties can be dependent upon the collagen source, extraction method, amount of thermal denatured employed, and the degree of cross-linking, thereby making gelatin a very versatile polymer (Foox and Zilberman 2015). Furthermore, its ability to produce a thermoreversible gel makes it a very good candidate as a targeted drug-delivery carrier and, as a result, gelatin can be utilized to develop specific drug-release profiles, allowing for a broad range of applications in drug delivery (Foox and Zilberman 2015).

The primary advantages of gelatin are its biodegradability, availability, and cheap cost (Karim and Bhat 2009). Porcine skin-derived gelatin is the most popular source, followed by bovine skin, and bone (Bigi et al. 2002). However, there are health concerns over transmission of pathogenic vectors such as prions (Bigi et al. 2002). Recombinant gelatins, however, can be utilized to overcome the disadvantages with animal tissue-derived materials (Rajangam and Ssa 2013).

Research into improving gelatin's mechanical properties will need to be investigated in the future for gelatin to have a promising future as a primary biomaterial, but, until then, the most likely incorporation of gelatin will be as a composite with other natural or synthetic biomaterials, or as a carrier for drug delivery.

2.3 Fibrin

Fibrin is a 360-kD fibrinogen-derivative biopolymer involved in the natural blood-clotting process, enhancing cell adhesion and proliferation (Ahmad et al. 2008). In addition to possessing excellent biodegradability and biocompatibility, fibrin exhibits high elastic and viscous properties; stiffens in response to shear, tension, or compression; and has excellent deformability (Kjaergard et al. 2000). One of the first developed fibrin-based products was fibrin glue (fibrin sealant). Today, an expansive variety of products possessing different compositions and adhesive properties are available in the market. These products are broadly used for hemostasis and tissue-sealing applications in various surgical procedures, including neurosurgery and plastic and reconstructive surgery (Tredwell and Sawatzky 1990, Mcgill et al. 1997, de La Puente et al. 2013).

Fibrin is natural, highly available, implantable, inexpensive, easy to use, and has low fibrinogen concentrations. Due to its porous morphology, a fibrin scaffold system is adequate for cell attachment, proliferation, differentiation (de La Puente et al. 2011, Linnes et al. 2007), and has a release system of growth factors such as vascular endothelial growth factor and basic fibroblast growth factor (bFGF) (Toole 2004, El Maradny et al. 1997). Fibrin-immobilized growth factors have been shown to be continuously released for several days in a controlled manner, making it optimal for numerous tissue-engineering purposes (Ahmad et al. 2008). Thus, autologous fibrin could prevent the complications of techniques derived from the use of current commercially available fibrin products and should be further investigated. Fibrin-based scaffolds do have some limitations, such as weak mechanical strength and quick degradation rates; however, these properties have been shown to be improved by incorporating stronger natural and synthetic polymers, utilizing various cross-linking methods, and utilizing micro/nanospheres (Ahmad et al. 2008).

2.4 Hyaluronic Acid (HA)

HA is an essential component of the ECM and its structural and biological properties mediate cellular signaling, morphogenesis, matrix organization, and wound repair (Meyer and Palmer 1934, Mori et al. 2004) (Scheme 2). In 1943, HA was first isolated from the vitreous humor by Meyer and Palmer (Allison et al. 2006). HA is a member of the GAG family, which involves linear polysaccharides consisting of alternating units of N-acetyl-d-glucosamine and glucuronic acid ranging in size from 5,000 to 20,000,000 Daltons *in vivo*, and is present in almost every tissue in vertebrates. Native sources of HA include rooster combs, bovine vitreous humor, and synovial fluids (Prestwich 2008). HA degradation occurs in the body through free radicals, such as nitric oxide and matrix metalloproteinases found in the ECM, and then undergoes endocytosis. Further digestion of HA by lysosomal enzymes results in mono- and disaccharides, which are then converted into ammonia, carbon dioxide, and water (Prestwich 2008).

Recently, HA has become recognized as an important building block for the creation of new biomaterials for use in cell therapy, 3D cell culture, and tissue engineering (Burdick and Prestwich 2011, Yu et al. 2013, Niiyama and Kuroyanagi 2014). As HA is secreted at the early stage of wound healing, it has been extensively researched for wound-dressing applications (Prestwich 2011, Kumar et al. 2004). HA can be recognized by receptors on a variety of cells associated with tissue repair, and thus presents the capability to stimulate angiogenesis and to regulate injury-induced inflammation as a free radical scavenger (Raveendran et al. 2017).

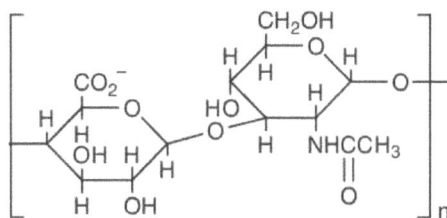

Scheme 2: Chemical structure of hyaluronic acid.

2.5 *Chitin / Chitosan*

Chitosan is another naturally derived biodegradable polysaccharide commonly used in tissue engineering (Akman et al. 2010) (Scheme 3). Chitosan is a derivative of chitin—the second most abundant natural polymer commonly found in the exoskeletons of crustacean and insects as well as the cell walls of fungi (Akman et al. 2010). Chitin is partially deacetylated to form chitosan, which is composed of glucosamine and N-acetyl glucosamine linked in a $\beta(1–4)$ manner (Gobin and Mathur 2006). The molecular weight and the degree of deacetylation, which are essential in assessing the characteristics of chitosan, are reliant on the source and production process. Specifically, chitin/chitosan-based materials have demonstrated potential with regard to connective, nerve, adipose, and vascular tissue-engineering applications (Wu et al. 2007, Madhumati et al. 2010, Kumar et al. 2010). For example, a silk fibroin (SF)/chitin-based scaffold was used to repair a musculo-fascial defect in the abdominal wall, displaying continuous integration with adjacent native tissue and mechanical strength similar to native tissue (Hillyard et al. 1964).

Furthermore, chitin/chitosan-based materials have been incorporated successfully into cutaneous wound management. A number of studies have reported the use of chitin/chitosan scaffolds and membranes to treat patients with deep burns (Sugano et al. 1988, Kozen et al. 2008). Recently, novel α-chitin/silver nanoparticles (AgNPs) and β-chitin/AgNP composite scaffolds were tested for wound-healing applications (Sugano et al. 1988, Kozen et al. 2008). These chitin/AgNPs composite scaffolds were found to possess excellent antibacterial activity against *Staphylococcus aureus* and *E. coli*, combined with good blood clotting ability (Sugano et al. 1988, Kozen et al. 2008). In addition, β-chitin/AgNP composite scaffolds functioned as promising matrices capable of providing good cell attachment apart from their antibacterial activity, which suggests that these composite scaffolds are ideal for wound-healing applications (Kozen et al. 2008).

Chitin/chitosan-based materials are widely studied for several different tissue-engineering applications, regenerative medicine, wound healing, and drug delivery for good reason. They have excellent biodegradability and biocompatibility and have been known to have antiulcer, anti-acid, hypocholesterolemic action, wound-healing, antitumor, and hemostatic properties (Ueno and Fujinaga 2011, Kaur et al. 2007, Calvert 1997, Betancur et al. 1997). Although chitin/chitosan-based materials tend to lack mechanical properties, they possess functional reactive side groups that can be cross-linked to make bridges between polymeric chains, optimizing the resistance and elasticity of these materials (Vmd et al. 2017).

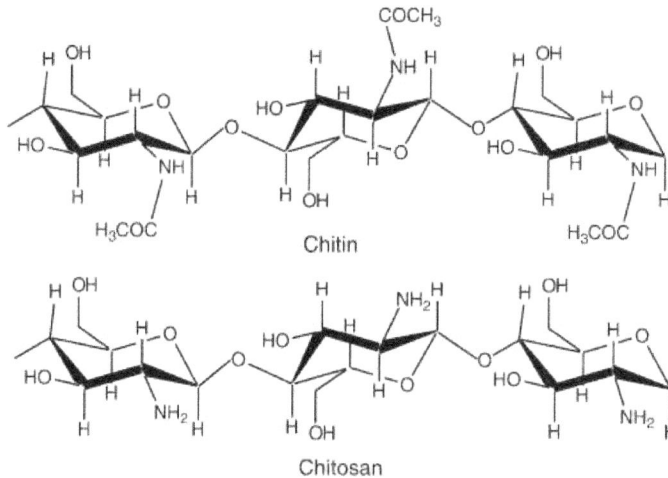

Scheme 3: Chemical structure of chitosan.

2.6 Starch

Starch is the primary energy reserve polysaccharide in plants, and it is present in the form of granules composed of amylose and amylopectin (Dragan 2014). Amylose is a linear polymer composed of glucose monomers linked through α-D-(1–4) glycosidic linkages, whereas amylopectin molecules are huge, branched polymers of glucose known to be one of the highest molecular-weight natural polymers (Saboktakin et al. 2011). Different plants have slightly different granule sizes, amylose/amylopectin ratio, mineral contents, and amount of phosphorous and phospholipid contents that lead to varying starch properties (Dragan 2014). The specific characterization of starch is particularly important due to different swelling, solubility, gelatinization, mechanical behavior, enzymatic digestibility, rheological characteristics, and surface characteristics, which affect the way it needs to be processed to convert it to a more usable form such as hydrogels, pastes, and nanoparticles (Pawar and Edgar 2012).

Starch has been proposed as a possible drug-delivery system (Lee and Mooney 2012). In its hydrogel form, it can efficiently entrap drugs in interstitial spaces, thus protecting them from undesirable conditions in the human body (Altman et al. 2003). In addition, starch hydrogels are resistant to gastric juices, allowing for potential oral drug-delivery systems, and they can be modified to be degraded in very specific portions of the gastrointestinal tract, thus allowing for site-specific delivery (Cao and Zhang 2016). Moreover, physical modification of starch by retrogradation leads to the development of high levels of type three resistant starch—a very thermally stable, low solubility form of starch that makes it suitable for colon-specific delivery systems (Lee and Mooney 2012). However, in practice, studies on starch as a possible drug-delivery system have been very limited. Most of the research on starch as part of a drug-delivery system is theoretical, and scientists are still exploring the possible consequences of physical modification of starch on the mechanical and structural properties of hydrogels.

2.7 Alginate

Alginate is a polysaccharide derived from the cell wall of brown seaweed and extracellularly in some bacteria. It is an anionic polymer that is biocompatible, nontoxic, and noninflammatory, as long as it undergoes multi-purification steps, but it is primarily known for its mild gelation conditions, low cost, and relatively simple modifications in making alginate derivatives with new properties (Jao et al. 2016). In particular, alginate hydrogels have been prepared by various chemical or physical cross-linking methods for diverse applications in wound healing, delivery of bioactive agents, and tissue engineering (Kundu et al. 2013). The primary drawbacks of alginate are its generalized lack of strong mechanical properties, poor cell adhesion, and its lack of degradability in mammals (Jao et al. 2016). However, by combining alginate with other biomaterials such as agarose and chitosan, and by partially oxidizing alginate with molecules like sodium periodate, scientists have managed to enhance its mechanical properties and degradability, conferring significant promise on alginate-based biomaterials (Kundu et al. 2013).

Alginate is a widely utilized biomaterial, especially in regenerative medicine and in tissue engineering, due to its biocompatibility, mild and physical gelation process, chemical and physical cross-linking abilities, non-thrombogenic nature, and the resemblance of its hydrogel matrix texture to that of the ECM (Wenk et al. 2011). Moreover, alginate happens to be easily modified into any form, such as microspheres, sponges, foams, elastomers, fibers, and hydrogels, thereby broadening the scope of application of alginate-based biomaterials, and it can be combined with other natural biomaterials to create and enhance new and existing properties (Mottaghitalab et al. 2015).

The future of alginate-based wound dressings hinges upon establishing more control over the delivery of one or more drugs, as well as their duration and sequence of release while considering external environmental changes (Kundu et al. 2013). Furthering our understanding of the fundamentals of alginate properties will help researchers take advantage of the remarkable properties and bioavailability of alginate and utilize genetic engineering techniques to control the bacterial synthesis of alginate with new and improved properties, thus revolutionizing the use of this material (Kundu et al. 2013).

2.8 Silk

Silk fibers are natural biopolymers derived primarily from the silkworm Bombyx mori (Hofer et al. 2015). The silk fiber consists of two parallel SF proteins, held together by a layer of silk sericin protein glue on the surface (Pritchard et al. 2013). Until recently, silk sericin has been deemed to be immunologically incompatible with the human body and has, therefore, been largely neglected as a biopolymer (Seib et al. 2013). However, SF has been used as a biomedical suture material for centuries (Hofer et al. 2015). It is a semi-crystalline structure that has an incredible combination of mechanical properties, possessing very high tensile strength, coupled with excellent elasticity and flexibility (Park et al. 2013).

Besides the properties already mentioned, SF has many other unique and standout properties that make it excellent for use as a drug carrier (Mottaghitalab et al. 2015). It can allow for loading of even the most sensitive of drugs, such as proteins and nucleic acids, due to its mild, all-aqueous processing conditions (Seib et al. 2013). Moreover, SF has a diverse range of amino acids with several functional groups that can simplify the attachment of different types of biomolecules or antibodies, giving it a wide degree of functionalization. Finally, SF naturally has an intrinsic response to pH changes, making it easy to control drug-release kinetics, and the mechanism of elimination from the body can easily be done by degradation via proteolytic enzymes in the body, leaving no likely side effects. Recently, most research groups have been investigating the mechanistic component of incorporating SF nanoparticulate for protein delivery, small-molecules delivery, and even anticancer delivery.

SF-based biomaterials have immense potential as one of the preeminent natural biopolymers studied today. Its ease of structural modification, controllable degradability, high tensile strength, elasticity and flexibility, potential to introduce physical cross-links, hemostatic and self-healing attributes, and its ability to be processed into numerous different forms, such as sponges, films, and hydrogels, make SF a polymer with various biomedical applications (Hofer et al. 2015).

2.9 Polyhydroxyalkanoates (PHA)

PHA is a class of natural, biodegradable polyesters synthesized by microorganisms as intracellular carbon, and energy storage compounds in uneven growth conditions (Chen et al. 2009) (Scheme 4). They have exceptional biodegradability and biocompatibility, and produce nontoxic degradation products, making them excellent for use in biomedical applications such as drug delivery, tissue engineering, and substitute for implantable devices (Li and Loh 2017). A very important character of PHAs is that they can be developed in a way to have various physicochemical behaviors in properties such as amphiphilicity, crystallinity, and mechanical properties by simply choosing an appropriate production strain, cultivation conditions, and carbon sources (Saito and Doi 1994).

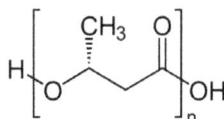

Scheme 4: Chemical structure of polyhydroxyalkanoates.

3. Synthetic Biopolymers

The logic behind employing polymers in the design of therapeutic agents have been widely investigated for a number of decades. In 1975, a rational model for pharmacologically active polymers was first proposed by Helmut Ringsdorf (Ringsdorf 1975). His concept of covalently bound polymer-drug conjugates still forms the basis for much of the work in this area. The Ringsdorf model (Figure 1) primarily consists of a biocompatible polymer backbone bound to three

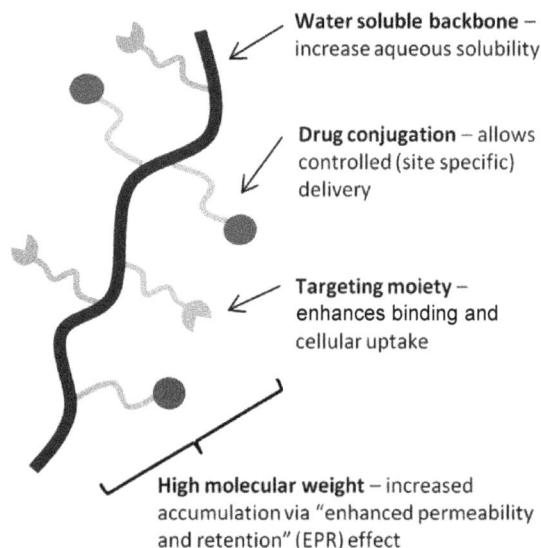

Water soluble backbone – increase aqueous solubility

Drug conjugation – allows controlled (site specific) delivery

Targeting moiety – enhances binding and cellular uptake

High molecular weight – increased accumulation via "enhanced permeability and retention" (EPR) effect

Figure 1: Concept of drug-delivery using polymer-drug conjugates. Reproduced with Permission from ref 121, copyright 2011 American Chemical Society.

components: (1) a solubilizer, which serves the purpose of imparting hydrophilicity and ensuring water solubility; (2) a drug, usually bound to the polymeric backbone via a linker; and (3) a targeting moiety whose function is to provide transport to a desired physiological destination or bind to a particular biological target.

A substantial amount of effort is directed toward developing anticancer polymer drug conjugates. Anticancer agents are often limited by poor water solubility and metabolic instability, and their clinical use is often limited by dose dependent toxicity. The therapeutic index of a given drug is defined as the ratio between its toxic and therapeutic dose. For the clinician, the goal is to deliver an anticancer agent at a dose high enough to achieve cytotoxicity within tumor tissues. However, the actual dose administered is very often limited by toxicity to other vital organs. Thus, any improvement in the therapeutic index for such drugs that allows the clinician the ability to deliver higher drug concentrations to tumor tissue while maintaining manageable side effects can yield benefits for cancer patients. One of the primary ways in which polymer-drug conjugates can increase the therapeutic index of anticancer agents is via the "enhanced permeability and retention (EPR) effect", first described by Matsumura and Maeda in 1986 (Kaneda et al. 2004). They proposed that increased uptake of macromolecules by solid tumors can occur as a result of a combination of poor lymphatic drainage and increased vascular permeability present within the tumor microenvironment (Figure 2).

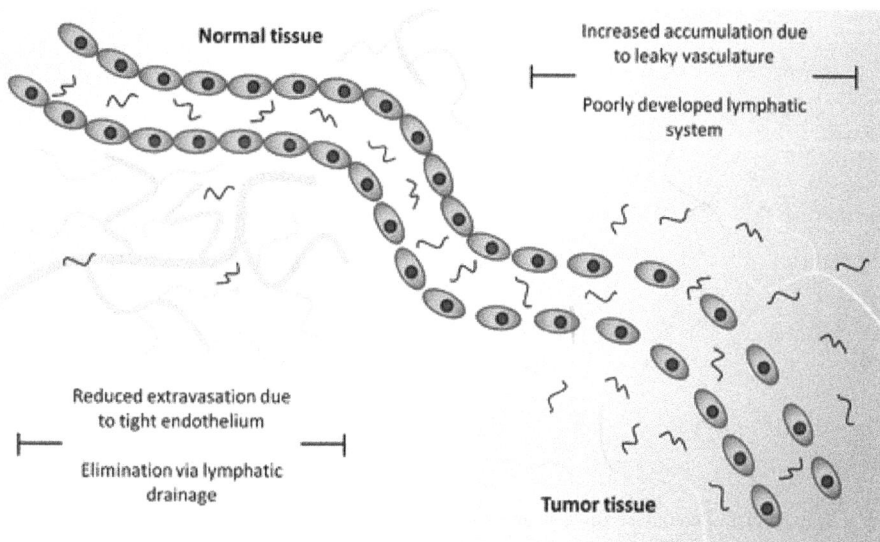

Figure 2: The enhanced permeability and retention, via a combination of increased extravasation and reduced lymphatic drainage in tumor tissues. Reproduced with Permission from ref 121, copyright 2011 American Chemical Society.

3.1 Linear Polymers

Many different drug conjugates have been synthesized utilizing water-soluble linear polymers. While many polymeric carriers have been described such as poly(vinyl pyrrolidone) (PVP) (Kamada et al. 2000, Yasukawa et al. 1999), poly(vinyl alcohol) (Chpman et al. 2006), polyglutamic acid (PGA), and poly(malic acid) (Ljubimova et al. 2008), two of the most widely investigated chemistries are those based on poly(ethylene glycol) (PEG) (Pasut and Veronese 2009a) and N-(2-hydroxypropyl) methacrylamide (HPMA) copolymers (Pasut et al. 2008).

3.1.1 Poly(ethylene glycol)

Although numerous different polymer compositions have been synthesized and studied (Figure 3), some of the simplest polymers, such as poly(ethylene glycol) (PEG), maintain widespread use and versatility. PEG-protein conjugates have gained particular importance because of the ability of PEG to protect against protein enzymatic degradation and reduce uptake by the reticulo-endothelial system (RES) (Veronese et al. 2002, New and Brechbiel 2009), both properties imparted via simple steric hindrance. Numerous functionalized PEGs are available to aid in conjugation (Figure 3A). Whereas some functionalities allow conjugation to biomolecules such as proteins and antibodies, others can be more generally applied in the synthesis of novel biomaterials. One such approach widely utilized involves "click" chemistry (Van Dijk et al. 2009). A click reaction is a highly specific, high yield conjugation reaction wherein mild conditions are commonly used, and byproducts are easily removed. The most prevalent example is the 1,3-dipolar cycloaddition of alkynes and azides (Joralemon et al. 2010). A number of other conjugation chemistries have also been investigated (Pasut et al. 2009, Pasut and Veronese 2009b, Subr et al. 2009).

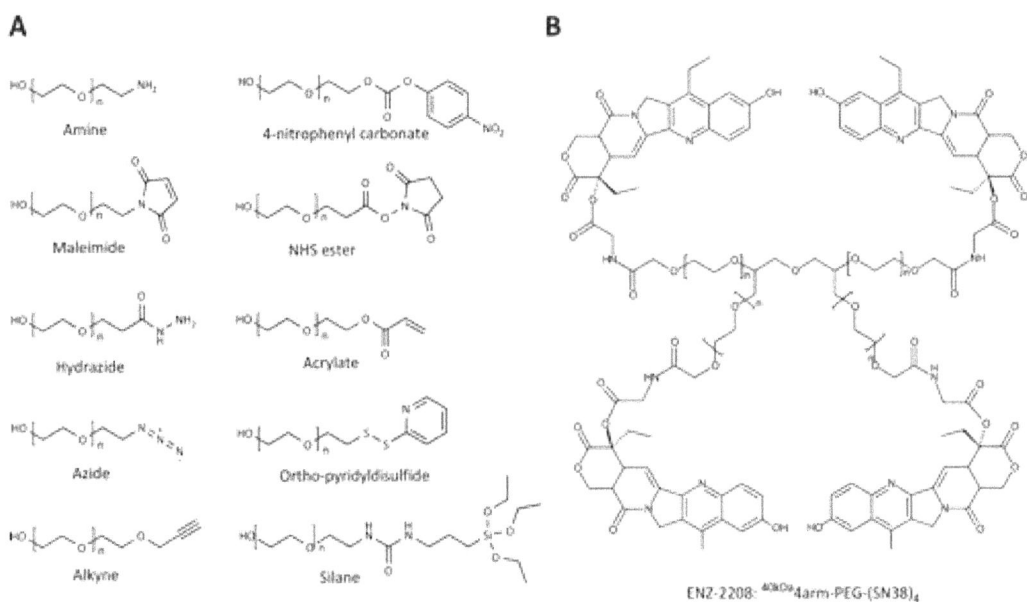

Figure 3: (A) Examples of commercially available functionalized PEGs. (B) multiarm PEG conjugates. Adapted and reprinted with permission from ref 78, copyright 2009 Elsevier B.V.

3.1.2 N-(2-Hydroxypropyl)methacrylamide (HPMA) Copolymers

A major driving force behind the continued development of HPMA copolymers as drug carriers was the development of oligopeptide sequences as drug linkers (Putnam and Kopecek 1995a,b). These sequences were specifically designed to ensure hydrolytic stability during systemic transport and the ability to be enzymatically cleaved by lysosomal enzymes following cellular internalization (Vicent et al. 2004). In developing such a system, early studies with model enzymes demonstrated that factors such as peptide sequence structure and length, drug loading, drug structure, and steric hindrance play important roles in stability and drug release kinetics (Kasuya et al. 2011). Studies evaluating release in the presence of the lysosomal enzyme cathepsin B resulted in the isolation of the tetrapeptide sequence glycylphenylalanylleucylglycine (GFLG). Numerous HPMA copolymer-drug conjugates utilizing this lysosomally cleavable linker have been reported to date, including several HPMA copolymers used in clinical trials.

3.2 Dendrimers

Dendrimers are branched polymeric macromolecules forming a star-like structure (Figure 4A). Such unique structures allow the conjugation of drugs to the surface, thus maximizing the potential for biological interactions. A wide array of chemistries can be employed in the synthesis of dendrimers, where the core, monomer units, and surface functionality determine physiochemical characteristics. However, for use in drug delivery applications, it is necessary to maintain biocompatibility. Physiochemical properties such as solubility, surface group functionality, surface charge density, and stability must therefore be considered (Figure 4). Tomalia et al. first described the synthesis of poly(amido amine) (PAMAM) dendrimers in 1985 (Tomalia et al. 1985, 1986).

In the field of drug delivery, much of the work with dendrimers has focused on their use in the encapsulation and formulation of drugs (Gajbhiye et al. 2009). Due to their hyper branched structure, dendrimers often possess open cavities between adjacent branches, thus allowing encapsulation of drugs (Cheng et al. 2008). This can aid in the solubilization of poorly water-soluble drugs. In addition, dendrimers formulated (physically mixed) with drugs have been investigated as both transdermal (Goldberg et al. 2010) and oral (Xu et al. 2010) delivery systems. Dendrimers with positively charged surface functionalities, such as poly(ethyleneimine) and PAMAM dendrimers, have also been investigated as gene carriers (Xu et al. 2010), as a result of their ability to complex with negatively charged DNA.

The covalent attachment of drugs to dendrimers has been widely investigated. Numerous chemotherapeutics have been attached to the surface of dendrimers in an attempt to increase aqueous solubility and provide specific delivery to tumor tissues. For example, multifunctional drug delivery platforms utilizing dendrimers conjugated with imaging agents, drugs, and targeting moieties have been investigated (Strickley 2004).

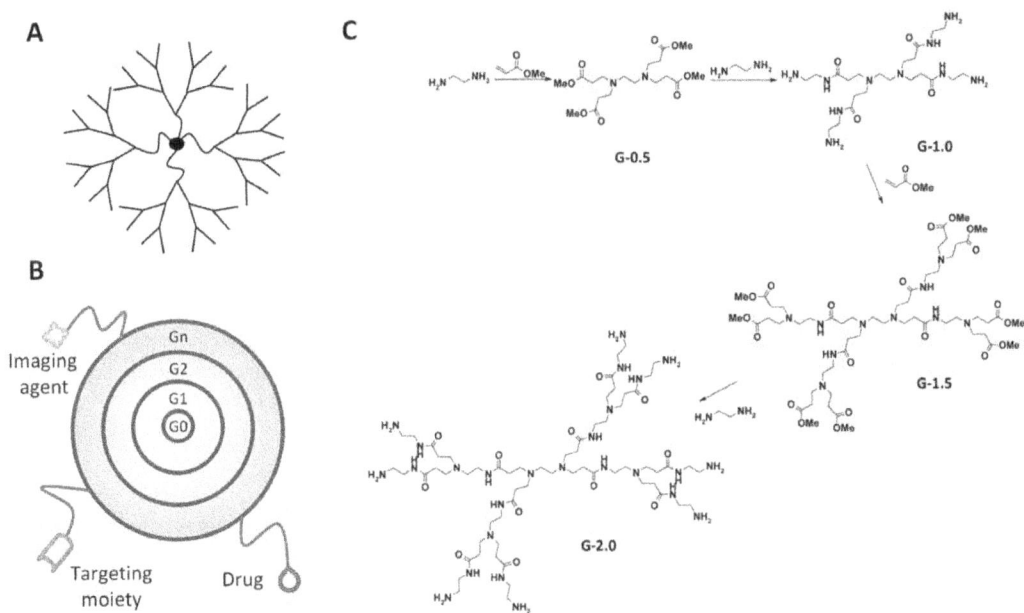

Figure 4: (A) Star-like polymers, where drugs can be either conjugated to the dendrimer surface or encapsulated within center. (B) Dendrimers grow linearly in size and exponentially in surface area. (C) Synthesis of poly(amido amine) (PAMAM) dendrimers occurs from an ethylenediamine core with alternating reactions with methyl acrylate and ethylenediamine to produce each generation. Reproduced with Permission from ref 121, copyright 2011 American Chemical Society.

3.3 Polymeric Micelles

Micelles are colloidal particles with a size of about 5–150 nm that consist of self-assembled aggregates of amphiphilic molecules or surfactants. Amphiphiles, at low concentrations in aqueous media, exist as unimers in solution. However, as their concentration is increased, thermodynamic processes drive the formation of aggregates, which sequester hydrophobic regions into core like structures surrounded by a hydrophilic corona or shell (Figure 5). The concentration at which aggregation occurs is commonly referred to as the critical micelle concentration (CMC). Traditionally, low molecular weight surfactants (i.e., polysorbates, sodium dodecyl sulfate, etc.) with relatively high CMCs in the range of 10^{-3} to 10^{-4} M have been used extensively in pharmaceutical formulations, primarily as excipients to increase the aqueous solubility of poorly watersoluble drugs (Mikhail and Allen 2010). Hydrophobic drugs are contained within and associate with the hydrophobic regions of the micelle. However, following administration, dilution of a given pharmaceutical formulation occurs rapidly, and as the micelle concentration drops below its CMC, its stability is compromised.

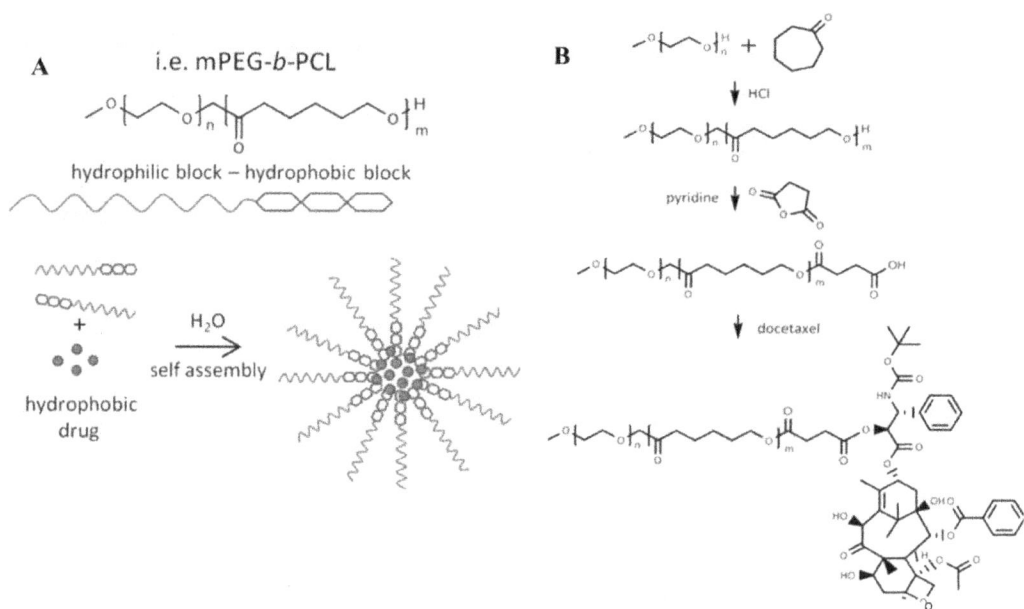

Figure 5: (A) A typical example of polymeric micelle unimer structure composed of both hydrophilic (mPEG) and hydrophobic (PCL) blocks. (B) Synthetic scheme for mPEG-b-PCL-docetaxel micelle unimer. Adapted and reprinted with permission from ref 97, copyright 2010 American Chemical Society.

3.4 Stimuli Responsive Polymers

So called "smart polymers" have been engineered to contain a vast array of properties, including the ability to respond to changes in environmental stimuli such as pH, ionic strength, temperature or externally applied heat, magnetic or electric fields, or ultrasound (Torchilin and Eur 2009). Such polymers commonly respond via conformational and/or electrostatic changes, which can be exploited to help facilitate a particular function (i.e., drug release, endosomal escape, etc.). Carriers that respond to variations in pH and temperature have found the greatest versatility in drug delivery and will be reviewed in brief as follows.

3.4.1 pH-Sensitive Systems

Exploiting physiological variations in pH has been widely investigated as a means to obtain site specific delivery. The pH of diseased areas such as tumors, infarction, and sites of inflammation may drop to around 6.5, almost one full pH unit below that of normal blood (pH 7.4), as a result of hypoxic conditions and extensive cell death (Wike-Hooley et al. 1984, Bae et al. 2003). In addition, following cellular uptake via endocytosis, the pH of late endosomes may reach values as low as 5.0, further providing a gradient over which release may be triggered (Bae et al. 2005).

A number of pH responsive polymeric micelles have also been described, including systems in which doxorubicin was conjugated to the side chains of the micelle core-forming blocks via hydrazone bonds (Yin et al. 2006). The micelles demonstrated both time and pH dependent release, with increased release under endosomal low pH conditions (5.0–5.5) (Yin et al. 2006). Biodistribution studies showed minimal signs of premature drug release, and selective accumulation in tumors and the antitumor efficacy of these pH sensitive micelles was significantly higher than that achieved with comparable doses of free doxorubicin (Kamada et al. 2004).

3.4.2 Temperature-Sensitive Systems

The concept of using temperature to control drug delivery is partially due to the observation that elevated temperature can be associated with diseased tissues. In addition, the external application of hyperthermia can be utilized as a trigger to induce changes in polymer structure resulting in drug release. Water-soluble temperature sensitive polymers such as those based on poly(Nisopropylacrylamide) (poly(NIPAAM)) undergo a lower critical solution temperature (LCST) phase transition, wherein polymer chains collapse and aggregate at temperatures above their LCST, as a result of the reversible dehydration of hydrocarbon side chains (Yin et al. 2006). The LCST for poly(NIPAAM) is approximately 32°C. However, the LCST of such polymers can be adjusted by changing the N-substituted carbon chain or via copolymerization (Kamada et al. 2004).

3.4.3 Enzyme-Sensitive Conjugates

Another powerful method for targeted drug release exploits the enzymatic cleavage of linkers in polymer-drug conjugates (Straley and Heilshorn 2009). In an attempt to increase the rate and maximum extent of side-chain hydrolysis by lysosomal enzymes, Duncan and co-workers developed polymer-drug conjugates using *N*-(2-hydroxypropyl)methacrylamide copolymers and *p*-nitroaniline drug analogues, bearing oligopeptidyl-*p*-nitroanilide side chains, which are specific to certain lysosomal proteinases, yielding a potential delivery system.

4. Polymeric Biomaterials for Controlled Release

All controlled release systems aim to improve the effectiveness of drug therapy (Langer 1998, Brouwers 1996). This improvement can take the form of increasing therapeutic activity compared to the intensity of side effects, reducing the number of drug administrations required during treatment, or eliminating the need for specialized drug administration (e.g., repeated injections). Two types of control over drug release can be achieved-temporal and distribution control. In temporal control, drug delivery systems aim to deliver the drug over an extended duration or at a specific time during treatment. Controlled release over an extended duration is highly beneficial for drugs that are rapidly metabolized and eliminated from the body after administration, the benefit is shown schematically in Figure 6. A diverse range of mechanisms have been developed to achieve both temporal and distribution-controlled release of drugs using polymers. This diversity is a necessary consequence of different drugs imposing various restrictions on the type of delivery system employed. From a

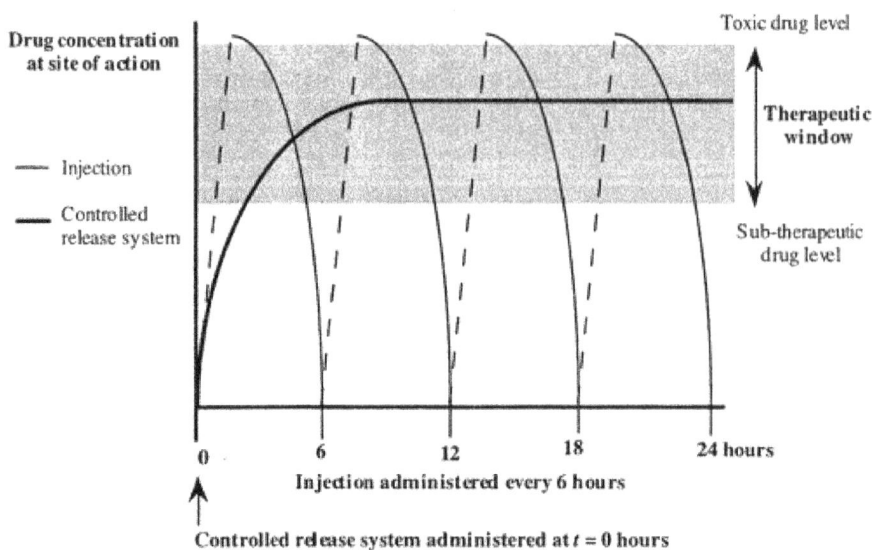

Figure 6: Drug concentrations at site of therapeutic action after delivery as a conventional injection (thin line) and as a temporal controlled release system (bold line). Adapted and reprinted with permission from ref 122, copyright 1999 American Chemical Society.

polymer chemistry perspective, it is important to appreciate that different mechanisms of controlled release require polymers with a variety of physicochemical properties (Figure 6).

4.1. Polymers used for Controlled Drug Release

4.1.1. Poly(esters)

Poly(esters) have been extensively employed in drug delivery applications and comprehensively reviewed (Wong and Mooney 1997, Lanza et al. 1997, Shalaby 1994, Shields et al. 1968). The predominant synthetic pathway for production of poly(esters) is from ring-opening polymerization of the corresponding cyclic lactone monomer. The more prominent poly(esters) and their starting materials are shown in Figure 7.

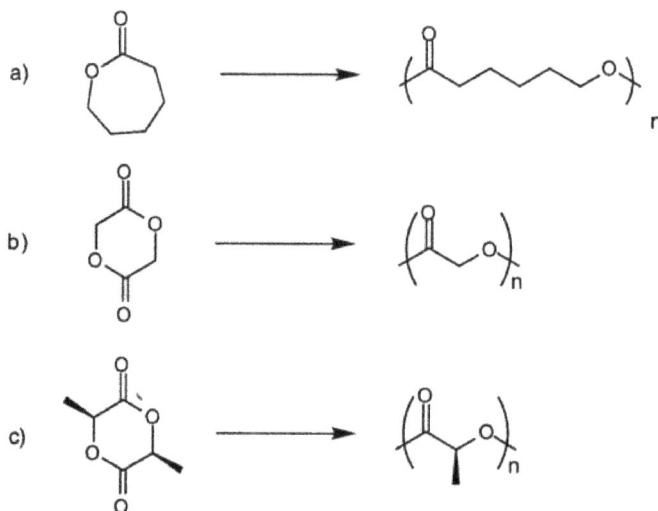

Figure 7: Ring-opening polymerization of selected cyclic lactones to give different lactic acids.

4.1.2 Poly(ethylene glycol) Block Copolymers

Poly(ethylene glycol) (PEG) is also referred to as poly(ethylene oxide) (PEO) at high molecular weights. Biocompatibility is one of the most noted advantages of this material. Typically, PEG with molecular weights of 4000 amu is 98% excreted in man (Andrade et al. 1996). One of the emerging uses for inclusion of PEG in a controlled release system arises from its protein resistivity (Zalipsky 1995). The hydrophilic nature of PEG is such that water hydrogen bonds tightly with the polymer chain and thus excludes, or inhibits, protein adsorption. Many research groups are investigating attachment of PEG chains to therapeutic proteins; PEG chains at the surface allow for longer circulation of the protein in the body by prolonging biological events such as endocytosis, phagocytosis, liver uptake and clearance, and other adsorptive processes (Nucci et al. 1991, Natre 1993, Gaertner and Offord 1996, Delgado 1992, Johnson et al. 1996) (Figure 8).

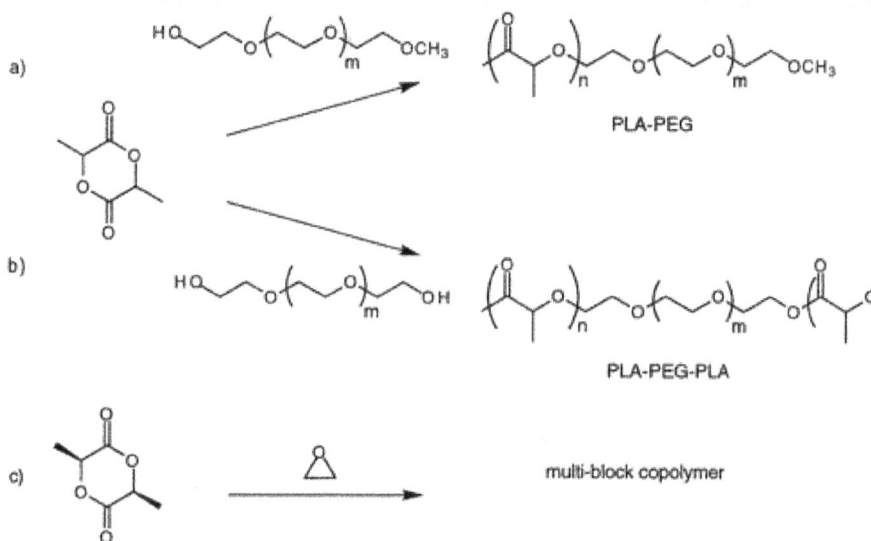

Figure 8: Synthesis of polylactic acid-PEG copolymers.

4.1.3 Poly(iminocarbonates)

Poly(amino acids) are highly insoluble, non-processible, and antigenic when the polymers contain three or more amino acids (Silver et al. 1992). To circumvent these problems, "pseudo"-poly(amino acids) synthesized from tyrosine dipeptide were investigated (Engelberg and Kohn 1999). These de-gradable polymers are derived from the polymerization of desaminotyrosyl tyrosine alkyl esters. The general structure of these polymers is shown in Figure 9. Tyrosine-derived poly(carbonates) are readily processible polymers that support the growth and attachment of cells and have also shown a high degree of tissue compatibility (Daniels et al. 1990, Nate and Hamidreza 2012, Scott and Langer 1999). Tyrosine-derived poly(carbonates) are characterized by their relatively high strength and stiffness exceeding poly(esters) such as poly(ortho esters) but not poly(lactic acid) or poly-(glycolic acid).

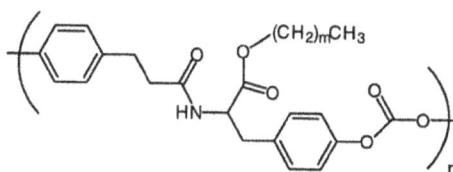

Figure 9: Degradable polymers derived from the polymerization of desaminotyrosyl tyrosine alkyl esters. Adapted and reprinted with permission from ref 122, copyright 1999 American Chemical Society.

5. Conclusion and Future Outlook

Many therapeutics based on nanoparticulate polymer and metal nanoparticle drug carriers are currently under development. These include polymer-drug conjugates, single-crystal nanoparticles, magnetic nanoclusters, micelles, dendritic polymer carriers, polymerosomes, polymer-coated liposomes, magnetoliposomes and niosomes, nanocapsules and nanoemulsions, and sophisticated stimuli-responsive systems, nanorobots, and diagnostic devices. In addition, a growing number of polymer conjugates and other organic or inorganic DDS are entering clinical development and evaluation. New synthetic methods in polymer and materials chemistry and technology have enabled the preparation of well-defined nanostructures that can exploit passive, active, and even magnetic tumor targeting simultaneously. Abilities to manipulate the architecture of polymeric and nanoscale systems and to conjugate these systems to drugs in sophisticated controllable ways (covalent or noncovalent) mean that it is becoming possible to create systems with tunable drug release profiles and preprogrammed responses to specific external or physiological stimuli. With the continued convergence and synergies in tailoring structure, dynamic function, and biological complexity, the prospects are promising for biomaterial systems that allow exquisite structural assembly, enable active interactions, and perform tailored, high-level functions at complex biological interfaces. Such function will enable the tuning of cell-materials' interactions that will be important for expanding our knowledge of the underlying mechanisms that control cell and tissue fate and that result in new and lasting contributions to the biomedical sciences.

References

Ahmed, T.A.E., E.V. Dare and M. Hincke. 2008. Fibrin: A versatile scaffold for tissue engineering applications. Tissue Eng. Part B Rev. 14(2): 199–215.

Akman, A.C., R.S. Tigli, M. Gumusderelioglu and R.M. Nohutcu. 2010. bFGF-loaded HA-chitosan: A promising scaffold for periodontal tissue engineering. J. Biomed. Mater. Res. A 92(3): 953–962.

Allison, D.D. and K.J. Grande-Allen. 2006. Review G-AKJ. Review. Hyaluronan: A powerful tissue engineering tool. Tissue Eng. 12(8): 21310–2140.

Altman, G.H., Frank Diaz, Caroline Jakuba, Tara Calabro, Rebecca L. Horan, Jingsong Chen, Helen Lu, John Richmond and David L. Kaplan. 2003. Silk-based biomaterials. Biomaterials 24(3): 401–416.

Andrade, J.D., V. Hlady and S.I. Jeon. 1996. Poly(ethylene oxide) and protein resistance principles, problems, and possibilities. Adv. Chem. Ser. 248: 51.

Bae, Y., S. Fukushima, A. Harada and K. Kataoka. 2003. Design of environment-sensitive supramolecular assemblies for intracellular drug delivery: Polymeric micelles that are responsive to intracellular pH change. Angew. Chem. Int. Ed. Engl. 42(38): 4640–3.

Bae, Y., N. Nishiyama, S. Fukushima, H. Koyama, M. Yasuhiro and K. Kataoka. 2005. Preparation and biological characterization of polymeric micelle drug carriers with intracellular pH-triggered drug release property: Tumor permeability, controlled subcellular drug distribution, and enhanced *in vivo* antitumor efficacy. Bioconjugate Chem. 16(1): 122–30.

Barbucci, R. 2002. Integrated Biomaterials Science. New York: Kluwer Academic/Plenum Publishers.

Betancur, A.D., G.L. Chel and H.E. Cañizares. 1997. Acetylation and characterization of Canavalia ensiformis starch. J. Agri. Food Chem. 45(2): 378–382.

Bigi, A., G. Cojazzi, S. Panzavolta, N. Roveri and K. Rubini. 2002. Stabilization of gelatin films by crosslinking with genipin. Biomaterials 23(24): 4827–4832.

Brechbiel, M.W. 2009. Growing applications of "click chemistry" for bioconjugation in contemporary biomedical research. Cancer Biother. Radiopharm. 24(3): 289–302.

Brouwers, J.R.B. 1996. Advanced and controlled drug delivery systems in clinical disease management. J. Pharm. World Sci. 18: 153.

Burdick, J.A. and G.D. Prestwich. 2011. Hyaluronic acid hydrogels for biomedical applications. Adv. Mater. 23(12): H41–H56.

Burke, C.J. and T.A. Hsu. 1999. Formulation VDB. stability, and delivery of live attenuated vaccines for human use. Crit. Rev. Ther. Drug Carrier Syst. 16(1): 1–83.

Calvert, P. 1997. The structure of starch. Nature 389: 338.

Cao, T.-T. and Y.-Q. Zhang. 2016. Processing and characterization of silk sericin from Bombyx mori and its application in biomaterials and biomedicines. Mater. Sci. Eng. C 61: 940–952.

Chen, G.Q. 2009. A microbial polyhydroxyalkanoates (PHA) based bio- and materials industry. Chem. Soc. Rev. 38(8): 2434–2446.

Cheng, Y., Z. Xu, M. Ma and T. Xu. 2008. Dendrimers as drug carriers: Applications in different routes of drug administration. J. Pharm. Sci. 97(1): 123–43.

Chipman, S.D., F.B. Oldham, G. Pezzoni and J.W. Singer. 2006. Biological and clinical characterization of paclitaxel poliglumex (PPX, CT-2103), a macromolecular polymer-drug conjugate. Int. J. Nanomed. 1(4): 375–83.

Daniels, A., M. Chang, K. Andriano and J. Heller. 1990. Mechanical properties of biodegradable polymers and composites proposed for internal fixation of bone. J. Appl. Biomater. 1: 57.

de La Puente, P., D. Ludeña, A. Fernández, J.L. Aranda, G. Varela and J. Iglesias. 2011. Autologous fibrin scaffolds cultured dermal fibroblasts and enriched with encapsulated bFGF for tissue engineering. J. Biomed. Mater. Res. A 99A(4): 648–654.

de La Puente, P., D. Ludeña, M. López, J. Ramos and J. Iglesias. 2013. Differentiation within autologous fibrin scaffolds of porcine dermal cells with the mesenchymal stem cell phenotype. Exp. Cell Res. 319(3): 144–152.

Delgado, C. 1992. The uses and properties of PEG-linked proteins. Crit. Rev. Ther. Drug Carrier Syst. 9: 249.

Dragan, E.S. 2014. Design and applications of interpenetrating polymer network hydrogels. A review. Chem. Eng. J. 243: 572–590.

Duncan, R., H.C. Cable, J.B. Lloyd, P. Rejmanová and J. Kopecek. 1983. Makromol Chem. 184(10): 1997–2008.

Dvir, T., B.P. Timko, D.S. Kohane and R. Langer. 2011. Nanotechnological strategies for engineering complex tissues. Nat. Nanotechnol. 6(1): 13–22.

El Maradny, E., N. Kanayama, H. Kobayashi, B. Hossain, S. Khatun, S. Liping, T. Kobayashi and T. Terao. 1997. The role of hyaluronic acid as a mediator and regulator of cervical ripening. Hum. Reprod. 12(5): 1080–1088.

Engelberg, I. and J. Kohn. 1991. Physico-mechanical properties of degradable polymers used in medical applications: A comparative study. Biomaterials 12: 292.

Foox, M. and M. Zilberman. 2015. Drug delivery from gelatin-based systems. Expert. Opin. Drug Deliv. 12(9): 1547–1563.

Freyman, T.M., I.V. Yannas, R. Yokoo and L.J. Gibson. 2001. Fibroblast contraction of a collagen–GAG matrix. Biomaterials 22(21): 2883–2891.

Gaertner, H.F. and R.E. Offord. 1996. Site-specific attachment of functionalized poly(ethylene glycol) to the amino terminus of proteins. Bioconjugate Chem. 7: 38.

Gajbhiye, V., V.K. Palanirajan, R.K. Tekade and N.K.J. Jain. 2009. Pharm. Dendrimers as therapeutic agents: A systematic review. Pharmacol. 61(8): 989–1003.

Gelse, K., E. Pöschl and T. Aigner. 2003. Structure C-. function, and biosynthesis. Advanced Drug Delivery Reviews 55(12): 1531–1546.

Gobin, A.S., C.E. Butler and A.B. Mathur. 2006. Repair and regeneration of the abdominal wall musculofascial defect using silk fibroin-chitosan blend. Tissue Eng. 12(12): 3383–3394.

Goldberg, D.S., H. Ghandehari and P.W. Swaan. 2010. Cellular entry of G3.5 poly (amido amine) dendrimers by clathrin and dynamin-dependent endocytosis promotes tight junctional opening in intestinal epithelia. Pharm. Res. 27(8): 1547–57.

Gollwitzer, H., K. Ibrahim, H. Meyer, W. Mittelmeier, R. Busch and A. Stemberger. 2003. Antibacterial poly(D,L-lactic acid) coating of medical implants using a biodegradable drug delivery technology. J. Antimicrob. Chemother. 51(3): 585–591.

Gou, Y., D. Miao, M. Zhou, L. Wang, H. Zhou and G. Su. 2018. Bio-inspired protein-based nanoformulations for cancer theranostics. Front. Pharmacol. 9: 421.

Hillyard, I.W., J. Doczi and P.B. Kiernan. 1964. Antacid and antiulcer properties of the polysaccharide chitosan in the rat. Proc. Soc. Exp. Biol. Med. 115: 1108–1112.

Hofer, M., G. Winter and J. Myschik. 2015. Recombinant spider silk particles for controlled delivery of protein drugs. Biomaterials 33(5): 1554–1562.

Huang, Y., Y.J. Wang, Y. Wang, Sijia Yi, Zhen Fan, Leming Sun, Derrick Lin, Nagaraju Anreddy, Hua Zhu, Michael Schmidt, Zhe-Sheng Chen and Mingjun Zhang. 2015. Exploring naturally occurring ivy nanoparticles as an alternative biomaterial. Acta Biomater. 25: 268–283.

Jao, D., X. Mou and X. Hu. 2016. Tissue regeneration: A silk road. J. Funct. Biomaterials 7(3): 22.

Johnson, O.L., J.L. Cleland, H.J. Lee, M. Charnis, E. Duenas, W. Jaworowicz, D. Shepard, A. Shahzamani, A.J.S. Jones and S.D. Putney. 1996. A month-long effect from a single injection of microencapsulated human growth hormone. Nat. Med. 2: 795.

Joralemon, M.J., S. McRae and T. Emrick. 2010. PEGylated polymers for medicine: From conjugation to self-assembled systems. Chem. Commun. (Cambridge, U.K.) 46(9): 1377–93.

Kamada, H., Y. Tsutsumi, Y. Yamamoto, T. Kihira, Y. Kaneda, Y. Mu, H. Kodaira, S.I. Tsunoda, S. Nakagawa and T. Mayumi. 2000. Antitumor activity of tumor necrosis factor-alpha conjugated with polyvinylpyrrolidone on solid tumors in mice. Cancer Res. 60(22): 6416–20.

Kamada, H., Y. Tsutsumi, Y. Yoshioka, Y. Yamamoto, H. Kodaira, S. Tsunoda, T. Okamoto, Y. Mukai, H. Shibata, S. Nakagawa and Tadanori Mayumi. 2004. Design of a pH-sensitive polymeric carrier for drug release and its application in cancer therapy. Clin. Cancer Res. 10: 2545–2550.

Kaneda, Y., Y. Tsutsumi, Y. Yoshioka, H. Kamada, Y. Yamamoto, H. Kodaira, T. Tsunoda Okamoto, Y. Mukai, H. Shibata, S. Nakagawa and T. Mayumi. 2004. The use of PVP as a polymeric carrier to improve the plasma half-life of drugs. Biomaterials 25(16): 3259–66.

Karim, A.A. and R. Bhat. 2009. Fish gelatin: Properties, challenges, and prospects as an alternative to mammalian gelatins. Food Hydrocoll. 23(3): 563–576.

Kasuya, Y., Z.R. Lu, P. Kopeckova, T. Minko, S.E. Tabibi and J. Kopecek. 2011. Synthesis and characterization of HPMA copolymer-aminopropylgeldanamycin conjugates. J. Controlled Release 74(1-3): 203–11.

Katre, N. 1993. Cojugation of proteins with polyethylene glycol and other polymers altering properties of proteins to enhance their therapeutic potential. Adv. Drug Delivery Rev. 10: 91.

Kaur, L., J. Singh and Q. Liu. 2007. Starch a potential biomaterial for biomedical applications. pp. 83–98. In: Mozafari, M.R. (ed.). Nanomaterials and Nanosystems for Biomedical ApplicationsDordrecht: Springer.

Kjaergard, H.K., J.L. Velada, J.H. Pedersen, H. Fleron and D.A. Hollingsbee. 2000. Comparative kinetics of polymerisation of three fibrin sealants and influence on timing of tissue adhesion. Thromb. Res. 98(2): 221–228.

Kohane, D.S. and R. Langer. 2008. Polymeric biomaterials in tissue engineering. Pediatr. Res. 63(5): 487–491.

Kopecek, J. and P. Kopeckova. 2010. HPMA copolymers: Origins, early developments, present, and future. Adv. Drug Delivery Rev. 62(2): 122–49.

Kozen, B.G., S.J. Kircher, J. Henao, F.S. Godinez and A.S. Johnson. 2008. An alternative hemostatic dressing: Comparison of CELOX, HemCon, and QuikClot. Acad. Emerg. Med. 15(1): 74–81.

Kumar, M.N.V.R., R.A.A. Muzzarelli, H. Muzzarelli Sashiwa and A.J. Domb. 2004. Chitosan chemistry and pharmaceutical perspectives. Chem. Rev. 104(12): 6017–6084.

Kumar, P.T.S., S. Abhilash, K. Manzoor, S.V. Nair, H. Tamura and R. Jayakumar. 2010. Preparation and characterization of novel-chitin/nanosilver composite scaffolds for wound dressing applications. Carbohydr. Polym. 80(3): 761–767.

Kundu, B., R. Rajkhowa, S.C. Kundu and X. Wang. 2013. Silk fibroin biomaterials for tissue regenerations. Adv. Drug Deliv. Rev. 65(4): 457–470.

Langer, R. and J.P. Vacanti. 1993. Tissue engineering. Science (New York, N. Y.) 260(5110): 920–926.

Langer, R. 1998. Drug delivery and targeting. Nature 392: 5.

Lanza, R.P., R. Langer and W.L. Chick. 1997. Principles of Tissue Engineering; R G Landes Co. and Academic Press: Austin, TX.

Lee, K.Y. and D.J. Mooney. 2012. Alginate: Properties and biomedical applications. Progress Polymer Science 37(1): 106–126.

Lenaghan, S.C., K. Serpersu, L. Xia, W. He and M. Zhang. 2011. A naturally occur ring nanomaterial from the Sundew (Drosera) for tissue engineering. Bioinspir. Biomim. 6(4): 046009.

Li, Z. and X.J. Loh. 2017. Recent advances of using polyhydroxyalkanoate-based nanovehicles as therapeutic delivery carriers. Wiley Interdiscip. Rev. Nanomed Nanobiotechnol. 9(3): e1429–n/a:e1429.

Linnes, M.P., B.D. Ratner and C.M. Giachelli. 2007. A fibrinogen-based precision microporous scaffold for tissue engineering. Biomaterials 28(35): 5298–5306.

Ljubimova, J.Y., M. Fujita, A.V. Ljubimov, V.P. Torchilin, K.L. Black and E. Holler. 2008. Covalent nanodelivery systems for selective imaging and treatment of brain tumors. Nanomedicine (London, U.K.) 3(2): 247–65.

Madhumathi, K., P.T. Sudheesh Kumar, S. Abhilash, V. Sreeja, H. Tamura, K. Manzoor, S.V. Nair and R. Jayakumar. 2010. Development of novel chitin/nanosilver composite scaffolds for wound dressing applications. J. Mater. Sci. Mater. Med. 21(2): 807–813.

Marijnissen, W.J.C.M., G.J.V.M. van Osch, J. Aigne, Simone W. van der Veen, Anthony P. Hollander, Henriëtte L. Verwoerd-Verhoef and Jan A.N. Verhaar. 2002. Alginate as a chondrocyte-delivery substance in combination with a non-woven scaffold for cartilage tissue engineering. Biomaterials 23(6): 1511–1517.

Mcgill, V., A. Kowal-Vern, M. Lee, D. Greenhalgh, E. Gomperts, G. Bray and R. Gamelli 1997. Use of fibrin sealant in thermal injury. J. Burn Care Rehabil. 18(5): 429–434.

Meyer, I.K. and J.W. Palmer. 1934. The polysaccharides of the vitreous humor. J. Biol. Chem. 107: 629–634.

Mikhail, A.S. and C. Allen. 2010. Poly(ethylene glycol)-b-poly(ε-caprolactone) micelles containing chemically conjugated and physically entrapped docetaxel: Synthesis, characterization, and the influence of the drug on micelle morphology. Biomacromolecules 11(5): 1273–80.

Mori, M., M. Yamaguchi, S. Sumitomo and Y. Takai. 2004. Hyaluronan-based biomaterials in tissue engineering. Acta Histochem. Cytochem. 37(1): 1–5.

Mottaghitalab, F., M. Farokhi, M.A. Shokrgozar, F. Atyabi and H. Hosseinkhani. 2015. Silk fibroin nanoparticle as a novel drug delivery system. J. Control. Release 206: 161–176.

Nate, L. and G. Hamidreza. 2012. Polymeric conjugates for drug delivery. Chem. Mater. 24: 840.853.

Niiyama, H. and Y. Kuroyanagi. 2014. Development of novel wound dressing composed of hyaluronic acid and collagen sponge containing epidermal growth factor and vitamin C derivative. J. Artif. Organs. 17(1): 81–87.

Nucci, M.L., R. Shorr and A. Abuchowski. 1991. The therapeutic value of poly(ethylene glycol)-modified-proteins. Adv. Drug Del. Rev. 6: 133.

Park, H., B. Choi, J. Hu and M. Lee. 2013. Injectable chitosan hyaluronic acid hydrogels for cartilage tissue engineering. Acta Biomater. 9(1): 4779–4786.

Pasut, G. and F.M. Veronese. 2009. PEGylation for improving the effectiveness of therapeutic biomolecules. Drugs Today (Barc.) 45(9): 687–95.

Pasut, G. and F.M. Veronese. 2009. PEG conjugates in clinical development or use as anticancer agents: An overview. Adv. Drug Delivery Rev. 61(13): 1177–88.

Pasut, G., F. Greco, A. Mero, R. Mendichi, C. Fante, R.J. Green and F.M. Veronese. 2009. Polymer-drug conjugates for combination anticancer therapy: Investigating the mechanism of action. J. Med. Chem. 52(20): 6499–502.

Pasut, G., M. Sergi and F.N. Veronese. 2008. Anti-cancer PEG-enzymes: 30 years old, but still a current approach. Adv. Drug Deliv. Rev. 60: 69–78.

Pawar, S.N. and K.J. Edgar. 2012. Alginate derivatization: A review of chemistry, properties and applications. Biomaterial 33(11): 3279–3305.

Piskin, E. 1995. Biodegradable polymers as biomaterials. J. Biomater. Sci. Polym. Ed. 6(9): 775–795.

Prestwich, G.D. 2008. Engineering a clinically useful matrix for cell therapy. Organogenesis 4(1): 42–47.

Prestwich, G.D. 2011. Hyaluronic acid-based clinical biomaterials derived for cell and molecule delivery in regenerative medicine. J. Control. Release 155(2): 193–199.

Pritchard, E.M., T. Valentin, B. Panilaitis, F. Omenetto, D.L. Kaplan, M. Pritchard Eleanor and L. Kaplan David. 2013. Antibiotic-releasing silk biomaterials for infection prevention and treatment. Adv. Funct. Mater. 23(7): 854–861.

Putnam, D. Kopecek. 1995. Enantioselective release of 5-fluorouracil from N-(2- hydroxypropyl)methacrylamide-based copolymers via lysosomal enzymes. J. Bioconjugate Chem. 6(4): 483–492.

Putnam, D. and J. Kopeček. 1995. Polymer conjugates with anticancer activity. In Biopolymers II, Springer: Berlin 122: 55–123.

Rajangam, T. and A. Ssa. 2013. Fibrinogen and fibrin based micro and nanoscaffolds incorporated with drugs, proteins, cells and genes for therapeutic biomedical applications. Int. J. Nanomedicine 8: 3641–3662.

Raveendran, S., A. Rochani, T. Maekawa and D. Kumar. 2017. Smart carriers and nanohealers: A nanomedical insight on natural polymers. Materials 10(8): 929.

Ringsdorf, H.J. 1975. Structure and properties of pharmacologically active polymers. Polym. Sci. Polym. Symp. 51(1): 135–153.

Saboktakin, M.R., R.M. Tabatabaie, A. Maharramov and M.A. Ramazanov. 2011. Synthesis and *in vitro* evaluation of carboxymethyl starch–chitosan nanoparticles as drug delivery system to the colon. Int. J. Biol. Macromolecule 48(3): 381–385.

Saddler, J.M. and P.J. Horsey. 1987. The new generation gelatins. A review of their history, manufacture and properties. Anaesthesia 42(9): 998–1004.

Saito, Y. and Y. Doi. 1994. Microbial synthesis and properties of poly(3-hydroxybutyrate-co-4-hydroxybutyrate) in Comamonas acidovorans. Int. J. Biol. Macromol. 16(2): 99–104.

Scott, M.C. and S.L. Robert. 1999. Polymeric systems for controlled drug release. Chem. Rev. 99: 3181–3198.

Seib, F.P., G.T. Jones, J. Rnjak-Kovacina, Y. Lin, D.L. Kaplan, T. Jones Gregory and L. Kaplan David. 2013. pH-dependent anticancer drug release from silk nanoparticles. Adv. Health Mater. 2(12): 1606–1611.

Shalaby, S.W. 1994. In Biomedical Polymers: Designed-to-Degrade Systems; Shalaby, S.W. (ed.). Hanser/Gardner: Cincinnati, OH.

Shields, R., J. Harris and M. Davis. 1968. Suitability of polyethylene glycol as a dilution indicator in the human colon. Gastroenterology 54: 331–333.

Silver, F., M. Marks, Y. Kato, C. Li, S. Pulapura and J. Kohn. 1992. Tissue compatibility of tyrosine-derived polycarbonates and polyaminocarbonates and initial evaluation. J. Long-Term Effects Med. Implants. 1: 329.

Straley, K.S. and S.C. Heilshorn. 2009. Dynamic, 3d-pattern formation within enzyme-responsive hydrogels. Adv. Mater. 21(41): 4148–4152.

Strickley, R.G. 2004. Solubilizing excipients in oral and injectable formulations. Pharm. Res. 21(2): 201–30.

Subr, V., J. Kopecek, J. Pohl, M. Baudys and V. Kostka. 1988. Cleavage of oligopeptide side-chains in N-2(hydroxpropyl) meth-acrylamide copolymers by mixtures of lysosomal enzymes. J. Controlled Release 8(2): 133–140.

Sugano, M., S. Watanabe, A. Kishi, M. Izume and A. Ohtakara. 1988. Hypocholes-terolemic action of chitosans with different viscosity in rats. Lipids 23(3): 187–191.

Taylor, P. 2015. Global Markets for Implantable Biomaterials [market report on the Internet]. Massachusetts: BCC Research. 2015 Jan [cited March 1, 2016]. Available from: http://www.bccresearch.com/market-research/advanced-materials/implantable-biomaterials-markets-report-avm118a. html . Accessed March 2019.

Tomalia, D.A., H. Baker, J. Dewald, M. Hall, G. Kallos, S. Martin, J. Roeck, J. Ryder and P. Smith. 1985. A new class of polymers: Starburst-dendritic macromolecules. Polym. J. 17(1): 117–132.

Tomalia, D.A., H. Baker, J. Dewald, M. Hall, G. Kallos, S. Martin, J. Roeck, J. Ryder and P. Smith. 1986. Dendritic macromolecules: Synthesis of starburst dendrimers. Macromolecules 19(9): 2466–2468.

Toole, B.P. 2004. Hyaluronan: From extracellular glue to pericellular cue. Nat. Rev. Cancer 4(7): 528–539.

Torchilin, V. and J. Eur. 2009. Micellar nanocarriers: Pharmaceutical perspectives. Pharm. Biopharm. 71(3): 431–44.

Tredwell, S.J. and B. Sawatzky. 1990. The use of fibrin sealant to reduce blood loss during cotrel-dubousset instrumentation for idiopathic scoliosis. Spine 15(9): 913–915.

Ueno, H., T. Mori and T. Fujinaga. 2011. Topical formulations and wound healing applications of chitosan. Advanced Drug Delivery Reviews 52(2): 105–115.

Vacanti, J.P. and R. Langer. 1999. Tissue engineering: The design and fabrication of living replacement devices for surgical reconstruction and transplantation. Lancet 354: S32–S34.

van Dijk, M., D.T. Rijkers, R.M. Liskamp, C.F. van Nostrum and W.E. Hennink. 2009. Synthesis and applications of biomedical and pharmaceutical polymers via click chemistry methodologies. Bioconjugate Chem. 20(11): 2001–16.

Veronese, F.M. and J.M. Harris. 2002. Introduction and overview of peptide and protein pegylation. Adv. Drug Delivery Rev. 54(4): 453–606.

Veronese Francesco M., Paolo Caliceti, Oddone Schiavon and Mauro Sergi. 2002. Polyethylene glycol–superoxide dismutase, a conjugate in search of exploitation. Advanced Drug Delivery Reviews 54: 587–606.

Vert, M. 2005. Aliphatic polyesters: Great degradable polymers that cannot do everything. Biomacromolecules 6(2): 538–546.

Vicent, M.J., S. Manzanaro, J.A. de la Fuente and R. Duncan. 2004. Investigating the mechanism of enhanced cytotoxicity of HPMA copolymer-Dox-AGM in breast cancer cells. J. Drug Targeting 12(8): 503–15.

Vmd, O.C., B. Stringhetti Ferreira Cury, R.C. Evangelista and M.P. Daflon Gremião. 2017. Development and characterization of cross-linked gellan gum and retrograded starch blend hydrogels for drug delivery applications. J. Mech. Behav. Biomed. Mater. 65: 317–333.

Wenk, E., H.P. Merkle and L. Meinel. 2011, Silk fibroin as a vehicle for drug delivery applications. J. Control. Release 150(2): 128–141.

Wike-Hooley, J.L., J. Haveman and H.S. Reinhold. 1984. The relevance of tumour pH to the treatment of malignant disease Radiother. Oncol. (4): 343–66.

Wong, W.H. and D.J. Mooney. 1997. In Synthetic Biodegradable Polymer Scaffolds. Atala Mooney (ed.). Birkhauser: Boston, MA.

Wu, X., L. Black, G. Santacana-Lafftte and C.W. Patrick. 2007. Preparation and assessment of glutaraldehyde-crosslinked collagen–chitosan hydrogels for adipose tissue engineering. J. Biomed. Mater. Res. A 81A(1): 59–65.

Xu, Q., C.H. Wang and D.W. Pack. 2010. Polymeric carriers for gene delivery: Chitosan and poly(amidoamine) dendrimers. Curr. Pharm. Des. 16(21): 2350–68.

Yasukawa, T., H. Kimura, Y. Tabata, H. Miyamoto, Y. Honda, Y. Ikada and Y. Ogura. 1999. Targeted delivery of anti-angiogenic agent TNP-470 using water-soluble polymer in the treatment of choroidal neovascularization. Invest. Ophthalmol. Visual Sci. 40(11): 2690–6.

Yin, X., A.S. Hoffman and P.S. Stayton. 2006. Poly(N-isopropylacrylamide-co-propylacrylic acid) copolymers that respond sharply to temperature and pH. Biomacromolecules 7(5): 1381–5.

Yu, A., H. Niiyama, S. Kondo, A. Yamamoto, R. Suzuki and Y. Kuroyanagi. 2013. Wound dressing composed of hyaluronic acid and collagen containing EGF or bFGF: Comparative culture study. J. Biomater. Sci. Polym. Ed. 24(8): 1015–1026.

Zalipsky, S. 1995. Functionalized poly(ethylene glycols) for preparation of biologically relevant conjugates. Bioconjugate Chem. 6: 150.

10

Therapeutic Applications of Nanomaterials

Sreejan Manna[1] and *Sougata Jana*[2,3,]*

1. Introduction

In recent times, nanotechnology based drug delivery systems have emerged as one of the flourishing field in the delivery of therapeutics. Possessing a huge advantage of size and surface charge moderation, the nanoparticles offer spectrum of opportunity in disease management. In spite of being extensively investigated in pharmaceutical research in recent past, nanotechnology based delivery systems are still regarded to be in the initial phase of development (Sandhiya et al. 2009). Nanomaterials play a major role in delivering a wide range of therapeutics with excellent tissue targeting ability. Nanoparticles are extensively used as useful tool to resolve the solubility problem of drugs having poor solubility. Having the advantage of smaller particle size, nanoparticles offer prolonged circulation time *in vivo*. Both phagocytosis and pinocytosis method take part in the uptake of nanoparticles into cellular targets (Jiacheng and Martina 2018). Depending on the targeting requirements, distinctive surface modification can be done to localize nanoparticles at target tissue. An increased surface area is another attracting feature of nanomaterials which facilitate biotargeting to a specific receptor for enhancing therapeutic efficacy. Along with it, a quick initial release with excellent absorption behaviour is one of its highlighted characteristics (Wang and Thanou 2010, Moghimi et al. 2005).

With remarkable research work on different nanoparticulate system, polymeric nanomaterials remain the most investigated for delivering therapeutic agents. Higher percentage of drug encapsulation, better *in vivo* targeting, controllable drug release from matrix and prolonged storage ability makes polymeric nanoparticles an attractive delivery system for therapeutics (Alexis et al. 2008, Soppimath et al. 2001, Jean-Christophe et al. 1996). The selection of polymers and other components in the preparation of polymeric nanoparticles depends on the pharmacokinetic properties of drug, site of absorption, specification of target tissue, physicochemical characteristics of the polymer and duration of the therapy (D'Mello et al. 2009).

---

[1] Department of Pharmaceutical Technology, Brainware University, Barasat, Kolkata, West Bengal 700125, India.
[2] Department of Pharmaceutics, Gupta College of Technological Sciences, Ashram More, G.T. Road, Asansol-713301, West Bengal, India.
[3] Department of Health and Family Welfare, Directorate of Health Services, Kolkata, India.
* Corresponding author: janapharmacy@rediffmail.com

Polymeric nanocarriers can deliver potential drugs to specific targets due to the unique properties of polymers. The application of different polymers in the preparation of nanostructured substances has revolutionized the therapeutic approach of nanomaterials. Initially, polymeric nanoparticles were mostly made of non-biodegradable polymers, namely polyacrylamide, polyacrylates and poly(methyl methacrylate) (Vijayan et al. 2013, Shastri 2003). The non-biodegradable characteristic of the polymers has enforced a quick clearance assuring minimal tissue distribution and accumulation. Having the advantages of reduced toxicity and better biocompatibility, biodegradable polymers have become one of the focused areas in nanoparticulate research. Many synthetic polymers such as poly(lactide-*co*-glycolide), poly(lactide) and different polysaccharides are extensively used in the development of nanomaterials (Zhang et al. 2013). Polysaccharide based nanoparticles are one of the most investigated nanoparticles which has gained a lot of attention among scientists.

Alginate is one such polysaccharide which has shown promising characteristics for fabrication of nanoparticles (Jana et al. 2016). The mucoadhesive property along with the network forming ability has made it an appealing candidate for the preparation of nanomaterials (Hamidi et al. 2008). It also possesses the ability to form gelatinous layer under certain conditions which acts as a barrier for drug diffusion, facilitating the controlled release of drug at desired site (Douglas et al. 2006). Another cationic polysaccharide based biopolymer, chitosan which is obtained from deacetylation of chitin, exhibits excellent biocompatibility (Jana et al. 2015a) . Due to its non-toxicity, mucoadhesive and antimicrobial property, chitosan is favoured by many scientists for developing nanomaterials (Jana et al. 2013, Elgadir et al. 2015, Hamedi et al. 2018). Biodegradable polyester such as polylactic acid and poly(lactic-co-glycolic acid) acid are extensively used in nanoparticulate drug carriers due to their adaptability, flexibility and other physicochemical properties. PLA and PLGA offer excellent tumour targeting ability with required modification of surface characteristics (Soppimath et al. 2001, Danhier et al. 2012, Byung et al. 2016).

2. Therapeutic Applications of Nanomaterials

2.1 Delivery of Antihypertensive Therapeutics

A polyionic nanoparticulate novel hybrid system was developed based on chitosan and sodium alginate for delivering antihypertensive drugs. Amlodipine, captopril and valsartan are the three hydrophobic drugs which were encapsulated individually with encapsulation efficiency varying between $42 \pm 0.9\%$ to $96 \pm 1.9\%$. The nano-scale polyionic structure between carboxylate and amino group of alginate and chitosan was confirmed by FTIR study. Zeta potential study revealed a stable nature in suspension form with positive zeta potential value. *In vitro* drug release showed fewer than 8% drug release in physiological buffer solution in initial 24 hr. This drug delivery system can reduce the dosing frequency and thus improve the antihypertensive therapy for cardiovascular patient (Taskeen et al. 2016).

Ramipril loaded chitosan based hollow nanospheres were developed using single emulsification technique. Poly-D, L-lactide-co-glycolide was used as template developed by solvent evaporation technique. The drug entrapment in chitosan matrix was facilitated by interaction between the carboxylic group present in ramipril and amino group present in chitosan. A change in zeta potential and size indicated chitosan layer adsorption on PLGA template. A higher encapsulation of ramipril was observed when ramipril concentration was maintained 5 mg/mL in polymeric matrix. The release characteristics of ramipril were found to be influenced by pH of the buffer solution. 86% and 73% of drug release was reported in pH 3.3 acetate buffer and pH 6.3 phosphate buffers, respectively, while in pH 8.0 Tris buffer only 48% drug was released. The drug release was found to follow Korsemeyer-Peppas model indicating a swelling controlled release from the biodegradable hollow microspheres (Tanushree et al. 2018).

A potent calcium channel blocker, isradipine, was successfully entrapped by nanoprecipitation technique using polymers like poly(d,l-lactide), poly(epsilon-caprolactone) and poly(d,l-lactide-co-glycolide). The average particle diameter was reported between 110 nm to 208 nm. The zeta potential was found to be about –25 mV, which imparts good stability. Isradipine was found in amorphous form in formulation, indicating a molecular dispersion in polymer matrix. Study findings indicated that the nanosphere formulation is effective in reduction of initially obtained hypotensive peak and to prolong the effect of the drug (Martine et al. 1998).

A widely used calcium channel blocker felodipine was incorporated in nanoparticulate system by emulsion solvent evaporation method using poly-D, L-lactic-co-glycolic acid. Differential scanning calorimetry study confirmed absence of intermolecular interaction between felodipine and PLGA. The *in vitro* release of felodipine from PLGA nanoparticles was found to decrease initial burst release and prolong up to 96 hr. Toxicity study in albino mice ensured the oral and cellular safety of developed felodipine loaded PLGA nanoparticles (Utpal et al. 2014).

2.2 Delivery of Anticancer Therapeutics

A widely used antineoplastic drug 5-Flurouracil was successfully developed into hierarchical nanoflower formulation comprising a polymeric core of cationic-β-cyclodextrin with petals of chitosan and alginate. Ionic gelation method was employed to develop the nanoflower formulation. β-cyclodextrin, choline chloride and epichlorohydrin was used to prepare polymeric core of cationic-β-cyclodextrin via a single step polycondensation technique. The average size of the prepared nanoflowers was found to be 300 nm with a positive zeta potential of 9.90 mV. An initial burst release followed by controlled release of drug up to 24 hr was observed. The pH dependent, non-toxic nanoflower formulation of Cat-β- CD/Alg-Chi was found to suppress the growth of L929 cells depending on concentration (Jaya et al. 2017).

An anticancer drug doxorubicin was incorporated with galactosylated chitosan on graphene oxide and was investigated for anticancer therapy. Initially, graphene oxide-doxorubicine complex was formed, which was then interacted with galactosylated chitosan to obtain GC-GO-DOX nanoparticles. The highest drug loading was found to be 1.08 mg/mg of polymer. The system was found to be a pH responsive system as the drug release was targeted at an environment of low pH. The formulations were assessed for cellular uptake and cell proliferation study and the results revealed a higher cytotoxicity towards HepG2 and SMMC-7721 cells compared to chitosan/graphene oxide/doxorubicin formulations. The fluorescence intensity and tumour inhibiting ability was also found better for GC-GO-DOX than CS-GO-DOX nanoparticles' formulation (Chen et al. 2018).

A novel hydrogel system of stimuli-responsive nanocomposite was synthesized by N-isopropyl acrylamide and acrylic acid copolymerization onto chemically modified chitosan with the help of Fe_3O_4 nanoparticles. Co-precipitation method was employed to prepare magnetic nanoparticles of Fe_3O_4. The morphology and structure of the prepared nanocomposites was characterized by FTIR, XRD, SEM and TEM. The temperature and pH sensitive magnetic nanocomposite was employed as an eminent carrier, facilitating controlled release of an anticancer drug doxorubicin. The study findings revealed the highest drug loading of 89%, facilitating a temperature and pH dependant drug release. A maximum of 82% of drug release was reported within 48 hr (Soleyman et al. 2019).

Bevacizumab-a monoclonal antibody—was successfully encapsulated in polymeric nanocarrier system composed of chitosan grafted poly(ethylene glycol) methacrylate. A double crosslinking method including ionic and covalent method in reverse emulsion provides mechanical stability to the polymeric nanocarriers. The SEM images revealed nanoparticles of different diameters depending on the stirring speed and polymer ratio. The prepared nanoparticles were examined for their drug release profile and the results showed promising characteristics for the posterior therapy treatment of eye. The *in vitro* study showed a controlled release of bevacizumab for several days with no signs of cytotoxicity (Corina-Lenuța et al. 2019).

Quercetin loaded chitosan-quinoline nanoparticles were developed by O/W nanoemulsification method using 2-chloro-3-formylquinoline and 3-formylquinolin-2(1H)-one. The morphological study showed a nanorod shaped nanoparticles with a size range between 141–174.8 nm. The encapsulation efficiency of quercetin in the nanoparticle matrix was found between 65.8–77%. The *in vitro* drug release study displayed a burst release at low pH environment due to the presence of cleavage in imine linkage between chitosan-quinoline derivatives. The morphology study under acidic environment showed that the nanorod shape of nanoparticles has been changed into tiny nanocube shaped crystals leading to faster drug release. *In vitro* cytotoxicity was performed against HeLa cells and the study report indicated a better cytotoxic profile in comparison to free quercetin (Shahnaz et al. 2019).

A chitosan based nanomicelle was synthesized for delivering an anticancer drug methotrexate. Chitosan was initially modified by 4-cyano, 4-[(phenylcarbothioyl) sulfanyl] pentanoic acid to develop a microinitiator. Then, in presence of the microinitiator, reversible addition-fragmentation chain transfer polymerization occurs to form nanomicelles. Fluorescein dye entrapment into the nanomicelle core helped to trace cellular uptake in cancer cells present in breast. The conjugated methotrexate also acts as a targeting ligand by attaching to the surface of the nanomicelles. The flow cytometry and imaging technology confirmed the cellular uptake of methotrexate loaded nanomicelles (Farideh et al. 2019).

Sudipta et al. (2019) have reported a safe and effective drug carrier for targeting mitochondria. Glycol chitosan and dequalinium are combined to form an amphiphilic polymer self assembled into nanocomposite holding the drug dequalinium on its surface. The prepared nanoparticles were found nontoxic in comparison to positive control assessed in carcinogenic and non-carcinogenic cells. Flow cytometry and confocal microscopy method confirmed the cellular uptake and mitochondria targeting. Curcumin was successfully encapsulated after promising *in vitro* test results (Sudipta et al. 2019).

A chemotherapeutic drug docetaxel was successfully encapsulated in a chitosan based nanocarrier system. A targeting ligand GX1 is incorporated for vascular targeting of nanoparticles. The average size of the obtained nanoparticles is found to be 150.9 nm with smooth surface and spherical shape. A sustained drug release was observed *in vitro* with an accelerated release in acidic medium. The cytotoxic study showed a better cytotoxicity against cancerous cells of gastric and human umbilical vein endothelial cells. The confocal microscopy study confirmed an enhanced cellular uptake. The *in vivo* tumour inhibiting ability was tested for tumour bearing mice and the tumour inhibition rate was reported to be 67.05% (Enhui et al. 2019).

2.3 *Delivery of Antiviral Therapeutics*

Zidovudine, an antiviral drug, was successfully entrapped into nanoparticulate formulation developed by a covalent conjugation between amino acid and sodium alginate through amidation reaction. Alginate based nanoparticles were developed via chemical crosslinking without a chemical cross linker. The spherical shape nanoparticles were confirmed from microscopic images. The loading efficiency of the nanoparticles was determined by dialysis method and reported as $29.5 \pm 3.2\%$. A sustained release of drug was reported in phosphate buffer solution pH 7.4. Glioma cell lines were used to study cell viability and cellular uptake (Joshy et al. 2018). A novel heparin nanoassemblies were developed to combat majorly HSV type 1, HSV type 2, human papilloma virus and HPV-16 and respiratory syncytial virus. Heparin was hydrophobically modified by using various esterification methods. Crystal like hexagonal shaped nanoassemblies were synthesized by auto-association of cyclodextrin with O-palmitoyl-heparin in aqueous environment. The average hydrodynamic diameters of nanoassemblies were reported between 340 to 659 nm based on the concentration and type of heparin or α-cyclodextrin used. The study findings showed that the concentration of the components did not affect the antiviral activity. The antiviral activity was found to be dependent on the degree of sulfation (David et al. 2014).

Two antiretroviral agents, tenofovir disoproxil fumarate and maraviroc, were used in combination for delivering through a nanolipogel. The nanolipogel system was reported as a robust carrier for drug loading and with modulating drug release ability for two antiretroviral drugs. The drug loaded nanolipogel was evaluated for antiviral efficacy by using HIV-1 BaL in a cell culture system. In murine model, topical application in the vaginal mucosa of the prepared nano formulation showed promising result (Ramanathan et al. 2016).

A lipid based delivery system was developed by Praveen et al. to deliver a highly lipophilic antiretroviral drug efavirenz. Solid lipid nanoparticles were developed by using glyceryl monostearate and tween 80. The evaluation reports showed particles with a mean size of 124.5 ± 3.2 with a negative zeta potential. The highest drug entrapment reported was 86%. The *in vitro* drug release results confirmed a sustained release of drug up to 24 hr. A 5.32 fold higher C_{max} and 10.98 fold higher AUC was observed for the developed nanoparticles when compared to efavirenz suspension (Praveen et al. 2014).

A poorly bioavailable drug lopinavir was successfully entrapped in solid lipid nanoparticle core to target lymphatic vessel of intestine. The mean size of SLNs was reported to be 230 nm with a polydispersity index < 0.27. The zeta potential of the formulated nanoparticles was found negative (–27 mV) which indicates good stability in suspension. The atomic force microscopy and DSC study has confirmed the homogenous distribution and solid nature of drug in SLNs matrix. The intestinal lymphatic transport study report showed a 4.91 fold higher lopinavir secretion in comparison to a pure drug in polymer solution. The AUC for lopinavir loaded SLN was found to be 2.13 fold higher with respect to the drug polymer solution (Alex et al. 2011). Self nano-emulsification method was employed to develop solid lipid nanoparticles. Polyethylene glycol, stearic acid and poloxamer were mixed and heated to initiate spontaneous self nano-emulsification which produced SLNs on subsequent cooling. Ternary phase diagram study was performed to assess the ability of self nano-emulsification. The mean particle size of the optimized formulation was found to be 180.6 ± 2.32 with a polydispersity index of 0.133 ± 0.001. An increase in oral bioavailability was observed for the drug loaded SLNs in comparison to bulk lopinavir due to the increased lymphatic transport of lopinavir (Jeetendra et al. 2013). Poly(d,l-lactic-co-glycolic acid) based biodegradable sustained release didanosine loaded nanoparticles were developed and studied for controlling HIV at early stage. Different evaluation results indicated that speed of homogenizing and drug polymer ratio played significant role in polydispersity index and drug loading for the nanoparticles. Formulations with PEG 400 showed slight increase in drug entrapment with respect to formulation with DMSO. The uptake of nanoparticles was found to be dependent on concentration. The *in vitro* study findings confirmed the transport of didanosine from the nanoparticle core into macrophages indicating a minimized toxicity and side effects (Gurudutta et al. 2012).

Christopher et al. has investigated the comparative *in vivo* pharmacokinetic profile of three anti retroviral drugs intraperitoneally injected with the same antiretroviral drugs nanoparticles. A solvent extraction method was employed to fabricate PLGA based nanoparticles. Efavirenz, lopinavir and ritonavir are the three model drugs used in this investigation. 20 mg per kg of antiretroviral nanoparticles were introduced in selected BALB/c mice. Anti-HIV efficacy was determined by *in vitro* assay. The animals were euthanized at different time intervals both for free drugs and antiretroviral nanoparticles. Different organs along with serum were collected for determination of drug concentrations by HPLC. Free drug in serum showed C_{max} after 4 hr of injection and eliminated by 72 hr. Nanoparticle based antiretroviral drugs showed detectable quantity of efavirenz, lopinavir and ritonavir in different tissues after 28 d of administration *in vivo* (Christopher et al. 2010). Poly(iso-butylcyanoacrylate) nanocapsules were investigated for improving the delivery of a nucleoside reverse transcriptase inhibitor azidothymidine-triphosphate. The challenge of drug encapsulation due to low molecular weight has been overcome by associating azidothymidine-triphosphate with a cationic polymer poly(ethyleneimine). The developed hybrid nanocapsules showed slow *in vitro* drug release. A 30-fold increase in cellular uptake was observed for the prepared nanocapsules with respect to the free drug satisfying the intracellular concentration required for the treatment of HIV affected patients (Hervé et al. 2006).

2.4 Delivery of Antidiabetic Drug

Delivery of an antidiabetic drug liraglutide was investigated by developing nanocapsules coated with chitosan. Calcium alginate beads were prepared and were evaluated with a different microscopic method which showed smooth and spherical particles. The investigation revealed that the bead diameter was dependent on alginate concentration and method of encapsulation. The drug loading was reported as 92.5%. The *in vitro* drug release was performed in simulated gastrointestinal buffer solution in sequential method which showed a sustained release of drug. The release of liraglutide was restricted in acidic pH and a controlled release in higher pH was observed (Fatemeh et al. 2018). Self assembled spherical nanoparticles were synthesized using trimethyl chitosan (TMC) and a sulphated polysaccharide fucoidan (FD). Fucoidan possesses hypoglycaemic activity while trimethyl chitosan exhibits mucoadhesive properties. Insulin was encapsulated into the nanoparticle core. The TMC/FD nanoparticles were found to decrease the barrier properties of Caco-2 cells present as a monolayer in intestinal linings. The paracellular transport across the intestinal membrane was found to increase for insulin. The glucosidase inhibitory effect was also observed for developed nanoparticle formulation and at 2 mg/mL, the ratio of inhibition was reported as 33.2% (Li-Chu et al. 2019).

An investigation was made for oral delivery of insulin with a nano-carrier system consisting chitosan and lecithin. Solvent injection technique was followed to develop insulin loaded lecithin chitosan nanoparticles by preparing insulin-phospholipid complex. The compatibility between drug and polymers was established by FTIR study. The transmission electron microscopy confirmed a multilamellar structure consisting a hollow core surrounded by a number of bilayers. The loading efficiency of insulin was determined by rapid ultrafiltration and RP-HPLC technique and was found to be 4.5%. The results showed a mean size of nanoparticles of 180 nm. The *in vivo* hypoglycaemic activity was studied in streptozotocin-induced rats which showed significant reduction of glucose level in diabetic rats (Liyao et al. 2016).

2.5 Delivery of Protein or Peptide Drugs

Ultrasonication technique was used to synthesize magnetite nanoparticles by co-precipitation followed by coating with chitosan. Polyvinyl alcohol was attached to the surface of the prepared nanoparticles without the application of cross-linking agent. The magnetite nanoparticles were investigated to examine the adsorption of protein drug bovine serum albumin. A mean size of 10.62 nm was confirmed by transmission electron microscopy. FT-IR results have confirmed the coating of the nanoparticles. The chitosan coated nanoparticles were reported to have highest zeta potential of +48.1 mV. The adsorptive nature of the nanoparticles was studied by UV-visible spectroscopy technique. The PVA modified nanoparticles were reported to have minimum protein adsorption (Hamidreza et al. 2015).

Novel polybutylcyanoacrylate nanoparticles were synthesized for delivering functional proteins into neuronal cells. The uptake of the prepared nanoparticles was found to be dependent on endocytosis process via lipoprotein receptor. Three different functional proteins were selected for demonstrating the delivery to the neurons. Toxicity study on cultured neurons displayed marginal toxicity produced by the nanoparticles. A continuous and positive enzyme activity is observed on the treatment of neural cultures with *E. coli* beta-galactosidase. Small GTPase rhoG, when delivered with nanoparticle formulation, was found to induce neuron growth with differentiation in PC12 cell line. Delivery of a monoclonal antibody via nanoparticles was found to interact with alpha synuclein in cultured neuronal cells. PBCA nanoparticles were considered suitable as a potential *in vivo* carrier for therapeutic proteins to treat neuronal disease (Linda et al. 2009). Sang et al. has investigated the delivery of cytochrome C—a therapeutic protein by nanoparticulate carrier system composed of lipid and apolipoprotein for targeting lung tumour cells. The green fluorescent protein and cytochrome C were individually coupled with a membrane permeable sequence peptide before incorporating into nanoparticle formulations.

A xenograft model was followed to design tumour specific protein loaded nanoparticles. The formulated nanoparticles were found within the range of 20–30 nm with a loading efficiency of 64–75%. After intravenous administration of the protein loaded nanoparticles in mice bearing H460 tumour, low accumulation in liver was reported. The nanoparticles were found to deliver the protein into tumour cells to induce apoptosis and retardation of tumour growth *in vivo* (Sang et al. 2012). Magnetic nanoparticles of iron oxide were successfully synthesized from tumour tatgeting. Epichlorohydrin was used to crosslink the hydroxyl group onto the parent nanoparticles. PEGylation and heparin conjugation was performed to develop the nanoparticulate delivery system. Protamine was taken as a model drug for demonstrating the binding ability and loading content. The study result revealed that the loading is as high as 22.9 μg/mg Fe. The pharmacokinetic study displayed a 37.5 fold longer half-life than heparin. A superior targeting ability to the tumour cells was confirmed from the study findings along with enhanced plasma stability on exposure to tumour cells (Jian et al. 2013).

2.6 *Delivery of Antibiotics*

One of the most commonly used antifungal drug amphotericin B is encapsulated in a nanostructured lipid carrier (NLC) system. Nanostructured lipid carriers are suitable carrier system for delivering hydrophobic drugs other than intravenous route. High pressure homozenization technique was employed to develop the NLCs. The prepared NLCs possessed low cytotoxicity along with high pH selectivity. TEM study revealed a well maintained NLC structure even at dehydrated condition. The drug release was found to be dependent on the ratio of alginate and cross-linking agent. The NLCs were reported to maintain their structure after rehydration which indicates a suitable oral delivery system for amphotericin B (Juliana et al. 2018). Rifampicin loaded hybrid nanoparticles were synthesized to develop oral controlled release delivery system by using sodium alginate and cellulose nanocrystals.

Cellulose nanocrystals were extracted from banana fibres. The developed nanoparticles were reported to have a size range below 100 nm. The alginate based nanoparticles showed pH regulated swelling behaviour *in vitro*. Non-toxic behaviour of the nanoparticles was confirmed by MTT assay. 15% of rifampicin release was reported after 2 hr which ensured an effective protection of drug from gastric surroundings (Deepa et al. 2018). The adhesion effect of clindamycin loaded charged nanoparticles of poly (lactic-*co*-glycolic acid)-polyethylenimine was investigated by Hasan et al. Both positively and negatively charged nanoparticles were prepared and the adhesion property of the prepared NPs was studied on methicillin resistant *S. aureous* infected wounds. Sustained drug release over 48 h was observed for the prepared NP formulations. Harmless nature of both the NPs was confirmed when exposed to fibroblast cells. The positively charged nanoparticles showed higher degree of adhesion compared to other group. The study reports showed a better re-epithelialization and wound healing property using mouse model in favour of the positively charged NPs (Hasan et al. 2019).

A widely used antibiotic doxycycline was incorporated in poly(DL-lactic-co-glycolic acid) nanoparticles to prevent the abnormal aortic wall expansion known as abdominal aortic aneurysms.

Matrix metalloproteases are responsible for decreasing elasticity of abdominal aorta. The released doxycycline was found to increase the accumulation and synthesis of the elastic matrix. The positively charged nanoparticles were also found to be decreasing the synthesis and activity of matrix metalloproteases –2. The NPs were also reported to enhance the elastogenic nature by specifically binding to elastin and by inhibiting elastolysis (Balakrishnan and Anand 2013).

2.7 Delivery of Genes

PLGA based nanoparticles were synthesized for targeted delivery of siRNA *in vivo*. Being a biodegradable polymer, PLGA undergo hydrolysis to slowly release SiRNA. The study findings showed eight functions including endosomal escape, cell penetration and tumour targeting. The SiRNA resulted in a sustained knockdown of the targeted PLK1 gene and retarded tumour growth. Three ligands were conjugated with nanoparticle surface which imparted flexibility to the NP formulations. The surface bound PEG was also reported to increase the circulatory time (Jiangbing et al. 2012).

Polymeric nanoparticles were used as a carrier system to deliver plasmid DNA *in vivo* for restoring the function of p53 gene. A sustained intracellular delivery was achieved by using poly(lactic-co-glycolic acid) nanoparticles. A p53 null, PC-3 human prostate cancer cells were subcutaneously injected in mouse for *in vivo* testing. Local and systemic intake of p53 NPs resulted in a reduced tumour growth with improved survival for mice. Imaging technology revealed accumulation of p53 NPs after IV injections. A greater and sustained gene expression was reported for tumours exposed to p53 NPs than p53 DNA only (Sharma et al. 2011).

The antiproliferative efficacy of wild-type p53 gene encapsulated nanoparticles was investigated in breast tumour cell lines. Biodegradable nanoparticles of poly(D,L-lactide-co-glycolide) were developed for sustained and efficacious gene delivery. Multiple emulsion solvent evaporation technique was employed to formulate NPs containing p53 genes. The intracellular trafficking was monitored for understanding the actual mechanism of prolonged gene expression. The study findings indicated a significantly distinguished antiproliferative activity than with wild type p53 gene alone. The intracellular accumulation of p53 NPs was confirmed by fluorescent labeled confocal microscopy indicating a slow release of plasmid DNA (Swayam and Vinod 2004).

Nanoparticle based gene transfection was investigated by plasmid DNA encapsulation by using poly(D,L-lactide-co-glycolide) and polylactide. Polyvinyl alcohol was used as an emulsifier during nanoparticle preparation by multiple emulsion solvent evaporation technique. The gene expression of the DNA loaded nanoparticles was determined by using PC-3 cell lines for prostate cancer and MCF-7 cell lines for breast cancer. The PLGA based NPs showed greater gene expression than PLA based formulation. PLGA, with higher molecular weight, was found to increase the DNA loading which resulted in a greater gene transfection than low molecular weight PLGA. PVA in lower concentration resulted in higher gene expression and better intracellular uptake (Swayam and Vinod 2003).

2.8 Delivery of Anti-parkinson Drugs

An effective anti-parkinson drug ropinirole hydrochloride was incorporated into chitosan nanoparticles prepared by emulsification crosslinking method. Polysorbate 80 was used to form a coating on the surface of the nanoparticles. The encapsulation efficiency, zeta potential, and average particle size of the nanoparticles was found satisfactory. The release pattern indicated an initial burst release followed by sustained drug release over 10 hr. The *in vivo* study demonstrated a higher concentration of ropinirole hydrochloride in brain in comparison to liver, kidney and spleen for polysorbate 80 coated NPs with respect to uncoated NPs and pure drug (Ray et al. 2018). Cerium oxide NPs were developed and assessed against the biochemical and behavioural changes which occur in Parkinson's disease. Rat model was selected for the necessary animal testing in this study. The control group received intrastriatal injection of saline and a low dose of cerium oxide NPs were injected in test group. Behavioural assessment was conducted after completion of experimental period and then the animals were sacrificed for biochemical study.

The biochemical study mainly focused on striatal dopamine, activity of caspase-3 and stress markers. Partial neuroprotection was confirmed by the results along with partially decreased oxidative stress. The striatal dopamine level was found unaffected with respect to untreated group (Maha et al. 2017). Chitosan-sodium tripolyphosphate nanoparticles were prepared to deliver pramipexole dihydrochloride through nasal route. The average particle size was reported as 292.5 nm ± 8.80. Diffusion of drug was studied in goat nasal mucosa and compared with an artificial membrane. Microscopic study revealed spherical shape of nanoparticles with rough surfaces. An improved photoactometer score with reduced motor deficit was reported for the group treated with pramipexole dihydrochloride chitosan nanoparticles in comparison to its oral tablets or nasal solution. The study findings also indicated a higher status of antioxidant along with a significant increase in dopamine in brain (Ruhi et al. 2018).

2.9 Delivery of NSAIDs

Celecoxib loaded NPs were prepared by using a two step pulverization using hydroxypropyl cellulose. The freeze dried nanoparticles were physicochemically evaluated which exhibited an improved dissolution characteristics. Buffer solution pH 1.2 was used to characterize the dissolution behaviour and the study result demonstrated a 21.8 fold better dissolution profile in comparison to crystalline celecoxib. The dissolution behaviour was found unchanged after storing of NPs for 4 wk at 40°C. The oral absorption was evaluated in rat model and the results indicated a 4.7 hr faster T_{max} and 4.3 times higher C_{max} than crystalline celecoxib (Keisuke et al. 2018). An investigation was made for oral delivery of poorly water soluble NSAIDs such as ketoprofen, ibuprofen and nabumetone through sonication by forming a nano emulsion of lipids in aqueous phase. Sustained release of NSAIDs from solid lipid nanoparticles was achieved by using capmul GMS as lipoidal matrix and gelucire as surfactant. The entrapment efficiency for ketoprofen, ibuprofen and nabumetone was reported as 74, 53 and 69%, respectively. Raw 264.7 cells were used for cytotoxicity study which confirmed the non-toxic nature of the developed SLNs (Raj et al. 2018).

A topical approach for piroxicam delivery was investigated to avoid the side effects associated with its extended oral use. Solid lipid nanoparticle based formulations were developed through solvent evaporation method to improve the topical permeation rate. A nanolipidic topical gel was formulated for assessing the percuteneous permeation of drug. Skin permeation study of piroxicam loaded nanolipidic gel and commercial gel of the same drug showed a better skin permeation for the SLN based nanolipidic gel (Soliman et al. 2018). Nanogel formulations were developed by using carbopol 940 incorporating chitosan and egg albumin based nanoparticles for delivering aceclofenac. Heat coagulation method was employed to prepare the nanoparticulate formulations (Figure 1). A sustained release of aceclofenac was observed over 8 hr of *in vitro* dissolution. Excised mouse abdominal skin was used for skin permeation study which revealed a sustained drug release from nanogel formulations. Anti inflammatory activity was evaluated in carrageenan induced rat model. The study indicated a higher percentage of swelling inhibition of paw edema in rats treated with carbopol 940 based aceclofenac nanogel in comparison to commercial aceclofenac gel after 4 hr (Figure 2) (Jana et al. 2014).

Another investigation of chitosan (CS) and locust bean gum (LBG) IPN nanocomposites delivery of aceclofenac by Jana and Sen. The nanocomposites were synthesized by glutaraldehyde cross-linking (Figure 3). *In vitro* result of drug loaded nanocomposites showed sustained release of aceclofenac (Jana and Sen 2017).

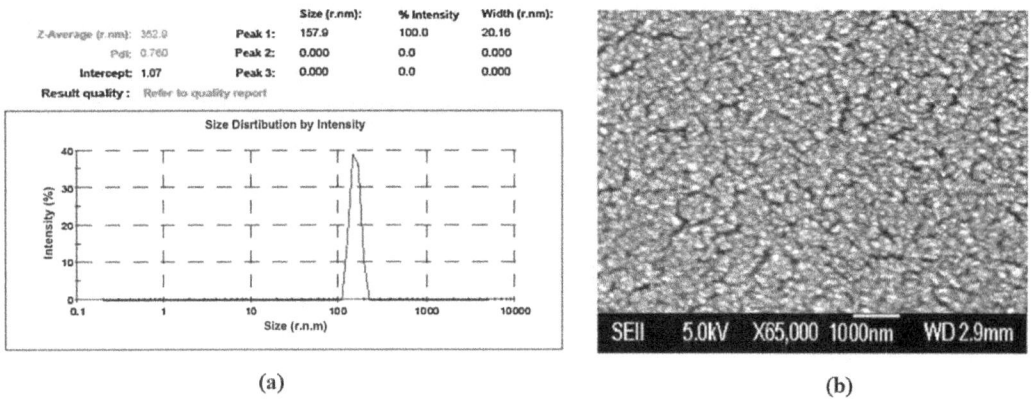

(a)   (b)

Figure 1: (a) Particle size distribution of 2% (w/v) NaTPP cross-linked aceclofenac-loaded chitosan-egg albumin nanoparticles (F-6) prepared using 200 mg chitosan and 500 mg egg albumin; (b) FE-SEM image of 2% (w/v) NaTPP cross-linked aceclofenac-loaded chitosan-egg albumin-nanoparticles (F-6) at 65 000× magnification.

Source: Jana, S., S. Manna, A.K. Nayak, K.K. Sen and S.K. Basu. 2014. Carbopol gel containing chitosan-egg albumin nanoparticles for transdermal aceclofenac delivery. Colloid. Surface. B 114: 36–44. Copyright (2014), with permission from Elsevier.

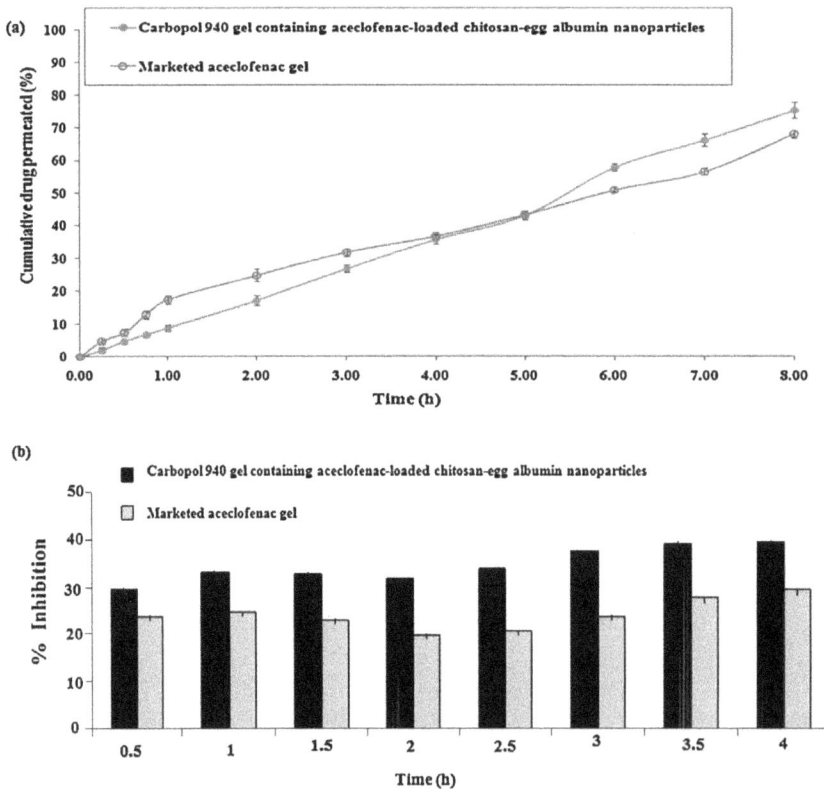

Figure 2. (a) The comparative *ex vivo* drug permeation from Carbopol 940 gel containing aceclofenac-loaded chitosan-egg albumin nanoparticles and a marketed aceclofenacgel through excised mouse skin (mean ± S.D.; n = 3); (b) comparative percentage inhibition profile of paw edema for Carbopol 940 gel containing aceclofenac-loaded nanoparticles and marketed aceclofenac gel at various time intervals in carrageenan-induced rat model for anti-inflammatory activity evaluation.

Source: Jana, S., S. Manna, A.K. Nayak, K.K. Sen and S.K. Basu. 2014. Carbopol gel containing chitosan-egg albumin nanoparticles for transdermal aceclofenac delivery. Colloid. Surface. B 114: 36–44. Copyright (2014), with permission from Elsevier.

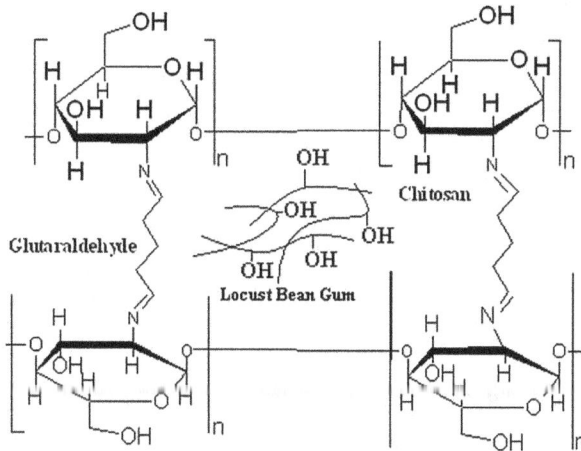

Figure 3: The reaction mechanism of GA crosslink CS-LBG IPN.
Source: Jana, S. and K.K. Sen. 2017. Chitosan—Locust bean gum interpenetrating polymeric network nanocomposites for delivery of aceclofenac. International Journal of Biological Macromolecules 102(2017): 878–884. Copyright (2017), with permission from Elsevier.

2.10 Delivery of Vaccines

PLGA based nanoparticulate delivery system was developed for delivering an inactivated antigen SwIV H1N2 for preventing swine influenza viruses. Intranasal route was preferred for vaccination to pigs which showed an antigen specific proliferation of lymphocytes. An increased frequency of cytotoxic T-cells was reported in peripheral blood cells. The clinical symptoms of swine flu were not found in the vaccinated pigs while the pigs of controlled group were reported to have fever. Reduced lung pathology was reported for vaccinated pigs with subsequent clearance of heterologous virulent (Santosh et al. 2017). An intranasal painless nanomaterial based system was fabricated to deliver Hepatitis B vaccines. Chitosan and tripolyphosphate based nanoparticles were synthesized which were combined with the antigen protein of Hepatitis B.

The developed chitosan based gene vaccine was subjected to *in vivo* test using rats. The study results confirmed good absorption of chitosan in nasal mucosal layer which facilitate uninterrupted antigen supply in body (Kim and Kang 2008). Intradermal administration of polymeric nanoparticles was investigated. The antigen loaded nanoparticles resulted in a burst transit via lymph nodes along with a restricted systemic exposure. Monophosphoryl Lipid A along with ovalbumin and imiquimod was incorporated in PLGA based NP formulation to use for vaccine delivery. Ovalbumin loaded PLGA NPs offered quick maturation kinetics for antibody affinity in comparison to soluble vaccine of ovalbumin. The microneedle delivery system resulted in a higher antibody response with respect to intramascular injection of similar vaccine (Lin et al. 2019).

Biodegradable polymeric nanoparticles were synthesized for delivering oral protein vaccine targeted to the area of Peyer's patches. A model vaccine, bovine serum albumin, was incorporated in chitosan-tripolyphosphate nanoparticles prepared by ionic gelation technique. The nanoparticles were further subjected to coating by electrostatic interaction using Eudragit L100. The stability and release behaviour of the NPs was reported as satisfactory. The targeting characteristic was evaluated in rats by fluorescence visualization technique which reported specific accumulation of NPs after dissolvation of eudragit L100 coating in intestinal fluid (Xu et al. 2018).

2.11 Delivery of Antibacterial Agent

A widely used antibacterial agent chlorhexidine was encapsulated in a nanostructured composite of sodium alginate and hydroxyapatite. The nanocarrier system was developed specifically for dental application. The solubility of alginate-hydroxyapatite composite was determined by using phosphate buffer which controlled release of chlorhexidine. The drug release was also found dependent on alginate content. The maximum chlorhexidine concentration after 72 hr was reported to be 0.04 mg/l. XRD study indicated an increased alginate concentration in nanocomposite, which reduced the average crystal size for hydroxyapatite and increased the lattice deformation (Leonid et al. 2018). Povidone–iodine nanoparticles were successfully synthesized by following two step method.

Poly(N-vinyl-2-pyrrolidone-co-methyl methacrylate) NPs were synthesized initially followed by complex formation, with iodine resulting in the development of water insoluble povidone–iodine nanoparticles. The feed ratio of NVP was found to influence the chemical composition, morphology, molecular weight and other properties. The antibacterial efficacy was studied by *E. coli*, *S. aureus* and *P. aeruginosa* by colony count technique. The developed NPs were introduced in three different products such as ink, dye and glue which showed significant antimicrobial effect (Gao et al. 2017). Copper iodide nanoparticles were synthesized by co-precipitation technique. The NPs were reported with a negative zeta potential of –21.5 mV. Gram positive and gram negative bacteria are found sensitive when tested with the NP formulations. DH5α was reported as the most sensitive bacteria and *Bacillus subtilis* was found more resistant to the nanoparticles. The study inferred that copper iodide NPs generate reactive oxygen species causing suppressing the DNA transcription for both gram positive as well as gram negative bacteria. The atomic force microscopy study indicated membrane damage which acts as a major bactericidal mechanism (Pramanik et al. 2012).

2.12 Hormone Delivery

Jana et al. (2015b) developed alginate nanocapsules containing testosterone for hormone delivery. Drug loaded alginate nanocapsules were further characterized by different instrumentation techniques. Pharmacokinetic evaluations drug loaded nanocapsules were performed in a rat model and different parameters were evaluated in comparison with control, drug loaded nanoformulation, pure and commercial formulation. *In-vivo* data showed successful delivery of hormone from drug loaded alginate nanocapsules (Figure 4).

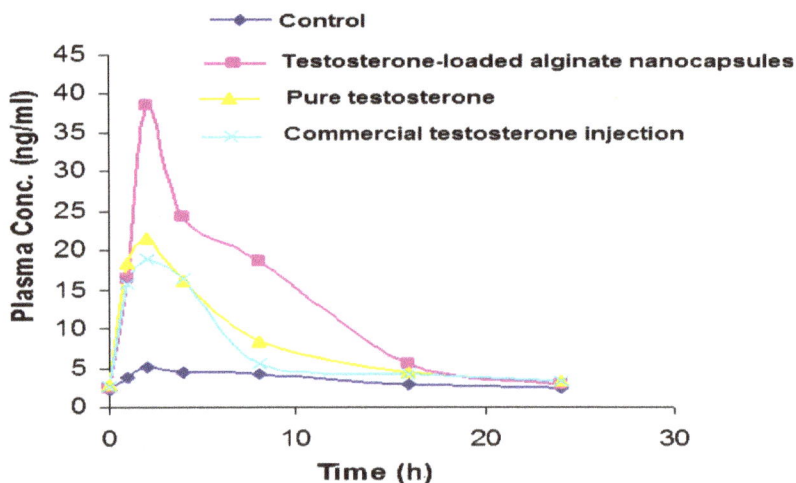

Figure 4: Testosterone plasma concentration vs. time curve.
Source: Jana, S., A. Gangopadhaya, B.B. Bhowmik, A.K. Nayak and A. Mukherjee. 2015b. Pharmacokinetic evaluation of testosterone-loaded nanocapsules in rats. International Journal of Biological Macromolecules 72: 28–30. Copyright (2015b), with permission from Elsevier.

3. Conclusion

Nanoparticulate drug delivery system has gained serious attention during the past few decades due to the low toxicity and biodegradability of biopolymers. Protein and polysaccharide based biopolymers played a significant role in the development of nanomaterials. Continuous progress has been noted in different techniques such as crosslinking and surface modification which has evolved over the years and contributed in the development process. Though many drugs are successfully delivered through nanoparticles in laboratory scale, a very few approved nanoparticulate formulations are available. One of the major reasons of it is the unmatched *in vivo* response in clinical trial, immunogenicity and impulsive initial drug release. The unpredictability in degradation pathway and fate of the polymer matrix adds further complexity in the procedure. However, there is a constant increase in the FDA approval of nanomaterials based drug delivery system since 1990s. With the rapidly progressive research and advances in the field of nanoscience, the challenges will be resolved to revolutionize nanotherapeutics.

References

Alex, M.R.A., A.J. Chacko, S. Jose and E.B. Souto. 2011. Lopinavir loaded solid lipid nanoparticles (SLN) for intestinal lymphatic targeting. Eur. J. Pharm. Sci. 42: 11–18.

Alexis, F., E. Pridgen, L.K. Molnar and O.C. Farokhzad. 2008. Factors affecting the clearance and biodistribution of polymeric nanoparticles. Mol. Pharm. 5: 505–515.

Balakrishnan, S. and R. Anand. 2013. Multifunctional nanoparticles for doxycycline delivery towards localized elastic matrix stabilization and regenerative repair. Acta Biomater. 9: 6511–6525.

Byung, K.L., Y. Yeonhee and P. Kinam. 2016. PLA micro- and nano-particles. Adv. Drug Deliv. Rev. 107: 176–191.

Chen, W., Z. Zhiqiang, Binbin Chen, G. Liuqiong, L. Yuan and Y. Shuwei. 2018. Design and evaluation of galactosylated chitosan/graphene oxide nanoparticles as a drug delivery system. J. Colloid. Interf. Sci. 516: 332–341.

Christopher, J.D., B. Todd, G. Michael, S. Annemarie and A.B. Michael. 2010. Antiretroviral release from poly(DL-lactide-co-glycolide) nanoparticles in mice. J. Antimicrob. Chemother. 65: 2183–2187.

Corina-Lenuța, S., P. Marcel, D. Christelle, C. Marcel, C. Dănuț and A.P. Cătălina. 2019. Chitosan grafted-poly(ethylene glycol) methacrylate nanoparticles as carrier for controlled release of bevacizumab. Mater. Sci. Eng. C 98: 843–860.

Danhier, F., E. Ansorena, J.M. Silva, R. Coco, A.L. Breton and V. Preat. 2012. PLGA-based nanoparticles: An overview of biomedical applications. J. Control. Release 161: 505–522.

David, L., D. Manuela, L. Claire, C. Valeria, C. Andrea, P.B. Elsa, Z. Narimane and B. Kawthar. 2014. Auto-associative heparin nanoassemblies: A biomimetic platform against the heparan-sulfate-dependent viruses HSV-1, HSV-2, HPV-16 and RSV. Eur. J. Pharm. Biopharm. 88: 275–282.

Deepa, T., M.S. Lathab and K.K. Thomas. 2018. Synthesis and *in vitro* evaluation of alginate-cellulose nanocrystal hybrid nanoparticles for the controlled oral delivery of rifampicin. J. Drug Deliv. Sci. Tec. 46: 392–399.

Douglas, K.L., C.A. Piccirillo and M. Tabrizian. 2006. Effects of alginate inclusion on the vector properties of chitosan-based nanoparticles. J. Control. Release 115: 354–361.

D'Mello, S.R., S.K. Das and N.G. Das. 2009. Polymeric nanoparticles for small molecule drugs: Biodegradation of polymers and fabrication of nanoparticles. pp. 16–34. *In*: Yashwant, P. and T. Deepak (eds.). Drug Delivery Nanoparticles Formulation and Characterization. Informa Healthcare. New York, NY, USA.

Elgadir, M.A., U.M. Salim, S. Ferdosh, A. Adam, A.J.K. Chowdhury and M.Z.I. Sarker. 2015. Impact of chitosan composites and chitosan nanoparticle composites on various drug delivery systems: A review. J. Food Drug Anal. 23: 619–629.

Enhui, Z., X. Ronge, L. Song, L. Kecheng, Q. Yukun, Y. Huahua and L. Pengcheng. 2019. Vascular targeted chitosan-derived nanoparticles as docetaxel carriers for gastric cancer therapy. Int. J. Biol. Macromol. 126: 662–672.

Farideh, M., J. Behrooz and G. Marjan. 2019. Chitosan-based nanomicelle as a novel platform for targeted delivery of methotrexate. Int. J. Biol. Macromol. 126: 517–524.

Fatemeh, S., T. Elnaz and K. Khosro. 2018. Development of chitosan coated calcium alginate nanocapsules for oral delivery of liraglutide to diabetic patients. Int. J. Biol. Macromol. 120: 460–467.

Gao, T., H. Fan, X. Wang, Y. Gao, W. Liu, W. Chen, A. Dong and Y.J. Wang. 2017. Povidone-iodine-based polymeric nanoparticles for antibacterial applications. ACS Appl. Mater. Interfaces 9: 25738–25746.

Gurudutta, P., S. Biswadip, M. Biswajit, G. Saikat, B. Sandip, M. Subhasish and B. Tanmoy. 2012. Submicron-size biodegradable polymer-based didanosine particles for treating HIV at early stage: An *in vitro* study. J. Microencapsul. 29: 666–676.

Hamedi, H., S. Moradi, S.M. Hudson and A.E. Tonelli. 2018. Chitosan based hydrogels and their applications for drug delivery in wound dressings: A review. Carbohyd. Polym. 199: 445–460.

Hamidi, M., A. Azadi and P. Rafiei. 2008. Hydrogel nanoparticles in drug delivery. Adv. Drug Deliv. Rev. 60: 1638–1649.

Hamidreza, S., M.G. Sayed and M. Mohammad. 2015. Improvement of interaction between PVA and chitosan via magnetite nanoparticles for drug delivery application. Int. J. Biol. Macromol. 78: 130–136.

Hasan, N., C. Jiafu, L. Juho, P.H. Shwe, A.O. Murtada, N. Muhammad, K. Min-Hyo, L.L. Bok, J. Yunjin and Y. Jin-Wook. 2019. Bacteria-targeted clindamycin loaded polymeric nanoparticles: Effect of surface charge on nanoparticle adhesion to MRSA, antibacterial activity, and wound healing. Pharmaceutics 11: 236. 10.3390/pharmaceutics11050236.

Hervé, H., L.D. Trung, A. Martine and C. Patrick. 2006. Hybrid polymer nanocapsules enhance *in vitro* delivery of azidothymidine-triphosphate to macrophages. J. Control. Release 116: 346–352.

Jana, S., N. Maji, A.K. Nayak, K.K. Sen and S.K. Basu. 2013. Development of chitosan-based nanoparticles throughinter-polymeric complexation for oral drug delivery. Carbohydrate Polymers 98: 870–876.

Jana, S., S. Manna, A.K. Nayak, K.K. Sen and S.K. Basu. 2014. Carbopol gel containing chitosan-egg albumin nanoparticles for transdermal aceclofenac delivery. Colloid. Surface. B 114: 36–44.

Jana, S., B. Laha and S. Maiti. 2015a. Boswellia gum resin/chitosan polymer composites: Controlled delivery vehicles for aceclofenac. Int. J. Biol. Macromol. 77: 303–306.

Jana, S., A. Gangopadhaya, B.B. Bhowmik, A.K. Nayak and A. Mukherjee. 2015b. Pharmacokinetic evaluation of testosterone-loaded nanocapsules in rats. International Journal of Biological Macromolecules 72: 28–30.

Jana, S., K.K. Sen and A. Gandhi. 2016. Alginate based nanocarriers for drug delivery applications. Current Pharmaceutical Design 22: 3399–3410.

Jana, S. and K.K. Sen. 2017. Chitosan—Locust bean gum interpenetrating polymeric network nanocomposites for delivery of aceclofenac. International Journal of Biological Macromolecules 102: 878–884.

Jaya, R.L., M. Thabo and W.M.K. Rui. 2017. Cationic cyclodextrin/alginate chitosan nanoflowers as 5-fluorouracil drug delivery system. Mater. Sci. Eng. C 70: 169–177.

Jean-Christophe, L., A. Eric, D.J. Fanny, D. Eric and G. Robert. 1996. Biodegradable nanoparticles from sustained release formulations to improved site specific drug delivery. J. Control. Release 39: 339–350.

Joshy, K.S., M.A. Susan, S. Snigdha, K. Nandakumar, A.P. Laly and T. Sabu. 2018. Encapsulation of zidovudine in PF-68 coated alginate conjugate nanoparticles for anti-HIV drug delivery. Int. J. Biol. Macromol. 107: 929–937.

Jeetendra, S.N., C. Pronobesh, K.S. Ashok and R. Veerma. 2013. Development of solid lipid nanoparticles (SLNs) of lopinavir using hot self nano-emulsification (SNE) technique. Eur. J. Pharm. Sci. 48: 231–239.

Jiacheng, Z. and H.S. Martina. 2018. Entry of nanoparticles into cells: The importance of nanoparticle properties. Polym. Chem. 9: 259–272.

Jian, Z., C.S. Meong, E.D. Allan, Z. Jie, L. Kyuri, H. Huining and C.Y. Victor. 2013. Long-circulating heparin-functionalized magnetic nanoparticles for potential application as a protein drug delivery platform. Mol. Pharm. 10: 3892–3902.

Jiangbing, Z., R.P. Toral, F. Michael, P.B. James and W.M. Saltzman. 2012. Octa functional PLGA nanoparticles for targeted and efficient siRNA delivery to tumors. Biomaterials 33: 583–591.

Juliana, P.S., N.B. Thaís, C. Stephani, C.C. Talita, J.S. Michael, D.H.E.S. Kattya Gyselle and R.E.M. Claudia. 2018. Dual alginate-lipid nanocarriers as oral delivery systems for amphotericin B. Colloid. Surface. B 166: 187–194.

Keisuke, Y., O. Mizuki, S. Hiroki, S. Yoshiki, S. Hideyuki and Satomi Onoue. 2018. Physicochemical and biopharmaceutical characterization of celecoxib nanoparticle: Avoidance of delayed oral absorption caused by impaired gastric motility. Int. J. Pharm. 552: 453–459.

Kim, B.G. and I.J. Kang. 2008. Evaluation of the effects of biodegradable nanoparticles on a vaccine delivery system using AFM, SEM, and TEM. Ultramicroscopy 108: 1168–1173.

Leonid, F.S., B.S. Liudmyla, L. Olena and P. Yuriy. 2018. Synthesis and characterization of hydroxyapatite-alginate nanostructured composites for the controlled drug release. Mater. Chem. Phys. 217: 228–234.

Li-Chu, T., C. Chien-Ho, L. Cheng-Wei, H. Yi-Cheng and M. Fwu-Long. 2019. Development of mutlifunctional nanoparticles self-assembled from trimethyl chitosan and fucoidan for enhanced oral delivery of insulin. Int. J. Biol. Macromol. 126: 141–150.

Lin, N., Y.C. Leonard, A.B. Scott, J.H. Kris and P. Jayanth. 2019. Intradermal delivery of vaccine nanoparticles using hollow microneedle array generates enhanced and balanced immune response. J. Control. Release 294: 268–278.

Linda, H., K. Jörg, H. Hiroaki, Tadao Iwasaki and M.G. Julia. 2009. Functional protein delivery into neurons using polymeric nanoparticles. J. Biol. Chem. 284: 6972–6981.

Liyao, L., Z. Cuiping, X. Xuejun and L. Yuling. 2016. Self-assembled lecithin/chitosan nanoparticles for oral insulin delivery: Preparation and functional evaluation. Int. J. Nanomed. 11: 761–769.

Maha, A.E.H., M.M. Hala, A.A.E. Doaa, Y.E. Fatma, A.A. Malik, A.A. Fisal, M.A. Obaid and M.A. Mahdi. 2017. The possible role of cerium oxide (CeO2) nanoparticles in prevention of neurobehavioral and neurochemical changes in 6-hydroxydopamine induced parkinsonian disease. Alex. J. Med. 53: 351–360.

Martine, L.V., F. Laurence, K. Young-II, H. Maurice and M. Philippe. 1998. Preparation and characterization of nanoparticles containing an antihypertensive agent. Eur. J. Pharm. Biopharm. 46: 137–143.

Moghimi, S.M., A.C. Hunter and J.C. Murray. 2005. Nanomedicine: Current status and future prospects. FASEB J. 19: 311–330.

Pramanik, A., D. Laha, D. Bhattacharya, P. Pramanik and P. Karmakar. 2012. A novel study of antibacterial activity of copper iodide nanoparticle mediated by DNA and membrane damage. Colloid. Surface. B 96: 50–55.

Praveen, K.G., M. Shikha, B. Meenakshi and M. Anushika. 2014. Enhanced oral bioavailability of efavirenz by solid lipid nanoparticles: *In vitro* drug release and pharmacokinetics studies. BioMed. Res. Int. 363404. 10.1155/2014/363404.

Raj, K., S. Ashutosh, G. Neha and F.S. Prem. 2018. Solid lipid nanoparticles for the controlled delivery of poorly water soluble nonsteroidal anti-inflammatory drugs. Ultrason. Sonochem. 40: 686–696.

Ramanathan, R., Y. Jiang, B. Read, S. Golan-Paz and K.A. Woodrow. 2016. Biophysical characterization of small molecule antiviral-loaded nanolipogels for HIV-1 chemoprophylaxis and topical mucosal application. Acta Biomater. 36: 122–131.

Ray, S., P. Sinha, B. Laha, S. Maiti, U.K. Bhattacharyya and A.K. Nayak. 2018. Polysorbate 80 coated crosslinked chitosan nanoparticles of ropinirole hydrochloride for brain targeting. J. Drug Deliv. Sci. Tec. 48: 21–29.

Ruhi, R., W. Sarika, S. Vinay and G. Ram. 2018. Pramipexole dihydrochloride loaded chitosan nanoparticles for nose to brain delivery: Development, characterization and *in vivo* anti-Parkinson activity. Int. J. Biol. Macromol. 109: 27–35.

Sandhiya, S., S.A. Dkhar and A. Surendiran. 2009. Emerging trends of nanomedicine—An overview. Fundam. Clin. Pharmacol. 23: 263–269.

Sang, K.K., B.F. Michael and Leaf Huang. 2012. The targeted intracellular delivery of cytochrome C protein to tumors using lipid-apolipoprotein nanoparticles. Biomaterials 33: 3959–3966.

Santosh, D., H. Jagadish, B. Kathryn, S.L. Yashavanth, S. Duan-Liang, O. Kang, O. Kyung-il, B. Basavaraj, G. Jonathan, T. Kairat, K. Steven, N. Balaji, L. Chang-Won and J.R. Gourapura. 2017. Biodegradable nanoparticle delivery of inactivated swine influenza virus vaccine provides heterologous cell-mediated immune response in pigs. J. Control. Release 247: 194–205.

Shahnaz, R., K. Sepideh and G. Mehdi. 2019. Preparation and characterization of rod like chitosan–quinoline nanoparticles as pH-responsive nanocarriers for quercetin delivery. Int. J. Biol. Macromol. 128: 279–289.

Sharma, B., M. Wenxue, M.A. Isaac, P. Jayanth, D. Sanja and L. Vinod. 2011. Nanoparticle-mediated p53 gene therapy for tumor inhibition. Drug Deliv. Transl. Res. 1: 43–52.

Shastri, V.P. 2003. Non-degradable biocompatible polymers in medicine: Past, present and future. Curr. Pharm. Biotechnol. 4: 331–337.

Soleyman, H., H. Hossein, P. Shahryar and K. Zahra. 2019. Synthesis of stimuli responsive chitosan nanocomposites via RAFT copolymerization for doxorubicin delivery. Int. J. Biol. Macromol. 121: 677–685.

Soliman, M.S., Z. Shirin and E.A. Elaheh. 2018. Piroxicam loaded solid lipid nanoparticles for topical delivery: Preparation, characterization and *in vitro* permeation assessment. J. Drug Deliv. Sci. Tec. 47: 427–433.

Soppimath, K.S., T.M. Aminabhavi, A.R. Kulkarni and W.E. Rudzinski. 2001. Biodegradable polymeric nanoparticles as drug delivery devices. J. Control. Release 70: 1–20.

Sudipta, M., J.S. Su, B. Yoonhee and S.C. Joon. 2019. Self-assembled nanoparticles composed of glycol chitosan-dequalinium for mitochondria-targeted drug delivery. Int. J. Biol. Macromol. 132: 451–460.

Swayam, P. and L. Vinod. 2003. Critical determinants in PLGA/PLA nanoparticle mediated gene expression. Pharm. Res. 21: 354–364.

Swayam, P. and L. Vinod. 2004. Nanoparticle-mediated wild-Type p53 gene delivery results in sustained antiproliferative activity in breast cancer cells. Mol. Pharm. 1: 211–219.

Tanushree, B., P. Bonamali and S. Satnam. 2018. Hollow chitosan nanocomposite as drug carrier system for controlled delivery of ramipril. Chem. Phys. Lett. 706: 465–471.

Taskeen, N., N. Habib, S. Saima, R. Asma and I. Muhammad. 2016 Polyionic hybrid nano engineered systems comprising alginate and chitosan for antihypertensive therapeutics. Int. J. Biol. Macromol. 91: 180–187.

Utpal, J., K.M. Anjan, L.P. Sovan, K.M. Prabal and P.M. Guru. 2014. Felodipine loaded PLGA nanoparticles: Preparation, physicochemical characterization and *in vivo* toxicity study. Nano Convergence 1: 31.

Vijayan, V., K.R. Reddy, S. Sakthivel and C. Swetha. 2013. Optimization and charaterization of repaglinide biodegradable polymeric nanoparticle loaded transdermal patchs: *In vitro* and *in vivo* studies. Colloid. Surface. B 111: 150–155.

Wang, M. and M. Thanou. 2010. Targeting nanoparticles to cancer. Pharmacol. Res. 62: 90–99.

Xe, B., W. Zhang, Y. Chen, Y. Xu, B. Wang and L. Zong. 2018. Eudragit® L100-coated mannosylated chitosan nanoparticles for oral protein vaccine delivery. Int. J. Biol. Macromol. 113: 534–542.

Zhang, Z., P.-C. Tsaib, T. Ramezanlib and B.B. Michniak-Kohn. 2013. Polymeric nanoparticles-based topical delivery systems for the treatment of dermatological diseases. WIRES Nanomed. Nanobi. 5: 205–21.

11

Chitosan Based Nanobioceramics in Bone Tissue Engineering Application

J. Venkatesan,[1] *S. Revathi,*[2] *R. Sasirekha,*[2] *P.N. Sudha,*[2]
P. Pazhanisamy[3] and *T. Gomathi*[2],*

1. Introduction

Through the logical integration of cells, biomimetic matrices, biological signals, and biophysical cues, tissue engineering aims to recreate or restore the normal biological activities of the tissues or organs. Tissue engineering, on the other hand, uses scaffolds to attract endogenous cells for tissue repair while utilising the body's inherent capacity to mend itself. The potential of engineers to replicate the microenvironment *ex vivo* in order to create constructs that resemble tissue and can be transplanted is essential to the success of *in vitro* tissue engineering.

Although the ultimate "engineers" in the end are the cells, the significance of the scaffolding components cannot be over emphasised. By delivering specialised cell recognition sites and signalling molecules in a temporal and spatial manner, polymeric scaffolds are intended to regulate cell functions and enhance tissue formation rather than serving as simple physical templates (Place et al. 2009, Langer et al. 1993).

Because of how closely they resemble the natural extracellular matrix (ECM) in terms of structure and functionality, hydrogels are the most alluring tissue engineering scaffolds. The term "hydrogel" refers to a group of interconnected, macroscopic-scale networks made of hydrophilic (or amphiphilic) building blocks that are insoluble due to the presence of crosslinks. Insoluble tiny substances like nanoparticles and nanofibrils as well as macromonomer polymers and soluble monomers can all be used to create scaffolds. Crosslinks can be either chemical or physical intersections where even more than two microscopic or polymeric chains come together.

For tissue engineering, scaffold should generally be biocompatible, biodegradable, provide mechanical support and effectively transmit forces from the environment to the developing tissue over an extended period of time, and include structural and temporal information from the modern biological world (Langer et al. 2004, 2009). In order to produce cells with a biologically relevant

[1] Yenepoya Research Centre, Yenepoya University, Mangalore-575018, Karnataka, India. Email: venkatjchem@gmail.com
[2] PG and Research Department of Chemistry, D.K.M. College for Women (Autonomous), Affiliated to Thiruvalluvar University, Vellore, Tamil Nadu, India.
[3] Department of Chemistry, Sir Theagaraya College, Affiliated to University of Madras, Chennai-600021, India.
* Corresponding author: drgoms1@gmail.com

micro environment that promotes cell proliferation, migration, and ECM production, the intrinsic properties of the scaffolds are discussed in this chapter along with how they can be rationally designed and processed to improve the development of functional tissues.

2. Bone Tissue Engineering

Tissue engineering is a contemporary area of interdisciplinary research that integrates the understanding of systems engineering and biology with the objective of rebuilding or replicating restored body tissue in clinical or pathological instances caused by tumours or systemic disorders. As a result, efforts were made to enhance biological substitutes that might restore, maintain, or grow organ and tissue function. In a typical tissue engineering technique, cells, bioactive compounds, and biomaterials are used in combination.

The process of preparing the matrix of new tissue is carried out by the cells. Cell migration, differentiation, and proliferation can all be induced by bioactive substances. The new tissue is supported by biomaterials, which provide an atmosphere that encourages cell proliferation and differentiation (Lanza et al. 2014).

Every tissue in a healthy state is made up of cell components and extracellular matrix (ECM). The ECM serves as the foundation upon which cells develop and arrange themselves. It creates the structure and microenvironment necessary for cell development and proliferation, as well as a storage area for vital growth factors, nutrients, and water for cells (Chen et al. 2015, Schoen and Levy 2005). In this aspect, the need for tissue replication necessitates the presence of a constitution that acts as a temporary matrix and mimics the ECM. This permits tissue growth and cell proliferation until full regeneration has occurred (Salgado et al. 2004, Cartmell et al. 2009). According to Rodriguez-Vázquez et al. (2015), this structure, known as the scaffold, which functions as a temporary ECM during tissue regeneration, may be a porous and three-dimensional solid biomaterial. In order to replicate tissue, biomaterials are essential and can decide whether a tissue engineering strategy is successful or not (Miranda et al. 2011, Keane and Badylak 2014). The biomaterial mimics ECM, to improve vascularization with three-dimensional framework. Additionally, it should process as a carter of molecules that are biologically connected and, in a similar manner, modify their biological reactions and the reproduction process (Figure 1).

Consequently, a scaffold biomaterial's usage has specific properties that make it analogous to ECM (Hutmache 2000, Rinaudo 2006). The scaffold must possess a transitory shape which gradually degrades over duration at a rate equal to the rate at which new tissue is forming, enabling

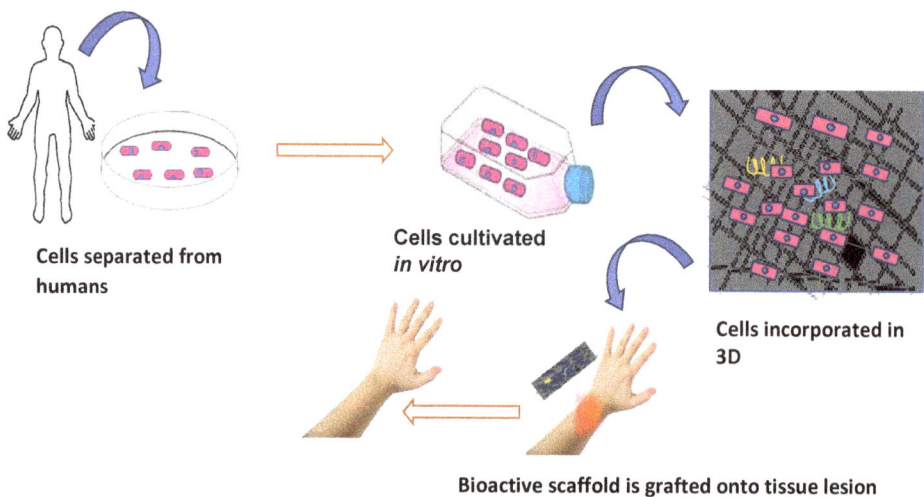

Cells separated from humans

Cells cultivated *in vitro*

Cells incorporated in 3D

Bioactive scaffold is grafted onto tissue lesion

Figure 1: Diagram of the concept of tissue engineering (Adapted from Dvir et al. 2011, Lanza et al. 2014).

regenerated tissue to replicate it. In order to do this, it is required to have a number of crucial characteristics, including biocompatibility, biodegradability, mechanical performance, and surface characteristics that promote cell attachment, such as porosity (Nalwa et al. 2004, 2000). In general, calcium phosphate ceramics such as hydroxyapatite (HA) and beta-tricalcium phosphate (β-TCP), synthetic polymers such as polyglycolic acid (PGA) and poly(lactic-co-glycolic) acid (PLGA), and naturally occurring biodegradable polymers such as collagen, hyaluronic acid, silk fibroin, gelatin, and chitosan are used for tissue regeneration (Shabafrooz et al. 2014, Obregon et al. 2014, Zhou and Lee 2011).

It has been determined that the polysaccharide chitosan derived from chitin (sources - exoskeletons of crustaceans, mollusks, and insects as well as on the cell walls of fungus), which comes from plant sources, has a immense prospective for usage as a scaffold for tissue regeneration (Li 2006). A linear chain of D-glucosamine and N-acetyl-glucosamine units connected by β(1-4) glycosidic connections makes up the chemical structure of chitosan (CS).

3. Nanobioceramics

Chitosan is can be created as a scaffold for the renewal of alveolar bone because those regions are very vulnerable to bacterial infection (Jie et al. 2004, 2004). Chitosan's chemical characteristics are thought to be responsible for its ability to convey cell bonds and multiplication. Chitosan has a polysaccharide backbone that resembles glycosaminoglycans, which make up the majority of the extracellular matrix in bone and cartilage. Efforts are currently being made to enhance the mechanical and biological properties of chitosan scaffolds by adding bio ceramics like hydroxyapatite (HA) (Zakaria et al. 2013, Jiang 2009), tricalcium phosphate (Cai 2008), and calcium phosphate (Di Martino et al. 2005), as well as biomaterials like gelatin (Zhang et al. 2014), alginate (Abbah et al. 2013 (Rabea et al. 2003). The resorption kinetics of the chitosan scaffolds have been known to be reduced by the inclusion of bio ceramics (Rabea et al. 2003). When bio ceramics are incorporated into chitosan scaffolds, the mechanical properties improve.

In order to promote cell adhesion, migration, differentiation, and proliferation, gelatin, a partial imitation of collagen, has already been made into composites via combining with some other materials, such as chitosan, tricalcium phosphate (TCP), and hydroxyapatite (HA). A certain amino acid sequence found in gelatin promotes cell adhesion by including amino acids including glycine, proline, and hydroxyproline (Jayakumar et al. 2011). The scaffold's characteristics are known to be affected by gelatin absorption.

Biodegradable and osteoconductive biomaterials utilised for bone service are bioactive glasses. Francis (Suh et al. 2000) enhanced it as a material for bone repair. Recent research has demonstrated that the bioactive glass products can promote gene expression of osteoblast, induce the synthesis of growth-promoting molecules, and increase cell proliferation (Venkatesan et al. 2010, Chan et al. 2008). Both hard and soft tissues can be attached to bioactive glass (Fergal et al. 2011, Langer et al. 2004). The glasses' propensity to create a coating of hydroxycarbonate apatite (HCA) on their surface has been credited with improving their ability to connect with bone (Ana Claudia et al. 2013). Various studies have demonstrated the superiority of bioactive glasses over hydroxyapatite coating for implant surfaces (Skallevold et al. 2019). *In vivo* implantation result reveals that these compositions don't cause any local or systemic toxicity, inflammation, or a reaction to a foreign body (Skallevold et al. 2019). Bioactive glasses use the sol-gel process and have a mesoporous texture with simpler compositions than those derived from melts and display higher bioactivity and resorb capacity. In the present, sol-gel method, nano-bioactive glass ceramic were created (Braghirolli et al. 2014, Dalton et al. 2013), and demonstrating the recent development of novel three-dimensional (3D) scaffolds with multi-scale hierarchical architectures from electrospun materials for tissue engineering applications. Given that this has never been done previously, it is exciting to study the impact of nBGC in nano-composite scaffolds. This report concentrated on the way the coexistence of various micro- and nano-features in the scaffolds affects the cellular behaviour (attachment, migration, and differentiation).

3.1 Preparation of Bioactive Glass Ceramic Nanoparticles (nBGC)

Following is a quick explanation of the steps involved in creating nBGC nanoparticles (SiO_2: CaO: P_2O_5 (mol%)55:40:5): 9.84 cc of TEOS was combined with 7.639 g of calcium nitrate to create combination A. Subsequently, the mixture A was dissolved in a solution containing water and ethanol (1:2 (mol)). Citric acid was added to the combination to bring the pH within the range of 1–2, and the mixture was blended until a translucent mixture B was produced. A second combination C was created by mixing 1500 ml of deionized water with 1.078 g of diammonium hydrogen phosphate and 15 g of poly(ethylene glycol) (PEG Mw 20,000), and then bringing the pH 10 with ammonium water. After vigorously mixing mixtures B and C together, the resulting mixture is matured for 24 hours at room temperature to produce a white gel precipitate. The above precipitate underwent filtering and washing. Following lyophilization to create a fine powder, the filtrate was then heated to 700°C to create.

3.2 Preparation of Nano-composite Scaffolds

In a solution of 1% acetic acid, 2% (w/v) chitosan was digested and homogenized with gelatin for 12 hours at 37°C. Next, after stirring for 24 hours, nBGC (1 weight percent) has been incorporated to that same chitosan-gelatin (CG) combination. To disperse the resulting mixture and reduce particle size, ultrasonication was substituted (Vibra cell VC 505), and 0.25% glutaraldehyde was added. The prepared mixture was put into 12 well culture plates, pre-frozen at –20°C for 12 hours, and then freeze-dried (Christalpha LD plus) at –80°C for 48 hours. The scaffolds were then rinsed with ion-free water after being neutralised for two hours with 2% NaOH and 5% NaBr. Then the scaffolds

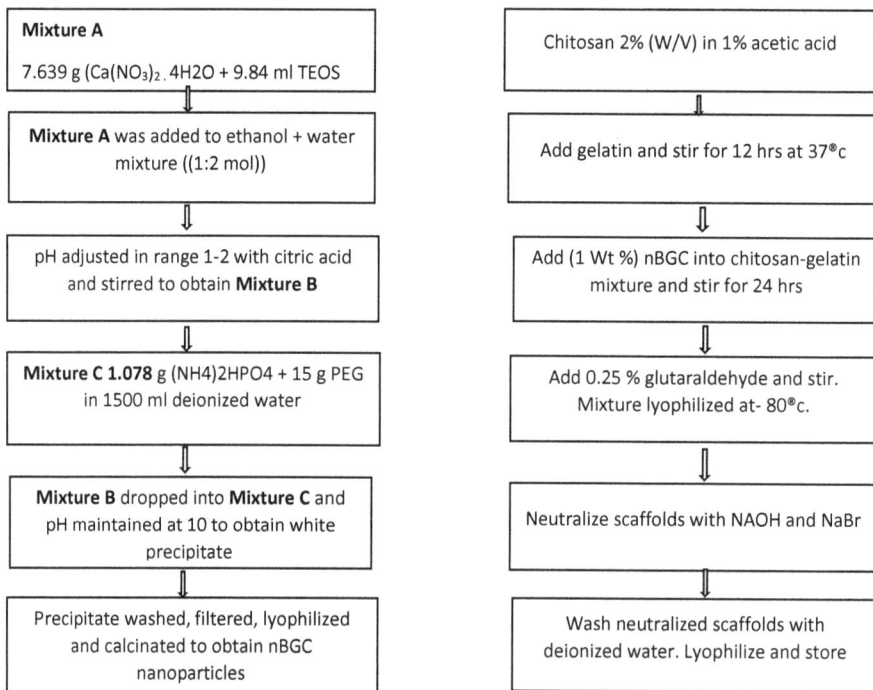

Figure 2: (a) Flow chart showing the steps in the synthesis of nBGC nanoparticles and (b) shows the steps in preparation of composite scaffolds (Harris et al. 1998).

were neutralized by 2% NaOH and 5% NaBr for 2 h and next washed with ion free water. The scaffolds were then freeze-dried and kept in storage for further usage. To create the nano-composite scaffolds, three different concentrations (0.5, 1 and 2%, w/w) of CG mixture were created and mixed with 1% nBGC before being freeze dried (Figure 2b).

3.3 *Characteristics of Nanobioceramics*

The morphology and dimension of nBGC was estimated using TEM. nBGC was distributed in ethanol and TEM images were taken. The shape morphology of the nano-composite scaffolds was examined using scanning electron microscope (SEM). Scaffolds were segmented into thin section with razor blade. The section was put on aluminum stub and covered with platinum 2 min at 10 mA before imaging. The standard pore size was determined by estimating the size of 30 pores. FTIR spectra of dried nBGC and nano-composite scaffolds were ground and mixed entirely with potassium bromide at a ratio of 1:5 (sample:KBr) and pelleted. The IR spectrum of the prepared pellet was taken using Perkin-Elmer RX1 operating at the range of 400–4000cm^{-1}. XRD patterns of nano-composite scaffolds and nBGC were available at room temperature using analytical XPERT PRO powder diffractometer (Cu Kα radiation) employed at a voltage of 40 kV. XRD were taken at 2θ angle range of 5–60° and the process parameters were scan step size 0.02 (2θ) and scan step time 0.05 s.

3.4 *Nanocomposites Featuring Nanobioceramics*

Current advances in the multidisciplinary field of tissue engineering or regenerative medicine rely heavily on the use of several types of scaffolds. According to the improved emphasis on tissue engineering as described at a National Science base workshop, scaffolds are the ideal materials for promoting healing, maintaining, and preserving the tissue process (Ho et al. 2004). They play a unique role in repair and more importantly regeneration of tissues by providing a suitable platform, allowing essential supply of different components accumulated with survival, proliferation and variation of cells (Hollister 2005).

Scaffolds can be prepared by synthetic or absorbable, naturally occurring, biological, degradable or non-degradable polymeric materials (Okano et al. 1995). Various techniques have been used to construct scaffolds (Okano et al. 1993) but the four major scaffolding approaches (Takezawa et al. 1990) include the use of ECM secreting cell sheets (Figure 3) (Boccaccini et al. 2005), premade porous scaffolds of synthetic, natural and biodegradable biomaterials (Figure 4) (Chan et al. 2007), de-cellularized ECM scaffolds (Figure 5) (Chan et al. 2005), and cells entrapped

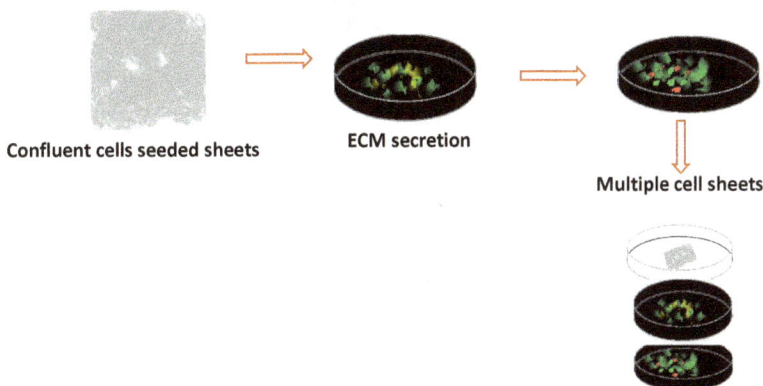

Figure 3: Cell sheets that secrete extracellular matrix (ECM). Cells are coated on the sheets and permit to secret ECM that facilitates the growth and proliferation. Multiple cell seeded sheet have the capacity to secret ECM that are used for implantation at the wound site.

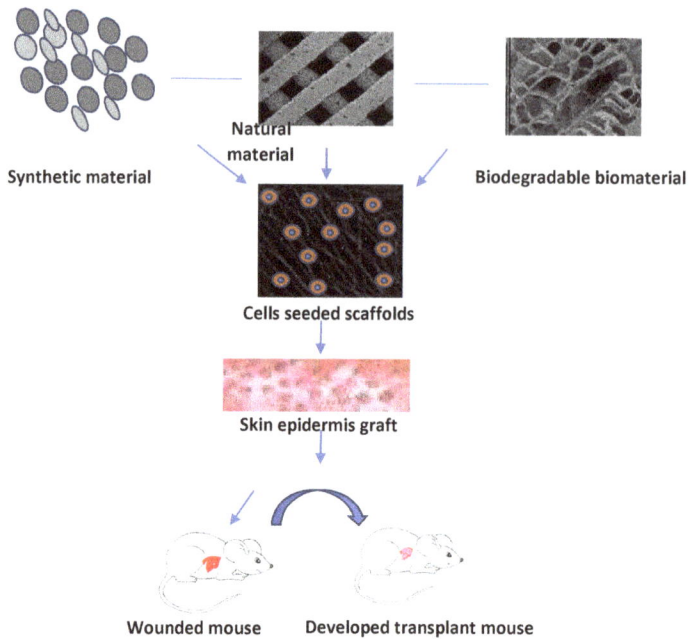

Figure 4: Porous scaffolding using several biomaterials. Many natural, synthetic and biodegradable materials are used for formation of highly porous scaffolds. These scaffolds give a favorable environment for cell development and proliferation. The porous nature of such scaffolds facilitates the continuous flow of nutrients and oxygen for the skin cells, such as keratinocytes and fibroblasts. The thick skin that developed on such scaffolds is used for wound transplant.

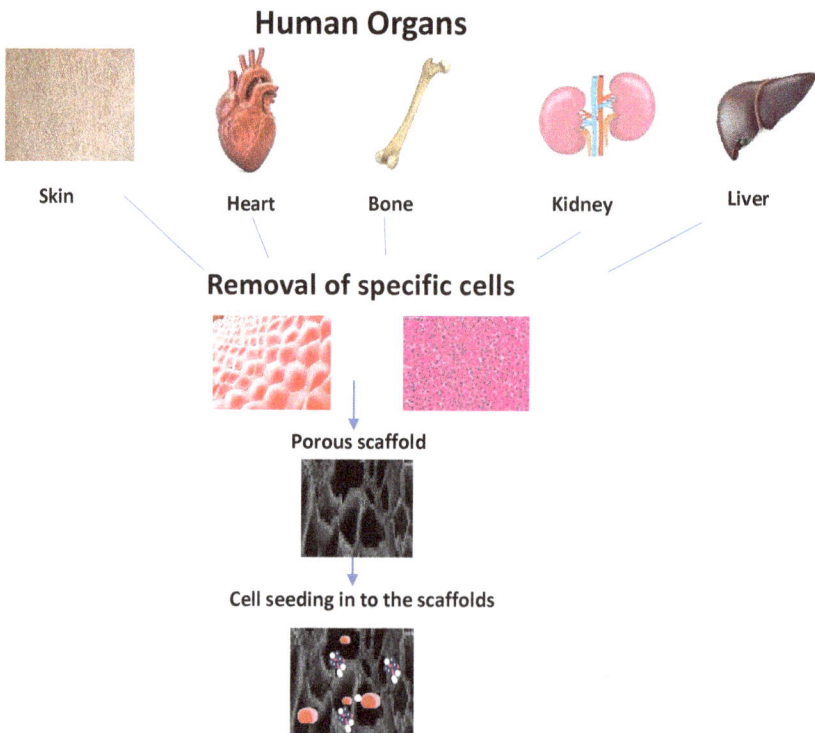

Figure 5: Acellular scaffolding approach. In this method, the organ is completely decellularized to form extracellular (ECM) beneath matrices. On such scaffolds, the target cells, can be effectively produced (Chan et al. 2005).

in hydrogels (Chevalier et al. 2008). All of these methods have benefits and shortcomings. Based on the biomaterial designs of the various types of scaffolds and their benefits and drawbacks, we want to concentrate our investigation on those scaffolds that are frequently employed for skin tissue regeneration (Table 1). Figure 3 shows that cells are planted on thermally controlled polymeric materials based on poly(N-isopropylacrylamide), which have a lower critical solution temperature (LCST) of 25°C. The ECM-secreting cell sheets that result from allowing cells to reach complete confluence are then produced. Then, without the requirement for trypsinization, these cell sheets can be easily removed from the polymer surface by simply dropping the incubation temperature below 25°C.

Table 1: Advantages and disadvantages of different types of scaffold.

Scaffold Types	Advantages	Disadvantages	Future Prospects
Porous scaffolds	Extracellular matrix (ECM) secretion can occur in an environment with high porosity, which also supplies the cells with nutrients. Some cell types' hole diameters prevent the cells from aggregating, which stops necrotic centres from forming.	A porous environment prevents a uniform supply of cells. For the different cell types, different pore diameters are necessary, which takes time.	Improvement in the connectivity of pores and thereby the shape of the scaffolds is required.
Fibrous scaffolds	For cell adhesion, proliferation, and differentiation, a very microporous form is ideal. Upon implantation, little inflammatory reaction.	The nanofibers used to build these scaffolds need to be functionalized in their surroundings.	Drugs and biological molecules such as proteins, genes, growth factors, etc., can be incorporated in fibrous scaffolds for release applications.
Hydrogel scaffolds	Highly biocompatible and biodegradation rate inhibition.	Due to soft structures, mechanical strength is constrained.	Degradation character of the hydrogels and tenability should be well-defined. Hydrogels incorporating growth factors to facilitate cell differentiation.
Microsphere scaffolds	Simple fabrication with regulated physical qualities improves cell adhesion and migratory properties and is advantageous for slow or quick drug delivery.	Sometimes, microsphere sintering techniques are incompatible with cells and reduce cell viability.	These scaffolds can be used as a target particular delivery vehicle for the drugs such as antibiotics, anticancer, etc.
Composite scaffolds	High absorbability and mechanical strength with a high biodegradability.	During deterioration, acidic byproducts are created. Mediocre cell affinity. Creating composite scaffolds requires time-consuming work.	Nano-bioceramic and polymer composites with speed degradation are at present being developed.
Acellular scaffolds	Native ECM is preserved, maintaining fundamental anatomical characteristics. Low immunological and inflammatory response and higher mechanical strength.	In order to avoid immunological reactions, incomplete de-cellularization is necessary.	Such scaffolds hold promise towards improving artificial organs.

3.4.1 Ceramic Matrix Nanocomposites (CMCs)

A multiphase solid material known as a nanocomposite has nanoscale constant distances between its constituent phases and includes at least one phase with 1D, 2D, or 3D less than 100 nm. Colloids, porous media, copolymers, and gels are included in this description; nevertheless, it is typically the solid mixture of a bulk matrix and a nanodimensional phase with diverse properties resulting properties brought on by modifications to its chemistry and structure (Ajayan et al., 2003).

In a nanocomposite material, the nanoscale phase typically exhibits an unusually high surface to volume ratio and/or an abnormally high aspect ratio. Sheets, particles, or fibres can make up the reinforcing material. Compared to conventional composite materials, the intermediate region between the matrix and the reinforcing phase is an order of high magnitude. The characteristics of the matrix material close to the reinforcing phase are dramatically affected. Numerous kinds of nanoparticulate may lead to better heat resistance, optical characteristics, dielectric properties, or mechanical properties like strength, stiffness, and wear resistance. A metal makes up the second component in maximal ceramic-matrix nanocomposites. Similar to this, the metallic and ceramic components are uniformly distributed to produce particular nanoscopic characteristics. Zhao et al. (2014) examined the mechanical properties of SiC/AlON because AlON (aluminium oxynitride) is a promising ceramic material in structural applications due to its strong corrosion and wear resistance, high rigidity, and good chemical stability, but low fracture toughness and strength. Benavente et al. discovered a significant increase in the fracture toughness of Al_2O_3-ZrO_2 composites with varying ZrO_2. Al_2O_3-ZrO_2 nanocomposites is suitable for a range of biological and technical applications due to their high wear resistance, good electrical resistivity, low friction, excellent corrosion resistance, and good biocompatibility. Estili and Kawasaki [15] established a method for mass-producing CNTs/alumina nanocomposites and stated that the CNT concentration must be in the range of 2.4–16 vol% for homogeneous dispersion. Walker et al. demonstrated good mechanical performances with a determined 235% increase in toughness compared to the monolith by merely inserting 1.5% volume graphene platelets into Si3N3 ceramic particles using water colloidal methods to create homogeneous platelet dispersion.

3.4.2 Metal-Matrix Nanocomposites

Metal matrix nanocomposites (MMNCs) have gained popularity in recent years, owing to their exceptional properties that make them appropriate for a variety of applications, including functional and structural applications. The development of new materials has attracted a lot of focus over past 20 years the last two decades in order to meet new criteria for working conditions for electronic devices as well as broaden their industrial applications in fields such as automotive, aerospace, and electronic packaging (Rashad et al. 2014, Wang et al. 2016, Perez-Bustamante et al. 2014, Saboori et al. 2018, Tabandeh-Khorshid et al. 2016). Metal matrix nanocomposites (MMNC) are made of a ductile metal or alloy matrix with reinforcing elements implanted at the nanoscale. These materials combine the ductility and toughness of metal with the high strength and modulus of ceramic, respectively. Metal matrix nanocomposites can therefore be used to produce materials with strong shear/compression strength and high service temperature capabilities. They exhibit tremendous application potential in a variety of fields, including the automotive and aerospace sectors and the creation of structural materials (Tjong and Wang 2004). Although both MMNC and CMNC with CNT nanocomposites have potential, they also present difficulties for actual success. Pérez-Bustamante (2014) and colleagues created the Al/GNPs nanocomposites by mechanical alloying followed by traditional press-sintering, and they discovered that increasing the graphene content increased the composite's hardness. In 2011, Bartolucci et al. looked at how graphene affected the mechanical properties of pure aluminium. They used hot isostatic pressing (HIP), hot extrusion, and ball-milling to create the Al/GNPs nanocomposite. According to their findings, the mechanical characteristics of the composites were worse than monolithic aluminium due to the production of aluminium carbide. Gao et al. (2016) recently investigated the impact of graphene nanoplatelets on the mechanical characteristics of pure aluminium and discovered that when the graphene concentration increases, the composite's fracture mechanism shifts from ductile fracture to brittle fracture. The energetic nanocomposite is the following type of nanocomposite that often takes the form of a hybrid sol-gel with a silica foundation that, when combined with metal oxides and nanoscale aluminium powder, creates super thermite materials.

3.4.3 Polymer Nanocomposite

The ease of production, light weight, and frequently ductile character of polymer materials make them widely used in industry. They are less strong and have a lower modulus than metals and ceramics, for example, which are drawbacks. In this situation, reinforcing the polymer matrix with fibres, whiskers, platelets, or other particles is a very effective way to increase mechanical qualities. Polymers have been filled with a variety of inorganic chemicals, either synthetic or natural, to increase their heat and impact resistance, flame retardancy, and mechanical strength, and to decrease their electrical conductivity and gas permeability with regard to oxygen and water vapour, for example (Fischer 2003).

Additionally, metal and ceramic reinforcements provide striking pathways to specific inorganic nanoparticle-derived magnetic, electrical, optical, or catalytic capabilities that add to other polymer qualities including processibility and film forming capacity (Athawale et al. 2003). This method allows polymers to be enhanced while retaining their ductility and light weight (Jordan et al. 2005, Akita and Hattori 1999). Another significant factor is that, as will be seen later, nanoscale reinforcements have a remarkable capacity to produce novel phenomena. This results in unique features in these materials. It should be noted that these composites have reinforcing efficiencies that are comparable to fibres in microcomposites at 40–50%, even at low volume fractions (Ray and Bousmina 2005).

A polymer matrix's performance is improved by adding nanoparticulate by utilising the characteristics and makeup of the nanoscale filler. This process produces the best polymer nanocomposite when the filler properties at the nanoscale are the best and differ from the matrix reinforcement and when the filler is well disseminated. It's possible that improving mechanical qualities won't help to maintain strength or stiffness. Time-dependent characteristics can be generated by using nanofillers. On the other hand, high-performance nanocomposites' emerging properties are a result of the fillers' maximum surface areas and high aspect ratios, as nanoparticulates have improved surface area to volume ratios when there is better dispersion.

4. Chitosan as a Biomaterial for Bone Tissue Engineering

4.1 Physiochemical Properties of Chitosan

Chitosan consists of β-(1–4) linked D-glucosamine with irregularly distributed N-acetylglucosamine groups, depending on the degree of deacetylation of the chitin (Kim et al. 2008). Strong alkaline hydrolysis is essential for this version due to the barrier given by the trans arrangement of the C2–C3 substituents in the sugar ring (Knight et al. 2008). Chitosan is rigid and simple to convert into films with great mechanical strength thanks to hydrogen bonding in its molecular structure. However, the structure of chitosan differs from that of chitin due to the inclusion of amino groups, giving chitosan more impressive properties. Because of its low solubility, chitin has little practical utility. The amino groups of chitosan D-glucosamine, on the other hand, may be protonated and are soluble in diluted acidic water solutions (pH 6) (Lanza et al. 1996).

Chitosan is soluble in mineral acids such as hydrochloric acid; and generates insoluble chitosan sulphate in sulfuric acid. Chitosan's acidic solubility allows it to be processed even in moderate circumstances, which opens the door to a wide range of applications. Furthermore, due of the presence of amino groups, chitosan effectively gives numerous complexes with metal ions that are effectively used for heavy metal removal in wastewater treatment (Orive et al. 2004). Chitosan is also used as a polyelectrolyte in the layer-by-layer deposition of multilayered films. Chitosan is the only naturally occurring positively charged polysaccharide, and as such, it interacts with other negatively charged synthetic polymers like poly(acrylic acid) to form complexes, and it also forms films on negatively charged surfaces like lipids, cholesterol, proteins, and macromolecules. Amino and hydroxyl functional groups on chitosan chains work with other functional groups to create stable covalent bonds.

Due to the hydrophilic amino groups and hydrophobic acetyl groups in its molecular structure, chitosan improves amphiphilic qualities; this has an impact on its physical properties in liquids and solid states. In both people and animals, chitosan improves a number of qualities, including the ability to bind fat and its antibacterial, antifungal, mucoadhesive, analgesic, hemostatic, and wound-healing properties (Uludag et al. 2004). It is capable of decomposing into harmless leftovers. Because of this, chitosan can be used in many different types of medical procedures such topical ocular application, implantation, and injection. Chitosan is an excellent choice for tissue engineering applications due to all of these distinctive qualities.

4.2 Biological Properties of Chitosan

The term "bioactive material" refers to compounds that can affect their surroundings by causing a biological reaction. These substances can be classified as either osteoconductive, osteoproductive, and osteoinductive. Osteoconductive materials are those that can adhere to dense tissue, such bone, and encourage the growth of that tissue along the implant surface. Osteoproductive materials can adhere to cartilage and soft tissue and stimulate the formation of new bone on their surface that is not in touch with an implant or bone. According to Chitosan-based nanomaterials' unique chemical features, including desired biodegradability, compatibility, and nontoxicity, a lot of attention has been dedicated to many biomedical disciplines. In tissue engineering, chitosan is a practical biomaterial for creating extracellular tissue matrixes (Sivashankari and Prabaharan 2016).

For bioactive materials to be successful in tissue engineering, the fundamental properties of gases, metabolites, nutrients, and signal molecules need to be enhanced both within and across various materials in addition to the host environment (Kim et al. 2008). Different kinds of bioactive materials based on metals, polymers, ceramics, and their composites play a vital part in tissue engineering sectors such bone, neuron, muscle, and ligament regeneration (Stratton et al. 2018). However, bioactive mechanisms made of synthetic and/or natural polymer-based materials have more advantages than any of those made of metal or ceramic due to the fact that degraded materials are bioresorbable, which negates the need for a second surgery to remove the implanted devices once the injured tissue has fully healed.

4.3 Chitosan-based Bio Nanocomposites

Biomaterials for tissue engineering, medication delivery systems, and dressings for burns and wounds that are based on chitosan are widely used. Due to the hydrophilic (D-glucosamine) and hydrophobic (N-acetylated residues) parts of its molecules that allow it to generate emulsion systems, chitosan has been used as an emulsifier in the food industry and has emerged as a promising candidate for anti-microbial packaging materials.

Additionally, bio nanocomposites exhibit outstanding qualities like superior mechanical, barrier, and increased transparency. These qualities are a result of the functionality of the surface and the makeup of the nanofillers. Higher aspect ratios and surface areas lead to specific bio nanocomposites being dispersed at the nanoscale. By marginally decreasing reinforcing efficiency compared to standard nanocomposites, these nanoscale bio composites with chitosan display more durable and stable bio nanocomposites matrices materials. Therefore, chitosan-based bio nanocomposites also resemble bioactive and compatible matrices with remarkable mechanical properties (Coakley et al. 2015). As a result, under regulated circumstances, biopolymeric matrix is combined with nanoscale inorganic fillers to create bio nanocomposites, which are identified as the most practical hybrid products. The prefix "bio" denotes that things are environmentally friendly and biodegradable.

Chitosan has emerged as a desirable replacement for environmentally dangerous synthetic plastic polymers since it is a natural biopolymer. Polysaccharides, polynucleic acids, proteins, aliphatic polyesters, and nanofillers are the most widely used biopolymer matrices. Typically, nanofillers consist of nanotube-hydroxyapatite (HA), clay nanoparticles, metal nanoparticles, and nanofibers.

5. Methods and Techniques for Producing Chitosan-based Bio Nanocomposites

Typical methods are commonly used to prepare the chitosan-based bio nanocomposites.

5.1 Solution-casting Method

The simplest approach for creating polymer nanocomposites is solution casting. There are three steps in this approach for creating nanocomposites. Initially, mechanical agitation or sonication are used to dissolve or distribute the fillers in an appropriate solvent. After that, the resulting polymer is melted in a related solvent. Finally, the polymer solution and the filler solution are combined at the proper temperature. The creation of bio nanocomposites occurs when this combination is cast or precipitated. Using a solution-casting method, Regiel-Futyra and his research group created chitosan-gold bio nanocomposites in 2015 (Kuo et al. 2015).

5.2 In situ Technique

The best way to create bio nanocomposites with evenly distributed filler components is using the *in situ* process. In this method, filler chemicals are spread across monomers whether a solvent is present or not. The curing ingredient is then added after, at the precise temperature needed for optimal polymerization. Hebeish's research team used this method in 2014 to successfully create nanocomposites based on chitosan. Silver ions were chemically reduced in graft copolymerization of acrylonitrile onto chitosan films to create poly acrylonitrile silver nanocomposites.

5.3 Electrospinning Technique

As thin as tens of nanometers in diameter, superior and thin fibres can be produced using the electrospinning technology. The three essential parts of an electrospinning instrument are a high-voltage electrical source, a spinneret with several needles, and a grounded conductor that serves as a collector. The electrospinning liquid is loaded into a syringe at a predetermined rate controlled by the syringe pump. Immobilizing charges on the surface of a liquid droplet results in a continuous jet. Nanofibers are created when the fluid filament is whipped and stretched quickly. By strengthening chitin nanocrystals, Naseri's research team was able to create chitosan/polyethene oxide-based fibre mats (Xia et al. 2004).

5.4 Freeze-drying Technique

A homogenous mixture of polymer and filler solution is poured into a copper mould and cooled to extremely low temperatures by quenching in liquid nitrogen. The solvents are then removed from the polymer scaffold by freezing-drying it. The porosity of the developed scaffold is 90%, and the pore sizes range from 15 to 35 m. The use of these scaffolds in tissue engineering has skyrocketed. In 2013, Liu's research team created a chitosan-based bio nanocomposite with halloysite nanotubes by combining the freeze-drying and solution-casting techniques (Wu et al. 2008).

6. Benefits and Drawbacks of Chitosan based Bionanocomposites

6.1 Benefits

6.1.1 Biocompatibility

The biocompatibility of such chitosan-based bionanocomposites represents its most salient quality, which favours their use in pharmaceuticals, cosmetics, food packaging, and wastewater purification. By forming hydrogen bonds with electrostatic attraction, the positively charged ammonium groups ($-NH_3^+$) of chitosan and the negatively charged groups have this property. Thus, the strong linkages that have been created prove that chitosan-based bionanocomposites are highly biocompatible (Huang et al. 2016).

6.1.2 Nontoxicity

Low/non-toxicity is another advantageous quality of chitosan-based bionanocomposites. Numerous research on animals have shown that chitosan derivatives don't have any long-term harmful effects. Additionally, rats treated orally with several chitosan derivatives showed no evidence of toxicity or death (Hu et al. 2017).

6.1.3 Biodegradability

Polymers can be regarded as biodegradable if they break down caused by the actions of microorganisms or enzymes in either aerobic or anaerobic conditions. In the same way, the enzyme N-acetyl-b-D-glycosaminidase (NAGase) lysozyme is responsible for degrading chitosan-based bionanocomposites in humans.

6.1.4 Excellent Film-forming Ability

Bionanocomposites based on chitosan are able to produce thin films that can be employed as food packaging materials. A crucial component of film formation is tensile strength; weak films are more likely to shatter and contaminate food products. It was discovered that chitosan-based bionanocomposite films give packaging material strength and shield the goods from deterioration (Gilarska et al. 2018).

6.1.5 Thermal Stability

Chitosan was previously found to be unable to endure high temperature settings between 200 and 220°C. This was caused by the pure polymer or biopolymer's low thermal expansion coefficient for dimensional stability. The chitosan bionanocomposites, however, exhibit exceptional heat stability. For instance, nanoclay, nanofibrillated cellulose, hydroxyapatite, calcium carbonate, and other components made sure that the chitosan-based bionanocomposite could exert outstanding thermal stability (Huang et al. 2016).

6.1.6 Excellent Mechanical and Barrier Facility

The mechanical characteristics (rigidity and tensile strength) of the bionanocomposites are enhanced by the incorporation of nanoparticles throughout the biopolymer matrix. The barrier and mechanical properties of the polymer films were strengthened in bionanocomposites based on chitosan using a variety of techniques. Researchers discovered that the mechanical properties of chitosan-based bionanocomposites were successfully improved by the hybridization of nanofillers and surface-functionalized nanofillers (Arakawa et al. 2017). Collagen, cellulose, carboxymethyl cellulose, sulfonated cellulose fibres, poly N-vinyl pyrrolidone, polyvinyl alcohol, polyethylene glycol, and polyethylene oxide, among others, may help to improve the mechanical property (Aliramaji et al. 2017).

6.1.7 Strong Anti-microbial Activity

Chitosan bionanocomposites exhibit anti-microbial activity against a wide range of microorganisms. Grass-negative and gram-positive bacteria, yeast, filamentous fungus, and other microorganisms are all susceptible to the anti-microbial properties of chitosan bionanocomposites. The kind of chitosan used, its concentration, MW, and DD all affect its action. As it pertains to the quantity of free amino groups in the molecule, it is thought that DD is the most important element that promotes or retards the antibacterial activity of chitosan-based bionanocomposites. Additionally, Rodrigues' work team discovered that because it can convert the proton to ammonium (NH_3^+), it has significant potential to demonstrate strong anti-microbial activity in an acid media (Zhang et al. 2019).

6.1.8 Anti-oxidant Potential

The high antioxidant qualities of chitosan-based bionanocomposites are another noteworthy advantage. It is utilised to treat oxidative stress and some disorders as a result of its capacity to scavenge free radicals. Researchers discovered that the radical scavenging activity and anti-oxidant capabilities of chitosan film with rosehip seed oil were dramatically improved (Kim et al. 2015).

6.1.9 Anti-fungal Potential

Chitosan's biocidal effect on microorganisms demonstrates excellent economic potential in the healthcare and agricultural industries. Additionally, numerous studies have demonstrated that chitosan's anti-microbial activity is carried out by concentrating on the cell surface and interfering with cellular energy, suggesting that chitosan is a good alternative course of action for antifungal medication. The anti-fungal capabilities of chitosan-based bionanocomposites aid in the management of plant diseases brought on by fungi in the agricultural industry. Chitosan-based bionanocomposites' antifungal qualities aid in the management of crop and fruit diseases. These characteristics resulted from a wide variety of pH levels that prevented the microbe strain from growing at the desired rate. As a result, if fruits are stored at room temperature, their shelf life may be increased by two to three times.

6.2 Drawbacks and Challenges

Despite the molecule's exceptional biological and physiochemical characteristics, a few flaws prevent it from being widely used in some important fields and present serious difficulties. Below is a list of the most significant drawbacks that require immediate attention.

Due to physiological pH, it has low solubility, which could provide a problem for the medical area.

1. Despite being approved as a food-contact substance by the FDA, the European Food Safety Authority disagrees with it (Kim et al. 2015). Therefore, the issue of cleanliness and safety remains unresolved.

2. The low colloidal stability of chitosan-based bionanocomposites is another significant flaw that prevents their widespread use for drug delivery (Tan et al. 2009).

3. Another disadvantage that restricts the use of chitosan-based bionanocomposites is their high degree of elasticity.

4. For various medical applications, particularly drug delivery, the chitosan-based bionanocomposite exhibits satisfactory results. However, the researchers continue to run across problems with drug release efficiency, drug loading capacity, bionanocomposites material degradation rate, delivery time, and many other issues (Ruel-Gariepy et al. 2004).

5. Lastly, industrial processing facilities still have a financial challenge in demonstrating the viability of sustainable biopolymers on the open market. The examination of the potential uses and advantages of chitosan-based bionanocomposites appears to be a viable biopolymer for potential future global concerns. However, some of the portrayed issues and difficulties need to be resolved.

7. Applications of Chitosan-based Nanocomposites in Bone Tissue Engineering

7.1 Chitosan based Scaffolds

The scaffolds for bone tissue engineering were created utilising chitosan and had a highly interconnected porosity structure. The ability of these scaffolds to replicate the qualities of th etissue is their primary necessity. In order to replace lost or damaged bones, these scaffolds must be biocompatible, osteoconductive, osteoinductive, and mechanically strong (Sivashankari and Prabaharan 2016). The creation of scaffolds for bone tissue engineering has been the subject of extensive research using both natural and synthetic materials. In the construction of scaffolds, chitosan is highly valued. However, compared to regular bone, chitosan scaffolds' mechanical characteristics are poorer (Islam et al. 2018, Biswas et al. 2018). They cannot meet the load-bearing demands of bone implants as a result. Furthermore, chitosan itself is not osteoconductive, making it unable to mimic the characteristics of real bones. To improve the mechanical strength and structural integrity of chitosan biocomposites for bone tissue engineering applications, biopolymers such as chitin, silk, Alg, GL, PCL, PLA, and HA, as well as bioactive nano ceramics such as HAp, SiO_2, TiO_2, ZrO_2, etc., are created (Saravanan et al. 2016, Leena et al. 2017).

7.1.1 Chitosan/Nano Hydroxyapatite Scaffold

Numerous studies have shown that HAp $[Ca_{10}(PO_4)_6(OH)_2]$, one of the most stable forms of calcium phosphate and an important component of bone (60–65%), can enhance the mechanical characteristics and osteoconductivity of implants (Kong et al. 2005). HAp also interacts with the living system and promotes the production of new bone without resorption. Additionally, nanostructured HAp has greater bioactivity and a higher surface area. The organic and inorganic components of actual bone can therefore be mimicked by composites consisting of chitosan and HAp, which are now being further researched (Venkatesan and Kim 2010, Li et al. 2002). Collagen-chitosan-polyethylene-glycol-HAp was combined with freeze drying and dehydrothermal cross-linking procedures to create a porous 3D scaffold (Kozlowska et al. 2019). In this study, the impact of HAp in chitosan-based materials was examined. After HAp and PEG were added to the composites, the mechanical properties of the chitosan-collagen scaffolds were improved. This suggested that the deformation rate and compression resistance had increased (Kozlowska et al. 2019). However, composites made from HAp and chitosan exhibit inferior mechanical strength due to increased brittleness. Kar et al. created a composite by incorporating clay minerals into a HAp-chitosan composite to address this issue (Kar et al. 2016). Clay has a layered silicate structure, so adding clay of the montmorillonite (MMT) variety can improve the mechanical qualities of a composite (Islam et al. 2017). To enhance polymer matrix miscibility, the scientists added alkyl ammonium salts to the MMT clay (Kar et al. 2016).

7.1.2 Chitosan/Nano Bioactive Glass Ceramics (BGC)

When creating scaffolds for bone tissue engineering, chitosan is combined with synthetic polymers such as poly methyl methacrylate (PMMA), polyethylene glycol, polycaprolactone, poly lactic acid, etc. to affect the mechanical properties and biocompatibility of the finished composite (Pereira et al. 2019, Shakir et al. 2015, Ghaee et al. 2017, Jing et al. 2015, Cui et al. 2010). Due to their low hydrophilicity and lack of cell recognition sites, exclusively synthetic polymer scaffolds have poor cell affinity; as a result, mixtures of synthetic and natural polymers have gained popularity in research (Venugopal et al. 2005). Due to qualities like mechanical strength and moldability, PMMA has been utilised as bone cement for a long time (Hide et al. 2004). PMMA polymerizes at high temperatures, but this might result in thermal injury to the tissue around it. PMMA is coupled with chitosan, bioactive glass (BG), calcium phosphate ($Ca_3(PO4)_2$, and certain bioinorganic compounds to solve this issue.

7.1.3 Chitosan/Nano SiO$_2$

To create mesoporous materials, nanobio-ceramics including hydroxyapatite, silica, and bioglass are frequently employed. Incorporating such a nanobio-ceramic not only results in great drug loading capacity but also good cytocompatibility and implant osteointegration. Increased drug loading and sustained drug release characteristics are the goals of the functionalization of nanobio-ceramics within the pore walls. This is accomplished by swapping out the hydrogen atom in silanol for an organic group R that can form a covalent bond with the oxygen atom. Thus, it is conceivable to transplant specific biomolecules that act as signals to improve or trigger the desired *in vivo* response, such as peptides, proteins, or growth factors.

7.1.4 Chitosan/Nano TiO$_2$

A study on the impact of various TiO$_2$ nanostructures on the surface of Ti was undertaken by Kumbar et al. in 2014. Numerous surface nanomorphologies, including mesoporous nanoscaffolds, nanoflowers, nanoneedles, nanorods, and octahedral bipyramids, were produced utilising the hydrothermal approach. This was accomplished by changing the hydrothermal conditions, including the composition of the reaction medium, its concentration, its duration, and its temperature. All Ti plates with nanostructures demonstrated improved fibronectin and vitronectin adsorption when compared to polished Ti plates, with the octahedral bipyramid adsorbing the greatest amount of adhesion proteins. The vitality and proliferation capabilities of osteoblasts *in vitro* and *in vivo* were then examined in a subsequent study using such nanomorphologies on the surface of Ti screws.

7.1.5 Chitosan/Nano ZrO$_2$

Implants for prosthetic limbs have articulating surfaces made of zirconia toughened alumina (ZTA) (Lao et al. 2008). Zirconia nanoparticles' nanocomposite properties are spread throughout an alumina matrix. After 8 million cycles, the material showed good cytocompatibility and less wear debris than alumina (Wu et al. 2008). ZTA's high hardness, excellent elasticity, and resistance to age deterioration make it an attractive material for tissue engineering applications involving significant loading loads (Islama et al. 2020). Numerous studies have used nanocomposites scaffolds constructed of nano bioceramics in these chitosan-based composites for bone tissue engineering. In Table 2, a few of the nanocompsites were listed.

7.2 Chitosan based Hydrogels

The bones that make up the human skeleton are incredibly useful connective tissues that mechanically support the body, protect internal organs, and participate in a number of physiological processes. When damaged, bone tissues can heal themselves by inducing MSCs to differentiate into osteoblasts and creating new blood vessels for small-sized bone defects, but for larger abnormalities, bone transplants should be used in clinical operations (Gómez-Barrena et al. 2015). Because of low immunogenicity, physiological inertia, osteoconductivity and osteogenesis, CS based hydrogels have excellent application possibilities in bone tissue regeneration (Jiang et al. 2019).

Table 2: Nanocomposites featuring nano bio ceramics applied in bone tissue engineering.

Scaffolds	Components
Aligned nanofibrous multi-component scaffolds (Subramanian et al. 2005)	PLGA/collagen/nano-HA
Mineralized nanofibers (Kuo et al. 2009)	Calcium phosphate contained collagen fibrils
Nanocomposites (Tan et al. 2009)	Collagen/nano-HA/nano-ACP
Double membrane (Chong et al. 2007)	ACP/collagen/PLGA
Biodegradable composite scaffolds (Mallick et al. 2016)	Nano-HA/chitosan/CMC
Nanofibrous scaffolds	PCL/nano-HA or PCL/TCP
Nanofibrous membranes (Mallick et al. 2018)	PCL/nano-HA
Nanofibrous composite scaffolds (Keikhaei et al. 2019)	Nano-sized demineralized bone powders
(DBPs)/PLA Composite scaffolds (Aranaz et al. 2017)	Chitosan/nano-HA
Nanostructured mesoporous silicon fibrous scaffolds (Subia et al. 2010)	Silicon (PSi)/PCL fibers
Electrospun fibre calcium phosphate cements	PLGA/CPCs (tetra calcium phosphate, dicalcium phosphate anhydrous and chitosan lactate)
Solid casted composite film (Liu 2017)	Nano-PLGA-HA

Additionally, in order to build multifunctional biomaterials scaffolds, CS-based hydrogels must constantly be coupled with other synthetic or natural polymers, as well as bioactive pharmaceutical compounds, to increase their mechanics and numerous functions. For instance, Oudadesse et al. (2020) suggested employing the freeze-gelation approach to create a nBG-loaded hybrid CS-based hydrogel. By adjusting the BGN proportion, it is possible to modify the surface area, porosity, and mechanical properties of the BGN/CH composite hydrogel. This could encourage the formation of an apatite layer on the hydrogel's surface, which would facilitate direct bone bonding with the hydrogels after implantation. By interspersing a methacrylate gelatin network into a nanocomposite hydrogel made of methacrylate chitosan and polyhedral oligomeric silsesquioxane, Zhang et al. (2019) created a biodegradable hybrid DN hydrogel. Using a rat with calvarial abnormalities, hybrid DN hydrogels were able to more easily manipulate the MSCs toward osteogenic differentiation in vitro and speed up new bone regeneration *in situ* (Tang et al. 2020).

8. Advances in Chitosan based Nanocomposites for Bone Tissue Engineering

Designing, building, and improving three-dimensional (3D) scaffolds with a highly interconnected porous structure requires the application of knowledge from the fields of chemistry, engineering, and biological science. This discipline is known as "bone tissue engineering." The ability of these scaffolds to match the characteristics of the tissue that they will replicate is their most important prerequisite. Additionally, these scaffolds should be strong enough to heal weakened or injured bones, osteoconductive, osteoinductive, and biocompatible (Middleton et al. 2000). In-depth research has been done to develop scaffolds for bone tissue engineering using both natural and synthetic materials. More emphasis has been placed on chitosan in the manufacture of scaffolds.

But chitosan scaffolds' mechanical qualities fall short of those of regular bone (Pereira et al. 2015). They therefore are unable to support the load-bearing demands of bone implants. Additionally, chitosan is not osteoconductive, making it unable to mimic the characteristics of real bones. To improve the mechanical strength and structural integrity of chitosan biocomposites for bone tissue engineering applications, biopolymers including of natural bones. Biopolymers such as chitin, silk,

Alg, GL, PCL, PLA, and HA, as well as bioactive nano ceramics such as HAP, SiO_2, TiO_2, ZrO_2, etc., have been created.

Alkylammonium salts are substituted for MMT clay by the authors in order to enhance polymer matrix miscibility (Islam et al. 2018). The mechanical properties and bioactivity of this composite were improved, and it was also non-toxic. Cell viability was low, but it was still within a manageable range. A recent study reported the use of biomimetic chitosan/hydroxyapatite-zinc oxide nanocomposites to aid in the organic modification of montmorillonite clay (OMMT) (Islam et al. 2017). The coating bioactive trace elements like Sr^{2+}, Zn^{2+}, Mg^{2+}, Cu^{2+}, and Si^{4+} into HAP by ion replacement on the other hand, can promote osteogenic differentiation of mesenchymal stem cells and speed up angiogenesis (MSCs). Due to its ability to promote the formation of new bone and prevent bone resorption, strontium (Sr) has emerged as a desirable option in bone tissue engineering (Prabhakaran et al. 2017).

The criteria that influence the fabrication of scaffold for bone tissue engineering applications are good mechanical properties and biocompatibility of the composites (Cui et al. 2010). Due to the absence of cell recognition sites and reduced hydrophilicity of exclusively synthetic polymer scaffolds, combinations of natural and synthetic polymers have gained popularity in research (Venugopal et al. 2005). Additionally, the hydrogel's ability to be injected has the benefit of filling small, uneven flaws and transporting cells and medicinal substances.

9. Recent Innovation and Future Prospects

As nanotubes and nanopowders' capabilities grow and are integrated with cutting-edge processing methods, new multifunctional materials are being improved. Rapid improvements in commercial scale nanotube production have resulted from carbon nanotubes' distinct mechanical, electrical, and thermal properties.

Alumina nanocomposites containing CNTs were enhanced in the most recent study. Despite the fact that ceramic matrices are prone to CNT aggregation, dimethylformamide was designed to successfully disperse the CNTs throughout the material. The produced materials had outstanding electrical and mechanical properties. It was discovered that alumina-doped CNTs were more effective at achieving intermediate adhesion with the ceramic matrix. This hydrogel might introduce a novel and unconventional method for the engineering of bone tissue. Two naturally occurring polymers with polysaccharide bases are chondroitin sulphate and hyaluronic acid. Additionally, due of the synergetic impact between Ca^{2+} and Sr^{2+} ions, Sr-HAp chitosan scaffolds increase very good osteoinductivity, making them an outstanding choice for bone engineering application.

10. Conclusion

Excellent physio-chemical characteristics and particular interactions with proteins, cells, and biological things are displayed by chitosan. Primary amines are found along the backbone of its structure, which aids in the formation of polycation (following protonation in acidic circumstances), allowing this material to be processed with other anionic polymers in a variety of shapes. The successful usage of chitosan-based bioactive materials in a variety of tissues and organs, including bone, blood vessels, cornea, cartilage, and skin, suggests, according to the researchers, that these materials have a bright future for applications in fixing, repair, and regeneration. Because chitosan-based matrices naturally mimic host environments in terms of structure, chemistry, and functionalities, it is envisaged that their application in bone tissue engineering will have a significant impact on the quality of the tissue outputs. The process of modifying materials to improve cell-specific interactions, the development of more complex tissue structures, and the long-term biocompatibility impact of using these materials within the human body all require further study despite recent technological advancements.

References

Abbah, S.A., J. Liu, J.C. Goh and H.K. Wong. 2013. Enhanced control of *in vivo* bone formation with surface functionalized alginate microbeads incorporating heparin and human bone morphogenetic protein-2. Tissue Eng. Part A 19: 350–359.

Aliramaji, S., A. Zamanian and M. Mozafari. 2017. Super-paramagnetic responsive silk fibroin/chitosan/magnetite scaffolds with tunable pore structures for bone tissue engineering applications. Mater. Sci. Eng. C 70: 736–744.

Ana Claudia M. Renno, Paulo Sérgio Bossini, Murilo C. Crovace, Ana Candida M. Rodrigues, Edgar Dutra Zanotto and Nivaldo Antonio Parizotto. Characterization and *in vivo* biological performance of biosilicate. BioMed Research International, vol. 2013, Article ID 141427, 7 pages, 2013. https://doi.org/10.1155/2013/141427.

Arakawa, C., R. Ng, S. Tan, S. Kim, B. Wu and M. Lee. 2017. Photopolymerizable chitosan–collagen hydrogels for bone tissue engineering. J. Tissue Eng. Regen. Med. 11(1): 164–174.

Aranaz, I., E. Martinez-Campos, C. Moreno-Vicente, A. Civantos, S. Garcia-Arguelles and F. del Monte. 2017. Macroporous calcium phosphate/chitosan composites prepared via unidirectional ice segregation and subsequent freeze-drying. Materials 10(5): 516.

Athawale, A.A., S.V. Bhagwat, P.P Katre, A.J. Chandwadkar and P. Karandikar. 2003. Aniline as a stabilizer for metal nanoparticles. Materials Letters 57(24-25): 3889–3894.

Ajayan, P.M., L.S. Schadler and P.V. Braun (eds.). 2003. Nanocomposite Science and Technology. Wiley.

Bartolucci, S.F., J. Paras, M.A. Rafiee, J. Rafiee, S. Lee, D. Kapoor and N. Koratkar. 2011. Graphene–aluminum nanocomposites. Mater. Sci. Eng. A 528: 7933–7937.

Benavente, R., M. Salvador, F. Penaranda-Foix, E. Pallone and A. Borrell. 2014. Mechanical properties and microstructural evolution of alumina-zirconia nanocomposites by microwave sintering. Ceramics International 40(7): 11291–11297.

Biswas, S., M.M. Islam, M. Hasan, S. Rimu, M. Khan, P. Haque and M. Rahman. 2018. Evaluation of Cr (VI) ion removal from aqueous solution by bio-inspired chitosan clay composite: Kinetics and isotherms. Iran. J. Chem. Eng. 15(4).

Boccaccini, A.R. and J.J. Blaker. 2005. Bioactive composite materials for tissue engineering scaffolds. Expert. Rev. Med. Devices 2: 303–317.

Braghirolli, D.I., D. Steffens and P. Pranke. 2014. Electrospinning for regenerative medicine: A review of the main topics. Drug Discov. Today 19: 743–753.

Cai, Y. and R. Tang. 2008. Calcium phosphate nanoparticles in biomineralization and biomaterials. J. Mater. Chem. 18: 3775.

Cartmell, S. 2009. Controlled release scaffolds for bone tissue engineering. Journal of Pharmaceutical Sciences 98(2): 430–44.

Chan, B.P. and K.F. So. 2005. Photochemical crosslinking improves the physicochemical properties of llagen scaffolds. J. Biomed. Mater. Res. Part A 75: 689–701.

Chan, B.P. and K.W. Leong. 2008. Scaffolding in tissue engineering: General approaches and tissue-specific considerations. Eur. Spine J. 17: 467–479.

Chan, B.P., T.Y. Hui, O.C. Chan, K.F. So, W. Lu, K.M. Cheung, E. Salomatina and A. Yaroslavsky. 2007. Photochemical cross-linking for collagen-based scaffolds: A study on optical properties, mechanical properties, stability, and hematocompatibility. Tissue Eng. 13: 73–85.

Chen, Q. and G.A. Thouas. 2015. Biomaterials, A Basic Introduction, CRC Press, Boca Raton, USA.

Chevalier, E., D. Chulia, C. Pouget and M. Viana. 2008. Fabrication of porous substrates: A review of processes using pore forming agents in the biomaterial field. J. Pharm. Sci. 97: 1135–1154.

Chong, E.J., T.T. Phan, I.J. Lim, Y. Zhang, B.H. Bay, S. Ramakrishna and C.T. Lim. 2007. Evaluation of electrospun PCL/gelatin nanofibrous scaffold for wound healing and layered dermal reconstitution. Acta Biomater. 3(3): 321–330.

Chudhuri, B., D. Bhadra, S. Dash, G. Sardar, K. Pramanik and B. Chaudhuri. 2013. Hydroxyapatite and hydroxyapatite-chitosan composite from crab shell. J. Biomater. Tissue Eng. 3: 653.

Coakley, D.N., F.M. Shaikh, K. O'Sullivan, E.G. Kavanagh, P.A. Grace and T.M. McGloughlin. 2015. *In vitro* evaluation of acellular porcine urinary bladder extracellular matrix—A potential scaffold in tissue engineered skin. Wound Med. 10: 9–16.

Cui, W., Y. Zhou and J. Chang. 2010. Electrospun nanofibrous materials for tissue engineering and drug delivery. Sci. Technol. Adv. Mater. 11(1): 014108.

Dalton, P.D., C. Vaquete, B.L. Farrugia, T.R. Dargaville, T.D. Brown and D.W. Hutmacher. 2013. Electrospinning and additive manufacturing: Converging technologies. Biomater. Sci. 1: 171–185.

Dhandayuthapan, B., Y. Yoshida, T. Maekawa and D.S. Kumar. 2011. Polymeric scaffolds in tissue engineering application: A review. International Journal of Polymer Science, vol. 2011, Article ID 290602, 19 pages.

Di Martino, A., M. Sittinger and M.V. Risbud. 2005. Chitosan: A versatile biopolymer for orthopaedic tissue-engineering. Biomaterials 26: 5983.

Dvir, T.B., P. Timko, D.S. Kohane and R. Langer. 2011. Nanotechnological strategies for engineering complex tissues. Nature Nanotechnology 6(1): 13–22.

Estili, M. and A. Kawasaki. 2008. An approach to mass-producing individually alumina-decorated multi-walled carbon nanotubes with optimized and controlled compositions, Scripta Materialia, 58(10): 906–909.

Fergal, O.B.J. 2011. Biomaterials and scaffolds for tissue engineering. Mater. Today 14: 88–95.

Francis Suh, J.-K. and H.W. Matthew. 2000. Application of chitosan based polysaccharide biomaterials in cartilage tissue engineering: A review. Biomaterials 21: 2589.

Gao, X., H. Yue, E. Guo, H. Zhang, X. Lin, L. Yao and B. Wang. 2016. Preparation and tensile properties of homogeneously dispersed graphene reinforced aluminum matrix composites. Mater. Des. 94: 54–60.

Ghaee, A., J. Nourmohammadi and P. Danesh. 2017. Novel chitosan-sulfonated chitosan polycaprolactone-calcium phosphate nanocomposite scaffold. Carbohydr. Polym. 157: 695–703.

Giannitelli, S.M., D. Accoto, M. Trombetta and A. Rainer. 2014. Current trends in the design of scaffolds for computer-aided tissue engineering. Acta Biomater. 10: 580–594.

Gilarska, A., J. Lewandowska-Łańcucka, W. Horak and M. Nowakowska. 2018. Collagen/chitosan/hyaluronic acid–based injectable hydrogels for tissue engineering applications–design, physicochemical and biological characterization. Colloids Surf. B Biointerfaces 170: 152–162.

Hide, I. and A. Gangi. 2004. Percutaneous vertebroplasty: History, technique and current perspectives. Clin. Radiol. 59(6): 461–467.

Ho, M.H., P.Y. Kuo, H.J. Hsieh, T.Y. Hsien, L.T. Hou, J.Y. Lai and D.M. Wang. 2004. Preparation of porous scaffolds by using freeze-extraction and freeze-gelation methods. Biomaterials 25: 129–138.

Hollister, Scott J. 2005. Porous scaffold design for tissue engineering. Nat. Mater. 4.7: 518–524.

Hu, Y., J. Chen, T. Fan, Y. Zhang, Y. Zhao, X. Shi and Q. Zhang. 2017. Biomimetic mineralized hierarchical hybrid scaffolds based on in situ synthesis of nano-hydroxyapatite/chitosan/chondroitin sulfate/hyaluronic acid for bone tissue engineering. Colloids Surf. B Biointerfaces 157: 93–100.

Huang, Y., X. Zhang, A. Wu and H. Xu. 2016. An injectable nano-hydroxyapatite (n-HA)/glycol chitosan (G-CS)/hyaluronic acid (HyA) composite hydrogel for bone tissue engineering. RSC Adv. 6(40): 3529–33536.

Hutmacher, D.W. 2000. Scaffolds in tissue engineering bone and cartilage. Biomaterials 21(24): 2529–2543.

Islam, M.M., M.N. Khan, S. Biswas, T.R. Choudhury, P. Haque, T.U. Rashid and M.M. Rahman. 2017. Preparation and characterization of bijoypur clay-crystalline cellulose composite for application as an adsorbent. Adv. Mater. Sci. 2: 1–7.

Islam, M.M., S. Biswas, M.M. Hasan, P. Haque, S.H. Rimu and M.M. Rahman. 2018. Studies of Cr (VI) adsorption on novel jute cellulose-kaolinite clay biocomposite. Desalination Water Treat. 123: 265–276.

Islam, M.M., Md. Shahruzzaman, Shanta Biswas, Md. Nurus Sakib and Taslim Ur Rashid. 2020. Chitosan based bioactive materials in tissue engineering applications—A review. Bioactive Materials 5(1): 164–183.

Jayakumar, R., M. Prabaharan, P. Sudheesh Kumar, S. Nair and H. Tamura. 2011. Biomaterials based on chitin and chitosan in wound dressing applications. Biotechnol. Adv. 29: 322.

Jiang, D. and J. Zhang. 2019. Calcium phosphate with well controlled nanostructure for tissue engineering. Curr. Appl. Phys. 9: S252.

Jie, W. and L. Yubao. 2004. Tissue engineering scaffold material of nanoapatite crystals and polyamide composite. Eur. Polym. J. 40: 509.

Jing, X., H.Y. Mi, X.C. Wang, X.F. Peng and L.S. Turng. 2015. Shish-kebab-structured poly (ε-caprolactone) nanofibers hierarchically decorated with chitosan–poly (ε-caprolactone) copolymers for bone tissue engineering. ACS Appl. Mater. Interfaces 7(12): 6955–6965.

Jordan, J., K.I. Jacob, R. Tannenbaum, M.A. Sharaf and I. Jasiuk. 2005. Experimental trends in polymer nanocomposites: a review. Materials Science and Engineering: A 393(1-2): 1–11.

Kar, S., T. Kaur and A. Thirugnanam. 2016. Microwave-assisted synthesis of porous chitosan–modified montmorillonite–hydroxyapatite composite scaffolds. Int. J. Biol. Macromol. 82: 628–636.

Keane, T.J. and S.F. Badylak. 2014. Biomaterials for tissue engineering applications. Seminars in Pediatric Surgery 23(3): 112–118.

Keikhaei, S., Z. Mohammadalizadeh, S. Karbasi and A. Salimi. 2019. Evaluation of the effects of β-tricalcium phosphate on physical, mechanical and biological properties of poly (3-hydroxybutyrate)/chitosan electrospun scaffold for cartilage tissue engineering applications. Mater. Technol. 1–11.

Kim, I.Y., S.J. Seo, H.S. Moon, M.K. Yoo, I.Y. Park, B.C. Kim and C.S. Cho. 2008. Chitosan and its derivatives tissue engineering applications. Biotechnology Advances 26(1): 1–21.

Kim, B.S., C.E. Baez and A. Atala. 2000. Biomaterials for tissue engineering. World J. Urol. 8: 2–9.

Knight, R.L., H.E. Wilcox, S.A. Korossis, J. Fisher and E. Ingham. 2008. The use of acellular matrices for the tissue engineering of cardiac valves. Proc. Inst. Mech. Eng. 22: 129–143.

Kong, L., Y. Gao, W. Cao, Y. Gong, N. Zhao and X. Zhang. 2005. Preparation and characterization of nano-hydroxyapatite/chitosan composite scaffolds. J. Biomed. Mater. Res. Part A: Off. J. Soc. Biomater. Jpn. Soc. Biomater. Aust. Soc. Biomater. Kor. Soc. Biomater. 75(2): 275–282.

Kozlowska, J., N. Stachowiak and A. Sionkowska. 2019. Preparation and characterization of collagen/chitosan poly (ethylene glycol)/nanohydroxyapatite composite scaffolds. Polym. Adv. Technol. 30(3): 799–803.

Kuo, Y.C. and Y.R. Hsu. 2015. Tissue-engineered polyethylene oxide/chitosan scaffolds as potential substitutes for articular cartilage. J. Biomed. Mater. Res. Part A: Off. J. Soc. Biomater. Jpn. Soc. Biomater. Aust. Soc. Biomater. Kor. Soc. Biomater. 91(1): 277–287.

Langer, R. and D.A. Tirrell. 2004. Designing materials for biology and medicine. Nature 428: 487–492.

Langer, R. and J.P. Vacanti. 1993. Tissue engineering. Science 260(5110): 920–926.

Lanza, R., R. Langer and J. Vacanti. 2014. Principles of Tissue Engineering. Elsevier, San Diego, Calif, USA.

Lanza, R.P., J.L. Hayes and W.L. Chick. 1996. Encapsulated cell technology. Nat. Biotechnol. 14: 1107–1111.

Leong, K.F., C.M. Cheah and C.K. Chua. 2003. Solid free form fabrication of three-dimensional scaffolds for engineering replacement tissues and organs. Biomaterials 24(13): 2363–2378.

Li, B., Q. Hu, X. Qian, Z. Fang and J. Shen. 2002. Bioabsorbable chitosan/hydroxyapatite composite rod prepared by *in situ* precipitation for internal fixation of bone fracture. Acta Polym. Sin. (6): 828–833.

Li, X., Q. Feng, X. Liu, W. Dong and F. Cui. 2006. Collagen-based implants reinforced by chitin fibres in a goat shank bone defect model. Biomaterials 27: 1917.

Liu, H. 2017. Chitosan/hydroxyapatite composite scaffolds for articular cartilage injury, Chin. J. Tissue Eng. Res. 21(2): 244–248.

Lutolf, M.P., P.M. Gilbert and H.M. Blau. 2009. Designing materials to direct stem-cell fate. Nature 462: 433–441.

Mallick, S.P., B.N. Singh, A. Rastogi and P. Srivastava. 2018. Design and evaluation of chitosan/poly (L-lactide)/ pectin based composite scaffolds for cartilage tissue regeneration. Int. J. Biol. Macromol. 112: 909–920.

Mallick, S.P., K. Pal, A. Rastogi and P. Srivastava. 2016. Evaluation of poly (L-lactide) and chitosan composite scaffolds for cartilage tissue regeneration. Des. Monomers Polym. 19(3): 271–282.

Minas, T. and S. Nehrer. 1997, June. Current concepts in the treatment of articular cartilage defects. Orthopedics 20(6): 525–38. Doi: 10.3928/0147-7447-19970601-08. PMID: 9195635.

Miranda, S.C.C.C.G., A.B. Silva, R.C.R. Hell, M.D. Martins, J.B. Alves and A.M. Goes. 2011. Three-dimensional culture of rat BMMSCs in a porous chitosan-gelatin scaffold: A promising association for bone tissue engineering in oral reconstruction. Archives of Oral Biology 56(1): 1–15.

Mostafa, N.Y. 2005. Characterization, thermal stability and sintering of hydroxyapatite powders prepared by different routes. Mater. Chem. Phys. 94: 333.

Nalwa, H.S. (ed.). 2000. Handbook of Nanostructured Materials and Nanotechnology, Academic Press, San Diego, CA, 1–5.

Obregon, R., J. Ramon-Azcon, S. Ahadian, H. Shiku, H. Bae, M. Ramalingam and T. Matsue. 2014. The use of microtechnology and nanotechnology in fabricating vascularized tissues. J. Nanosci. Nanotechnol. 14: 487.

Okano, T., N. Yamada, H. Sakai and Y. Sakurai. 1993. A novel recovery system for cultured cells using plasma-treated polystyrene dishes grafted with poly (N-isopropylacrylamide). J. Biomed. Mater. Res. 27: 1243–1251.

Okano, T., N. Yamada, M. Okuhara, H. Sakai and Y. Sakurai. 1995. Mechanism of cell detachment from temperature-modulated, hydrophilic-hydrophobic polymer surfaces. Biomaterials 16: 297–303.

Orive, G., R.M. Hernández, A.R. Gascón, R. Calafiore, T.M. Chang, P. De Vos, G. Hortelano, D. Hunkeler, I, Lacík, A.M. Shapiro and J.L. Pedraz. 2003. Cell encapsulation: Promise and progress. Nat. Med. 9: 104–107.

Orive, G., R.M. Hernandez, A. Rodriguez Gascon, R. Calafiore, T.M. Chang, P. de Vos, G. Hortelano, D. Hunkeler, I. Lacik and J.L. Pedraz. 2004. History, challenges and perspectives of cell microencapsulation. Trends Biotechnol. 22: 87–92.

Pereira, I.C., A.S. Duarte, A.S. Neto and J. Ferreira. 2015. Chitosan and polyethylene glycol based membranes with antibacterial properties for tissue regeneration. Mater. Sci. Eng. C 96(2019): 606–615.

Place, E.S., N.D. Evans and M.M. Stevens. 2009. Complexity in biomaterials for tissue engineering. Nat. Mater. 8: 457–470.

Pon-On, W., P. Suntornsaratoon, N. Charoenphandhu, J. Thongbunchoo, N. Krishnamra and I.M. Tang. 2018. Synthesis and investigations of mineral ions-loaded apatite from fish scale and PLA/chitosan composite for bone scaffolds. Mater. Lett. 221: 143–146.

Rabea, E.I., M.E.T. Badawy, C.V. Stevens, G. Smagghe and W. Steurbaut. 2003. Chitosan as antimicrobial agent: Applications and mode of action. Biomacromolecules 4: 1457.

Rinaudo, M. 2006. Chitin and chitosan: Properties and applications. Progress in Polymer Science (Oxford) 31(7): 603–632.

Rodríguez-Vázquez, M., B. Vega-Ruiz, R. Ramos-Zúñiga, D.A. Saldaña-Koppel and L.F. Quiñones-Olvera. 2015. Chitosan and its potential use as a scaffold for tissue engineering in regenerative medicine. BioMed Research International. Article ID 821279, 15 pages.

Rogina, A., L. Pribolšan, A. Hanžek, L. Gómez-Estrada, G.G. Ferrer, I. Marijanović, M. Ivanković and H. Ivanković. 2016. Macroporous poly (lactic acid) construct supporting the osteoinductive porous chitosan-based hydrogel for bone tissue engineering. Polymer 98: 172–181.

Salgado, A.J., O.P. Coutinho and R.L. Reis. 2004. Bone tissue engineering: State of the art and future trends. Macromolecular Bioscience 4(8): 743–765.

Schoen, F.J. and R.J. Levy. 2005 March. Calcification of tissue heart valve substitutes: Progress toward understanding and prevention. Ann. Thorac. Surg. 79(3): 1072–80.

Shabafrooz, V., M. Mozafari, D. Vashaee and L. Tayebi. 2014. Electrospun nanofibers: From filtration membranes to highly specialized tissue engineering scaffolds. J. Nanosci. Nanotechnol. 14: 522.

Shakir, M., R. Jolly, M.S. Khan, N.E. Iram, T.K. Sharma and S.I. Al-Resayes. 2015. Synthesis and characterization of a nano-hydroxyapatite/chitosan/polyethylene glycol nanocomposite for bone tissue engineering. Polym. Adv. Technol. 26(1): 41–48.

Sivashankari, P.R. and Prabaharan, M. 2016, Dec. Prospects of chitosan-based scaffolds for growth factor release in tissue engineering. Int. J. Biol. Macromol. 93(Pt B): 1382–1389. Doi: 10.1016/j.ijbiomac.2016.02.043. Epub 2016 Feb 17. PMID: 26899174.

Skallevold, H.E., D. Rokaya, Z. Khurshid and M.S. Zafar. 2019, Nov. Bioactive glass applications in dentistry. Int. J. Mol. Sci. 27; 20(23): 5960. Doi: 10.3390/ijms20235960. PMID: 31783484; PMCID: PMC6928922.

Subramanian, A., D. Vu, G.F. Larsen and H.-Y. Lin. 2005. Preparation and evaluation of the electrospun chitosan/PEO fibers for potential applications in cartilage tissue engineering. J. Biomater. Sci. Polym. Ed. 16(7): 861–873.

Takezawa, T., Y. Mori and K. Yoshizato. 1990. Cell culture on a thermo-responsive polymer surface. Biotechnology 8: 854–856.

Tan, H., J. Wu, L. Lao and C. Gao. 2009. Gelatin/chitosan/hyaluronan scaffold integrated with PLGA microspheres for cartilage tissue engineering. Acta Biomater. 5(1): 328–337.

Uludag, H., P. de Vos and P.A. Tresco. 2000. Technology of mammalian cell encapsulation. Adv. Drug Deliv. Rev. 42: 29–64.

Upadhyaya, L., J. Singh, V. Agarwal and R.P. Tewari. 2013. Recent progress in antimicrobial applications of nanostructured materials. J. Nanopharmaceutics Drug Delivery 1: 4.

Venkatesan, J. and S.K. Kim. 2010. Chitosan composites for bone tissue engineering—An overview. Mar. Drugs 8(8): 2252–2266.

Venugopal, J., L. Ma, T. Yong and S. Ramakrishna. 2005. *In vitro* study of smooth muscle cells on polycaprolactone and collagen nanofibrous matrices. Cell Biol. Int. 29(10): 861–867.

Wang, X., J. Li and Y. Wang. 2016. Improved high temperature strength of copper-graphene composite material. Mater. Lett. 181: 309–312.

Williams, D.F. 2009. On the nature of biomaterials. Biomaterials 30(30): 5897–5909.

Wu, H., Y. Wan, X. Cao and Q. Wu. 2008. Proliferation of chondrocytes on porous poly (DL-lactide)/chitosan scaffolds. Acta Biomater. 4(1): 76–87.

Zakaria, S.M., S.H.S. Zein, M.R. Othman, F. Yang and J.A. Jansen. 2013. Nanophase hydroxyapatite as a biomaterial in advanced hard tissue engineering: A review. Tissue Eng. PT B-Rev. 19: 431.

Zhang, J., J. Nie, Q. Zhang, Y. Li, Z. Wang and Q. Hu. 2014. Preparation and characterization of bionic bone structure chitosan/hydroxyapatite scaffold for bone tissue engineering. J. Biomater. Sci. Polym. Ed. 25: 61.

Zhang, X.Y., Y.P. Chen, J. Han, J. Mo, P.F. Dong, Y.H. Zhuo and Y. Feng. 2019. Biocompatible silk fibroin/carboxymethyl chitosan/strontium substituted hydroxyapatite/cellulose nanocrystal composite scaffolds for bone tissue engineering. Int. J. Biol. Macromol. 136: 1247–1257.

Zhou, H. and J. Lee. 2011. Nanoscale hydroxyapatite particles for bone tissue engineering. Acta Biomater. 7: 2769.

12

Nanobiomaterials in Cancer Therapeutics

Noel Vinay Thomas,[1] *Nithya R.*[2,]* and *Nirubama K.*[3]

1. Introduction

In olden days, cancer was a deadly disease and due to ineffective treatments, there was no way out than cruel death. Today also, cancer is a dreadful disease in accordance with its advanced phases. The difference between the former and the present day spread is that it was a rare phenomenon in olden days and widespread nowadays. However, the disease remains the same and the treatment has become improved. The conceptions in finding drugs from natural resources which do not adversely affect health rather than taking recourse to chemical methods are essentially needed today. Resorting to nanobiotechnology is one such idea of preparing drugs from nature.

Nanobiotechnology can be explained as "nanotechnology applied to biological systems". Nanotechnlogy is a conglomeration of several aspects of science, i.e., physics, chemistry, biology, clinical science and technology. In the field of medicine, it is potentially active to overcome the shortcomings of traditional drug delivery system by reducing side effects and improving efficiency. Nanotechnology is efficiently helpful in cancer therapeutics with its potential for diagnosis, for imaging and for drug delivery (Lungu et al. 2019).

Nanobiotechnology involves the engineering of new materials at a nano scale level for various biological usages. As such, nanobiomaterials form a bridge between biology and nanotechnolgy. As already defined by many authors, nano materials are those whose length measures between 1 and 100 nm. At these standards, they sustain clear optical, mechanical and electrical properties (Avti et al. 2011) which at length cater to the needs of cancer therapeutics.

1.1 Key Properties of Nanobiomaterials

Several nanobiomaterials are in vogue, namely metallic, ceramic, organic, inorganic, carbon based, polymer nanobiomaterials and nanocomposites. Their key properties are their size, shape and surface features. These aspects assume significant part in targeted drug delivery.

---

[1] Department of Medical Laboratory Science, Komar University of Science and Technology, Sulaymaniyah, Kurdistan Region, Iraq. Email: noel.thomas@komar.edu.iq

[2] Department of Chemistry, Dr. M.G.R. Educational and Research Institute, Chennai-95.

[3] Department of Biochemistry, Kongunadu Arts and Science college, Coimbatore, Tamil Nadu, India.

* Corresponding author: nithyar.22@gmail.com

In order to gain a prolonged period of drug suspension, nanoparticles of small size are preferable for circulating through the blood stream (Ankamwar 2012). When designed in proper manner, the polymeric nanoparticles ranging from 20–100 nm act efficiently *in vivo* owing to their sustained journey in blood stream. As regards shape, it decides the cell absorption and targeted delivery of drug. Nanoparticles of spherical shape enter into the cells faster than cylindrical ones (Elsabahy and Wooley 2012). The surface of functionalized nanomaterials decide whether they are hydrophobic or hydrophilic, lipophobic or lipophilic, and also the range of the molecular interaction (Ankamwar 2012). The rough surface area of nanoparticles helps promote biomedical applications, drug release in particular (Lin and Gu 2014).

2. Types and Applications of Nanobiomaterials in Cancer Therapy

Numerous studies on cancer theraupeutics have been conducted by nanomaterials because they overcome the biological barriers, such as enzymatic degradation or non-specific delivery during circulation and improved intracellular penetration. The usage of biodegradable polymers can be targeted to a precise location which would make the drug much more effective and less harmful (Sinha et al. 2006). This present chapter will try to provide a broad understanding about different types of bionanomaterials towards cancer therapeutic applications.

2.1 Chitosan Nanomaterials

To be successfully used in cancer treatment, the materials used should be inert, biodegradable, biocompatible and should be free of impurities. Chitosan, a biodegradable biopolymer, mainly consists of β-(1→4)-linked 2-amino-2-deoxy-β-D-glucopyranose (GlcN) units. It is due to these amino groups that the pH alters significantly, which in turn alters its property, thereby influencing biomedical applications of chitosan (Kumirska et al. 2011).

The drug release study of chitosan encapsulated piperlongumine (PL) (PL–CSNPs) which was prepared *via* ionic gelation method revealed the sustainable release of PL from the CSNPs and are pH-dependent. The PL–CSNPs showed efficient cytotoxicity against human gastric carcinoma (Venkatesan et al. 2016). Magnetic nanoparticles are noteworthy as a therapeutic agent for cancer treatment. Chitosan oligosaccharide-stabilized ferrimagnetic iron oxide nanocubes (Chito-FIONs) with 30-nm-sized FIONs was developed as an effective heat nanomediator for cancer hyperthermia. The results showed excellent antitumor efficacy (Bae et al. 2012). The antitumor nanovaccines consisting of self-assembled chitosan (CS) nanoparticles and two-component mucin1 (MUC1) glycopeptide antigens prepared by physically mixing showed that it is a useful and safe adjuvant system for MUC1 glycopeptide based cancer vaccines (Chena et al. 2017).

The identification of novel anticancer substances derived from natural sources is of great interest. Using breast cancer cell line (MCF 7) to test the anticancer extract's ability to inhibit the proliferation of cancer cells, Indigofera intricata crude alcoholic extract and chitosan were combined, and the findings were encouraging (Shahiwala et al. 2019). Curcumin was successfully encapsulated in chitosan-gum arabic nanoparticles via an emulsification solvent diffusion method with 136 nm size, which showed superior anti-colorectal cancer activity (Udompornmongkol and Chiang 2015). The evaluation of chitosan loaded nanoparticle of *Allivum sativum* by using ionic gelation method was prepared and the *in vitro* release of the drug was found to be the highest, i.e., up to 90.65% and was more stable. Hence, it can be used to treat cancer (Gupta et al. 2019).

Because of their stability, low toxicity, and straightforward and gentle manufacturing processes, silk fibroin-modified chitosan nanoparticles are excellent candidates for chemotherapeutic drug delivery in the treatment of cancer (Yang et al. 2015). Peripheral blood mononuclear cells (PBMC) were used to create a cancer-associated fibroblast (CAF) cell line, which was then subjected to an in vitro treatment with 100 μg/mL of chitosan nanoparticle. The results indicated that Ch-Np could be used as a delivery system for targeting cancer cells for the treatment of esophageal cancer (Potdar and Shetti 2016).

2.2 Alginate Nanomaterials

Natural biopolymer alginate is a linear copolymer made up of blocks of (1,4)-linked d- and l-mannuronate residues. The blocks are made up of alternating M and G residues (GMGMGM), followed by consecutive G residues (GGGGGG), and then consecutive M residues (MMMMMM). It allows efficient cell penetration into matrix and cell encapsulation. It is used as anticancer drug because of its biocompatiblity, nontoxicity, and biodegradability.

Inverse-micelle process was used to prepare calcium-crosslinked alginate matrix with 90 nm size. These nanoalginate (NALG) carriers were assessed by a doxorubicin (DOX) formulation (NALG-DOX) and examined their potency on breast cancer cells (4T1-luc2-GFP) (Rosch et al. 2018). The alginate-conjugated folic acid nanoparticles (AF NPs) were encapsulated with 5-aminolevulinic acid (5ALA) through a water-in-oil (W/O) emulsion method and was accurately delivered to cancer cells without any degradation (Lee and Lee 2020).

The side effects of an oral chemotherapeutic drug Exemestane (EXE) was reduced by loading with alginate nanoparticles by gelation method. *In vitro* drug release studies showed that ALG-NPs loaded EXE was used to treat breast cancer effectively (Jayapal and Dhanaraj 2017).

Curcumin, polyphenolic compound of turmeric, has excellent biomedical applications. But, its poor solubility, short half-life and low bioavailability make its application limited in cancer therapy. Hence, to overcome these drawbacks, curcumin is loaded with biopolymer. Curcumin loaded magnetic alginate/chitosan nanoparticles were fabricated and tested the cytotoxicity of loaded curcumin to Human Caucasian Breast Adenocarcinoma cells (MDA-MB-231) (Song et al. 2018). Curcumin diglutaric acid (CG), a prodrug of curcumin, was loaded with chitosan/alginate nanoparticles and revealed better anticancer activity against Caco-2, human hepatocellular carcinoma (HepG2) and human breast cancer (MDA-MB-231) cells (Sorasitthiyanukarn et al. 2018).

2.3 Carbon Nanomaterials

Carbon nanomaterials like graphene, carbon nanotubes, carbon quantum dots, and carbon nanospheres are highly used in biomedical applications because of their high surface to volume ratio, unique size and excellent mechanical and electrical properties. Carbon nanotubes are hydrophobic and hence they cannot be used as such. The surface modification or functionalization of CNTs allows them to disperse into aqueous solutions which makes them most advantageous in biomedical applications (Shao et al. 2013).

Supramolecular π–π stacking to load doxorubicin (DOX) onto branched polyethylene glycol (PEG) functionalized SWNTs for *in vivo* drug delivery applications was carried out by Zhuang Liu et al. (2009) and found that the new drug formulation SWNT–DOX revealed an enhanced therapeutic efficacy with less toxicity compared with free DOX. Zhang et al. (2011) studied doxorubicin-loaded PEGylated nanographene oxide (NGO-PEG-DOX) to facilitate combined chemotherapy and photothermal therapy in one system both *in vivo* and *in vitro*. This combined treatment resulted in higher therapeutic efficacy with less toxicity than DOX alone.

The movement of single functional Qdots in tumors of mice from a capillary vessel to cancer cells was observed by Hiroshi Tada et al. (2007) and the *in vivo* processes of delivery were successfully analyzed. Indium phosphide (core)-zinc sulphide (shell) quantum dots (InP/ZnS QDs) were created and discovered to be highly effective and non-toxic optical probes for imaging live pancreatic cancer cells (Yong et al. 2009).

2.4 Hydroxyapatite Nanoparticles

Hydroxyapatite nanoparticles (HANP) are known to be biocompatible and non-immunogenic and are accumulated in bone tissues. The magnetic hydroxyapatite nanoparticles were prepared by adding Fe^{2+} by co-precipitation method. Then this magnetic-HAP powder (mHAP) or pure HAP powder (HAP) was mixed with phosphate buffer solution (PBS) and injected around the tumor cells of a mouse and the results revealed good biocompatibility and less toxicity (Hou et al. 2009).

Mesoporous hydroxyapatite nanoparticles containing significant quantities of chemotherapeutic drug vincristine was found to be a potential nanocarrier for bone cancer treatment (Maia et al. 2018). Mesoporous silica/hydroxyapatite (MSNs/HAP) hybrid carrier loaded with anticancer drug doxorubicin (DOX) (DOX@MSNs/HAP) fabricated. Oligo HA (oHA) and hyaluronan (HA) were coated onto the nanoparticles (HA-DOX@MSNs/HAP, oHA-DOX@MSNs/HAP). oHA-DOX@MSNs/HAP showed much higher efficiency cellular uptake and drug release in tumor regions, which proved that the anticancer effect of oHA-DOX@MSNs/HAP is better than HA-DOX@MSNs/HAP (Kang et al. 2019).

HAP nanoparticle (nHAP) possessed the ability for inhibiting cancer cell growth *in vitro* for 3 days and *in vivo*. *In vitro* studies revealed that the proliferation of human cancer cells was inhibited by more than 65% and *in vivo* injection of nHAP about 50% (Han et al. 2014). HAP nanoparticles were used to treat human breast cancer cells (MCF-7) in culture at different doses, and the results showed that HAP both slows cell growth and promotes apoptosis (Meena et al. 2012).

2.5 Gold Nanoparticles

GNPs exhibit unique properties like large surface area, penetrating deep into the biological tissues, the ability to bind amine and thiol groups and allows surface modification which makes them suitable for biomedical applications. The colloidal gold also exhibits plasmon surface resonance (LPSR), which means that it can absorb light at specific wavelengths, which makes them suitable for hyperthermic cancer treatments (Vines et al. 2019).

Gold nanoparticles (GNPs) can be used in radiation as well as chemotherapy together. GNP-mediated chemoradiation has the potential to improve cancer care treatment (Yang et al. 2018). The detection of oral epithelium alive cancer cells *in vivo* and *in vitro* was demonstrated via surface plasmon resonance (SPR) scattering imaging or SPR absorption spectroscopy produced from antibody-conjugated gold nanoparticles (El-Sayed et al. 2005).

3. Future Perspectives and Conclusion

Nanotechnology is a fast-growing area of science. The size and shape of nanoparticles plays an important role in biomedical applications. Even though there are many *in vitro* studies on cancer therapy, *in vivo* studies have to be conducted to check the viability. Then, they will reach their full potential as common drug delivery methods for regular cancer chemotherapy. For example, though there are many studies on the toxicity of carbon nanotubes, uncertainty on CNT toxicity prevails. If this uncertainty is resolved, the CNT-based therapeutics can be possibly applied clinically.

One of the directions for the best results is the integration of therapies and diagnosis like phototherapy, photodynamic therapy (PDT), photoacoustic therapy, chemoradiation therapy, etc., along with conventional therapies. Good example is iron oxide coated hyaluronic acid as a biopolymeric material in cancer therapy. More attention should be paid to improve the basic theories of drug release from nanocarriers and to improve biocompatibility in future investigations.

References

Aliasgar Shahiwala, Naglaa G. Shehab, Maryam Khider and Rawoof Khan. 2019. Chitosan nanoparticles as a carrier for indigofera intricata plant extract: Preparation, characterization and anticancer activity. Current Cancer Therapy Reviews 15: 162.

Bae, K.H., M. Park, M.J. Do, N. Lee, J.H. Ryu, G.W. Kim, C. Kim, T.G. Park and T. Hyeon. 2012. Chitosan oligosaccharide-stabilized ferrimagnetic iron oxide nanocubes for magnetically modulated cancer hyperthermia. ACS Nano. 6(6): 5266–5273.

Balaprasad Ankamwar. 2012. Size and shape effect on biomedical applications of nanomaterials, biomedical engineering-technical applications in medicine. Radovan Hudak, Marek Penhaker and Jaroslav Majernik, IntechOpen.

Celina Yang, Kyle Bromma, Caterina Di Ciano Oliveira, Gaetano Zafarana, Monique van Prooijen and Devika B. Chithrani. 2018. Gold nanoparticle mediated combined cancer therapy. Cancer Nano. 9: 4.

Chun-Han Hou, Sheng-Mou Hou, Yu-Sheng Hsueh, Jinn Lin, Hsi-Chin Wu and Feng-Huei Lin. 2009. The *in vivo* performance of biomagnetic hydroxyapatite nanoparticles in cancer hyperthermia therapy. Biomaterials 30: 3956–3960.

El-Sayed, I.H., X. Huang and M.A. El-Sayed. 2005. Surface plasmon resonance scattering and absorption of anti-EGFR antibody conjugated gold nanoparticles in cancer diagnostics: Applications in oral cancer. Nano Lett. 5(5): 829–834.

Feuangthit Niyamissara Sorasitthiyanukarn, Chawanphat Muangnoi, Pahweenvaj Ratnatilaka Na Bhuket, Pornchai Rojsitthisak and Pranee Rojsitthisak. 2018. Chitosan/alginate nanoparticles as a promising approach for oral delivery of curcumin diglutaric acid for cancer treatment. Materials Science and Engineering: C 93: 178–190.

Gupta, D.K., S. Kesharwani, N.K. Sharma and M.K. Gupta. 2019. Formulation and evaluation of herbal extract of allivum sativum (garlic) loaded chitosan nanoparticle. Journal of Drug Delivery and Therapeutics 9: 715–718.

Hiroshi Tada, Hideo Higuchi, Tomonobu M. Wanatabe and Noriaki Ohuchi. 2007. *In vivo* real-time tracking of single quantum dots conjugated with monoclonal Anti-HER2 antibody in tumors of mice. Cancer Res. 67(3).

Iulia Ioana Lungu, Alexandru Mihai Grumezescu, Adrian Volceanov and Ecaterina Andronescu. 2019. Nanobiomaterials used in cancer therapy: An up-to-date overview. Molecules 24: 3547.

Jayachandran Venkatesan, Moch Syaiful Alam, Eun Ji Hong, Se-Kwon Kim and Min Suk Shim. 2016. Preparation of piperlongumine-loaded chitosan nanoparticles for safe and efficient cancer therapy. RSC Adv. 6: 79307–79316.

Jayapal, J.J. and S. Dhanaraj. 2017. Exemestane loaded alginate nanoparticles for cancer treatment: Formulation and *in vitro* evaluation. Int. J. Biol. Macromol. 105: 416–421.

Jolanta Kumirska, Mirko X. Weinhold, Jorg Thöming and Piotr Stepnowski. 2011. Biomedical activity of chitin/chitosan based materials—influence of physicochemical properties apart from molecular weight and degree of N-Acetylation. Polymers 3: 1875–1901.

Justin G. Rosch, Anna L. Brown, Allison N. DuRoss, Erin L. DuRoss, Gaurav Sahay and Conroy Sun. 2018. Nanoalginates via inverse-micelle synthesis: Doxorubicin-encapsulation and breast cancer cytotoxicity. Nanoscale Research Letters 13: 350.

Ken-Tye Yong, Hong Ding, Indrajit Roy, Wing-Cheung Law, Earl J. Bergey, Anirban Maitra and Paras N. Prasad. 2009. Imaging pancreatic cancer using bioconjugated InP quantum dots. ACS Nano. 24. 3(3): 502–510.

Mahmoud Elsabahy and Karen L. Wooley. 2012. Design of polymeric nanoparticles for biomedical delivery applications. Chem. Soc. Rev. 41(7): 2545–2561.

Maia, A.L.C., C.A. Ferreira, A.L.B. Barros, A.T.M.E. Silva, G.A. Ramaldes, A.D. Silva Cunha Júnior, D.C.P. Oliveira, C. Fernandes and D.C. Ferreira Soares. 2018. Vincristine-loaded hydroxyapatite nanoparticles as a potential delivery system for bone cancer therapy. J. Drug Target 26(7): 592–603.

Ming-Hui Yang, Tze-Wen Chung, Yi-Shan Lu, Yi-Ling Chen, Wan-Chi Tsai, Shiang-Bin Jong, Shyng-Shiou Yuan, Pao-Chi Liao, Po-Chiao Lin and Yu-Chang Tyan. 2015. Activation of the ubiquitin proteasome pathway by silk fibroin modified chitosan nanoparticles in hepatic cancer cells. Int. J. Mol. Sci. 16: 1657.

Potdar, P.D. and A.U. Shetti. 2016. Evaluation of anti-metastatic effect of chitosan nanoparticles on esophageal cancer associated fibroblasts. J. Cancer Metasta. Treat. 2: 259–67.

Pramod K. Avti, Sunny C. Patel and Balaji Sitharaman. 2011. Nanobiomaterials: Current and Future Prospects. In Nanobiomaterials Handbook, Taylor and Francis Group, LLC.

Pu-Guang Chena, Zhi-Hua Huanga, Zhan-Yi Sun, Yue Gao, Yan-Fang Liu, Lei Shi, Yong-Xiang Chen, Yu-Fen Zhao and Yan-Mei Li. 2017. Chitosan nanoparticles based nanovaccines for cancer immunotherapy. Pure Appl. Chem. USA 89(7): 931–939.

Rajni Sinha, Gloria J. Kim, Shuming Nie and Dong M. Shin. 2006. Nanotechnology in cancer therapeutics: Bioconjugated nanoparticles for drug delivery. Mol. Cancer Ther. 5(8).

Ramovatar Meena, Kavindra Kumar Kesari, Madhu Rani and R. Paulraj. 2012. Effects of hydroxyapatite nanoparticles on proliferation and apoptosis of human breast cancer cells (MCF-7). J. Nanopart. Res. 14: 712.

Sara Lee and Kangwon Lee. 2020. pH-sensitive folic acid conjugated alginate nanoparticle for induction of cancer-specific fluorescence imaging. Pharmaceutics 12: 537.

Shao, W., P. Arghya, M. Yiyong, L. Rodes and S. Satya Prakash. 2013. Carbon nanotubes for use in medicine: Potentials and limitations, 285–311.

Udompornmongkol, P. and B.H. Chiang. 2015. Curcumin-loaded polymeric nanoparticles for enhanced anti-colorectal cancer applications. Journal of Biomaterials Applications 30(5): 537–546.

Vines, J.B., J.-H. Yoon, N.-E. Ryu, D.-J. Lim and H. Park. 2019. Gold nanoparticles for photothermal cancer therapy. Front. Chem. 7: 167.

Wenxing Song, Xing Su, David Alexander Gregory, Wei Li, Zhiqiang Cai and Xiubo Zhao. 2018. Magnetic alginate/chitosan nanoparticles for targeted delivery of curcumin into human breast cancer cells. Nanomaterials 8: 907.

Xubo Lin and Ning Gu. 2014. Surface properties of encapsulating hydrophobic nanoparticles regulate the main phase transition temperature of lipid bilayers: A simulation study. Nano Research 7(8): 1195–1204.

Yao Kang, Wen Sun, Shuyi Li, Mingle Li, Jiangli Fan, Jianjun Du, Xing-Jie Liang and Xiaojun Peng. 2019. Oligo hyaluronan-coated silica/hydroxyapatite degradable nanoparticles for targeted cancer treatment. Adv. Sci. 6: 1900716.

Yingchao Han, Shipu Li, Xianying Cao, Lin Yuan, Youfa Wang, Yixia Yin, Tong Qiu, Honglian Dai and Xinyu Wang. 2014. Different inhibitory effect and mechanism of hydroxyapatite nanoparticles on normal cells and cancer cells *in vitro* and *in vivo*. Sci. Rep. 4: 7134.

Zhang, W., Z. Guo, D. Huang, Z. Liu, X. Guo and H. Zhong. 2011. Synergistic effect of chemo-photothermal therapy using PEGylated graphene oxide. Biomaterials 32(33): 8555–8561.

Zhuang Liu, Alice C. Fan, Kavya Rakhra, Sarah Sherlock, Andrew Goodwin, Xiaoyuan Chen, Qiwei Yang, Dean W. Felsher and Hongjie Dai. 2009. Supramolecular stacking of doxorubicin on carbon nanotubes for *in vivo* cancer therapy. Angew. Chem. Int. Ed. Engl. 48(41): 7668–7672.

13

Nanoporous Silicon Materials for Bioapplications by Electrochemical Approach

Kelvii Wei GUO

1. Introduction

In recent years, huge scientific research effort has been put into nanotechnology, and with the development of nanotechnology, the technology related to human is dramatically changed, even including an extensively profound influence on our daily life. It is well known that nanoporous materials are a subset of nanotechnology, which is also a significant class with captivating applications such as sensors (Adams-McGavin et al. 2017, Weremfo et al. 2017, Wisser et al. 2016), drug delivery (Lee et al. 2018, Pawlik et al. 2018, Yallappa et al. 2018, Zeleňák et al. 2018), catalysis (Clough et al. 2017, Sharma et al. 2018), electrodes (Fu et al. 2018, Kapusta-Kołodziej et al. 2018, Shobana and Kim 2017) and molecular separation (Gholampour and Yeganegi 2014, Gomis-Berenguer et al. 2017, Smith et al. 2017, Sumer and Keskin 2017), etc. To date, the applications of nanoporous materials related to the biomedical field have been extensively explored owing to their unique properties (such as tunable size of pores, large volume of pores, high specific surface area, feasible surface modification and chemical stability, etc.). Meanwhile, the structures of nanoporous materials also have fascinating conducting, magnetic and fluorescent properties resulting in attracting the abovementioned biomedical applications, for instance, optical sensors, electrochemical sensors, biomolecule determination, targeted therapy, drug encapsulation, controlled drug release, drug solubility improvement, theranostics, magnetic resonance imaging, fluorescent imaging, enzyme immobilization, gene transfer, nucleic acid protection, proteome analysis, adjuvants, implants, regeneration medicine, tissue engineering, etc.

It often defines the highly porous nanostructure with pore sizes ranging from a few nanometers to one hundred nanometers as nanoporous material. In line with the requirements of the International Union of Pure and Applied Chemistry (IUPAC), it can be categorized into three different types of pore on the basis of the diameter of pore: nanopores (< 2 nm), mesopores, 2 nm < pore size < 50 nm and macropores (> 50 nm). On such nanoscale, the nanoporous materials possess a series of unique properties (such as quantum confinement effect (Kim and Kim 2017, Notario et al. 2016,

Department of Mechanical and Biomedical Engineering, City University of Hong Kong, 83 Tat Chee Avenue, Kowloon Tong, Kowloon, Hong Kong. Email: kelviiguo@yahoo.com

Tripathy et al. 2015), plasmonic (Mueller et al. 2017, Norek et al. 2014, Ruffino et al. 2017, Watkins and Borensztein 2018), high surface to volume area (Hosseini et al. 2018, Mehrafzoon et al. 2018, Sadeghi et al. 2017, Sharma et al. 2018) and photonic (Abbasimofrad et al. 2018, Gomis-Berenguer et al. 2017, Likodimos 2018), etc.), which are closely related to the materials. Up till now, many efforts have been devoted to exploring different porous materials such as metal (Gallican and Hure 2017, He and Abdolrahim 2018, Tran et al. 2018), ceramic (Subhapradha et al. 2018, Tovar et al. 2014), semi-conductor (Fic et al. 2018, González et al. 2016, Gregorczyk and Knez 2016, Tang et al. 2012) and organic (Lin et al. 2017, Wang et al. 2018, Wichmann et al. 2014).

Because of the contribution to the state-of-art synthesis strategies, bulk materials can be taken for preparing the porous materials with the cutting-edge techniques. Numerous types of pore morphologies along with the well-developed nanostructures have been proposed. Generally, pore morphologies consist of open pores which display a connection throughout the structure to the surface of the materials. Moreover, many different attractive pore shapes have been developed such as spherical (Haller et al. 2016, Shiba et al. 2013), triangular (Mangipudi et al. 2016, Wierzbicka et al. 2016), cylindrical (Brach et al. 2017, Chen et al. 2018) and sponge-like (Feng et al. 2018, Xu et al. 2018), etc. Furthermore, some nanostructured pores with the special characters, for instance, in the sinusoidal (Hernández-López et al. 2016) and wavy (Bargmann et al. 2018) form can be explored with various controllable output waveforms fabricated by the electrochemical methods.

2. Methods for Synthesis of Nanoporous Materials

2.1 Etching-Dealloying

Etching such as dealloying process refers to a chemical process in which the alloy is partially dissolved by the selective etching (Farid et al. 2018). In the alloys' system, a less noble element is dissolved by the etchants and leaves behind a noble alloy constituent and an open nanoporous structure. The evolution of nanoporosity during the dealloying has been explored early by Erlebacher et al. in 2001 with the relevant results published in Nature (Erlebacher et al. 2001). Study shows that the gold atoms are not dissolved and tend to cluster together to form Au islands, it opens up the pore and etches continuously throughout the bulk structure. Finally, the sponge-like porous Au is obtained after etching. Recently, Ruffino et al. (Ruffino et al. 2017) fabricated a kind of nanoporous Au structures by dealloying Au/Ag. By HNO_3 dealloying etching, the particles nanoporous keep the shape and the density of the surface density successfully along with the particles volume shrinking to some extent, resulting in the lattice defects and the plastic deformation of Au crystal structure. It also indicates that the dealloying process is more efficient on the particles obtained by the liquid state process to obtain higher homogeneity of the AuAg alloy forming the particles.

2.2 Etching-Electrochemical Etching

In general, electrochemical etching is a common top-down approach to fabricate nanoporous materials, the bulk material is usually electrochemically etched, the pore is formed by an applied voltage or current. The surface of the bulk materials reacts with the electrolyte (the etchants) to generate the pore structure and such reaction usually begins in the defect sites of the surface. Literature related to nanoporous materials by the electrochemical etching has been published: porous silicon (pSi) (Amri et al. 2016, Ervin et al. 2018, Tsai et al. 2018), porous Ni (Du et al. 2018, Lin and Chou 2017), porous titania (Kapusta-Kołodziej et al. 2018, Mei et al. 2015, Rahman et al. 2018) and porous alumina (Abbasimofrad et al. 2018, Ateş et al. 2018, Fukutsuka et al. 2016, Masuda et al. 2018).

2.3 Templating Method

Templating method is a technique adopted to prepare porous materials by a sacrifice mold and fill with the target precursors into its void space. Generally, the electrochemical reduction or calcination is usually taken in order to turn it into the solid phase. Many materials such as porous silicon (pSi), and porous anodic alumina can be taken as the sacrificial template (Pérez-Page et al. 2016, Yan et al. 2017). In addition, due to the cylindrical pores in the porous anodic alumina, they can be filled with other materials to easily fabricate the well-defined nanorod or nanotube arrays (Schiavi et al. 2018). Also, the photonic porous silicon with the rugated structure can be replicated by filling other materials into the porous silicon, e.g., metal (Jaouadi et al. 2013, Romano et al. 2017) and polymer (Han et al. 2014, Vakifahmetoglu et al. 2016). It should be noted that the other templating method such as the nanosphere lithography (Lee et al. 2016, Purwidyantri et al. 2016) is a technique of the application of the nanospheres (silica or polystyrene) with diameter about 100 nm to 1 μm. Initially, the nanospheres are self-assembled by different methods involving the spincoating and the dipcoating on the substrate to form the ordered hexagonal structures. The structures can be a monolayer or a structure with three dimensions. The fabricated nanosphere nanostructure can be adopted as a template that can be filled into the void with various kinds of the materials afterwards (for instance, filled the metal by the electrodeposition processing). In the process of the electrodeposition, metal ions in an electrolyte are reduced to metal and filled into the interstices of the nanosphere template. Subsequently, the nanosphere can be removed by the dissolution or calcination to obtain porous materials with the desired characteristics.

3. Porous Silicon (pSi)

Porous silicon (pSi) has attracted the intense scientific research significantly since the discovery of photoluminescence at room temperature due to its quantum confinement effects (Kumar et al. 2018a, Lu 2015). In addition to the photoluminescence of pSi (Azaiez et al. 2018), the applications (biosensing, *in vivo* imaging and gas sensing of other properties (high porosity, tailorable surface, biocompatibility and biodegradability)) of pSi have been well exploited. Moreover, in the reflectance spectra, the particular optical characteristics of pSi are extremely crucial to develop pSi-based sensor. It is well known that the single layer pSi displays Fabry-Pérot fringes and the modulated pSi multilayers with the waveform can fabricate into optical nanostructures, for example Bragg stacks and rugated filters (Bauer et al. 2013, Jaganathan and Godin 2012).

The top view and cross section images of the porous silicon etched by ozone oxidization at ozone of 1.5 SCFH for 20 min is shown in Figure 1. It shows that the pSi possesses high porosity with long and straight pores of the average diameter at 37 nm.

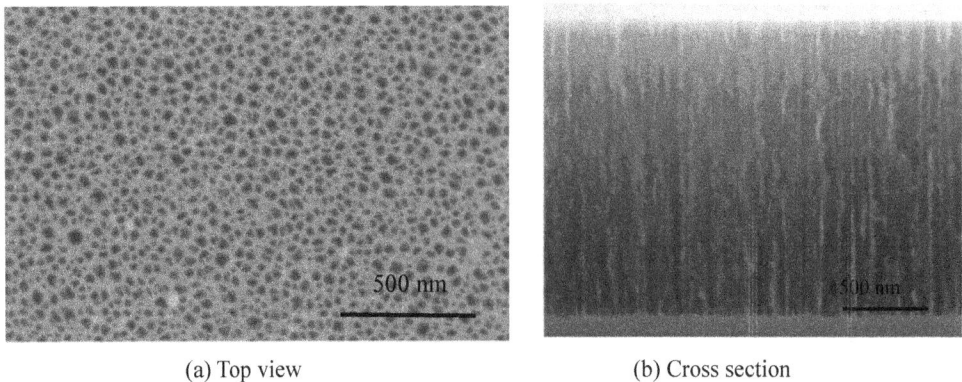

(a) Top view   (b) Cross section

Figure 1: FESEM images (a) top view (b) cross section of the porous silicon etched by ozone oxidization.

3.1 Synthesis Strategies for pSi

Porous silicon is usually fabricated by the electrochemical etching process on the crystalline silicon wafer in aqueous hydrofluoric acid (HF) connected to potentiostat as shown in Figure 2. Usually pSi samples are fabricated with the silicon wafer in a solution of 48% aqueous HF: ethanol (3:1). For increasing the wettability, ethanol is commonly added to the etchant to enhance the etchant infiltration and reduce the bubble formation. The mechanism of pore formation related to silicon etching is expressed as follows (Shcherban 2017).

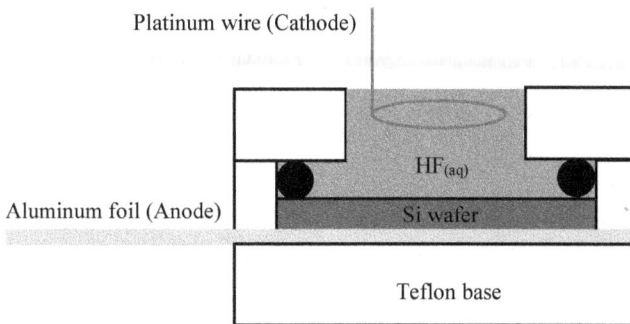

$$Si + 6HF + 2hole^+ \rightarrow H_2SiF_6 + 2H^+ + H_2$$

Figure 2: Schematic drawing of electrochemical etching process for porous silicon.

3.2 Porous Silicon for Biosensor

The attractive tunable pore sizes along with various optical nanostructures make the porous silicon a definitely promising candidate for bioengineering, especially in the field of biological sensing. One of the captivating biosensor is researched on the basis of the optical feature of pSi—an optical interferometric biosensor. The contribution of such pSi-based optical interferometric biosensor is based on Fabry–Pérot fringes which are the result of the peak maxima and minima of the reflection spectrum constructed by the constructive and destructive interference of the reflecting light from the top and the bottom of porous silicon layer. Results show that the change in the refractive index of the porous silicon matrix, as the peak shift of the effective optical thickness in the reflectance spectrum, can be easily detected by the charge-coupled device (CCD) (Chhasatia et al. 2018, Dhanekar and Jain 2013, Harraz 2014, Myndrul et al. 2017, RoyChaudhuri 2015, Tong et al. 2016).

3.3 Porous Silicon for Biomedical Applications

It is noted that the traditional silica NPs are widely utilized in the biomedical applications, as efficiently optical contrast agents in imaging, and highly effective drug delivery in drug releasing procedure. However, due to the limitations of silica NPs, it hinders its functionalization from the practical applications. Attractively, nanoporous silicon exhibits tunable size and volume of pores, and extremely high specific surface area to achieve the higher capacity of the therapeutic drug loading. The interior wall of nanoporous Si is amorphous and has the high density of the ordered framework with meso/nanoporous uniform distribution (Hernandez-Montelongo et al. 2014, Hou et al. 2014, Knieke et al. 2015, Jaganathan and Godin 2012, Salonen and Lehto 2008). Moreover, the cytocompatibility of the synthesized pSi was evaluated by *in vitro* testing with research on L132 cells adhesion after 24 h culture in the proposed method by Hernandez-Montelongo et al. (Hernandez-Montelongo et al. 2014). The relevant results are shown in Figure 3. It shows that cells are isolated and dispersed sparsely with a contracted semispherical morphology on npSi as shown in Figure 3a1, and only few short focal adhesions are discerned shown as white arrows in

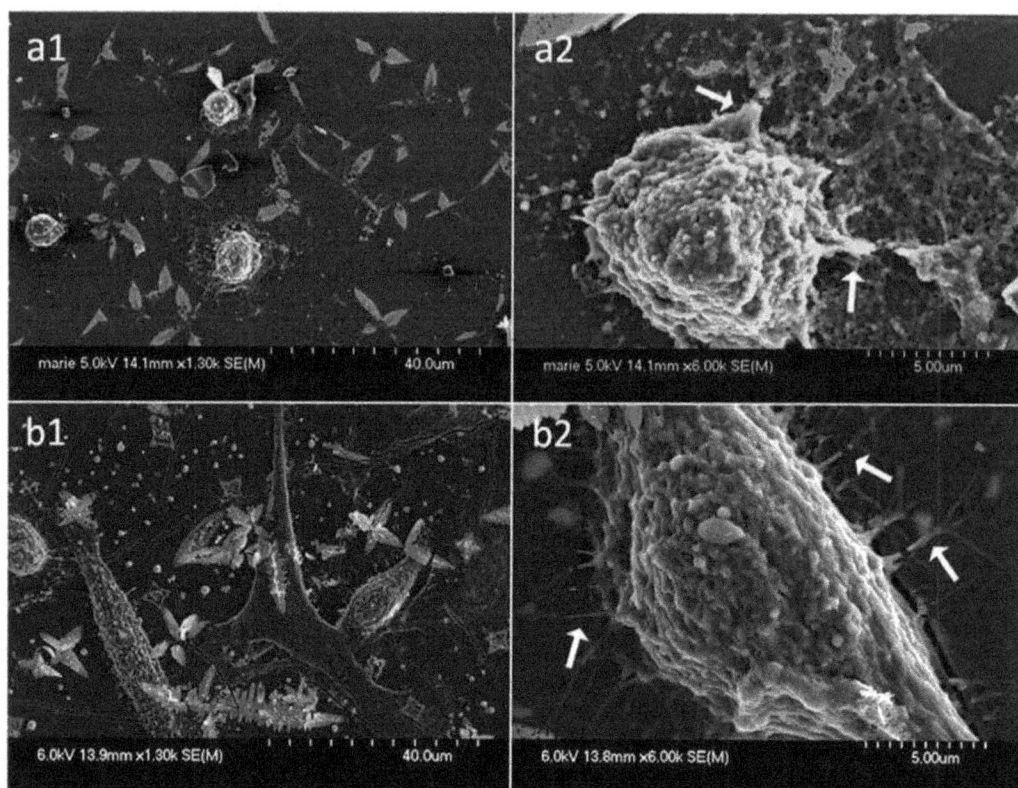

Figure 3: SEM images L132 cells in npSi (a1 and a2), npSi-CD (b1 and b2) after 24 h culture.

Figure 3a2. For cells on npSi-cyclodextrins (CD), the morphology of cells is oval along with the extended lamellipodia as shown in Figure 3b1. Moreover, numerous focal adhesions extending around the body effectively anchor the cell to the surface as white arrows shown in Figure 3b2. It indicates that a better adhesion to the surface with the elongated cytoskeleton is achieved successfully. At the meantime, it shows a good cytocompatibility of npSi and npSi-CD through the different cell growth behaviors on the testing samples.

Additionally, the pore size of pSi is able to be controlled at the certain nanometers, which is profitable for easy particles uptake by endocytosis in plant and animal cells without any significant cytotoxicity. In addition, the distribution of pore size in pSi is fascinatingly narrow and the diameter of pores can be tailored under 10 nm. Owing to these unique features, it is definitely beneficial to explore the relevant theory of drug release kinetics and various drug molecules loading (Bagheri et al. 2018, Bardhan et al. 2018, Correa et al. 2016, De Vitis et al. 2016, Diab et al. 2017, Hou et al. 2014, Jaganathan and Godin 2012, Ki et al. 2018, Knieke et al. 2015, Kumar et al. 2018b).

4. Conclusion

Due to the ease and quick fabrication by the electrochemical methods, porous silicon (pSi) has attractive optical properties with the controllable and tuneable porosity and pore size along with the enhanced morphological properties of the large internal surface area and the versatile surface chemistry. Owing to such unique properties of nanoporous materials (high porosity, modifiable surface, good biocompatibility and biodegradability), the nanoporous silicon materials prepared by the electrochemical methods will play more and more significant role in the field of catalysis, chemical, energy storage, gas sensing, biological sensing and *in vivo* imaging. Moreover, the abovementioned captivating properties of pSi fabricated by the electrochemical methods definitely make the porous silicon a promising candidate for bioapplications in the coming future. Meanwhile, for the future environmental risks and the sustainable development, the eco-friendly techniques shall be explored further because of the chemical usage.

References

Abbasimofrad, S., M.A. Kashi, M. Noormohammadi and A. Ramazani. 2018. Tuning the optical properties of nanoporous anodic alumina photonic crystals by control of allowed voltage range via mixed acid concentration. Journal of Physics and Chemistry of Solids 118: 221–231.

Adams-McGavin, R.C., Y.T. Chan, C.M. Gabardo, J. Yang, M. Skreta, B.C. Fung and L. Soleymani. 2017. Nanoporous and wrinkled electrodes enhance the sensitivity of glucose biosensors. Electrochimica Acta 242: 1–9.

Amri, C., R. Ouertani, A. Hamdi, R. Chtourou and H. Ezzaouia. 2016. Effect of porous layer engineered with acid vapor etching on optical properties of solid silicon nanowire arrays. Materials & Design 111: 394–404.

Ateş, S., E. Baran and B. Yazici. 2018. The nanoporous anodic alumina oxide formed by two-step anodization. Thin Solid Films 648: 94–102.

Azaiez, K., R.B. Zaghouani, S. Khamlich, H. Meddeb and W. Dimassi 2018. Enhancement of porous silicon photoluminescence property by lithium chloride treatment. Applied Surface Science 441: 272–276.

Bagheri, E., L. Ansari, K. Abnous, S.M. Taghdisi, F. Charbgoo, M. Ramezani and M. Alibolandi. 2018. Silica based hybrid materials for drug delivery and bioimaging. Journal of Controlled Release 277: 57–76.

Bardhan, M., A. Majumdar, S. Jana, T. Ghosh, U. Pal, S. Swarnakar and D. Senapati. 2018. Mesoporous silica for drug delivery: Interactions with model fluorescent lipid vesicles and live cells. Journal of Photochemistry and Photobiology B: Biology 178: 19–26.

Bargmann, S., B. Klusemann, J. Markmann, J.E. Schnabel, K. Schneider, C. Soyarslan and J. Wilmers. 2018. Generation of 3D representative volume elements for heterogeneous materials: A review. Progress in Materials Science 96: 322–384.

Bauer, S., P. Schmuki, K. von der Mark and J. Park. 2013. Engineering biocompatible implant surfaces: Part I: Materials and surfaces. Progress in Materials Science 58(3): 261–326.

Brach, S., S. Cherubini, D. Kondo and G. Vairo. 2017. Void-shape effects on strength properties of nanoporous materials. Mechanics Research Communications 86: 11–17.

Chen, Q., G.N. Wang and M.J. Pindera. 2018. Finite-volume homogenization and localization of nanoporous materials with cylindrical voids. Part 1: Theory and validation. European Journal of Mechanics—A/Solids 70: 141–155.

Chhasatia, R., M.J. Sweetman, B. Prieto-Simon and N.H. Voelcker. 2018. Performance optimisation of porous silicon rugate filter biosensor for the detection of insulin. Sensors and Actuators B: Chemical 273: 1313–1322.

Clough, M., J.C. Pope, L.T.X. Lin, V. Komvokis, S.Y.S. Pan and B. Yilmaz. 2017. Nanoporous materials forge a path forward to enable sustainable growth: Technology advancements in fluid catalytic cracking. Microporous and Mesoporous Materials 254: 45–58.

Correa, S., E.C. Dreaden, L. Gu and P.T. Hammond. 2016. Engineering nanolayered particles for modular drug delivery. Journal of Controlled Release 240: 364–386.

De Vitis, S., M.L. Coluccio, G. Strumbo, N. Malara, F.P. Fanizzi, S.A. De Pascali, G. Perozziello, P. Candeloro, E. Di Fabrizio and F. Gentile. 2016. Combined effect of surface nano-topography and delivery of therapeutics on the adhesion of tumor cells on porous silicon substrates. Microelectronic Engineering 158: 6–10.

Dhanekar, S. and S. Jain. 2013. Porous silicon biosensor: Current status. Biosensors and Bioelectronics 41: 54–64.

Diab, R., N. Canilho, I.A. Pavel, F.B. Haffner, M. Girardon and A. Pasc. 2017. Silica-based systems for oral delivery of drugs, macromolecules and cells. Advances in Colloid and Interface Science 249: 346–362.

Du, D.W., R. Lan, J. Humphreys, H. Amari and S.W. Tao. 2018. Preparation of nanoporous nickel copper sulfide on carbon cloth for high-performance hybrid supercapacitors. Electrochimica Acta 273: 170–180.

Erlebacher, J., M.J. Aziz, A. Karma, N. Dimitrov and K. Sieradzki. 2001. Evolution of nanoporosity in dealloying. Nature 410: 450–453.

Ervin, M.H., N.W. Piekiel and B. Isaacson. 2018. Process for integrating porous silicon with other devices. Sensors and Actuators A: Physical 280: 132–138.

Farid, S., R. Kuljic, S. Poduri, M. Dutta and S.B. Darling. 2018. Tailoring uniform gold nanoparticle arrays and nanoporous films for next-generation optoelectronic devices. Superlattices and Microstructures 118: 1–6.

Feng, L.Y., J.K. Sun, Y.H. Liu, X.X. Li, L. Ye and L.J. Zhao. 2018. 3D sponge-like porous structure of Mn_2O_3 tiny nanosheets coated on $Ni(OH)_2/Mn_2O_3$ nanosheet arrays for quasi-solid-state asymmetric supercapacitors with high performance. Chemical Engineering Journal 339: 61–70.

Fic, K., A. Platek, J. Piwek and E. Frackowiak. 2018. Sustainable materials for electrochemical capacitors. Materials Today 21(4): 437–454.

Fu, J.T., E. Detsi and J.T.M. De Hosson. 2018. Recent advances in nanoporous materials for renewable energy resources conversion into fuels. Surface and Coatings Technology 347: 320–336.

Fukutsuka, T., K. Koyamada, S. Maruyama, K. Miyazaki and T. Abe. 2016. Ion transport in organic electrolyte solution through the pore channels of anodic nanoporous alumina membranes. Electrochimica Acta 199: 380–387.

Gallican, V. and J. Hure. 2017. Anisotropic coalescence criterion for nanoporous materials. Journal of the Mechanics and Physics of Solids 108: 30–48.

Gholampour, F. and S. Yeganegi. 2014. Molecular simulation study on the adsorption and separation of acidic gases in a model nanoporous carbon. Chemical Engineering Science 117: 426–435.

Gomis-Berenguer, A., L.F. Velasco, I. Velo-Gala and C.O. Ania. 2017. Photochemistry of nanoporous carbons: Perspectives in energy conversion and environmental remediation. Journal of Colloid and Interface Science 490: 879–901.

González, A., E. Goikolea, J.A. Barrena and R. Mysyk. 2016. Review on supercapacitors: Technologies and materials. Renewable and Sustainable Energy Reviews 58: 1189–1206.

Gregorczyk, K. and M. Knez. 2016. Hybrid nanomaterials through molecular and atomic layer deposition: Top down, bottom up, and in-between approaches to new materials. Progress in Materials Science 75: 1–37.

Haller, X., Y. Monerie, S. Pagano and P.G. Vincent. 2016. Elastic behavior of porous media with spherical nanovoids. International Journal of Solids and Structures 84: 99–109.

Han, H., Z.P. Huang and W. Lee. 2014. Metal-assisted chemical etching of silicon and nanotechnology applications. Nano Today 9(3): 271–304.

Harraz, F.A. 2014. Porous silicon chemical sensors and biosensors: A review. Sensors and Actuators B: Chemical 202: 897–912.

He, L.J. and N. Abdolrahim. 2018. Deformation mechanisms and ductility enhancement in core-shell Cu@Ni nanoporous metals. Computational Materials Science 150: 397–404.

Hernández-López, J.M., A. Conde, J.J. de Damborenea and M.A. Arenas. 2016. Electrochemical response of TiO$_2$ anodic layers fabricated on Ti6Al4V alloy with nanoporous, dual and nanotubular morphology. Corrosion Science 112: 194–203.

Hernandez-Montelongo, J., N. Naveas, S. Degoutin, N. Tabary, F. Chai, V. Spampinato, G. Ceccone, F. Rossi, V. Torres-Costa, M. Manso-Silvan and B. Martel. 2014. Porous silicon-cyclodextrin based polymer composites for drug delivery applications. Carbohydrate Polymers 110: 238–252.

Hosseini, M., J. Azamat and H. Erfan-Niya. 2018. Improving the performance of water desalination through ultra-permeable functionalized nanoporous graphene oxide membrane. Applied Surface Science Part B 427: 1000–1008.

Hou, H.Y., A. Nieto, F.Y. Ma, W.R. Freeman, M.J. Sailor and L.Y. Cheng. 2014. Tunable sustained intravitreal drug delivery system for daunorubicin using oxidized porous silicon. Journal of Controlled Release 178: 46–54.

Jaganathan, H. and B. Godin. 2012. Biocompatibility assessment of Si-based nano- and micro-particles. Advanced Drug Delivery Reviews 64(15): 1800–1819.

Jaouadi, M., M. Gaidi and H. Ezzaouia. 2013. Effect of LiBr pore-filling on morphological, optical and electrical properties of porous silicon membrane. Superlattices and Microstructures 54: 172–180.

Kapusta-Kołodziej, J., A. Chudecka and G.D. Sulka. 2018. 3D nanoporous titania formed by anodization as a promising photoelectrode material. Journal of Electroanalytical Chemistry 823: 221–233.

Ki, M.R., T.K.M. Nguyen, H.S. Jun and S.P. Pack. 2018. Biosilica-enveloped ferritin cage for more efficient drug deliveries. Process Biochemistry 68: 182–189.

Kim, D. and H.S. Kim. 2017. Enhancement of fluorescence from one- and two-photon absorption of hemicyanine dyes by confinement in silicalite-1 nanochannels. Microporous and Mesoporous Materials 243: 69–75.

Knieke, C., M.A. Azad, D. To, E. Bilgili and R.N. Davé. 2015. Sub-100 micron fast dissolving nanocomposite drug powders. Powder Technology 271: 49–60.

Kumar, A., A. Kumar and R. Chandra. 2018a. Fabrication of porous silicon filled Pd/SiC nanocauliflower thin films for high performance H$_2$ gas sensor. Sensors and Actuators B: Chemical 264: 10–19.

Kumar, P., P. Tambe, K.M. Paknikar and V. Gajbhiye. 2018b. Mesoporous silica nanoparticles as cutting-edge theranostics: Advancement from merely a carrier to tailor-made smart delivery platform. Journal of Controlled Release 287: 35–57.

Lee, D., I.S. Shin, L. Jin, D. Kim, Y.J. Park and E.J. Yoon. 2016. Nanoheteroepitaxy of GaN on AlN/Si(111) nanorods fabricated by nanosphere lithography. Journal of Crystal Growth 444: 9–13.

Lee, H.J., N. Choi, E.S. Yoon and I.J. Cho. 2018. MEMS devices for drug delivery. Advanced Drug Delivery Reviews 128: 132–147.

Likodimos, V. 2018. Photonic crystal-assisted visible light activated TiO_2 photocatalysis. Applied Catalysis B: Environmental 230: 269–303.

Lin, J.D. and C.T. Chou. 2017. The influence of acid etching on the electrochemical supercapacitive properties of NiP coatings. Surface and Coatings Technology 325: 360–369.

Lin, T.C., K.C. Yang, P. Georgopanos, A. Avgeropoulos and R.M. Ho. 2017. Gyroid-structured nanoporous polymer monolith from PDMS-containing block copolymers for templated synthesis. Polymer 126: 360–367.

Lu, K. 2015. Porous and high surface area silicon oxycarbide-based materials—A review. Materials Science and Engineering: R: Reports 97: 23–49.

Mangipudi, K.R., V. Radisch, L. Holzer and C.A. Volkert. 2016. A FIB-nanotomography method for accurate 3D reconstruction of open nanoporous structures. Ultramicroscopy 163: 38–47.

Masuda, H., T. Yanagishita and T. Kondo. 2018. Fabrication of anodic porous alumina. Encyclopedia of Interfacial Chemistry, 226–235.

Mehrafzoon, S., S.A. Hassanzadeh-Tabrizi and A. Bigham. 2018. Synthesis of nanoporous Baghdadite by a modified sol-gel method and its structural and controlled release properties. Ceramics International 44(12): 13951–13958.

Mei, S., J. Yang, S. Christian, S.D. Yuan and J.M.F. Ferreira. 2015. Fabrication and characterisation of titania nanoporous thin film for photoelectrochemical (PEC) conversion of water. Energy Procedia 75: 2187–2192.

Mueller, A.D., L.Y.M. Tobing, Y. Luo and D.H. Zhang. 2017. Vertical growth of plasmonic nanostructures via electrodeposition on a conductive oxide. Procedia Engineering 215: 60–65.

Myndrul, V., R. Viter, M. Savchuk, M. Koval, N. Starodub, V. Silamiķelis, V. Smyntyna, A. Ramanavicius and I. Iatsunskyi. 2017. Gold coated porous silicon nanocomposite as a substrate for photoluminescence-based immunosensor suitable for the determination of Aflatoxin B1. Talanta 175: 297–304.

Norek, M., M. Włodarski and P. Matysik. 2014. UV plasmonic-based sensing properties of aluminum nanoconcave arrays. Current Applied Physics 14(11): 1514–1520.

Notario, B., J. Pinto and M.A. Rodriguez-Perez. 2016. Nanoporous polymeric materials: A new class of materials with enhanced properties. Progress in Materials Science 78-79: 93–139.

Pawlik, A., R.P. Socha, M.H. Kalbacova and G.D. Sulka. 2018. Surface modification of nanoporous anodic titanium dioxide layers for drug delivery systems and enhanced SAOS-2 cell response. Colloids and Surfaces B: Biointerfaces 171: 58–66.

Pérez-Page, M., E. Yu, J. Li, M. Rahman, D.M. Dryden, R. Vidu and P. Stroeve. 2016. Template-based syntheses for shape controlled nanostructures. Advances in Colloid and Interface Science 234: 51–79.

Purwidyantri, A., C.H. Chen, B.J. Hwang, J.D. Luo, C.C. Chiou, Y.C. Tian, C.Y. Lin, C.H. Cheng and C.S. Lai. 2016. Spin-coated Au-nanohole arrays engineered by nanosphere lithography for a Staphylococcus aureus 16S rRNA electrochemical sensor. Biosensors and Bioelectronics 77: 1086–1094.

Rahman, M.A., Y.C. Wong, G.S. Song, D.M. Zhu and C. Wen. 2018. Improvement on electrochemical performances of nanoporous titania as anode of lithium-ion batteries through annealing of pure titanium foils. Journal of Energy Chemistry 27(1): 250–263.

Romano, L., J. Vila-Comamala, K. Jefimovs and M. Stampanoni. 2017. Effect of isopropanol on gold assisted chemical etching of silicon microstructures. Microelectronic Engineering 177: 59–65.

RoyChaudhuri, C. 2015. A review on porous silicon based electrochemical biosensors: Beyond surface area enhancement factor. Sensors and Actuators B: Chemical 210: 310–323.

Ruffino, F., V. Torrisi, R. Grillo, G. Cacciato, M. Zimbone, G. Piccitto and M.G. Grimaldi. 2017. Nanoporous Au structures by dealloying Au/Ag thermal- or laser-dewetted bilayers on surfaces. Superlattices and Microstructures 103: 28–47.

Sadeghi, M.A., M. Aghighi, J. Barralet and J.T. Gostick. 2017. Pore network modeling of reaction-diffusion in hierarchical porous particles: The effects of microstructure. Chemical Engineering Journal 330: 1002–1011.

Salonen, J. and V.P. Lehto. 2008. Fabrication and chemical surface modification of mesoporous silicon for biomedical applications. Chemical Engineering Journal 137: 162–172.

Schiavi, P.G., P. Altimari, A. Rubino and F. Pagnanelli. 2018. Electrodeposition of cobalt nanowires into alumina templates generated by one-step anodization. Electrochimica Acta 259: 711–722.

Sharma, S., V. Saxena, A. Baranwal, P. Chandra and L.M. Pandey. 2018. Engineered nanoporous materials mediated heterogeneous catalysts and their implications in biodiesel production. Materials Science for Energy Technologies 1(1): 11–21.

Shcherban, N.D. 2017. Review on synthesis, structure, physical and chemical properties and functional characteristics of porous silicon carbide. Journal of Industrial and Engineering Chemistry 50: 15–28.

Shiba, K., S. Sato, T. Matsushita and M. Ogawa. 2013. Preparation of nanoporous titania spherical nanoparticles. Journal of Solid State Chemistry 199: 317–325.

Shobana, M.K. and Y.S. Kim. 2017. Improved electrode materials for Li-ion batteries using microscale and sub-micrometer scale porous materials—A review. Journal of Alloys and Compounds 729: 463–474.

Smith, K.J.P., M. May, R. Baltus and J.L. McGrath. 2017. A predictive model of separations in dead-end filtration with ultrathin membranes. Separation and Purification Technology 189: 40–47.

Subhapradha, N., M. Abudhahir, A. Aathira, N. Srinivasan and A. Moorthi. 2018. Polymer coated mesoporous ceramic for drug delivery in bone tissue engineering. International Journal of Biological Macromolecules 110: 65–73.

Sumer, Z. and S. Keskin. 2017. Molecular simulations of MOF adsorbents and membranes for noble gas separations. Chemical Engineering Science 164: 108–121.

Tang, Y.Y., C.L. Kao and P.Y. Chen. 2012. Electrochemical detection of hydrazine using a highly sensitive nanoporous gold electrode. Analytica Chimica Acta 711: 32–39.

Tong, W.Y., M.J. Sweetman, E.R. Marzouk, C. Fraser, T. Kuchel and N.H. Voelcker. 2016. Towards a subcutaneous optical biosensor based on thermally hydrocarbonised porous silicon. Biomaterials 74: 217–230.

Tovar, N., R. Jimbo, L. Witek, R. Anchieta, D. Yoo, L. Manne, L. Machado, R. Gangolli and P.G. Coelho. 2014. The physicochemical characterization and *in vivo* response of micro/nanoporous bioactive ceramic particulate bone graft materials. Materials Science and Engineering: C 43: 472–480.

Tran, H.T., J.Y. Byun and S.H. Kim. 2018. Nanoporous metallic thin films prepared by dry processes. Journal of Alloys and Compounds 764: 371–378.

Tripathy, M.K., N.K. Jena, A.K. Samanta, S.K. Ghosh and K.R.S. Chandrakumar. 2015. Theoretical investigations on Zundel cation present inside boron-nitride nanotubes: Effect of confinement and hydrogen bonding. Chemical Physics 446: 127–133.

Tsai, W.T., M.H. Nguyen, J.R. Lai, H.B. Nguyen, M.C. Lee and F.G. Tseng. 2018. ppb-level heavy metal ion detection by electrochemistry-assisted nanoporous silicon (ECA-NPS) photonic sensors. Sensors and Actuators B: Chemical 265: 75–83.

Vakifahmetoglu, C., D. Zeydanli and P. Colombo. 2016. Porous polymer derived ceramics. Materials Science and Engineering: R: Reports 106: 1–30.

Wang, Q., D. Wu and C.L. Liu. 2018. Electrostatic assembly of graphene oxide with Zinc-Glutamate metal-organic framework crystalline to synthesis nanoporous carbon with enhanced capacitive performance. Electrochimica Acta 270: 183–191.

Watkins, W.L. and Y. Borensztein. 2018. Ultrasensitive and fast single wavelength plasmonic hydrogen sensing with anisotropic nanostructured Pd films. Sensors and Actuators B: Chemical 273: 527–535.

Weremfo, A., S.T.C. Fong, A. Khan, D.B. Hibbert and C. Zhao. 2017. Electrochemically roughened nanoporous platinum electrodes for non-enzymatic glucose sensors. Electrochimica Acta 231: 20–26.

Wichmann, A., G. Schnurpfeil, J. Backenköhler, L. Kolke, V.A. Azov, D. Wöhrle, M. Bäumer and A. Wittstock. 2014. A versatile synthetic strategy for nanoporous gold-organic hybrid materials for electrochemistry and photocatalysis. Tetrahedron 70(36): 6127–6133.

Wierzbicka, E., K. Syrek, G.D. Sulka, M. Pisarek and M. Janik-Czachor. 2016. The effect of foil purity on morphology of anodized nanoporous ZrO_2. Applied Surface Science Part B 388: 799–804.

Wisser, F.M., J. Grothe and S. Kaskel. 2016. Nanoporous polymers as highly sensitive functional material in chemiresistive gas sensors. Sensors and Actuators B: Chemical 223: 166–171.

Xu, H.J., S.J. Pang and T. Zhang. 2018. Self-oxidized sponge-like nanoporous nickel alloy in three-dimensions with pseudocapacitive behavior and excellent capacitive performance. Journal of Power Sources 399: 192–198.

Yallappa, S., S.A.A. Manaf and G. Hegde. 2018. Synthesis of a biocompatible nanoporous carbon and its conjugation with florescent dye for cellular imaging and targeted drug delivery to cancer cells. New Carbon Materials 33(2): 162–172.

Yan, X.J., M. Sahimi and T.T. Tsotsis. 2017. Fabrication of high-surface area nanoporous SiOC ceramics using pre-ceramic polymer precursors and a sacrificial template: Precursor effects. Microporous and Mesoporous Materials 241: 338–345.

Zeleňák, V., D. Halamová, M. Almáši, L. Žid, A. Zeleňáková and O. Kapusta. 2018. Ordered cubic nanoporous silica support MCM-48 for delivery of poorly soluble drug indomethacin. Applied Surface Science 443: 525–534.

14

Bionanotechnology in Robust Biosolar Cells

Ashwini Ravi,[1] *J. Hemaprlya,*[2] *P.N. Sudhu,*[3] *S. Sugashini,*[3]
S. Pavithra[3] *and S. Vijayanand*[1,*]

1. Introduction

Energy is considered the main 'fuel' for social and economic development of a nation (World Energy Resources 2013). The US department of energy has divided energy consumers into different categories, viz., residential, commercial, industrial, electric power and transportation (NEED project 2018). While residential and commercial consumers use energy mainly for lighting purposes, the other sectors use them for socio-economic development. Therefore, sufficient and secure energy production has to be the main goal of any government in order to increase economy. The Energy Information Administration has stated that "There is a strong two-way relationship between economic development and energy consumption. On one hand, growth of an economy with its global competitiveness hinges on the availability of cost-effective and environmentally benevolent energy sources and on the other hand, the level of economic development has been observed to be dependent on the energy demand" (EIA 2010).

There are two factors that primarily determine the amount of energy production, i.e., the population and technology development. During the past two decades, there has been an increase in population by over 1.5 billion. In addition to this, the vast technology development over the past 2 decades has increased the electricity consumption of an individual. The increased population and the technological development have caused the reduction in number of people having access to energy which in turn increased the necessity for higher energy production (World Energy Resources 2013). As of 2017, the energy production in China and India has increased to 6,495 TW and 1,497 TW respectively. It has also been estimated that the energy consumption will further increase in 2050 to 21,000 TW in China and 4,700 TW in India. But in countries like US, European Union, Japan, Russian Federation and Canada, there have been downfall and some recovering. But still,

[1] Bioresource Technology Lab, Department of Biotechnology, Thiruvalluvar University, Serkkadu, Vellore, Tamil Nadu, India, 632115.
[2] Biomaterial Research Lab, Department of Chemistry, DKM College for Women (Autonomous), Vellore, Tamil Nadu, India, 632001.
[3] PG & Research Department of Microbiology, DKM College for Women (Autonomous), Vellore, Tamil Nadu, India, 632001.
* Corresponding author: vipni76@gmail.com

there has been increase of energy consumption in these countries by 2017 when compared with the previous years (World Energy 2017).

Therefore, there must be constant and sustainable energy sources for energy production for these growing population and growing needs. The main sources of energy production are coal, oil, natural gas, peat, geothermal, wind, solar, tidal and bioenergy from wastes (World Energy Resources 2016). Of the various sources, 80% of energy is produced using coal, natural gas and oil, 5% from the nuclear and 10% from the renewable sources. Two-thirds of energy from renewable sources is produced from hydroelectric power and the remaining is from wind, solar and bioenergy combined (Seger 2016). At present, awareness about the depletion of non-renewable source of energy and their social impacts on global warming has made government to look for more promising renewable source of energy production. This created several researchers to utilize the renewable underused source of energy production, the Sun. By using photovoltaic cells, nowadays highly advanced and developed Dye Sensitized Solar Cells (DSSC) and Biosensitized Solar Cells (BSSC) energy can be produced and can be utilized for several purposes. The DSSC or BSSC has three main components, viz., the photoanode (ITO, semiconductor, dye of choice), counter electrode and electrolyte (Thavasi et al. 2008). Several works have been performed on the choice of dye, semiconductor and electrolyte. Therefore, this chapter deals with the biosolar cells, their components and the involvement of nanotechnology in creating them.

2. Dye Sensitized Solar Cells

Solar energy is the most underused and unexplored source of energy production. It has been estimated that the sun rays strikes Earth with an energy of 167,000 TW which is highly sufficient to meet world's needs (Seger 2016). Therefore, production of energy from solar radiation has been concentrated by various countries for safe and economic energy production. In developing nations like India, installation of solar power plant has been increased and it has been shown that energy production from solar power plants has increased from 6.67 GW in 2015–16 to 12.28 GW in 2017–18 (Energy Statistics-India 2018). Globally, solar power production by photovoltaic cells has been found to be 227 GW at the end of 2015, contributing to 1% of energy production (IRENA 2016, World Energy Resources 2016).

Photovoltaic cells are those that convert solar energy directly in to electrical energy (Richhariya et al. 2017). These photovoltaic cells have been classified into three generations as first, second and third. The first generation PV cells are solely made up of silicon wafers and they have better efficiency among all three generations. But the main drawback with these types of cells is their higher fabrication cost (Tobnaghi et al. 2013). The second generation solar cells are the thin film solar cell mainly made up of amorphous silicon. Other components used instead of silicon are cadmium telluride and copper indium diselenide. The main drawback with these cells is their less efficiency of around 10–16% (Green 2007). The third generation photovoltaic cells are the Dye Sensitized Solar Cells (DSSC). While the first and second generation photovoltaic cells have been used commercially, the third generation photovoltaic cells (Gupta et al. 2018). Several researches have been conducted on these cells to find out the best combination in order to achieve maximum efficiency when compared with the other two generations.

The first Dye Sensitized Solar Cell was fabricated by Brian O' Regan and Michael Gratzel in the year 1991. Therefore, these cells are also called as Gratzel cells (O' Regan and Gratzel 1991). It has been proposed that these DSSCs were inspired by the photosynthetic process where light energy is converted to electrical energy (Guo et al. 2015, Pablo et al. 2016). The DSSCs fabricated by O' Regan and Gratzel was found to have energy conversion rate of 7.1% and incident photon electrical current conversion of approximately 80% (O' Regan and Gratzel 1991). Then on, several works on DSSCs have been performed since they have easy fabrication, simple structure, low cost technology and high efficiency (Gao et al. 2008, Gong et al. 2012).

3. Components of DSSC and their Functions

Dye Sensitized Solar Cells consist of three main components viz.

 (i) Photoanode

(ii) Counter electrode

(iii) Electrolyte

The schematic representation of working of solar cell and its components is given in Figure 1.

3.1 Photoanode

The photoanode consists of three main components: (i) a transparent anode made up of glass sheet treated with transparent conductive oxide which is mainly used for enhancing electrical conductivity and light transmittance. The main conductive glasses used in these cells are Fluorine doped Tin Oxide (FTO) and Indium doped Tin Oxide (ITO) (Sima et al. 2010); (ii) a mesoporous oxide layer which is a semiconductor metal, especially TiO_2 and ZnO (Sreekala et al. 2012). This semiconductor metal plays a major role in electron absorbance and transfer to the counter electrode for energy production; and (iii) dye of choice, which acts as a source of electron production. Both organic and natural dye molecules have been used in DSSC. When light falls on dye molecule, the photon gets excited from ground state to higher energy state injecting the electron into semiconductor which gets dispersed into the electrolyte and gets transferred to the counter electrode for the completion of energy production process (Narayan 2012).

3.2 Electrolyte

The electrolyte that is traditionally used in DSSC is iodide–triiodide (I^-/I_3^-) redox system which acts as a mediator between the counter electrode and photo anode. The lost electron of dye molecule will be replaced immediately by the Iodine I^- whereas the I^- is regenerated by I_3^- at the counter electrode. Apart from these, other electrolyte such as solid electrolyte, polymer electrolyte and gel electrolyte has been studied for its efficiency in DSSC (Gong et al. 2012).

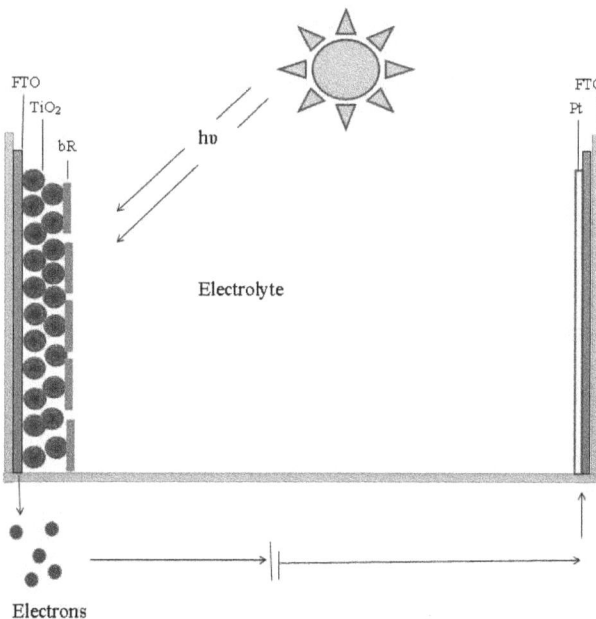

Figure 1: Representation of Dye Sensitized Solar Cells.

3.3 Counter Electrode

The counter electrode or the cathode is mainly coated with grapheme or platinum as catalyst. Apart from platinum, other materials such as graphite, activated carbon and poly(3,4-ethylenedioxythiophene)(PEDOT) have been used as counter electrode (Lindstrom et al. 2001, Saito et al. 2002, 2004).

4. Biosensitized Solar Cells (BSSC)

The Biosensitized Solar Cells (BSSC) have similar components and function as Dye Sensitized Solar Cell except the choice of dye. In DSSC, several organic and metal dyes have been used. Of various dyes used, ruthenium was found to have the highest efficiency. However, these dyes have several disadvantages such as high cost of production, complicated synthetic routes and low yields. In addition to these, they are not economically healthy to the environment. Therefore, natural dyes from plants and microorganisms have been now used in DSSCs (Mishra et al. 2009). Though the efficiency of natural dyes is comparatively lesser than the organic and metal dyes, they pose several advantages such as low cost of production, non-toxicity and biodegradability. Therefore, the use of natural dyes has been emphasized by various researchers around the world.

4.1 Chlorophyll

The main property of any photosensitizer to be used in DSSC is that it should have an active OH, C=O or COOH as a molecular bridge and the configuration to enable electron transfer between them and the semiconductor (Wongcharee et al. 2006, Narayan 2012). As far as chlorophyll is concerned, it is a natural photosensitizer with an absorption range of 400–450 nm and 600–700 nm in the visible range (Al-Alwani et al. 2015, Syafinar et al. 2015). All photosynthetic organisms have different types of chlorophyll but the higher plants only have two types, viz., Chlorophyll a and Chlorophyll b (Figure 2 and Figure 3). They both differ in their molecular structure and also in their absorption wavelengths (Rossi et al. 2017).

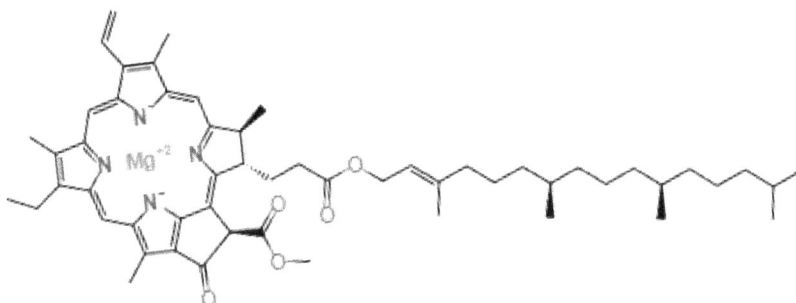

Figure 2: Structure of chlorophyll a (molview 1).

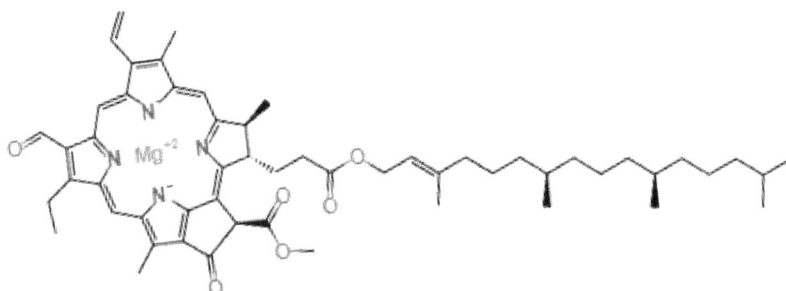

Figure 3: Structure of chlorophyll b (molview 2).

Though they both differ in their structure and wavelength, their primary function is to harvest sunlight and convert it into energy. Since they are abundant in nature and a universal photosensitizer, they have been abundantly studied for their application in DSSC. Some plants that have been used for the extraction of chlorophyll and its application as photosensitizer in DSSC have been listed in the table below:

S. No.	Name of the Plant	Photo Anode	Electrolyte	Counter Electrode	Reference
1.	*Spinach leaves*	TiO_2/TCO, TiO_2/FTO, TiO_2/ITO	I^-/I_3^-	Pt/TCO, Pt/FTO, Grpahite/ITO	Magsi et al. 2012, Cari et al. 2014, Syafinar et al. 2015, Rossi et al. 2017
2.	*Tectona grandis*	TiO_2/FTO	I^-/I_3^-	Pt/FTO	Kushwaha et al. 2013
3.	*Tamarindus indica*	TiO_2/FTO	I^-/I_3^-	Pt/FTO	Kushwaha et al. 2013
4.	*Eucalyptus globulus*	TiO_2/FTO	I^-/I_3^-	Pt/FTO	Kushwaha et al. 2013
5.	*Lawsoniainermis*	TiO_2/ITO	PEO I^-/I_3^-	Grpahite/ITO Platinum/ITO	Al Bathi et al. 2013, Ananth et al. 2014
6.	*Lemon*	TiO_2/FTO	I^-/I_3^-	Graphite/FTO	Maabong et al. 2015
7.	*Morula*	TiO_2/FTO	I^-/I_3^-	Graphite/FTO	Maabong et al. 2015
8.	*Euodia meliaefolia* (Hance) Benth	TiO_2/FTO	I^-/I_3^-	Pt/FTO	Liu et al. 2018
9.	*Matteuccia Struthiopteris* (L.) Todaro	TiO_2/FTO	I^-/I_3^-	Pt/FTO	Liu et al. 2018
10.	*Corylus heterophylla*	TiO_2/FTO	I^-/I_3^-	Pt/FTO	Liu et al. 2018
11.	*Filipendula intermedia*	TiO_2/FTO	I^-/I_3^-	Pt/FTO	Liu et al. 2018
12.	*Pteridium aquilinum* var. *latiusculum*	TiO_2/FTO	I^-/I_3^-	Pt/FTO	Liu et al. 2018
13.	*Populous L.*	TiO_2/FTO	I^-/I_3^-	Pt/FTO	Liu et al. 2018
14.	*Brassia oleracea*	TiO_2/FTO	KI/I_2	Carbon soot/FTO	Ikeogu et al. 2018
15.	*Tagetuspatula*	Steel/ZnS	S^{2-}/S_n^{2-}	Graphite	Panda et al. 2018
16.	*Anethum graveolus*	TiO_2/FTO	I^-/I_3^-	Pt/FTO	Taya et al. 2013
17.	*Arugula*	TiO_2/FTO	I^-/I_3^-	Pt/FTO	Taya et al. 2013
18.	*Parsley*	TiO_2/FTO	I^-/I_3^-	Pt/FTO	Taya et al. 2013
19.	*Pomogranate*	TiO_2/ITO	I^-/I_3^-	Pt/ITO	Chang and Lo 2010
20.	*Ocimumgrattisimum*	TiO_2/FTO	I^-/I_3^-	Pt/FTO	Eli et al. 2016
21.	*Jatropha*	TiO_2/ITO	I^-/I_3^-	C/ITO	Pramono et al. 2015
22.	*Papaya*	TiO_2/FTO	I^-/I_3^-	Pt/FTO	Suyitno et al. 2015
23.	*Azadirachta indica*	TiO_2/FTO	I^-/I_3^-	C/FTO	Swarnkar et al. 2015
24.	*Banana*	TiO_2/FTO	I^-/I_3^-	Pt/FTO	Taya et al. 2015
25.	*Peach*	TiO_2/FTO	I^-/I_3^-	Pt/FTO	Taya et al. 2015
26.	*Fig*	TiO_2/FTO	I^-/I_3^-	Pt/FTO	Taya et al. 2015
27.	*Hyophilainvoluta*	TiO_2/FTO	Polymer electrolyte	Pt/FTO	Hassan et al. 2016
28.	*Sargassum*	TiO_2/TCO	–	–	Ridwan et al. 2018

Apart from these, the derivatives of chlorophyll such as methyl 3-carboxy-3-devinyl-pyropheophorobide a, chlorins-1-5 possessing C32-carboxy and O174-esterified hydrocarbon groups including methyl, hexyl, dodecyl, 2-butyloctyl and cholesteryl were used in DSSCs (Wang et al. 2005, 2010). The stabilization of chlorophyll by Fe ions and the co-sensitiation of chlorophyll with betalin and anthocyanins have also been performed to increase the efficiency of chlorophyll in DSSCs (Arifin et al. 2017, Pratiwi et al. 2017, Sreeja and Pesala 2018).

4.2 Bacteriorhodopsin

Bacteriorhodopsin is a light driven proton pump that resides in the membrane of haloarchaea *Halobacterium salinarum* and was first identified by Oesterhelt and Stoeckenius in the year 1971 (Oesterhelt and Stoeckenius 1971, Ashwini et al. 2017). It is a purple coloured protein with a molecular weight of approximately 26.5 kDa and consists of seven membrane helical structures linked by short loops on either side of the cell membrane resembling the G-protein coupled receptor (Oesterhelt 1988, Kuhlbrandt 2000). The protein consists of 248 amino acids and is bound to chromophore at Lys 216 by a schiff's base linkage (Khorana et al. 1979). It acts as a green light driven photon pump that converts light energy into chemical energy which is utilized by the organism for production of ATP, amino acid uptake and locomotion. It also protects the organism from salt stress which is caused due to the extremely halophilic environment and ensures their survival even in deprived environmental conditions (Oesterhelt 1976, Oesterhelt and Stoeckenius 1973, Stoeckenius and Bogomolni 1982).

When the light range of 568 nm falls on to the bacteriorhodopsin, the photocycle of the protein initiates with photoisomeriaztion of the retinal chromophore where it gets deprotonated and passes through series of steps finally converting from all trans to cis conformation (Stoeckenius and Bogomolni 1982, Haupts 1997, Neutze et al. 2002). It passes to several stages from J to O at the moment of photon incidence liberating a proton as a product. This mechanism is a reversible cascade that gets completed in 3 ms to 1ps when the light hits on bacteriorhodopsin at a range of 250 mv, finally giving to 10,000 times of proton release inside the membrane. The final product O decays again to retain the bacteriorhodopsin getting ready for next cycle (Zimanyi et al. 1993, Kuhlbrandt 2000) (Figure 4).

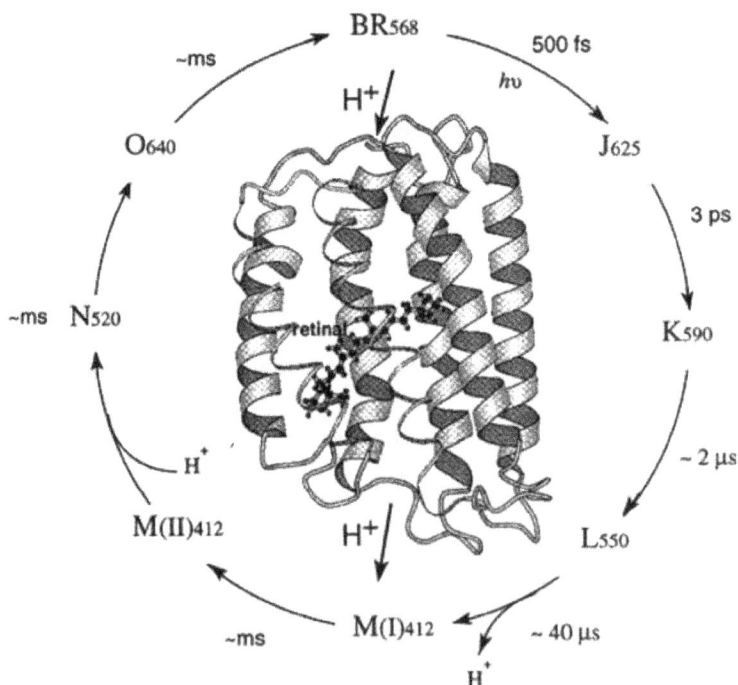

Figure 4: Photonic excitation of bacteriorhodopsin (Henderson et al. 1990).

Though the organism uses this light driven photon pump for their survival, it has attracted the attention of various researchers around the world since it is a natural mechanism of harvesting sunlight other than chlorophyll. Therefore, they have been applied in various optical studies, especially DSSC. Several works on bacteriorhodopsin and its mutants have been performed for its application in Dye sensitized solar cells. Thavasi et al. (2008) have studied the effect of bacteriorhodopsin and its three mutants in DSSC. In a study by Molaeirad et al. (2014), the bacteriorhodopsin was co-sensitized with bacterioruberin and was found to have better power converting efficiency of 0.18% when the two compounds were combined, which is comparatively higher than the bacteriorhodopsin (0.11%) and bacterioruberin (0.08%). In another study by Nesari and Molaeirad (2016), bacteriorhodopsin was made as a complex with N719, a synthetic dye, and was found to have better power converting efficiency.

4.3 Other Pigments

There are other natural pigments such as Beta Carotene, anthocyanin, etc., other than chlorophyll requiring limelight. Researchers around the world, despite using the chlorophyll from different sources, started studying variable dye sources of unalike compounds to be utilized in Dye Sensitized Solar Cells. The natural dye and its source of extraction have been listed in the table below:

S. No.	Source of Dye	Part of Extraction	Reference
1.	*Hehuang Safflowers*	Flower	Zhao et al. 2018
2.	*Callistemon citrinus*	Flower	Kushwaha et al. 2013
3.	*Amaranthus caudatus*	Flower	Godibo et al. 2015
4.	*Bougainvillea spectabilis*	Flower	Godibo et al. 2015
5.	*Delonix regia*	Flower	Godibo et al. 2015
6.	*Nerium oleander*	Flower	Godibo et al. 2015
7.	*Spathodeacompanulata*	Flower	Godibo et al. 2015
8.	*Dragon fruit*	Fruit	Isahkimpa et al. 2012
9.	*Beetroot*	Vegetable	Sathyajothi et al. 2017
10.	*Melinjo*	Skin	Cari et al. 2014
11.	*Shoe flower*	Flower	Cari et al. 2014
12.	*Blue pea*	Flower	Cari et al. 2014
13.	*Blue berry*	Fruit	Qin et al. 2018
14.	*Purple cabbage*	Peels	Qin et al. 2018
15.	*Onion*	Peels	Qin et al. 2018, Ammar et al. 2019
16.	*Tangerine*	Peel	Huizhi et al. 2011
17.	*Marigold*	Flower	Huizhi et al. 2011
18.	*Rose*	Flower	Huizhi et al. 2011
19.	*Raspberries*	Fruit	Alhamed et al. 2012
20.	*Grapes*	Fruit	Alhamed et al. 2012
21.	*Cherries*	Fruit	Jasim et al. 2011
22.	*Black carrot*	Vegetable	Tekerek et al. 2011
23.	*Black raspberry*	Fruit	Tekerek et al. 2011
24.	*Hibiscus sabdariffa*	Flower	Tekerek et al. 2011
25.	*Red cabbage*	Peels	Furkawa et al. 2009, Ammar et al. 2019

Table contd. ...

...Table contd.

S. No.	Source of Dye	Part of Extraction	Reference
26.	*Maqui*	Fruit	Cerda et al. 2016
27.	*Punica granatum*	Peel	Martinez et al. 2015
28.	*Clitoriaternatea*	Petals	Manna et al. 2017
29.	*Hibiscus rosasinensis*	Petals	Manna et al. 2017
30.	*Ixora chinensis*	Petals	Manna et al. 2017
31.	Carob	Fruit	Abdel-Latif et al. 2015
32.	Egg plant	Peel	Abdel-Latif et al. 2015
33.	Basil	Flower	Abdel-Latif et al. 2015
34.	Mint	Flower	Abdel-Latif et al. 2015
35.	*Royal poinciana*	Flower	Abdel-Latif et al. 2015
36.	Carob fruit	Fruit	Abdel-Latif et al. 2015

5. Role of Nanoparticles in Biosolar Cells

The "Quantum size effect" of nanoparticles provide them unique physical, chemical, electrical, mechanical and optical properties that have been applied in various fields. It has been potentially explored in several fields of science such as medicine, cosmetics, food and also in several consumables such as paints, paper, etc. (Janani et al. 2017). Now, nanoparticles have made their way in Dye Sensitized Solar cells, where they are used as semiconductor in conducting electrons produced by excitation of dye molecule and passing it to cathode via electrolyte.

The main property of nanomaterial to be used in solar cell is its surface to volume ratio which increases their chemical reactivity with the surroundings (Goldstein et al. 1992). This property of nanomaterial has been efficiently used in solar cells for light trapping since the photon path inside nanostructures is relatively higher than the bulk materials. Though different modes of nanoparticles such as nanowires, nanotubes, nanosheets, quantum dots (Garnet et al. 2005, Fan et al. 2011, Zhu et al. 2009, Peng and Peng 2001) and different types of nanoparticles such as titatinum dioxide, zinc oxide, etc. (Tributsch and Gerischar 1969, Spitler and Calvin 1977) have been used in Dye sensitized solar cells. In the first generation solar cells, silicon was used as nanowires (Kerlzenberg et al. 2008). Apart from this, cadmium telluride and cadmium sulphide were also used in first generation solar cells (Mia et al. 2012, Garnet et al. 2010). This made the way for replacement of bulk materials with nanomaterials in solar cells and with the advancement of solar cell to DSSC, usage of nanomaterials became predominant.

6. Nanoparticles in Biosolar Cells

Nanoparticles that are predominantly used in Dye Sensitized Solar Cells are discussed in the following sections.

6.1 TiO_2 Nanoparticles

The search made in Science direct search engine with keywords "TiO_2 and DSSC" shows 2,334 papers published in the topic from the year 2010 (Science direct). The efficiency of DSSC strongly depends on the wide band gap of semiconductor used (Gratzel 2003). Of the several semiconductors used, TiO_2 remains to be the highly utilized semiconductor in DSSC. The large band gap, suitable band edge levels for charge injection and extraction, long lifetime of excited electrons, resistance to photo corrosion, non-toxicity and low cost makes TiO_2 nanoparticle a perfect candidate for DSSC (Leung et al. 2010, Park et al. 2013). Apart from this, it has a large surface area which is very much essential for the absorption of dyes and it also has enhanced interaction with the electrolyte (Grunwald and Tributsch 1997). These characters of TiO_2 have made it an essential candidate as semiconductor in Dye Sensitized Solar Cells.

TiO_2 exists in three different forms as anatase, rutile and brookite (Figures 5a, 5b and 5c). Among these, anatase has been used usually in Dye sensitized solar cells due to its superior electron

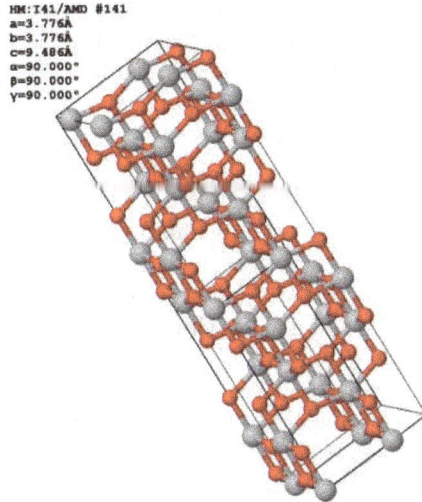

Figure 5a: Structure of Anatase (link ana).

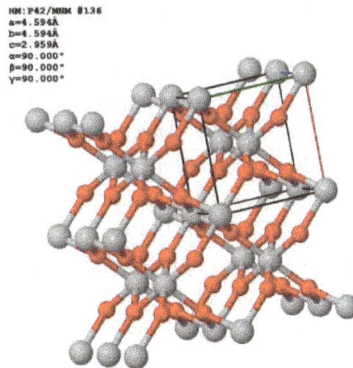

Figure 5b: Structure of Rutile (link rut).

Figure 5c: Structure of Brookite (Pauling blog).

transport (Chen and Mao 2007, Hsiao et al. 2007, Pavasupree et al. 2008). Some studies showed that a small amount of rutile form along with the anatase form increases the charge converting efficiency of solar cells. But rutile alone when used was found not to have better efficiency than anatase due to reduced conduction band, reduced dye adsorption and charge transport ability (Park et al. 2000).

In TiO_2 structure, shallow electron traps produced by the bulk oxygen vacancies, titanium interstitials and reduced crystal surfaces act as n type dopants. These dopants intensify the number of free electrons in TiO_2 which in turn increases the conductivity and current. These traps also have negative impact since they sometimes act as charge traps (Ardaani 1994, Asahi et al. 2000). Therefore, a check on the number of defects in TiO_2 structure is crucial for better device performance. TiO_2 for DSSC can be synthesized in different ways and the most commonly used methods include sol-gel synthesis (Karami 2010), hydrothermal synthesis (Xuelian and Junjiang 2012), direct oxidation (Peng and Chen 2004), electrospinning (Otieno et al. 2020), spin coating (Vasuki et al. 2014), atomic layer deposition (Edy et al. 2016), thermal oxidation (Ke et al. 2014), electrochemical deposition (Kavan et al. 1993) and pulsed laser ablation (Singh et al. 2016). In DSSC, TiO_2 has been used as nanoparticles instead of bulk particles in order to further increase the efficiency. Alongside, different structures of nanoparticles and doping with various metals have also been performed to study TiO_2 competence in DSSC.

Initially, the utilization of a compact layer of TiO_2 over FTO was first described by Burke et al. (2008). He and his co-workers proposed that a TiO_2 compact layer over FTO increases device efficiency even in low intensity lights, though his result was contrasted to other works proposed with ruthenium based dyes which proposed the introduction of TiO_2 had only marginal effects. This work of Burke concluded that different organic dyes have different functional properties other than ruthenium and dyes with more bulky structures function efficiently in DSSC. In that case, TiO_2 compact layer acts as a blocking layer in increasing the light harvesting efficiency making the cells function better.

The thickness of TiO_2 nanoparticles and its particle size are main criteria for functioning of Dye Sensitized Solar Cells. Thickness of 12.73 μm and particle size of 30 nm was found appropriate in increasing the efficiency of DSSC, though the size ranging from 20–30 nm did not affect cell's potency (Kumari et al. 2016, Son et al. 2018). TiO_2 nanoparticles were prepared in various forms such as nanotube membranes, nano dendrites, nanofibres, nanospindles, nanorods, nanowires and also as nanocomposites. TiO_2, when treated with chemicals such as titanium chloride, increases the efficiency of cell further (Kang et al. 2008, Bwana 2009, Qui et al. 2010, Tao et al. 2011, Song et al. 2013, Mohammadpour et al. 2015, Chou and Liu 2016, Ni et al. 2019). Apart from using it as a single element, TiO_2 is also doped for usage in DSSC with various compounds such as alkali metals (lithium (Submanian et al. 2012), magnesium (Kakiage et al. 2013)), metalloids (boron (Im et al. 2012), antimony (Wang et al. 2012)), non-metals (carbon (Hsu et al. 2013), nitrogen (Guo and Yin 2015)), transition metals (scandium (Latini et al. 2013), vanadium (Liu et al. 2013), iron (Eom et al. 2014)), post transition metals (aluminium (Manoharan and Venkatachalam 2015), gallium (Chandiran et al. 2011)) and lanthanides (lanthanum (Yahav et al. 2011), cerium (Zhang et al. 2014)). Apart from these, they are coated along with other nanostructures such as carbon nanotubes (Lee et al. 2009), MWCNT (Muduli et al. 2010), ZnO (John et al. 2016, Rahman et al. 2019), SnO (Ahn et al. 2014), etc. Though research on every doping metal or various forms of TiO_2 nanocrystals are carried out at large scale, it is evident that "without TiO_2 no DSSC exists".

6.2 ZnO Nanoparticle

While TiO_2 is predominantly used as the semiconductor in DSSC, other wide gap semiconductors have also been explored for their utility in the same. With this respect, ZnO was found to have excellent bulk particle mobility (Forro et al. 1994, Look et al. 1998). It also has high excitation binding energy, large saturation velocity, high optical gain, high radiative recombination efficiency, available at low cost and has stability against photocorrosion (Lee et al. 2011, Makkar and Bhatti 2011, Anta et al. 2012). ZnO was identified to be the first semiconductor to show irreversible electron injection from organic molecules into the conduction band (Tributsch and Calvin 1971, Anta et al. 2012). With these notable features, ZnO can be formulated as various types of nanoparticles and also in combinations for its promising use in DSSC.

The crystalline structure of ZnO is conductive to anisotropic growth (Baxter et al. 2006). This anisotropic characteristic of ZnO makes it to be produced as different nanostructures such as nanoparticles (Giannouli et al. 2018), nanograins (Rahman et al. 2013), nanotubes (Abd-Ellah et al. 2013), nanorods (Zhai et al. 2017), nanosheets (Lin et al. 2011), nanowires (Giannouli et al. 2018), nanoflowers (Lou et al. 2013), nanoflakes (Mou et al. 2011), nanodendrites (Lou et al. 2013) and tetrapods (Lee et al. 2012). In addition, they are flexible to be doped with other materials such as Cu (Selim et al. 2018, Fabbiyola and Kennedy 2019), gallium (Ghnimi et al. 2017), Co (Fabbiyola and Kennedy 2019), Fe (Fabbiyola and Kennedy 2019), Mn (Fabbiyola and Kennedy 2019), Carbon Nanotubes (Chang et al. 2012), Ag (Han et al. 2012), Al (Lee et al. 2010), etc. Several studies on deposition of these ZnO to photoanode were conducted and methods such as electrodeposition (Yoshida et al. 2009), spray pyrolysis (Shaban et al. 2016), chemical vapour deposition (Liang 2009), pulse laser deposition (Liu and Zhao 2007), hydrothermal synthesis (Ameen et al. 2012), etc., were found to be used. Because of these wide supple characters, they are considered more advantageous than TiO_2 nanoparticles.

6.3 Other Nanoparticles

Research on DSSC was more focussed on enhancing its efficiency and hence several metals have been studied for their efficiency to increase the efficacy. Therefore, the concept of surface plasmon resonance was utilized to increase the potency of DSSC by increasing its light harvesting efficacy (Zhao et al. 1997, Pillai and Green 2010). Hence, the metals and their nanoparticles, which are near field plasmonic such as Au (Mayumi et al. 2017), Ag (Song et al. 2017, Kazmi et al. 2017), Si (Brown et al. 2011), Cu (Prabhin et al. 2017), Ce/Gd (Hossain et al. 2018), Nb (Anuntahirunrat et al. 2017), etc., were studied with differing experimental designs in DSSC.

7. Conclusion

Microbes have been the most explored and exploited in several manufacturing industries. Apart from microbes, the nanoparticles slowly took over and it has now occupied almost all the manufacturing industries-right from food to non-consumable goods. The combination of these two widely used components was found to increase the quality of the product paving cost effectiveness, simple procedures and economic benefits. Biosolar cells, being a necessity in the present world to minimize the exploitation of non-renewable sources, are now utilizing the nanoparticles synthesized from biological entities, either microbes or plants. This utilization of bionanoparticles in biosolar cells will make it benign to the environment and efficiently use the solar cell for energy production. The continued researches in these DSSC in future will be the sole source of energy production in several developed countries which are already using photovoltaic cells for energy production and also in developing countries.

References

Abd-Ellah, M., N. Moghimi, L. Zhang, N.F. Heinig, L. Zhao, J.P. Thomas and K.T. Leung. 2013. Effect of electrolyte conductivity on controlled electrochemical synthesis of zinc oxide nanotubes and nanorods. J. Phys. Chem. 117-13: 6794–9.

Abdel-Latif, M.S., M.B. Abuiriban, N. Al Dahoudi, A.M. Al-Kahlout, S.A. Taya, T.M. El-Agez and H.S. El-Ghamri. 2015. Dye-sensitized solar cells using fifteen natural dyes as sensitizers of nanocrystalline TiO_2. Sci. Technol. Dev. 34: 135–139.

Ahn, S.H., D.J. Kim, W.S. Chi and J.H. Kim. 2014. Hierarchical double-shell nanostructures of TiO_2 nanosheets on SnO_2 hollow spheres for high-efficiency, solid-state, dye-sensitized solar cells. Adv. Funct. Mater. 24: 5037–5044.

Al-Alwani, M.A.M., A.B. Mohamad, A.A.H. Kadhum and N.A. Ludin. 2015. Effect of solvents on the extraction of natural pigments and adsorption onto TiO_2 for dye-sensitized solar cell applications. Spectrochim. Acta A Mol. Biomol. Spectrosc. 138: 130–137.

Al Bathi, S.A.M., I. Alaei and I. Sopyan. 2013. Natural photo-sensitizers for dye-sensitized solar cells. Int. J. Renew. Energy Res. 3: 138–143.

Alhamed, M., A.S. Issa and A.W. Doubal. 2012. Studying of natural dyes properties as photosensitizer for dye sensitized solar cells (DSSC). J. Electron. Devices 16: 1370–83.

Ameen, S., M.S. Akhtar, H.K. Seo, Y.S. Kim and H.S. Shin. 2012. Influence of Sn doping on ZnO nanostructures from nanoparticles to spindle shape and their photoelectrochemical properties for dye sensitized solar cells. Chem. Eng. J. 187: 351–6.

Ammar, A.M., H.S.H. Mohamed, M.M.K. Yousef, G.M. Abdel-Hafez, A.S. Hassanien and A.S.G. Khalil. 2019. Dye-Sensitized Solar Cells (DSSCs) based on extracted natural dyes. J. Nanomater. 1-10: 1867271.

Ananth, S., P. Vivek, T. Arumanayagam and P. Murugakoothan. 2014. Natural dye extract of lawsoniainermis seed as photo sensitizer for titanium dioxide based dye sensitized solar cells. Spectrochim. Acta A Molecul. Biomol. Spec. 128: 420–426.

Anta, J.A., E. Guillén and R. Tena-Zaera. 2012. ZnO-based dye-sensitized solar cells. J. Phy. Chem. C 116: 11413–25.

Anuntahirunrat, J., Y.M. Sung and P. Pooyodying. 2017. Efficiency of Nb-Doped ZnO nanoparticles electrode for dye-sensitized solar cells application. AMRMT Series Mater. Sci. Eng. 229: 012019.

Ardakani, H. 1994. Thin solid films, electrical and optical properties of *in situ* "hydrogen-reduced" titanium dioxide thin films deposited by pulsed excimer laser ablation. Thin Solid Films 248: 234–239.

Arifin, Z., S. Soeperman, D. Widhiyanuriyawan, S. Suyitino and A.T. Setyaji. 2017. Improving stability of chlorophyll as natural dye for dye-sensitized solar cells. J. Teknol. 80(1): 27–33.

Asahi, R., Y. Tagaand, W. Mannstadt and A.J. Freeman. 2000. Electronic and optical properties of anatase TiO_2. Phys. Rev. B 61: 7459–7465.

Ashwini, R., S. Vijayanand and J. Hemapriya. 2017. Photonic potential of Haloarchaeal pigment bacteriorhodopsin for future electronics. J. Curr. Microbiol. 74: 996–1002.

Baxter, J.B., A.M. Walker, K. Van Ommering and E.S. Aydil. 2006. Synthesis and characterization of ZnO nanowires and their integration into dye-sensitized solar cells. Nanotechnol. 17: 304–312.

Brown, M.D., T. Suteewong, R.S.S. Kumar, V. DInnocenzo, A. Petrozza, M.M. Lee, U. Wiesner and H.J. Snaith. 2011. Plasmonic dye-sensitized solar cells using core-shell metal-insulator nanoparticles. NanoLett. 11: 438–445.

Burke, A., S. Ito, H. Snaith, U. Bach, J. Kwiatkwoski and M. Gratzel. 2008. The function of a TiO_2 compact layer in dye-sensitized solar cells incorporating "planar" organic dyes. Nano. Lett. 8: 977–981.

Bwana, N.N. 2009. Comparison of the performances of dye-sensitized solar cells based on different TiO2 electrodes nanostructures. J. Nanoparticle Res. 1: 1917–23.

Cari, C., T.Y. Khairuddin, S.T.Y. Khairuddin, P.M. Suciatmoko, D. Kurniawan and A. Supriyanto. 2014. The preparation of natural dye for dye-sensitized solar cell (DSSC). AIP Conference Proceedings 1: 020106.

Cerda, B., R. Sivakumar and M. Paulraj. 2014. Natural dyes as sensitizers to increase the efficiency in sensitized solar cells. XIX Chilean Physics Symposium J. Phys. Conf. Series 720: 012030.

Chandiran, A.K., F. Sauvage, L. Etgar and M. Gratzel. 2011. Ga^{3+} and Y^{3+} cationic substitution in mesoporous TiO_2 photoanodes for photovoltaic applications. J. Phys. Chem. C 115: 9232–9240.

Chang, H. and Y.-J. Lo. 2010. Pomegranate leaves and mulberry fruit as natural sensitizers for dye-sensitized solar cells. J. Sol. Energy 84: 1833–1837.

Chang, W.C., Y.Y. Cheng, W.C. Yu, Y.C. Yao, C.H. Lee and H.H. Ko. 2012. Enhancing performance of ZnO dye-sensitized solar cells by incorporation of multiwalled carbon nanotubes. Nanoscale Res. 7: 166.

Chen, X. and S.S. Mao. 2007. Titanium dioxide nanomaterials: Synthesis, properties, modifications, and applications. Chem. Rev. 107: 2891–2959.

Chou, W.C. and W.J. Liu. 2016. Study of dye sensitized solar cell application of TiO_2 films by atmospheric pressure plasma deposition method. ICEP Proceedings 23: 664–668.

Edy, R., Y. Zhao, G.S. Huang, J.J. Shi, J. Zhang, A.A. Solovev and Y. Mei. 2016. TiO_2 nanosheets synthesized by atomic layer deposition for photocatalysis. Progress Nat. Sci. Mater. Int. 26: 493–497.

Eli, D., G.P. Musa and D. Ezra. 2016. Chlorophyll and betalain as light-harvesting pigments for nanostructured TiO_2 based dye-sensitized solar cells. J. Energy Nat. Res. 5: 53–58.

Energy Statistics. 2018. 25th issue. Central statistics office Ministry of statistics and programme implementation Government of India, New Delhi, 1–101.

Eom, T.S., K.H. Kim, C.W. Bark and H.W. Choi. 2014. Influence of Fe_2O_3 doping on TiO_2 electrode for enhancement photovoltaic efficiency of dye-sensitized solar cells. Mol. Cryst. Liq. Cryst. 600: 39–46.

Fabbiyola, S. and L.J. Kennedy. 2019. Bandgap engineering in doped ZnO nanostructures for dye sensitized solar cell applications. J. Nanosci. Nanotechnol. 19: 2963–2970.

Fan, Z.Y., H. Razavi, J.W. Do, A. Moriwaki, O. Ergen, Y.L. Chuch, P.W. Leu, J.C. Ho, T. Takahashi, L.A. Reichertz, S. Neale, K. Yu, M. Wu, J.M. Ager and A. Javey. 2009. Three-dimensional nanopillar-array photovoltaics on low-cost and flexible substrates. Nat. Mater. 8: 3756.

Forro, L., O. Chauvet, D. Emin, L. Zuppiroli, H. Berger and F.J. Levy. 1994. High mobility n-type charge carriers in large single crystals of anatase (TiO_2). Appl. Phys. 75: 633–635.

Furukawa, S., H. Iino, T. Iwamoto, K. Kukita and S. Yamauchi. 2009. Characteristics of dye-sensitized solar cells using natural dye. Thin Solid Films 518: 526–529.

Gao, F., Y. Wang, D. Shi, J. Zhang, M. Wang, X. Jing, R. Humphry-Baker, P. Wang, S.M. Zakeeruddin and M. Gratzel. 2008. Enhance the optical absorptivity of nanocrystalline TiO_2 film with high molar extinction coefficient ruthenium sensitizers for high performance dye sensitized solar cells. J. Am. Chem. Soc. 13: 10720–10728.

Garnett, E. and P. Yang. 2010. Light trapping in silicon nanowire solar cells. Nano Lett. 10: 1082.

Garnett, E.C. and D. Burthraw. 2005. Cost effectiveness of renewable electricity policies. Energy Econ. 27: 873.

Ghnimi, J., Z. Kadachi, M.B. Karoui and R. Gharbi. 2017. Performances enhancement of natural DSSC by Ga-doped ZnO layer used in the FTO/TiO2/ZnG photoanode. IEEE J. Photovoltaics 7: 1667–1673.

Giannouli, M., K. Govatsi, G. Syrrokostas, S.N. Yannopoulos and G. Leftheriotis. 2018. Factors affecting the power conversion efficiency in ZnO DSSCs nanowire vs nanoparticles. Materials 11: 411, 1–17.

Godibo, D.J., S.T. Anshebo and T.Y. Anshebo. 2015. Dye sensitized solar cells using natural pigments from five plants and quasi-solid state electrolyte. J. Braz. Chem. Soc. 26: 92–101.

Goldstein, A.N., C.M. Echer and A.P. Alivisatos. 1992. Melting in semiconductor nanocrystals. Science 256: 1425–1427.

Gong, J., J. Liang and K. Sumathy. 2012. Review on dye-sensitized solar cells (DSSCs) fundamental concepts and novel materials. Renew. Sustain. Energy Rev. 16: 5848–5860.

Grätzel, M. 2003. Dye-sensitized solar cells. J. Photochem. Photobiol. C 4: 145–53.

Green, M.A. 2005. Thin-film solar cells. Review of materials, technologies and commercial status. J. Mater. Sci. Mater. Electronics 18: 15–19.

Grünwald, R. and H. Tributsch. 1997. Mechanisms of instability in Ru-based dye sensitization solar cells. J. Phys. Chem. B 101: 2564–75.

Guo, E. and L. Yin. 2015. Nitrogen doped TiO2-Cu(x)O core-shell mesoporous spherical hybrids for high-performance dye-sensitized solar cells. Phys. Chem. Chem. Phys. 17: 563–574.

Gupta, M., S. Makda and R. Senthil. 2018. 2nd International conference on advances in mechanical engineering (ICAME). IOP Conf Series Mater. Sci. Eng. 402: 012176.

Han, Z., L. Ren, Z. Cui, C. Chen, H. Pan and J. Chen. 2012. Ag/ZnO flower heterostructures as a visible-light driven photocatalyst via surface plasmon resonance. Appl. Catal. B Environ. 126: 298–305.

Hassan, H.C., Z.H.Z. Abidin, F.I. Chowdhury and A.K. Arof. 2016. A high efficiency chlorophyll sensitized solar cell with quasi solid PVA based electrolyte. Int. J. Photoenergy 3685210: 1–9.

Haupts, U., J. Tittor, E. Bambergand and D. Oesterhelt. 1997. General concept for ion translocation by halobacterial retinal proteins: The isomerization/switch/transfer model. Biochem. 36: 78–85.

Henderson, R., J.M. Baldwin, T.A. Ceska, F. Zemlin, E. Beckmann and K.H. Downing. 1990. Model for the structure of bacteriorhodopsin based on high-resolution electron cryo-microscopy. J. Mol. Biol. 213: 899–929.

Hossain, M.A., C. Son and S. Lim. 2018. Improvement in the photovoltaic performance of a dye-sensitized solar cell by the addition of CeO_2 Gd nanoparticles in the photoanode. J. Ind. Eng. Chem. 65: 418–422.

Hsiao, P.T., K.P. Wang, C.W. Cheng and H. Teng. 2007. Nanocrystalline anatase TiO_2 derived from a titanate-directed route for dye-sensitized solar cells. J. Photochem. Photobiol. A Chem. 188: 19–24.

Hsu, C.W., P. Chen and J.M. Ting. 2013. Microwave-assisted hydrothermal synthesis of TiO_2 mesoporous beads having C and/or N doping for use in high efficiency all-plastic flexible dye-sensitized solar cells. J. Electrochem. Soc. 160: 160–165.

http://molview.org/?cid=11593175 (molview 2).

http://molview.org/?cid=12085802 (molview 1).

https://paulingblog.wordpress.com/2010/01/12/the-crystal-structure-of-brookite/.

https://www.chemtube3d.com/_anatasefinal/.

https://www.chemtube3d.com/_rutilefinal/.

https://www.sciencedirect.com/search?qs=TiO2%2C%20DSSC&years=2020%2C2019%2C2018%2C2017%2C2016%2C2015%2C2014%2C2013%2C2012%2C2011&lastSelectedFacet=years.

Huizhi, Z., L. Wu, Y. Gao and T. Ma. 2011. Dye-sensitized solar cells using 20 natural dyes as sensitizers. J. Photochem. Photobiol. A Chem. 219: 188–94.

Ikeogu, I., O.D. Adeniyi and A.A. Aboje. 2018. Photovoltaic solar cell based on chlorophyll dye pigments obtained from Brassica oleracea. IJERAT 2: 1–6.

Im, J.S., J. Yun, S.K. Lee and Y.S. Lee. 2012. Effects of multi-element dopants of TiO_2 for high performance in dye-sensitized solar cells. J. Alloys Compd. 513: 573–579.

International Renewable Energy Agency. 2016. https://www.irena.org/publications/2016/Jul/Renewable-Energy-Statistics-2016.

Isahkimpa, M., M. Momoh, K. UthmanIsah, H.N. Yahya and M.M. Damitso. 2012. Photoelectric characterization of dye sensitized solar cells using natural dye from pawpaw leaf and flame tree flower as sensitizers. Mater. Sci. Appl. 3: 281–286.

Janani, M., S.V. Nair and A.S. Nair. 2017. Photovoltaics: Role of nanotechnology in dye-sensitized solar cells. *In*: Raj, B., M.V. Voorde and Y. Mahajan. (eds.). Nanotechnology for Energy Sustainability, First Edition. Wiley-VCH Verlag GmbH and Co.KGaA.

Jasim, K.E., S. Aldallal and A.M. Hassan. 2011. Natural dye-sensitised photovoltaic cell based on nanoporous TiO_2. Int. J. Nanopart. 4: 359–682.

John, K.A., J. Naduvath, S. Mallick, J.W. Pledger, S.K. Remillard, P.A. De Young, M. Thankamoniamma, T. Shripathi and R.R. Philip. 2016. Electrochemical synthesis of novel Zn-doped TiO_2 nanotube/ZnO nanoflake heterostructure with enhanced DSSC efficiency. Nano-Micro Lett. 8: 381–387.

Kakiage, K., T. Tokutome, S. Iwamoto, T. Kyomen and M. Hanaya. 2013. Fabrication of a dye-sensitized solar cell containing a Mg-doped TiO_2 electrode and a Br_3^-/Br^- redox mediator with a high open-circuit photovoltage of 1.21 V. Chem. Commun. 49: 179–180.

Kang, S.H., S.H. Choi, M.S. Kang, J.Y. Kim, H.S. Kim, T. Hyeon and Y.E. Sung. 2008. Nanorod-based dye-sensitized solar cells with improved charge collection efficiency. Adv. Mater. 20: 54–58.

Karami, A. 2010. Synthesis of TiO_2 nano powder by the sol-gel method and its use as a photocatalyst. JICS 7: 154–160.

Kavan, L., B. ORegan, A. Kay and M. Gratzel. 1993. Preparation of TiO_2 anatase films on electrodes by anodic oxidative hydrolysis of $TiCl_3$. J. Electroanal. Chem. 346: 291–307.

Kazmi, S.A., S. Hameed and A. Azam. 2017. Efficiency enhancement in dye-sensitized solar cells using silver nanoparticles and $TiCl_4$. Energy Sources A Recovery, Utilization, Environ. Effects 39: 67–74.

Ke, W., G. Fang, J. Wang, P. Qin, H. Tao, H. Lei, Q. Liu, X. Dai and X. Zhao. 2014. Perovskite solar cell with an efficient TiO_2 compact film. ACS Appl. Mater. Interfaces 6: 15959–15965.

Kelzenberg, M.D., D.B. Turner-Evans, B.M. Kayes, M.A. Filler, M.C. Putnam, N.S. Lewis and H.A. Atwater. 2008. Photovoltaic measurements in single nanowire silicon solar cells. Nano Lett. 8: 710–714.

Khorana, H.G., G.E. Gerber, W.C. Herlihy, C.P. Gray, R.J. Anderegg, K. Nihei and K. Biemann. 1979. Amino acid sequence of bacteriorhodopsin. Proc. Natl. Acad. Sci. USA 76: 5046–5050.

Kuhlbrandt, W. 2000. Bacteriorhodopsin: The movie news and views. Nature 406: 569–570.

Kumari, J.M.K.W., N. Sanjeevadharshini, M.A.K.L. Dissanayake, G.K.R. Senadeera and C.A. Thotawatthage. 2016. The effect of TiO_2 photoanode film thickness on photovoltaic properties of dye-sensitized solar cells Ceylon. J. Sci. 45: 33–41.

Kushwaha, R., P. Srivastava and L. Bahadur. 2013. Natural pigments from plants used as sensitizers for TiO_2 based dye-sensitized solar cells. J. Energy 654953: 1–8.

Latini, A., C. Cavallo, F.K. Aldibaja, D. Gozzi, D. Carta, A. Corrias, L. Lazzarini and G. Salviati. 2013. Efficiency improvement of DSSC photoanode by scandium doping of mesoporous titania beads. J. Phys. Chem. C 117: 25276–25289.

Lee, J.J., M.M. Rahman, S. Sarker, N.C. Deb Nath, A.J.S. Ahammad and J.K. Lee. 2011. Metal oxides and their composites for the photoelectrode of dye sensitized solar cells. pp. 181–210. *In*: Attaf, B. (ed.). Composite Materials for Medicine and Nanotechnology.

Lee, K.M., W.H. Chiu, C.Y. Hsu, H.M. Cheng, C.H. Lee and C.G. Wu. 2012. Ionic liquid diffusion properties in tetrapod-like ZnO photoanode for dye-sensitized solar cells. J. Power Sources 216: 330–6.

Lee, S.H., S.H. Han, H.S. Jung, H. Shin, J. Lee, J.H. Noh, S. Lee, I.S. Cho, J.K. Lee, J. Kim and H. Shin. 2010. Al-doped ZnO thin film a new transparent conducting layer for ZnO nanowire-based dye-sensitized solar cells. J. Phys. Chem. C 114: 7185–7189.

Lee, W., J. Lee, S.K. Min et al. 2009. Effect of single-walled carbon nanotubes in PbS/TiO_2 quantum dots-sensitized solar cells. Mater. Sci. Eng. B 156: 48–51.

Leung, D.Y.C., X. Fu, C. Wang, M.K.H. Ni, L. Leung, X. Wang and X. Fu. 2010. Hydrogen production over titania-based photocatalysts. Chem. Sus. Chem. 3: 681–694.

Liang, Y. 2019. Chemical vapor deposition synthesis of Ge doped ZnO nanowires and the optical property investigation. Phy. Lett. 383: 2928–2932.

Lin, C.Y., Y.H. Lai, H.W. Chen, J.G. Chen, C.W. Kung, R. Vittal and K.C. Ho. 2011. Highly efficient dye-sensitized solar cell with a ZnO nanosheet-based photoanode. Energy Environ. Sci. 4: 3448–3455.

Lindstrom, H., A. Holmberg, E. Magnusson, S.E. Lindquist, L. Malmqvist and A. Hagfeld. 2001. A new method for manufacturing nanostructured electrodes on plastic substrates. Nano Lett. 1: 97–102.

Liu, J., Y. Duan, X. Zhou and Y. Lin. 2013. Influence of VB group doped TiO_2 on photovoltaic performance of dye sensitized solar cells. Appl. Surf. Sci. 277: 231–236.

Liu, Q., N. Gao, D. Liu, J. Liu and Y. Li. 2018. Structure and photoelectrical properties of natural photoactive dyes for solar cells. Appl. Sci. 8: 1–12 .

Liu, Y.D. and L. Zhao. 2007. Preparation of ZnO thin films by pulsed laser deposition. Zhongguo Jiguang/Chinese J. Lasers 34: 534–537.

Look, D.C., D.C. Reynolds, J.R. Sizelove, R.L. Jones, C.W. Litton, G. Cantwell and W.C. Harsch. 1998. Electrical properties of bulk ZnO. Solid State Commun. 105: 399−401.

Lou, Y.Y., Yuan, S., Zhao, Y., Wang, Z.Y. and Shi, L.Y. 2013. Influence of defect density on the ZnO nanostructures of dye-sensitized solar cells. Adv. Manuf. 1: 340–345. DOI 10.1007/s40436-013-0046-x.

Maabong, K., C.M. Muiva, P. Monowe, T.S. Sathiaraj, M. Hopkins, L. Nguyen, K. Malungwa and M. Thobega. 2015. Natural pigments as photosensitizers for dye-sensitized solar cells with TiO2 thin films. Int. J. Renew. Energy Res. 5: 54–60.

Magsi, K., P. Lee, Y. Kang, S. Bhattacharya and C.M. Fortmann. 2012. Enhanced chlorophyll A purification and dye sensitized solar cell performance. Mater. Res. Soc. Symp. Proc. 1390: 1 – 6.

Makkar, M. and H.S. Bhatti. 2011. Inquisition of reaction parameters on the growth and optical properties of ZnO nanoparticles synthesized via low temperature reaction route. Chem. Phys. Lett. 507: 122–127.

Manoharan, K. and P. Venkatachalam. 2015. Photoelectrochemical performance of dye sensitized solar cells based on aluminum-doped titanium dioxide structures. Mater. Sci. Semicond. Process 30: 208–217.

Mayumi, S., Y. Ikeguchi, D. Nakane, Y. Ishikawa, Y. Uraoka and M. Ikeguchi 2017. Effect of gold nanoparticle distribution in TiO_2 on the optical and electrical characteristics of dye-sensitized solar cells. Nanoscale Res. Lett. 12: 513.

Mc Gehee, M., Fan, S.H. and Cui, Y. 2009. Optical absorption enhancement in amorphous silicon nanowire and nanocone arrays. Nano Lett. 9: 279.

Mia, Y.U., Y.Z. Long, B. Sun and Z. Fan. 2012. Recent advances in solar cells based on one-dimensional nanostructure arrays. Nanoscale 4: 2783.

Mishra, A., M.K.R. Fischer and P. Bauerle. 2009. Metal free organic dyes for dye sensitized solar cells from structure: Property relationships to design rules. Angew. Chem. Int. Ed. 48: 2474–99.

Mohammadpour, F., M. Moradi, K. Lee, G. Cha, S. So, A. Kahnt, D.M. Guldi, M. Altomare and P. Schmuki. 2015. Enhanced performance of dye-sensitized solar cells based on TiO2 nanotube membranes using an optimized annealing profile. Chem. Commun. 51: 1631–1634.

Mou, J., W. Zhang, J. Fan, H. Deng and W. Chen. 2011. Facile synthesis of ZnOnanobullets/nanoflakes and their applications to dye-sensitized solar cells. J. Alloy Compd. 509: 961–5.

Muduli, S.K., V.V. Dhas, S. Hisamuddin Mujavar and S.B. Ogale. 2010. High efficient dye-sensitized solar cells using TiO2-multiwalled carbon nanotube (MWCNT) nanocomposite. WO079516.

Narayan, M.R. 2012. Review: Dye sensitized solar cells based on natural photosensitizers. Renew. Sustain. Energy Rev. 16: 208–215.

Nesari, M. and A. Molaeirad. 2016. Assembly of biosolar cells by using N719-bacteriorhodopsin complex. Journal of Paramedical Sciences (JPS) 7: 1–4.

Neutze, R., E.P. Peyroula, K. Edman, A. Royant, J. Navarro and E.M. Landau. 2002. Bacteriorhodopsin: A high-resolution structural view of vectorial proton transport. Biochim. Biophys. Acta 1565: 144–167.

Ni, S., F. Guo, D. Wang, S. Jiao, J. Wang, Y. Zhang, B. Wang, P. Feng and L. Zhao. 2019. Modification of TiO2 nanowire arrays with Sn doping as photoanode for highly efficient dye-sensitized solar cells. Crystals 9: 113.

O'Regan, B. and M. Gratzel. 1991. A low-cost, high-efficiency solar cell based on dye-sensitized colloidal TiO2 films. Nature 353: 737–740.

Oesterhelt, D. 1976. Bacteriorhodopsin as an example of a light driven proton pump. Angew. Chem. Int. Ed. Engl. 15: 17–24.

Oesterhelt, D. 1988. The structure and mechanism of the family of retinal proteins from Halophilic Archaea. Curr. Op. Struct. Biol. 8: 489–500.

Oesterhelt, D. and W. Stoeckenius. 1971. Rhodopsin-like protein from the purple membrane of Halobacterium halobium. Nat. New Biol. 233: 149–152.

Oesterhelt, D. and W. Stoeckenius. 1973. Functions of a new photoreceptor membrane. Proc. Natl. Acad. Sci. USA 70: 2853–2857.

Otieno, O.V., E. Csaki, O. Kéri, L. Simon, I.E. Lukas, K.M. Szecsenyi and I.M. Szilagyi. 2020. Synthesis of TiO2 nanofibers by electrospinning using water-soluble Ti-precursor. J. Therm. Anal. Calorim. 139: 57–66.

Panda, B.B., P.K. Mahapatra and M.K. Ghosh. 2018. Application of chlorophyll as sensitizer for ZnS photoanode in a dye-sensitized solar cell (DSSC). J. Electronic Materm. 1–9.

Park, H., Y. Park, W. Kim and W. Choi. 2013. Surface modification of TiO2 photocatalyst for environmental applications. J. Photochem. Photobiol. 15: 1–20.

Park, N.G., J.V. Lagemaat and A.J. Frank. 2000. Comparison of dye-sensitized rutile- and anatase-based TiO2 solar cells. J. Phys. Chem. B 104: 8989–8994.

Pavasupree, S., J. Jitputti, S. Ngamsinlapasathian and S. Yoshikawa. 2008. Hydrothermal synthesis, characterization, photocatalytic activity and dye-sensitized solar cell performance of mesoporous anatase TiO$_2$ nanopowders. Mater. Res. Bull. 43: 149–57.

Peng, X. and A. Chen. 2004. Aligned TiO$_2$ nanorod arrays synthesized by oxidizing titanium with acetone. RSC 14: 2542–2548.

Peng, Z.A. and X. Peng. 2001. Formation of high quality CdTe, CdSe, and CdN nanocrystals using CdO as precursor. J. Am. Chem. Soc. 123: 183.

Pillai, S. and M.A. Green. 2010. Plasmonics for photovoltaic applications. Solar Energy Materials and Solar Cells 94(9): 1481–1486.

Prabhin, V.S., K. Jeyasubramanian, N.R. Romulus and N.N. Singh. 2017. Fabrication of dye sensitized solar cell using chemically tuned CuO nanoparticles prepared by sol-gel method. Arch. Mater. Sci. Engn. 83: 5–9.

Pramono, S.H., E. Maulana, A.F. Prayogo and R. Djatmika. 2015. Characterization of dye-sensitized solar cell (DSSC) based on chlorophyll dye. Int. J. Appl. Engn. Res. 10: 193–205.

Pratiwi, D.D., F. Nurosyid, A. Supriyanto and R. Suryana. 2017. Performance improvement of dye-sensitized solar cells (DSSC) by using dyes mixture from chlorophyll and Anthocyanin. International Conference on Science and Applied Science, IOP Conf. Series: Journal of Physics: Conf. Series 909: 012025. Doi: 10.1088/1742-6596/909/1/012025.

Qin, C., D. Yang, P. Gu, X. Zhu and H. Sun. 2018. Natural dyes as sensitizers for dye-sensitized solar cells. Functional Mater. Lett. 11: 1850051, 1-4.

Qiu, Y., W. Chen and S. Yang. 2010. Double-layered photoanodes from variable-size anatase TiO2 nanospindles: A candidate for high-efficiency dye-sensitized solar cells. Angew. Chem. 122: 3757–3761.

Rahman, M.M., N.C.D. Nath, K.M. Noh, J. Kim and J.J. Lee. 2013. A facile synthesis of granular ZnO nanostructures for dye-sensitized solar cells. Int. J. Photoenergy Article ID 563170: 1–6.

Rahman, M.R., M. Wei, F. Xie and M. Khan. 2019. Efficient dye-sensitized solar cells composed of nanostructural ZnO doped with Ti. Catalysts 9: 273.

Richhariya, G., A. Kumar, P. Tekasakul and B. Gupta. 2016. Natural dyes for dye sensitized solar cell: A review. Renewable Sustainable Energy Reviews 69.

Ridwan, M.A., E. Noor and M.S. Rusli. 2018. Fabrication of dye-sensitized solar cell using chlorophylls pigment from sargassum, 1st International Conference on Tropical Studies and Its Application (ICTROPS). IOP Conf. Series: Earth and Environmental Science 144: 012039: 1–8.

Rossi, M., F. Matteocci, A. Di Carlo and C. Forni. 2017. Chlorophylls and xanthophylls of crop plants as dyes for Dye-Sensitized Solar Cells (DSSC). J. Plant Sci. Phytopathol. 1: 087–094.

Saito, Y., T. Kitamura, Y. Wada and S. Yanagida. 2002. Application of poly(3,4 ethylenedioxythiophene) to counter electrode in dye-sensitized solar cells. Chem. Lett. 31: 1060–1065.

Saito, Y., W. Kubo, T. Kitamura, Y. Wada and S. Yanagida. 2004. I−/I3−redox reaction behavior on poly(3,4-ethylenedioxythiophene) counter electrode in dye-sensitized solar cells. J. Photochem. Photobiol. A 164: 153–158.

Scanlon, D.O., C.W. Dunnill, J. Buckeridge, S.A. Shevlin, A.J. Logsdail, S.M. Woodley, C.R.A. Catlow, M. Powell, R.G. Palgrave, I.P. Parkin and G.W. Watson. 2013. Band alignment of rutile and anatase TiO$_2$. Nat. Mater. 12(9): 798–801.

Seger, B. 2016. Global Energy Consumption: The Numbers for Now and in the Future, 1–4.

Selim, H., A.A. Nada, M. El-Sayed, R.M. Hegazey, E.R. Souaya and M.F. Kotkata. 2018. The effect of ZnO and its nanocomposite on the performance of dye-sensitized solar cell. Nano Sci. Nano Technol. 12: 122.

Shaban, Z., M.H.M. Ara, S. Falahatdoost and N. Ghazyani. 2016. Optimization of ZnO thin film through spray pyrolysis technique and its application as a blocking layer to improving dye sensitized solar cell efficiency. Curr. Appl. Phys. 16: 131–4.

Sima, C., C. Grigoriu and S. Antohe. 2010. Comparison of the dye-sensitized solar cells performances based on transparent conductive ITO and FTO. Thin Solid Films 519: 595–7.

Singh, A., Vihinen, J., Frankberg, E., Hyvärinen, L., Honkanen, M. and Levänen, E. 2016. Pulsed laser ablation-induced green synthesis of TiO2 nanoparticles and application of novel small angle x-ray scattering technique for nanoparticle size and size distribution analysis. Nanoscale Res. Lett. 11(447): 1–9. Doi: 10.1186/s11671-016-1608-.

Son, Y.J., J.S. Kang, J. Yoon, J. Kim, J. Jeong, J. Kang, M.J. Lee, H.S. Park and Y.E. Sung. 2018. Influence of TiO2 particle size on dye-sensitized solar cells employing an organic sensitizer and a Cobalt(III/II) redox electrolyte. J. Phys. Chem. 122: 7051–7060.

Song, D.H., H.S. Kim, J.S. Suh, B.H. Jun and W.Y. Rho. 2017. Multi-shaped Ag nanoparticles in the plasmonic layer of dye-sensitized solar cells for increased power conversion efficiency. Nanomater. 7: 136.

Song, L., P. Du, X. Shao, H. Cao, Q. Hui and J. Xiong. 2013. Effects of hydrochloric acid treatment of TiO2 nanoparticles/nanofibers bilayer film on the photovoltaic properties of dye-sensitized solar cells. Mater. Res. Bull. 48: 978–982.

Spitler, M.T. and M. Calvin. 1977. Electron transfer at sensitized TiO2 electrodes. J. Chem. Phys. 66: 4294–4305.

Sreeja, S. and B. Pesala. 2018. Co₂ sensitization aided efficiency enhancement in betanin–chlorophyll solar cell. Mater. Renew. Sustain. Energy 7: 1–14.

Sreekala, C.O., I. Jinchu, K.S. Sreelatha, Y. Janu, N. Prasad, M. Kumar, A.K. Sadh and M.S. Roy. 2012. Influence of solvents and surface treatment on photovoltaic response of DSSC based on natural curcumin dye. IEEE J. Photovoltaics 2: 312–319.

Stoeckenius, W. and R.A. Bogomolni. 1982. Bacteriorhodopsin and related pigments of halobacteria. Annu. Rev. Biochem. 52: 587–616.

Subramanian, A., J.S. Bow and H.W. Wang. 2012. The effect of Li+ intercalation on different sized TiO2 nanoparticles and the performance of dye-sensitized solar cells. Thin Solid Films 520: 7011–7017.

Suyitno, S., T.J. Saputra, A. Supriyanto and Z. Arifin. 2015. Stability and efficiency of dye-sensitized solar cells based on papaya-leaf dye. Spectrochim. Acta A Mol. Biomol. Spec. 148: 99–104.

Swarnkar, A.K., S. Sahare, N. Chander, R.K. Gangwar, S.V. Bhoraskar and T.M. Bhave. 2015. Nanocrystalline titanium dioxide sensitised with natural dyes for eco-friendly solar cell application. J. Expt. Nanosci. 10: 1001–1011.

Syafinar, R., N. Gomesh, M. Irwanto, M. Fareq and Y.M. Irwan. 2015. Chlorophyll pigments as nature based dye for Dye-Sensitized Solar Cell (DSSC). Energy Procedia 79: 896–902.

Tao, P., Y. Li, A. Rungta, A. Viswanath, J. Gao, B.C. Benicewicz, R.W. Siegel and L.S. Schadler. 2011. TiO₂ nanocomposites with high refractive index and transparency. J. Mater. Chem. 21: 18623–18629.

Taya, S.A., T.M. El-Agez, H.S. El-Ghamri and M.S. Abdel-Latif. 2013. Dye-sensitized solar cells using fresh and dried natural dyes. Int. J. Mater. Sci. Appl. 2: 37–42.

Taya, S.A., T.M. El-Agez, K.S. Elrefi and M.S. Abdel-Latif. 2015. Dye-sensitized solar cells based on dyes extracted from dried plant leaves. Turk. J. Phys. 39: 24–0.

Tekerek, S., A. Kudret and U. Alver. 2011. Dye sensitized solar cells fabricated with black raspberry, black carrot and rosella juice. Indian J. Phys. 85: 1469–76.

Thavasi, V., T. Lazarova, S. Filipek, M. Kolinski, E. Querol, A. Kumar, S. Ramakrishna, E. Padrós and V. Renugopalakrishnan. 2008. Study on the feasibility of bacteriorhodopsin as bio-photosensitizer in excitonic solar cell: A first report. J. Nanosci. Nanotechnol. 8: 1–9.

The NEED project. Intermediate energy infobook 2018: 45–49.

Tobnaghi, D.M., R. Madatov and D. Naderi. 2013. The effect of temperature on electrical parameters of solar cells. Int. J. Adv. Res. Elect. Electro. Inst. Eng. 2: 6404–6407.

Tributsh, H. and H. Gerisher. 1969. Elektrochemische Untersuchungüber den Mechanismus der Sensibilisierung und Übersensibilisierung an ZnO-Einkristallen. Ber. Bunsenges. Phys. Chem. 73: 251–260.

Tributsch, H. and M. Calvin. 1971. Electrochemistry of excited molecules: Photo-electrochemical reactions of chlorophylls. Photochem. Photobio1. 4: 95–112.

Vasuki, T., M. Saroja, M. Venkatachalam and S. Shankar. 2014. Preparation and characterization of TiO₂ thin films by spin coating method. J. Nanosci. Nanotechnol. 2: 728–731.

Wang, M., S. Bai, A. Chen, Y. Duan, Q. Liu, D. Li and Y. Lin. 2012. Improved photovoltaic performance of dye-sensitized solar cells by Sb-doped TiO₂ photoanode. Electrochim. Acta 77: 54–59.

Wang, X.F., H. Tamiaki, L. Wang, N. Tamai, O. Kitao, H. Zhou and S. Sasaki. 2010. Chlorophyll-a derivatives with various hydrocarbon ester groups for efficient dye-sensitized solar cells: static and ultrafast evaluations on electron injection and charge collection processes. Langmuir 26: 6320–6327.

Wang, X.F., J. Xiang, P. Wang, Y. Koyama, S. Yanagida, Y. Wada, K. Hamada, S. Sasaki and H. Tamiaki. 2005. Dye-sensitized solar cells using a chlorophyll a derivative as the sensitizer and carotenoids having different conjugation lengths as redox spacers. Chem. Phys. Lett. 408: 409–414.

Wongcharee, K., V. Meeyoo and S. Chavadej. 2006. Dye-sensitized solar cell using natural dyes extracted from rosella and blue pea flowers. Sol. Energy Mater. Solar Cells 91: 566–571.

World energy resources. 2013. World Energy Council, London.

World energy resources. 2016. World Energy Council, London.

World energy resources. 2017. World Energy Council, London.

Xuelian, X. and Z. Junjiang. 2012. Hydrothermal synthesis of TiO_2 nanoparticles for photocatalytic degradation of ethane: Effect of synthesis conditions. Recent Patents on Chemical Engineering 5: 134–142.

Yahav, S., S. Ruhle, S. Greenwald, H.N. Barad, M. Shalom and A. Zaban. 2011. Strong Efficiency enhancement of dye-sensitized solar cells using a La modified TiCl4 treatment of mesoporous TiO2 electrodes. J. Phys. Chem. 115: 21481–21486.

Yoshida, T., J.B. Zhang, D. Komatsu, S. Sawatani, H. Minoura, T. Pauporte, D. Lincot, T. Oekermann, D. Schlettwein, H. Tada and D. Wohrle. 2009. Electrodeposition of inorganic/organic hybrid thin films. Adv. Funct. Mater. 19: 17–43.

Zhai, B.G., L. Yang and Y.M. Huang. 2017. Improving the efficiency of dye-sensitized solar cells by growing longer ZnO nanorods on TiO_2 photoanodes. J. Nanomater. Article ID 1821837: 1–8.

Zhang, J., J. Feng, Y. Hong, Y. Zhu and L. Han. 2014. Effect of different trap states on the electron transport of photoanodes in dye sensitized solar cells. J. Power Sources 257: 264–271.

Zhao, G., H. Kozuka and T. Yoko. 1997. Effects of the incorporation of silver and gold nanoparticles on the photoanodic properties of rose bengal sensitized TiO2 film electrodes prepared by sol-gel method. Sol. Energy Mater. Sol. Cell 46: 219–231.

Zhao, S.Q., Y.L. Xie, Z.X. Qi and P.C. Lin. 2018. Studies on photoelectric performance of natural dyes from safflower. Int. J. Electrochem. Sci. 13: 1945–1955.

Zimanyi, L., Y. Cao, R. Needleman, M. Ottolenghi and J.K. Lanyi. 1993. Pathway of proton uptake in the bacteriorhodopsin photocycle. Biochemistry 32(30): 7669–7678.

15

Nanobiomaterials in Tissue Engineering Applications

Jeeva Subbiah,[1,*] *G. Gowri,*[1] *K. Sangeethu,*[2] *P.N. Sudha*[2,*]
and *Srinivasan Latha*[3]

1. Introduction

The ultimate goal of tissue engineering is "to develop functional scaffolds to restore, repair or replace damaged tissue and/or organs."

Tissue engineering has definitely benefited much from the application of nanotechnologies, which follow the path of biomimicry (Abd El-Aziz et al. 2017). By combining the principles of materials science with cell transplantation, tissue engineering creates artificial tissues that encourage endogenous regeneration. The necessity to replace injured body components in order to return them to their original physiological states has always been the impetus behind research into the discovery and creation of new materials capable of carrying out this role as effectively as feasible. Materials that are biocompatible and biodegradable are used to create polymeric nanobiomaterials. They can range in size from 10 to 1000 nm, depending on how the nanoparticles were prepared. In accordance with their altered properties, it would also fulfil its purpose (Cheng et al. 2017).

1.1 Need and Role of Nanobiomaterials in Tissue Engineering

The primary objective of tissue engineering with biomedical nanomaterials is to achieve structural and functional restoration of damaged body parts by producing engineered constructs that are as similar to native tissue as feasible (Firouzi et al. 2017). Nano-biomaterial scaffolds for tissue engineering must be strong enough to support the skeleton, biocompatible to merge with the surrounding native tissues, and osteoconductive to draw osteoprogenitor cells into the porous structure. Materials that are nanosized can be good choices to meet a variety of needs for tissue-engineered scaffolds (Lotfi et al. 2016). Since (1) a stronger interaction between nanosized molecules and the implant material may improve the mechanical strength and (2) increased surface roughness may promote osteoblast adhesion and subsequent cellular functions, bone tissue substrate implants may be able to be easily integrated with surrounding tissues and support bone regeneration with the help of nanoscale materials (Ma et al. 2017).

[1] Center for inflammation, Immunity, and Infection, Institute for Biomedical Sciences, Georgia State University, Atlanta, GA 30303, USA.
[2] Department of Chemistry, D.K.M. College for Women, Vellore, Tamil Nadu, India.
[3] School of Advanced Science, Vellore Institute of Technology, Vellore, Tamil Nadu, India.
* Corresponding authors: jeevabio@gmail.com; drparsu8@gmail.com

Nanobiomaterials serve as the basic components of scaffolds in tissue engineering applications that can treat medical conditions like bleeding, infection, and fractures as well as support new bone tissue and promote the development of mature bone cells (Augustine and Hasan 2020). Nanobiomaterial scaffolds degraded more quickly because they can be used to release bioactive compounds from the scaffold or as a second conduit for cells to infiltrate (Aderibigbe and Mukaya 2016). Because it promotes cell motility, proliferation, and differentiation within the scaffold, proper cell infiltration is essential. Additionally, the scaffolds must maintain the extracellular environment while supporting cells to improve tissue alignment and cell-cell communication (Razavi et al. 2017).

Utilizing biomaterials, cells, and bioactive chemicals, tissue engineering blends engineering and biological principles (Konur 2017). A biomaterial's primary function in tissue regeneration is to act as a support system and scaffold for the development of cells. The extracellular matrix (ECM), a substance released by resident cells and supporting tissue and organs, serves as nature's model for biomaterial (Hannig and Hanning 2019). This offers cellular processes with not just physical support but also spatial organisation and a bioactive microenvironment.

There are numerous biomaterials for tissue engineering (TE) that can be used in a variety of clinical settings. In the creation of scaffolds for tissue engineering, three distinct categories of biomaterials—likely ceramics, synthetic polymers, and natural polymers—are typically utilised (Manna et al. 2018). The native vascular extracellular matrix may be able to be replicated by natural polymers, especially those that degrade more quickly (ECM). This highlights its essential biological and mechanical properties, which also include viscoelasticity, tensile strength, non-thrombogenicity, hemocompatibility, and biocompatibility with low cytotoxicity (Varga 2016; Ismail et al. 2018). This chapter discusses the tubes, sponges, solutions, powders, hydrogels, composites (Rajabi et al. 2016), and membranes that have been utilised for tissue engineering applications of skin, cartilage, bone, and other tissues taking into account these properties and the diversity of potential natural polymer modification processes.

1.2 *Chitosan as a Nanobiomaterial*

Chitosan is a natural polymer that can be used in tissue engineering since it is a cationic polysaccharide-based biomaterial that is similar to GAGs (glycosaminoglycans) found in native tissue and is primarily responsible for molecular water retention and tissue resistance to compression (Logith Kumar and Keshavnarayan 2016a). In terms of choosing a biomaterial for therapeutic usage, the metabolic destiny of chitosan in the body has the potential to play a significant role as a scaffolding material (Moreno-Vasquez et al. 2017). Chitosan is an excellent choice for tissue engineering and other biomedical applications due to its key qualities, including polymeric cationic character, high biocompatibility, biodegradability, antibacterial activity, non-antigenicity, and high adsorption-enhancing effects (Grifoll-Romero et al. 2018). The availability of amino groups makes chitosan accessible (Saranya et al. 2011, 2011), and CS's hydrophilic shape encourages cell adherence, proliferation, and differentiation of various cell types (Hoven et al. 2007).

Bone makes up the majority of the human body's interior framework. The structure of the human body is solidified and made physically stable by this biological component. Biomaterials for the regeneration of hard tissues, such as many parts of tissue engineering, are made from chitosan (He et al. 2011, Chen et al. 2017). Chitosan encourages calcium/phosphate ion accumulation, which causes apatite to deposit and improves biomineralization of composite materials (Tuzlakoglu and Reis 2007, Leonor et al. 2008). By turning chitosan into chitosan nanoparticles and mixing them with other materials, chitosan-based nanocomposite was created (Logith Kumar et al. 2016b). By ionic gelation of chitosan with sodium tripolyphosphate (STP) utilising the solution casting method, the vitality of chitosan nanoparticles gives functional therapeutic effect (Baboucarr 2019). The resulting nanoparticles were either the same size as the bioentities or smaller, and they were able to bind to other polymers to add new capabilities (Chandra et al. 2011, Mangal et al. 2015).

Scaffolds made of chitosan nanoparticles have been employed in tissue engineering. Snowflakes, circles, and spheres are only a few examples of the regular assemblage shapes that chitosan nanoparticles display (Puvaneswary et al. 2016). Chitosan can be used to create air-permeable films. It promotes cellular regeneration while defending tissues from microbial assaults. In addition to providing protection, it also speeds healing and reduces scarring.

1.3 Bioceramics as Nanobiomaterial

Basically, three-dimensional (3D) bone bioactive scaffolds for bone healing need to be built using biomaterials that have the right biocompatibility, bioactivity, osteoconduction, osteoinduction, and biodegradation characteristics. The subgroup of biomaterials known as bioceramics includes those bioactive ceramic materials that are also biocompatible. When utilised as an implant material, ceramics are used to repair and reconstruct bone tissue and skeletal systems that integrate the tissues at various phases of neotissue creation (Blunt et al. 2017). Bioceramics hold great promise as body-interactive materials since they have the ability to speed up the healing process or encourage tissue regeneration, which will return physiological functions. With intimate interaction surrounding the tissues existing in them, they are typically used as hard tissue replacement materials, mostly in the bone, cartilage, and teeth area of the body (Baino and Ferraris 2017).

Due to the way the tissue interacts with the implanted substance, the response seen at the implanted region of the body can be monitored at the nanoscale level (Mutsenko et al. 2017). They concentrate on the varied range of generating their materials that are suitable in the drug delivery, tissue engineering, implanting medical devices, and oxygen carriers, etc., based on the implanted bioceramics features and functions.

1.4 Bioceramic Scaffolds

Young's modulus, which measures mechanical stiffness, is frequently high in ceramic scaffolds, along with their extreme lack of elasticity and hard, brittle surface. They have high biocompatibility from the perspective of bone since they are chemically and structurally comparable to the mineral phase of natural bone (Mansouri et al. 2018). Because ceramics are known to promote osteoblast development and proliferation, interactions between osteogenic cells and ceramics are crucial for bone regeneration. In dentistry and orthopaedic surgery, several ceramics have been utilised to cover metallic implant surfaces and fill bone defects to enhance implant integration with the host bone (Cozza et al. 2018).

Bioceramics comprises the categorised groups such as calcium phosphates (Ca/P) as hydroxyapatite (HAp), the bioactive glasses, and tricalcium phosphate based on their potential and their tissue reaction (Zdarta et al. 2018). Due to these significantly greater levels of reactivity, hydroxyapatite has gained popularity as a bone tissue replacement material. It speeds up bone ingrowth around the defective area, which is thought to be a function of regulators including PO^{3-}, Ca^{2+}, and OH^- ions present in HAp (Tziveleka et al. 2017). Due to its structural resemblance to the inorganic components found in bone, artificial implants made of HAp are used in orthopaedic, dental, and maxillofacial applications. Collagen (20%), calcium phosphate, such as HAp (69%) for bone stiffness, and water (9%) make up the dynamic tissue that is bone (Grigore 2018). The most extensively used biomaterial for repairing and replacing flaws in bone tissue is hydroxyapatite.

Biomedical equipment and implants contain HAp because it is frequently used to coat the surfaces of metallic components, making the implants more bearable by the surrounding tissues (Lowe et al. 2016). HAp is also used as bone fillers in the form of powders, porous blocks, or beads to treat bone deformities or restore missing bone tissue in reconstructive procedures (Bhardwaj et al. 2018). Bone grafts, fillers, and coatings for metal implants are all made of this bioactive substance because bone cells develop there directly without competing with fibrous tissues (Ping

Liu et al. 2015). In particular, nanohydroxyapatite in tissue regeneration has been extensively used to strengthen Extracellular matrix (ECM) like polymer scaffolds due to their striking similarities to bone mineral component. It has also been demonstrated that nanocrystalline HAp (nHAp) gives better outcomes with regard to osteoblast adhesion, differentiation, and proliferation as well as biomineralization (Pal et al. 2017). In a study published in 2012, Curtin and colleagues found that a collagen nanoapatite composite scaffold containing HAp nanocrystals as an inorganic component had superior compressive stiffness and yield strength. Because it did not cause any inflammatory reaction in the skin, muscle, or hard tissue, it is increasingly used as a substitute for allogeneic or autologous bone grafts. The additional layer of calcium phosphate that the HAp nanocomposites add to the implanted site enhances integration and incorporation and avoids the development of a fibrous layer (Clarke et al. 2016).

The most crucial factors in selecting a tissue transplant are that the scaffold formed must be totally biocompatible and exhibit no harmful consequences. Comparatively to other polysaccharide scaffolds, the incorporation of HAp nanoparticles in carrageenan, alginate, cellulose, and chitosan (Pozzolini et al. 2018, Yuan et al. 2017, and Zolghadri et al. 2019) demonstrated a suitable location for bone cell attachment and tissue regeneration. The nano HAp based scaffolds will supply the material for cell growth as well as the delivery of oxygen and nutrients to the cells in the area (Komalakrishna et al. 2017).

1.5 Collagen as Nanobiomaterial

The predominant collagen type from the body is type I collagen, which can be found in bones, skin, ligatures, and even organs. Collagen is the main structural protein of mammalian fibrous tissues. Types I, II, III, V, and XI of the fibrillar collagen family are among the various classes of collagen; more recently, types XXIV and XXVII have been introduced to the family, expanding its size (Purcel et al. 2016). One of the main extracellular matrix components is collagen, and numerous studies have reported the use of collagen-based biomaterials in the development of the spinal cord, tendon regeneration, skeletal muscle engineering, and ocular tissue engineering. Myoblast proliferation and differentiation are aided by collagen's ability to replicate the natural extracellular matrix; a nanofiber matrix consisting of pure type I collagen also promotes adequate cell proliferation and preserves the myogenic phenotype (Das 2016). Nanofibrous collagen found in tendon has the capacity to direct and support neural cell proliferation because collagen may be helpful for the healing of nerves and the production of new connections.

Although collagen-based scaffolds have strong biocompatibility, biodegradability, and low immunogenicity, their weak mechanical qualities restrict their use. It has been established that synthetic polymers with outstanding mechanical properties include poly(-caprolactone) (PCL), poly(lactic acid), poly(glycolic acid), and their copolymer poly(lactic-co-glycolic acid) (PLGA). The most popular skin regeneration scaffolds are collagen-chitosan ones. By adding PLGA reinforcement to the hybrid collagen-chitosan scaffold, wound contraction was prevented, and defects in full-thickness mouse skin were successfully repaired (Govindharaj et al. 2019). Studying the osteoconductivity of collagen-PLA nanocomposite and bacterial cellulose (BC)-collagen nanocomposite, which can be employed as an alternative biomaterial for vascular tissue engineering, has been done using nanobiocomposite materials (Ghalia and Dahman 2016).

Crosslinking collagen with other substances or materials can make it stronger or change the degradation profile (Bouhout et al. 2019). The ability of collagen to generate strong and stable fibres as a result of its self-aggregation and crosslinking process is a key factor in the effectiveness of collagen-based nanobiomaterials in biomedical applications. Due of its hemostatic qualities, collagen can be used to create bioengineered tissues like blood arteries and heart valves (Liu and Fan 2019). The benefits of collagen, particularly its low antigenicity, its scaffolds' capacity to merge with surrounding tissues, and the fact that different collagen formulations are used in diverse applications (Guo et al. 2016).

2. Conclusion

Natural biomaterials offer wide range of mechanical properties and biodegradation features whereas inorganic nanoparticles have vital bioactivities. The extracellular matrix (ECM) of original tissues can be replicated in the form of scaffolds thanks to their integration capabilities, and modified scaffolds made of these biomaterials nanoparticles offer a more favourable environment for ingrowth and cell adhesion. While this is going on, the creation of nanoparticles like nHAP, bacterial cellulose nanofibers, and collagen-based implants shows significant roles in the storage and release of growth factors. It is a significant problem for material scientists and medical practitioners to design and create biomaterials for tissue engineering by properly matching materials with required bioactivities. Further research is required into the possibility of medication delivery and cell targeting using biomaterials made of degradable polymers and environmentally friendly inorganic nanoparticles.

.

References

Abd El-Aziz, A.M., R.M. El Backly, N.A. Taha, A. El-Maghraby and S.H. Kandil. 2017. Preparation and characterization of carbon nanofibrous/hydroxyapatite sheets for bone tissue engineering. Mater. Sci. Eng. 76: 1188–1195.

Aderibigbe, B.A. and H.E. Mukaya. 2016. Nanobiomaterials architectured for improved delivery of antimalaria drugs. Nanoarchitectonics for Smart Delivery and Drug Targeting, 169–200.

Augustine, R. and A. Hasan. 2020. Cellular response to nanobiomaterials. Handbook of Biomaterials Biocompatibility, 473–504.

Baboucarr Lowe. 2019. Marine-derived biomaterials for tissue engineering applications. Biomater. Sci. Eng. 14: 81–91.

Baino, F. and M. Ferraris. 2017. Learning from nature: using bioinspired approaches and natural materials to make porous bioceramics. Int. J. Appl. Ceram. Tec. 14: 507–520.

Bhardwaj, V.A., P.C. Deepika and S. Basavarajaiah. 2018. Zinc incorporated nano hydroxyapatite: A novel bone graft used for regeneration of intrabony defects. Contemp. Clin. Dent. 9: 427–433.

Blunt, J.W., B.R. Copp and R.A. Keyzers. 2017. Marine natural products. Nat. Prod. Rep. 34: 235–294.

Bouhout, S., S. Chabaud and S. Bolduc. 2019. Collagen hollow structure for bladder tissue engineering. Mat. Sci. Eng. C 102: 228–237.

Chandra, S., K.C.S. Barick and D. Bahadur. 2011. Oxide and hybrid nanostructures for therapeutic applications. Adv. Drug Del. Rev. 63(14): 1267–1281.

Chen, M., S. Zhang, X. Chen, Z. Wang, Y. Tian, P. Chen and L. Zhang. 2017. Peptide-modified chitosan hydrogels accelerate skin wound healing by promoting fibroblast proliferation, migration, and secretion. Cell Transplant 26: 1331–1340.

Cheng, D., T. Liu, X. Tang, X. Zhang and Q. Jia. 2017. Effects of Ca/P molar ratios on regulating biological functions of hybridized carbon nanofibers containing bioactive glass nanoparticles. Biomed. Mater. 12: 025019.

Clarke, S.A., S.Y. Choi and M. McKechnie. 2016. Osteogenic cell response to 3-D hydroxyapatite scaffolds developed via replication of natural marine sponges. J. Mater. Sci. Mater. Med. 27: 22.

Cozza, N., F. Monte and W. Bonani. 2018. Bioactivity and mineralization of natural hydroxyapatite from cuttlefish bone and Bioglass co-sintered bioceramics. Tissue. Eng. Regen. Med. 12: e1131–e1142.

Curtin, C.M., G.M. Cunniffe, F.G. Lyons, K. Bessho, G.R. Dickson, G.P. Duffy and F.J. O'Brien, 2012. Innovative collagen nano-hydroxyapatite scaffolds offer a highly efficient non-viral gene delivery platform for stem cell-mediated bone formation. Advanced Materials, 24(6): 749–754.

Das, R. 2016. Application of nanobioceramics in bone tissue engineering. Nanobiomaterials in Hard Tissue Engineering, 353–379.

Firouzi, R., H.H. Poursalehi, F. Delavari and M.A. Oghabian. 2017. Chitosan coated tungsten trioxide nanoparticles as a contrast agent for X-ray computed tomography. Int. J. Biol. Macromol. 98: 479–485.

Ghalia, M.A. and Dahman, Y. 2016. Advanced nanobiomaterials in tissue engineering. Nanobiomaterials in Soft Tissue Engineering, 141–172.

Govindharaj, M., U.K. Roopavath and S.N. Rath. 2019. Valorization of discarded Marine Eel fish skin for collagen extraction as a 3D printable blue biomaterial for tissue engineering. J. Clean. Prod. 230: 412–419.

Grifoll-Romero, S., H. Pascual, X. Aragunde, A. Biarnes and A. Planas. 2018. Chitin deacetylases: Structures, specificities, and biotech applications. Polymers 10: 352.

Grigore, M.E. 2018. Drug delivery systems in hard tissue engineering. Biotechnol. Biomed. Eng. 1: 1001.

Guo, C., J. Xue and Y. Dong. 2016. Hydroxyapatite-silver nanobiomaterial. Nanobiomaterials in Hard Tissue Engineering, 297–321.

Hannig, M. and C. Hannig. 2019. Nanobiomaterials in preventive dentistry. Nanobiomaterials in Clinical Dentistry, 201–223.

He, L.H., L. Yao, R. Xue, J. Sun and R. Song. 2011. *In-situ* mineralization of chitosan/calcium phosphate composite and the effect of solvent on the structure. Front. Mater. Sci. 5: 282–292.

Hoven, V.P., V. Tangpasuthadol, Y. Angkitpaiboon, N. Vallapa and S. Kiatkamjornwong. 2007. Surface-charged chitosan: Preparation and protein adsorption. Carbohydr. Polym. 68: 44–53.

Ismail, N.A., K.A. Mat Amin and M.H. Razali. 2018. Novel gellan gum incorporated TiO_2 nanotubes film for skin tissue engineering. Materials Letters 228: 116–120.

Komalakrishna, H., J.T.G. Shine and B. Kundu. 2017. Low temperature development of nanohydroxyapatite from Austromegabalanus psittacus, Star fish and Sea urchin. Mater. Today. Proc. 4: 11933–11938.

Konur, O. 2017. Recent citation classics in antimicrobial nanobiomaterials. Nanostructures for Antimicrobial Therapy, 669–685.

Leonor, E.T., M. Baran, R.L. Kawashita, Reis, T. Kokubo and T. Nakamura. 2008. Growth of a bone like apatite on chitosan microparticles after a calcium silicate treatment. Acta Biomater. 4: 1349–1359.

Liu, Y. and D. Fan. 2019. Novel hyaluronic acid-tyrosine/collagen-based injectable hydrogels as soft filler for tissue engineering. Int. J. Biol. Macromol. 141: 700–712.

LogithKumar, A.S. and D. Keshavnarayan. 2016a. A review of chitosan and its derivatives in bone tissue engineering. Carbohydr. Polym. 151: 172–188.

LogithKumar, A., S. KeshavNarayan, A. Dhivya, S. Chawla and N. Selvamurugan. 2016b. A review of chitosan and its derivatives in bone tissue engineering. Carbohydr. Polym. 151: 172–88.

Lotfi, G., M.A. Shokrgozar, R. Mofid, F.M. Abbas, F. Ghanavati, A.A. Baghban, S.K. Yavari and S. Pajoumshariati. 2016. Biological evaluation (*in vitro* and *in vivo*) of bilayered collagenous coated (nano electrospun and solid wall) chitosan membrane for periodontal guided bone regeneration. Ann. Biomed. Eng. 1–13.

Lowe, B., J. Venkatesan and S. Anil. 2016. Preparation and characterization of chitosannatural nano hydroxyapatite-fucoidan nanocomposites for bone tissue engineering. Int. J. Biol. Macromol. 93B: 1479–1487.

Ma, A., Z. Adayi, M. Liu, M. Li, L. Wu, Y. Xiao, Q. Sun, X. Cai and X. Yang. 2017. Asymmetric Collagen/chitosan membrane containing minocycline-loaded chitosan nanoparticles for guided bone regeneration. Sci. Rep. 6.

Mangal, R., S. Srivastava and L. Archer. 2015. Phase stability and dynamics of entangled polymer-nanoparticle composites. Nat. Commun. 6: 1–9.

Manna, S., D.A. Gopakumar, D. Roy, P. Saha and S. Thomas. 2018. Nanobiomaterials for removal of fluoride and chlorophenols from water. New Polymer Nanocomposites for Environmental Remediation, 487–498.

Mansouri, K., H. Fattahian, N. Mansouri, P.G. Mostafavi and A. Kajbafzadeh. 2018. The role of cuttlebone and cuttlebone derived hydroxyapatite with platelet rich plasma on tibial bone defect healing in rabbit: An experimental study. Kafkas. Univ. Vet. Fak. Derg. 24: 107–115.

Moreno-Vásquez, M.J., E.L. Valenzuela-Buitimea, M. Plascencia-Jatomea, J.C. Encinas-Encinas, F. Rodríguez-Félix, S. Sánchez-Valdes, E.C. Rosas-Burgos, V.M. Ocaño-Higuera and A.Z. Graciano-Verdugo. 2017. Functionalization of chitosan by a free radical reaction: Characterization, antioxidant and antibacterial potential. Carbohydr. Polym. 155: 117127.

Mutsenko, V.V., V.V. Bazhenov, O. Rogulska, D.N. Tarusin, K. Schütz, S. Brüggemeier, E. Gossla, A.R. Akkineni, H. Meißner, A. Lode, S. Meschke, A. Ehrlich, S. Petović, R. Martinović, M. Djurović, A.L. Stelling, S. Nikulin, S. Rodin, A. Tonevitsky, M. Gelinsky, A.Y. Petrenko, B. Glasmacher and H. Ehrlich. 2017. 3D chitinous scaffolds derived from cultivated marine demosponge Aplysina aerophoba for tissue engineering approaches based on human mesenchymal stromal cells. Int. J. Biol. Macromol. 104B: 1966–1974.

Pal, A., S. Paul and A.R. Choudhury. 2017. Synthesis of hydroxyapatite from Lates calcarifer fish bone for biomedical applications. Mater. Lett. 203: 89–92.

Ping Liu, Peng Chang, Yizao Wan, Zhiwei Yang, Gnangyao Xiong and Honglin Luo. 2015. Constructing a novel three-dimensional scaffold with mesoporous TiO2 nanotubes for potential bone tissue engineering. Journal of Material Chemistry B Issue 27, 2015, 3: 5595–5602.

Pozzolini, M., S. Scarfi and L. Gallus. 2018. Production, characterization and biocompatibility evaluation of collagen membranes derived from marine sponge Chondrosia reniformis Nardo. Mar. Drugs 16: E111.

Purcel, G., D. Melita, E. Andronescu and A.M. Grumezescu. 2016. Collagen-based nanobiomaterials. Nanobiomaterials in Soft Tissue Engineering, 173–200.

Puvaneswary, H.B., S. Raghavendran, M.R. Talebian, S.A. Murali, S. Mahmod and T. Kamarul. 2016. Incorporation of Fucoidan in β-Tricalcium phosphate-Chitosan scaffold prompts the differentiation of human bone marrow stromal cells into osteogenic lineage. Sci. Rep. 6.

Rajabi, M., M. Srinivasan and S.A. Mousa. 2016. Nanobiomaterials in drug delivery. Nanobiomaterials in Drug Delivery, 1–37.

Razavi, M., K. Zhu and Y.S. Zhang. 2017. Naturally based and biologically derived nanobiomaterials. Nanobiomaterials Science, Development and Evaluation, 61–86.

Saranya, A., S. Moorthi, M. Saravanan, P. Devi and N. Selvamurugan. 2011. Chitosan and its derivatives for gene delivery. Int. J. Biol. Macromol. 48: 234–238.

Tuzlakoglu, K. and R.L. Reis. 2007. Formation of bone-like apatite layer on chitosan fiber mesh scaffolds by a biomimetic spraying process. J. Mater. Sci. Mater. Med. 18: 1279–1286.

Tziveleka, L.A., E. Ioannou and D. Tsiourvas. 2017. Collagen from the marine sponges Axinella cannabina and Suberites carnosus: Isolation and morphological, biochemical, and biophysical characterization. Mar. Drugs 15: E152.

Varga, M. 2016. Self-assembly of nanobiomaterials. Fabrication and Self-Assembly of Nanobiomaterials, 57–90.

Yuan, F., M. Ma and L. Lu. 2017. Preparation and properties of polyvinyl alcohol (PVA) and hydroxylapatite (HA) hydrogels for cartilage tissue engineering. Cell. Mol. Biol. 63: 32–35.

Zdarta, J., K. Antecka and R. Frankowski. 2018. The effect of operational parameters on the biodegradation of bisphenols by Trametes versicolor laccase immobilized on Hippospongia communis spongin scaffolds. Sci. Total Environ. 615: 784–795.

Zolghadri, M., S. Saber-Samandari and S. Ahmadi. 2019. Synthesis and characterization of porous cytocompatible scaffolds from polyvinyl alcohol-chitosan. Bull. Mater. Sci. 42: 35.

16

Nanobiomaterials in the Fight against Mosquito Borne Diseases

Vijayalakshmi Ghosh,[1,2,]* *D. Sunita*[1] and *A.K. Gupta*[1]

1. Introduction

Mosquitoes are the key vectors of deadly diseases, i.e., chikungunya, dengue, Japanese encephalitis, lymphatic filariasis, malaria, St. Louis encephalitis, West Nile, yellow fever and Zika fever, etc. The control of these vector has been accomplished for a long time by using synthetic chemicals. Insecticides that can be used in a controlled way are increasingly becoming limited as the mosquitoes are becoming resistant to insecticides including DDT, dieldrin, fenitrothion, propoxur, organophosphorus and permethrins insecticides. Insecticide resistance, cytotoxicity, adverse effect on environment, high cost and elimination of non-target organisms have led researchers to develop alternative insecticides and repellents. In this regard, researchers have diverted their attention towards the plant world. Plant extracts or essential oils from plant may be an alternative source of mosquito control agents, since they constitute a rich source of bioactive compounds that are biodegradable.

With the emergence, rapid advancement and enormous application of nanotechnology in the academic and applied research in science and technology, nanoparticles' (NPs) production by various physical and chemical methods has become easier for their application in the desired field. Owing to the improved attributes of nanoparticles, their application in mosquito control has been investigated recently (Sap-Iam et al. 2010). However, excessive energy consumption by these methods, in terms of pressure and temperature (Salunkhe et al. 2011) and use of some harmful chemicals, has led to an increase of interest among scientific community to develop some biological methods for the nanoparticles' synthesis. Green nanotechnologies may help to boost the effectiveness of mosquito vector control by employing bio-nanomaterials such as green synthesized nanoparticles, polymer nanoparticles, plant essential oil based nanoemulsion, etc. The field of nanotechnology is one of the most active areas of research in modern material sciences. Nanoparticles are of 10–100 nm size and exhibit improved properties compared to bulk particle based on size, distribution and shape. Biological synthesis of nanoparticles has received attention due to a growing need of developing environmentally benign technologies in material synthesis. The bio-mediated biosynthesis of nanoparticles is advantageous over physical or chemical methods because it is a cost effective and ecofriendly method. It does not use high pressure, energy, temperature and toxic organic solvents.

---
[1] SoS in Life Science, Pt. Ravishankar Shukla University, Raipur, Chhattisgarh - 492010.
[2] G.D. Rungta College of Science & Technology, Kurud Rd, Kohka, Bhilai, Chhattisgarh - 490024.
* Corresponding author: vijayalakshmi.ghosh@gmail.com

The use of biological materials for the fabrication of nanoparticles is a single step method for the biosynthesis process, which is also rapid, low cost and eco-friendly. These green nanoparticles have demonstrated promising activity such as oviposition deterrent activity, ovicidal activity, larvicidal activity, pupicidal activity, adulticidal activity and repellent activity against mosquito vector, hence contributing to the fight against mosquito vector borne diseases. Nanobiomaterials are found safer to non-target organisms such as *Anisops bouvieri*, *Diplonychus indicus*, *Gambusia affinis*, *Mesocyclops thermocyclopoides* and *Poecilia reticulata* (Guppy fish) (Benelli and Govindarajan 2017, Gandhi et al. 2018, Yadav et al. 2019).

2. Green Nanoparticles

Now-a-days, there is a priority for development of eco-friendly tools for control of Culicidae vectors. Green nanotechnologies may help to boost the effectiveness of mosquito vector control. Synthesis of nanoparticle is a significant area of research with the search for the green materials and an eco-friendly method for current scenario. Several applications of synthesized metal nanoparticles acquired from different techniques have all been documented to be of diagnostic relevance *in-vitro*. Applications of nanoparticles in agriculture during crop production and protection, in food industry for food packaging against microbial spoilage and also in treatment of wastewater effluent, etc., are also reported. The applications of the metal nanoparticles depend on the particle size and also environmental factors such as pH and ionic strength.

2.1 Fabrication and Characterization of Green Nanoparticles

Nanoparticles can be synthesized using a wide array of biological materials as reducing agents such as plant extracts (whole plant extract, leaves, bark, seeds, flower, stem, etc.), mangroves, seaweeds, micro-organisms and arthropods, etc. (Figure 1).

Biogenic mode of green synthesis is useful owing to its non-toxic nature and reduced adverse effect on environment impact coupled with generation of a huge quantity of nanoparticles, which are energy conserving, cost effective and compatible for food and medical applications. The formation of nanoparticles is confirmed by color change from yellow to brown in case of silver

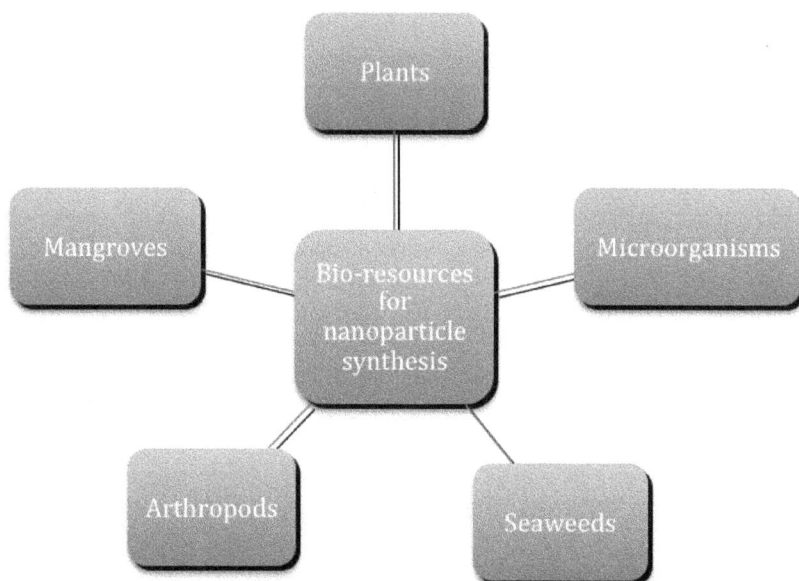

Figure 1: Different bio-resources employed for synthesis of nanoparticles.

nanoparticles and pale yellowish green to ruby red in case of gold nanoparticles due to reduction of the metal ions of the mixture to corresponding metal nanoparticle. The formation of nanoparticles in aqueous solution is also confirmed using UV-Visible spectrophotometer. Silver nanoparticle have characteristic absorbance at around 450 nm due to surface plasmon resonance (SPR) effect. The X-Ray Diffraction (XRD) studies are done to establish the crystalline nature of the biosynthesized metal nanoparticles whether the structure is cubic and/or hexagonal or consists of mixed phase. Fourier Transform Infra Red (FTIR) spectroscopy is done to understand the functional groups in the biological agents that reduces and efficiently stabilizes metal nanoparticles.

Morphology of the biosynthesized nanoparticles is observed through Atomic Force Microscope (AFM), Scanning Electron Microscope (SEM) and Transmission Electron Microscope (TEM). AFM is one of the primary tools for analyzing particle size, morphology and the agglomeration pattern. Additionally, it offers visualizations of the 3-Dimensional (z-axis) view of the nanoparticles, which can not be seen through the electron microscopes such as SEM and TEM. It also does not require any high-pressure vacuum conditions unlike the electron microscopes. SEM and TEM is done to investigate the shape and size distribution of the bio-fabricated nanoparticles. EDX profile predicts characteristic peaks corresponding to metal nanoparticles along with other peaks (if any), which may be due to the biomolecules adhering to the surface of bio-fabricated nanoparticles.

2.2 Mosquito Larvicidal Activity of Green Nanoparticles

Bio-fabricated nanoparticles have been used for mosquito vector control. One of the primary ways to achieve the Culicidae vector control is by killing the larva, which is one of the aquatic stages in the life cycle of the mosquito. Nanoparticles synthesized using plant extracts, mangroves, seaweeds, microorganisms and arthropods, etc., have demonstrated significant mosquito larvicidal activity.

2.2.1 Plants

Plant extracts rapidly reduce metal ions to metallic nanoparticle. Color of the solution turns from yellow to dark brown after the reaction in case of silver nanoparticle synthesis. This is due to presence of phytochemicals such as alkaloids, amino acids, enzymes, flavonoids, quinones, saponins, tannins, terpenoids and vitamins present in the plant extract which acts as reduction agent and capping agent during synthesis of nanoparticles. Apart from this, these plant extracts also stabilize the synthesized nanoparticles. Plant extracts are prepared from different plant parts like flower, fruits, leaves, seeds, stem, skin or whole plant extract for green synthesis of nanoparticles. The nanoparticles synthesized using plant extracts as reductants have major benefits over the biological organisms such as bacteria, fungi and algae, because of the easily availability, safety, nontoxicity and minimal number of steps required in downstream processing in case of plants. Though different types of nanoparticles like copper, titanium, magnesium, zinc, gold and silver have come up with different utilities, silver nanoparticles have proved to be the most efficient nanoparticles possessing good antimicrobial efficacy against bacteria, fungi, viruses and other harmful eukaryotic microorganisms, as anticancer agents, in diagnostics, and drug delivery.

Spherical silver nanoparticles (Ag NPs) with the size range of 35–60 nm are synthesized using *Eclipta prostrata* leaves extract as bio-reductant and tested its larvicidal activity against *Culex quinquefasciatus* and *Anopheles subpictus* (Rajakumar and Rahuman 2011). Antilarval silver nanoparticles have also been reported by using *Achyranthes aspera, Aganosma cymosa, Ammannia baccifera, Anisomeles indica, Annona muricata, A. reticulate, A. squamosal, Artemisia nilagirica, A. vulgaris, Belosy-napsis kewensis, Cinnamomum zeylanicum, Chrysanthemum, Cleistanthus collinus, Coccinia grandis, Cocos nucifera, Couroupita guianensis, Cynodon dactylon, Delphinium denudatum, Derris trifoliate, Eclipta prostrate, Ficus racemosa, Hyptis suaveolens, Holarrhena antidysenterica, Leucas aspera, Lippia citriodora, Melia azedarach, Merremia emarginatal,*

Pedalium murex, Piper longum, Pithecellobium dulce, Phyllanthus amarus, Ricinus communis, Sterculia foetida L., *Strychnos nux-vomica, Zornia diphylla*, etc. (Raman et al. 2012, Roopan et al. 2013, Suman et al. 2013, Velayutham et al. 2013, Rajasekharreddy and Rani 2014, Soni and Prakash 2014a, Suresh et al. 2014, Ramanibai and Velayutham 2015, Santhosh et al. 2015, Vimala et al. 2015, Bhuvaneswari et al. 2016, Elumalai et al. 2016, Govindarajan et al. 2016a, b, Ramanibai and Velayutham 2016, Azarudeen et al. 2017, Benelli and Govindarajan 2017, Elemike et al. 2017, Elumalai et al. 2017, Ishwarya et al. 2017, Kumar et al. 2017, Nalini et al. 2017, Soni and Dhiman 2017, Sundararajan and Kumari 2017, Ghramh et al. 2018, Jinu et al. 2018, Khader et al. 2018, Kumar et al. 2018, Parthiban et al. 2019, Yadav et al. 2019, Sharma et al. 2019). Green synthesized gold nanoparticles (Au NPs) (Balasubramani et al. 2015), nickel nanoparticles (Ni NPs) (Angajala et al. 2014, Elango et al. 2016a, b), selenium nanoparticles (Se NPs) (Yazhiniprabha and Vaseeharan 2019), titanium dioxide nanoparticles (TiO_2 NPs) (Gandhi et al. 2016, Udayabhanu et al. 2018, Gandhi et al. 2018) and zinc oxide nanoparticles (ZnO NPs) (Gandhi et al. 2017, Yazhiniprabha et al. 2019) have also been reported to have significant larvicidal activity against *An. stephensi, An. subpictus, Ae. aegypti* and *Cx. quinquefasciatus* (Table 1).

2.2.2 Microorganisms

Microorganisms such as bacteria and fungi have metal adsorbing and accumulating properties (Chen et al. 2003) and are thus known to form nanoparticles (Silver 2003, Sastry et al. 2003) by acting as reducing agents via their enzymatic machinery (Sanghi et al. 2011). These nanoparticles that are synthesized with the help of microbial cell free extracts have been examined for their larvicidal, pupicidal, adulticidal and/or repellent activity against various genera of mosquitoes. Silver (Ag), Copper (Cu), gold (Au) and cobalt (Co) nanoparticles have been synthesized using a wide array of fungi, e.g., *Aspergillus niger, Beauveria bassiana, Chrysosporium tropicum, C. keratinophilum, Cochliobolus lunatus, Isaria fumosorosea, Fusarium oxysporum, F. proliferatum* YNS2, *Metarhizium anisopliae, Phomopsis liquidambaris, Verticillium lecanii* and numerous bacteria such as *Bacillus megaterium, B. subtilis, B. thuringiensis, Listeria monocytogenes, Photorhabdus luminiscens, Streptomyces* sp. M25 and *Xenorhabdus indica*, which have excellent larvicidal activities against *Aedes, Anopheles* and *Culex* mosquitoes (Salunkhe et al. 2011, Soni and Prakash 2012a, b, c, Marimuthu et al. 2013, Soni and Prakash 2013, Banu et al. 2014, Banu and Balasubramaniam 2014a, b, Soni and Prakash 2014b, Banu and Balasubramaniam 2015, Soni and Prakash 2015, Amerasan et al. 2016, Prabakaran et al. 2016, Thammasittirong et al. 2017, El-Sadawy et al. 2018, Seetharaman et al. 2018, Vivekanandhan et al. 2018, Kalaimurugan et al. 2019a, b) (Table 2).

Ag NPs of 10–20 nm size synthesized via extract of fungus *Trichoderma harzianum* exhibited significant larvicidal activity against *Ae. aegypti* at minute concentration, i.e., at LC_{50} value of 0.09 ppm (Sundaravadivelan and Padmanabhan 2014). Kalaimurugan et al. (2019a) synthesized spherical Cu NPs of size 12.5 ± 0.21 nm using the fungus *Fusarium proliferatum* and observed its mosquito larvicidal activity indicating its use to be an effective substitute for chemical insecticides. A study was carried out to assess the cobalt (Co) nanoparticles synthesized with the help of bacteria *Bacillus thuringiensis* against the malaria vector *An. subpictus* and the dengue vector *Ae. aegypti*, where they observed better larvicidal efficacy of the *B. thuringiensis* bacterial extract synthesized Co NPs of an average size 85.30 nm with LC_{50} value of 3.59 and 2.87 ppm as compared to bacterial extract with LC_{50} value of 8.12 and 6.94 ppm L against *An. subpictus* and *Ae. aegypti*, respectively (Marimuthu et al. 2013). In another case, spherical Ag NPs of size 10–35 nm synthesized by microbial route using actinobacterium *Streptomyces* sp. M25 have been investigated for their promising larvicidal potential against mosquitoes of genera *Anopheles, Aedes* and *Culex* (Shanmugasundaram and Balagurunathan 2015).

In addition, magnetotactic bacteria like *Magnetospirillum magnetotacticum, M. magneticum* and *M. gryphiswaldense* form intracellular magnetosomes of nanometric size (35–120 nm), which

Table 1: Mosquito larvicidal activity of nanoparticles synthesized using plant extracts.

Plant Species	Plant Parts Used	NPs Synthesized	Size (nm)	Morphology	Target Mosquito Species	LC$_{50}$ (24 h) (ppm)	References
Eclipta prostrata	Leaves	Ag NPs	35–60	Spherical, elongated	*Cx. quinquefasciatus*	4.56	Rajakumar and Rahuman 2011
					An. subpictus	5.14	
Pithecellobium dulce	Leaves	Ag NPs	50–100	Spherical	*Cx. quinquefasciatus*	21.56	Raman et al. 2012
Ficus racemosa	Bark	Ag NPs	250	Cylindrical, uniform and rod	*Cx. quinquefasciatus*	12.00	Velayutham et al. 2013
					Cx. gelidus	11.21	
Ammannia baccifera	Aerial parts	Ag NPs	10–30	Spherical, triangle and hexagonal	*An. subpictus*	29.54	Suman et al. 2013
					Cx. quinquefasciatus	22.32	
Cocos nucifera	Coir	Ag NPs	23	Spherical	*An. stephensi*	87.24	Roopan et al. 2013
					Cx. quinquefasciatus	49.89	
Cinnamomum zeylanicum	Bark	Ag NPs	12	Spherical	*An. stephensi*	6.00	Soni and Prakash 2014a
					Cx. quinquefasciatus	1.50	
		Au NPs	47	Spherical	*An. stephensi*	1.00	
Delphinium denudatum	Root	Ag NPs	< 85	Spherical	*Ae. aegypti*	96.00	Suresh et al. 2014
Aegle marmelos	Leaves	Ni NPs	80–100	Triangular	*Ae. aegypti*	534.83	Angajala et al. 2014
					An. stephensi	595.23	
					Cx. quinquefasciatus	520.83	
Sterculia foetida L.	Seed	Ag NPs	7	Spherical	*Ae. aegypti, An. stephensi, Cx. quinquefasciatus*	–	Rajasekharreddy and Rani 2014
Annona muricata	Leaves	Ag NPs	30–45	Spherical	*An. stephensi*	37.70	Santhosh et al. 2015
					Cx. quinquefasciatus	31.29	
					Ae. aegypti	20.65	
Chloroxylon swietenia	Leaves	Au NPs	18–37	Spherical	*Ae. aegypti*	0.34	Balasubramani et al. 2015
					An. stephensi	0.19	
Couroupita guianensis	Leaves	Ag NPs	10–45	Cubic and hexagonal	*Ae. aegypti*	2.10	Vimala et al. 2015
	Fruits	Ag NPs	5–15	Cubic and hexagonal	*Ae. aegypti*	2.09	
Melia azedarach	Leaves	Ag NPs	3–31	Spherical	*A. aegypti*	4.27	Ramanibai and Velayutham 2015
Cocos nucifera	Coir	Ni NPs	47	Cubical	*Ae. aegypti*	259.24	Elango et al. 2016a
Cocos nucifera	Coir	Ni-Pd NPs	34	Spherical	*Ae. aegypti*	288.88	Elango et al. 2016b
Belosy-napsis kewensis	Leaves	Ag NPS	24	Spherical	*An. stephensi*	78.40	Bhuvaneswari et al. 2016
					A. aegypti	84.20	
Vitex negundo	Leaves	TiO$_2$ NPs	93	Spherical	*An. subpictus*	7.52	Gandhi et al. 2016
					Cx. quinquefasciatus	7.23	

Table 1 contd. ...

...Table 1 contd.

Plant Species	Plant Parts Used	NPs Synthesized	Size (nm)	Morphology	Target Mosquito Species	LC$_{50}$ (24 h) (ppm)	References
Achyranthes aspera	Leaves	Ag NPs	7–14	Cuboidal, Spherical	*Ae. aegypti*	3.68	Elumalai et al. 2016
					An. stephensi	3.89	
					Cx. quinquefasciatus	2.48	
Cynodon dactylon	Leaves	Ag NPs	14	Spherical	*Ae. aegypti*	3.02	Ramanibai and Velayutham 2016
					Cx. quinquefasciatus	3.21	
Anisomeles indica	Leaves	Ag NPs	50–100	Spherical	*An. stephensi*	31.56	Govindarajan et al. 2016a
					A. albopictus	35.21	
					Cx. quinquefasciatus	38.08	
Zornia diphylla	Leaves	Ag NPs	30–60	Spherical	*An. subpictus*	12.53	Govindarajan et al. 2016b
					Ae. albopictus	13.42	
					Cx. quinquefasciatus	14.61	
Aganosma cymosa	Leaves	Ag NPs	8–14	Spherical	*An. stephensi*	12.45	Benelli and Govindarajan 2017
					Ae. aegypti	13.58	
					Cx. quinquefasciatus	14.79	
Artemisia nilagirica	Leaves	Ag NPs	30	–	*Ae. aegypti*	0.33	Nalini et al. 2017
					An. stephensi	0.19	
Leucas aspera	Leaves	Ag NPs	7–22	Spherical, hexagonal, triangular and polyhedral	*Ae. aegypti*	4.02	Elumalai et al. 2017
					An. stephensi	4.69	
					Cx. quinquefasciatus	5.06	
Hyptis suaveolens	Leaves	Ag NPs	5–25	Spherical, hexagonal, triangular and polyhedral	*Ae. aegypti*	4.63	
					An. stephensi	4.04	
					Cx. quinquefasciatus	3.52	
Derris trifoliata	Leaves	Ag NPs	16	–	*Ae. aegypti*	5.87	Kumar et al. 2017
Artemisia vulgaris	Leaves	Au NPs	50–100	Spherical, triangular and hexagonal	*Ae. aegypti*	62.47	Sundararajan and Kumari 2017
Lippia citriodora	Leaves	Ag NPs	15–32	–	*Cx. quinquefasciatus*	138.62	Elemike et al. 2017
Momordica charantia	Leaves	ZnO NPs	21	Spherical	*An. stephensi*	5.42	Gandhi et al. 2017
					Cx. quinquefasciatus	4.87	
Ricinus communis	Leaves	Ag NPs	9	Spherical	*An. Stephensi*	0.75	Soni and Dhiman 2017
					Ae. aegypti	1.00	
Pedalium murex	Seed	Ag NPs	20–30	Hexagonal	*Ae. aegypti*	34.88	Ishwarya et al. 2017

Table 1 contd. ...

...Table 1 contd.

Plant Species	Plant Parts Used	NPs Synthesized	Size (nm)	Morphology	Target Mosquito Species	LC$_{50}$ (24 h) (ppm)	References
Merremia emarginatal	Leaves	Ag NPs	20	Spherical	*An. stephensi*	8.36	Azarudeen et al. 2017
					Ae. aegypti	9.20	
					Cx. quinquefasciatus	10.02	
Annona reticulata	Leaves	Ag NPs	8	Spherical	*Ae. aegypti*	4.43	Parthiban et al. 2019
Phyllanthus amarus	Leaves	Ag NPs	< 300	Spherical	*An. stephensi*	2.82	Khader et al. 2018
Annona squamosa	Leaves		1000	–		2.33	
Coccinia grandis	Leaves		100–200	Spherical		2.76	
Eclipta prostrata	Leaves		200	Cubic		2.51	
Holarrhena antidysenterica	Bark	Ag NPs	40–60	Spherical, hexagonal and triangular	*An. stephensi*	2.67	Kumar et al. 2018
Euphorbia hirta	Leaves	TiO$_2$ NPs	20–50	Spherical	*Ae. aegypti*	13.20	Udayabhanu et al. 2018
					Cx. quinquefasciatus	6.89	
Cleistanthus collinus	Leaves	Ag NPs	66–75	Triangular and pentagonal	*An. stephensi*	11.05	Jinu et al. 2018
					Ae. aegypti	11.38	
Strychnos nux-vomica	Leaves	Ag NPs	54–61	Irregular, spherical and round	*An. stephensi*	8.82	
					Ae. aegypti	7.75	
Chrysanthemum	Leaves	Ag NPs	40–100	–	*Ae. aegypti*	13.89	Ghramh et al. 2018
Momordica charantia	Leaves	TiO$_2$ NPs	70	Spherical	*An. stephensi*	3.29	Gandhi et al. 2018
Murraya koenigii	Berry extract	ZnO NPs	10–15	Hexagonal	*Cx. quinquefasciatus*	2.10	Yazhiniprabha et al. 2019
Murraya koenigii	Berry extract	Se NPs	50–150	Spherical	*Ae. aegypti*	3.54	Yazhiniprabha and Vaseeharan 2019
Piper longum L.	Leaves	Ag NPs	29	Spherical	*An. stephensi*	25.06	Yadav et al. 2019
					Ae. aegypti	46.35	
					Cx. quinquefasciatus	45.30	
Achyranthes aspera	Stem	Ag NPs	54.96	Spherical	*Ae. aegypti*	26.69	Sharma et al. 2019

contains crystalline magnetic iron minerals magnetite (Fe_3O_4) and/or greigite (Fe_3S_4) (Yan et al. 2012). These magnetic nanoparticles (MNPs) of *M. gryphiswaldense* have been exploited for their mosquito larvicidal and pupicidal potential (Murugan et al. 2017). The MNPs were highly toxic to the larvae as well as pupae of *An. stephensi* and *Ae. Aegypti*, indicating it to be a possible approach in vector control.

Table 2: Mosquito larvicidal activity of nanoparticles synthesized using microorganisms.

Microorganisms	NPs Synthesized	Size (nm)	Morphology	Target Mosquito Species	LC$_{50}$ (24 h) (ppm)	References
Bacteria						
Bacillus thuringiensis	Co NPs	85	Spherical and oval	*An. subpictus*	3.59	Marimuthu et al. 2013
				Ae. aegypti	2.87	
Bacillus thuringiensis	Ag NPs	44–143	–	*Ae. aegypti*	0.10	Banu et al. 2014
Bacillus megaterium	Ag NPs	46–81	–	*Cx. quinquefasciatus*	1.35	Banu and Balasubramanium 2015
				Ae. aegypti	0.49	
Listeria monocytogenes	Ag NPs	30	Rod, Spherical	*An. stephensi*	1.00	Soni and Prakash 2015
Bacillus subtilis	Ag NPs	76	Hexagonal, Spherical	*Cx. quinquefasciatus*	2.66	
Magnetospirillum gryphiswaldense	Magnetic NPs	–	–	*An. stephensi*	3.91	Murugan et al. 2017
				Ae. aegypti	4.85	
Bacillus thuringiensis subsp. *israelensis* K46 (Supernatant)	Ag NPs	120	Spherical	*Ae. aegypti*	0.01	Thammasittirong et al. 2017
Bacillus thuringiensis subsp. *israelensis* K46 (Inclusion Protein)	Ag NPs	100–300	Irregular	*Ae. aegypti*	0.01	
Xenorhabdus indica	Ag NPs	21–48	Spherical, triangular	*Cx. pipiens*	2.50	El-Sadawy et al. 2018
Photorhabdus luminescens laumondii HP88	Ag NPs	18–64	Spherical, triangular	*Cx. pipiens*	5.10	
Pseudomonas flourescens YPS3	Ag NPs	27	Spherical	*An. stephensi*	94.23	Kalaimurugan et al. 2019a
				Ae. aegypti	80.66	
				Cx. quinquefasciatus	113.24	
Actinobacteria						
Streptomyces sp. M25	Ag NPs	10–35	Spherical	*An. subpictus*	51.34	Shanmugasundaram and Balagurunathan 2015
				Ae. aegypti	60.23	
				Cx. quinquefasciatus	48.98	
Fungus						
Cochliobolus lunatus	Ag NPs	3–21	Spherical	*Ae. aegypti*	1.48	Salunkhe et al. 2011
				An. stephensi	1.30	
Chrysosporium tropicum	Ag NPs	20–50	Spherical	*Ae. aegypti*	4.00	Soni and Prakash 2012a
	Au NPs	2–15	Spherical	*Ae. aegypti*	12.00	
Aspergillus niger	Au NPs	10–30	–	*Ae. aegypti*	24.00	Soni and Prakash 2012b
				An. stephensi	1.69	
Chrysosporium tropicum	Ag NPs	20–50	Spherical	*An. stephensi*	4.00	Soni and Prakash 2012c
				Cx. quinquefasciatus	6.00	
	Au NPs	2–15	Spherical	*An. stephensi*	12.00	
				Cx. quinquefasciatus	24.00	

Table 2 contd. ...

...Table 2 contd.

Microorganisms	NPs Synthesized	Size (nm)	Morphology	Target Mosquito Species	LC$_{50}$ (24 h) (ppm)	References
Aspergillus niger	Ag NPs	20–70	Spherical	*Ae. aegypti*	4.67	Soni and Prakash 2013
				An. stephensi	2.00	
Beauveria bassiana	Ag NPs	37–61	Spherical	*Ae. aegypti*	0.79	Banu and Balasubramaniam 2014a
Isaria fumosorosea	Ag NPs	51–111	Spherical	*Cx. quinquefasciatus*	0.43	Banu and Balasubramaniam 2014b
				Ae. aegypti	0.09	
Chrysosporium keratinophilum	Ag NPs	24–51	Spherical	*Ae. aegypti*	2.00	Soni and Prakash 2014b
				An. stephensi	6.00	
	Au NPs	20–50	Spherical	*Cx. quinquefasciatus*	30.00	
				Ae. aegypti	30.00	
Verticillium lecanii	Ag NPs	20–50	Spherical	*Cx. quinquefasciatus*	2.00	
				An. stephensi	1.58	
				Ae. aegypti	30.00	
	Au NPs	1–40	Spherical	*Ae. aegypti*	30.00	
Trichoderma harzianum	Ag NPs	10–20	–	*Ae. aegypti*	0.09	Sundaravadivelan and Padmanabhan 2014
Metarhizium anisopliae	Ag NPs	28–38	Rod shaped	*An. culicifacies*	41.20	Amerasan et al. 2016
Beauveria bassiana	Ag NPs	20–34	Spherical	*Cx. quinquefasciatus*	13.19	Prabakaran et al. 2016
				Ae. aegypti	20.62	
				An. stephensi	22.87	
Fusarium oxysporum	Ag NPs	5–30	Spherical	*An. stephensi*	69.99	Vivekanandhan et al. 2018
				Ae. aegypti	42.94	
				Cx. quinquefasciatus	97.56	
Phomopsis liquidambaris	Ag NPs	19	Spherical	*Ae. aegypti*	1.02	Seetharaman et al. 2018
				Cx. quinquefasciatus	1.18	
Fusarium proliferatum (YNS2)	Cu NPs	13	Spherical	*An. stephensi*	39.25	Kalaimurugan et al. 2019b
				Ae. aegypti	81.34	
				Cx. quinquefasciatus	21.84	

2.2.3 Mangroves

Halophytic trees and shrubs that inhabit the coastal intertidal zone, estuarine and riverine areas in tropical and subtropical latitudes mainly characterize the mangroves. Mangroves play an immense ecological role in the marine and coastal environment by protecting shorelines from erosion caused by heavy winds, waves, and floods with their tangled root systems. They filter the pollutants and maintain the water quality. The zonation and distribution of distinctive mangrove species depends on salinity, nutrients available and other physico-chemical variations. Mangroves produce a wide array of biologically active compounds with immense medicinal potential for which fisher-folks have been traditionally using them to treat numerous diseases.

Rhizophora mucronata is a mangrove plant with a natural reservoir of various bioactive compounds such as alkaloids, flavonoids, polyphenols and terpenoids with antibacterial, antiplasmodial and antiviral activities. Anti-mosquito larval silver nanoparticles with size range of 60–95 nm have been bio-synthesized with *R. mucronata* leaf extract. These green nanoparticles have a dose-dependent mosquito larvicidal activity with LC_{50} value of 0.59 and 0.89 ppm and LC_{90} values of 2.62 and 6.29 ppm for *Ae. aegypti* and *Cx. quinquefasciatus* correspondingly (Gnanadesigan et al. 2011).

Suaeda is an important halophytic genus consisting of 110 species worldwide, covering the coastal areas of tropical and subtropical regions, of which *Suaeda maritima* (L.) Dumort is an annual herb belonging to Chenopodiaceae family that grows in alkaline and saline moist soils, and *S. maritima* is distributed throughout the east coast and west coast of India including Sunderbans mangroves in West Bengal, Mahanadi and Bhitarkanika mangroves in Orissa, Krishna and Godavari in Andhra Pardesh and Karangadu and Pichavaram in Tamil Nadu. Traditionally, *S. maritima* leaf extracts are used for the treatment of bacterial and viral infections. Silver nanoparticles synthesized using *Suaeda maritima* leaf extracts as bio-reductant are promising anti-larval agents for mosquito vector control with a LC_{50} value of 8.67 ppm, 10.10 ppm, 12.24 ppm and 14.89 ppm against I instar, II instar, III instar and IV instar larva of dengue vector *Ae. Aegypti*, respectively. Mosquito larval populations in the field, i.e., in water storage reservoirs, are also reduced in field after treatment with green nanoparticles. Significant *Ae. aegypti* pupicidal activity with LC_{50} value of 17.98 and LC_{90} value of 35.35 is obtained with these nanoparticles (Suresh et al. 2018) (Table 3).

Table 3: Silver nanoparticles synthesized using *Suaeda maritima* leaf extracts for control of *Ae. aegypti* mosquito (Suresh et al. 2018).

Activity	Concentration			
Ovicidal	**50 ppm**	**100 ppm**	**150 ppm**	**200 ppm**
Egg hatchability	53.00 ± 1.58	36.6 ± 1.14	25.2 ± 1.30	No hatchability
Larvicidal	**LC_{50} value (ppm)**		**LC_{90} value (ppm)**	
I instar	8.67		21.41	
II instar	10.10		24.32	
III instar	12.24		27.46	
IV instar	14.89		31.49	

2.2.4 Seaweeds

Seaweeds or marine macro algae are aquatic non-vascular plants (plants lacking xylem and phloem). Unfortunately compared to land plants, seaweeds remain largely unexplored and not much data is available on the extent of their diversity or endemism. Reasons for this include scarcity of seaweed taxonomists and comparative difficulty with exploring coastal regions. Seaweeds are very diverse in terms of the body size and habitatat (benthic/sub-tidal, inetrtidal zone, etc.). Macroalgae are classified into three groups based on the color: brown algae (Phaeophyta), red algae (Phodophyta) and green algae (Chlorophyta). Brown algae or seaweeds are brown in color due to the characteristic pigment fucoxanthin (a xanthophyll pigment), which masks the other xanthophylls, beta-carotene and chlorophyll. Red algae have the characteristic red color due to dominance of phycoerythrin and phycocyanin over the other pigments such as chlorophyll, beta-carotene and xanthophylls. Green seaweeds obtain characteristic green color due to rich proportion of chlorophyll a and b. The cell walls of marine macro algae or seaweeds are mainly composed of polysaccharides such as (e.g., alginate, carrageenan, laminarin and ulvan). These polysaccharides play vital role in synthesis of nanoparticles.

Table 4: Mosquito larvicidal activity of nanoparticles synthesized using seaweeds.

Seaweed Species	NPs Synthesized	Size (nm)	Morphology	Mosquito Species	LC_{50} (24 h) (ppm)	References
Caulrepa scalpelliformis	Ag NPs	20–35	Spherical, cubical	*Cx. quinquefasciatus*	4.64	Murugan et al. 2015a
Ulva lactuca	Ag NPs	20–35	Cubical	*An. stephensi*	4.63	Murugan et al. 2015b
Ulva lactuca	ZnO NPs	15	Triangles, hexagons, rods and rectangles	*Ae. aegypti*	22.38	Ishwarya et al. 2018
Sargassum wightii	ZnO NPs	20–62	Spherical	*An. stephensi*	5.69	Murugan et al. 2018

Nanoparticles synthesized using *Caulrepa scalpelliformis*, *Sargassum wightii* and *Ulva lactuca* have been used for mosquito larvicidal and pupicidal activity. Spherical silver nanoparticles synthesized from *Caulrepa scalpelliformis* with size range of 20–35 nm have demonstrated larvicidal activity against filariasis vector *Cx. quinquefasciatus* with a LC_{50} value of 3.08 ppm (I instar), 3.49 ppm (II instar), 4.64 ppm (III instar) and 5.86 ppm (IV instar) (Murugan et al. 2015a). Silver nanoparticles synthesized using *Ulva lactuca* exhibited larvicidal activity with LC_{50} value of 2.11 ppm (I instar), 3.09 ppm (II instar), 4.63 ppm (III instar) and 5.26 ppm (IV instar) against malaria vector *An. stephensi* (Murugan et al. 2015b), whereas *Ulva lactuca* synthesized zinc oxide nanoparticles killed the *Ae. aegypti* larva with LC_{50} value of 22.38 ppm (Table 4). Upon stereomicroscopic visualization, it is seen that *U. lactuca*-fabricated ZnO NPs disintegrated the epithelial layer and outer cuticle of the *Ae. aegypti* larva. The histopathological images showed the normal appearance in control group whereas the ZnO NPs treated larva experienced numerous histological modifications in the gastric caeca, muscles and nerve cord ganglia (Ishwarya et al. 2018) (Table 4).

2.2.5 Arthropods

Nanoparticles have also been synthesized using arthropod extracts as biological reducing agents. Antimicrobial peptide (AMP) crustin (Cr) from the blue crab, *Portunus pelagicus*, is used as a bioreductant for synthesis of TiO2NPs. Crustin can be extracted from blue crab haemolymph and further can be purified by matrix assisted affinity column chromatography. There are reports on crustin reduced titanium dioxide nanoparticles (Cr-TiO$_2$NPs) with tetragonal shape and particle size in the range of 10–50 nm. These synthesized Cr-TiO$_2$NPs killed third instar mosquito larvae of *An. stephensi* and *Cx. quinquefasciatus* within 48 h of exposure with LC_{50} value of 14.30 ppm and LC_{90} value of 94.10 ppm values for *An. stephensi* and LC_{50} value of 20.80 ppm and LC_{50} value of 148.10 ppm in case of *Cx. quinquefasciatus*, respectively (Rekha et al. 2019).

2.3 Mosquito Pupicidal Activity of Green Nanoparticles

There are a growing number of evidences about the larvicidal efficacy of bio-fabricated nano-larvicides, while only moderate efforts focused on their pupicidal activity (Table 5).

Antilarval silver nanoparticles have been synthesized using seaweeds such as *Caulrepa scalpelliformis* and *Ulva lactuca* with particle size ranging from 20–35 nm and proved to be significant pupicidal agent against *An. stephensi* and *Cx. quinquefasciatus* (Murugan et al. 2015a, b). *Sargassum wightii* synthesized ZnO NPs also kill the *An. stephensi* pupa with a LC_{50} value of 7.43 ppm (Murugan et al. 2018). Green nanoparticles synthesized from mangrove extract is also a promising mosquito pupicidal agent. *Suaeda maritima* synthesized spherical Ag NPs with particle size in the range of 20–60 nm killed the *Ae. aegypti* pupa and the LC_{50} and LC_{90}

Table 5: Mosquito pupicidal activity of green nanoparticles.

Scientific Name	Plant/ Organism	Nanoparticle	Size (nm)	Morphology	Mosquito Species	LC_{50} (24 h) (ppm)	References
Caulrepa scalpelliformis	Seaweed	Ag NPs	20–35	Spherical, Cubical	*Cx. quinquefasciatus*	4.64	Murugan et al. 2015a
Ulva lactuca	Seaweed	Ag NPs	20–35	Cubical	*An. stephensi*	6.86	Murugan et al. 2015b
Suaeda maritima	Mangrove	Ag NPs	20–60	Spherical	*Ae. aegypti*	17.98	Suresh et al. 2018
Magnetospirillum gryphiswaldense	Bacteria	Magnetic nanoparticles	–	–	*An. stephensi*	6.43	Murugan et al. 2017
					Ae. aegypti	7.55	
Momordica charantia	Plants	TiO_2 NPs	70	Spherical	*An. stephensi*	5.04	Gandhi et al. 2018
Sargassum wightii	Seaweed	ZnO NPs	20–62	Spherical	*An. stephensi*	7.43	Murugan et al. 2018

values are calculated to be 17.98 ppm and 35.35 ppm, respectively (Suresh et al. 2018). Magnetic nanoparticles from *Magnetospirillum gryphiswaldense* are reported to have significant pupicidal activity against malaria and dengue vector with LC_{50} value of 6.43 ppm and 7.55 ppm against *An. stephensi* and *Ae. Aegypti*, respectively (Murugan et al. 2017). Biosynthesized TiO_2 (NPs) using the *Momordica charantia* leaf aqueous extract as reducing and stabilizing agent showed the maximum mosquitocidal activity against *An. stephensi* Liston (Diptera: Culicidae) pupae with LC_{50} of 5.04 ppm (Gandhi et al. 2018).

3. Plant Essential Oil Based Nanoemulsion

3.1 Nanoemulsion Preparation

There are two different methods for nanoemulsion preparation, i.e., low-energy and high-energy methods in terms of energy consumption during the nanoemulsion preparation process.

3.1.1 Low-energy Emulsification Methods

The development of nanoemulsion by low-energy method depends on the spontaneous formation of nano-range droplets by altering composition of constituent phases and/or environment. The low-energy emulsification methods include emulsion inversion point (EIP) method, phase inversion temperature (PIT) method and spontaneous emulsification method. These methods are preferred in certain cases, especially in drug delivery systems, owing to their nondestructive nature. But the shortcoming lies in the selection of oil and emulsifier types and concentration, which will enable to achieve nano-range stable emulsion by using very low-energy. Also the requirement of high concentrations of emulsifiers in this case is not preferred in case of many food industries.

3.1.2 High-energy Emulsification Methods

High-energy emulsification methods use intense disruptive forces to reduce larger droplets size into smaller droplets size. These methods employ devices such as high-pressure homogenizer, microfluidizer and ultrasonicator to induce the required high sheer for production of tiny droplets required for nanoemulsion preparation. Initially, a coarse emulsion is prepared by adding all the constituent ingredients and then this mixture is subjected to high energy induced by these devices. The smaller droplet size can be produced by optimizing the type of device, type and concentration of oil, physico-chemical property of emulsifier (i.e., viscosity, hydrophile-lipophile balance value), and operating conditions such as energy input, temperature and time.

3.2 Characterization of Nanoemulsion

Nanoemulsions are characterized for their droplet size, droplet morphology, optical transparency and viscosity, etc. The droplet size is determined using particle size analyzer using dynamic light scattering (DLS) technique. Nanoemulsion is dependent on the process like surfactant type, surfactant concentration, oil–surfactant mixing ratio and emulsification time. Coarse emulsion is turbid or milky white in color due to high droplet size but nanoemulsions are optically transparent/ translucent, which can be measured by the optical density at 600 nm by an UV–Vis spectrophotometer. This optical transparency of submicron emulsions is due to the small droplets that scatter the light waves very weakly. Viscosity of the nanoemulsion depends on the type and concentration of oil and emulsifier. Increase in emulsifier concentration increases viscosity of the system. Morphology of nanoemulsion droplets is visualized by transmission electron microscope (TEM) and Atomic Force Microscope (AFM). In general, the nanoemulsion droplets are spherical is shape.

3.3 Nanoemulsions for Mosquito Vector Control

Nanoemulsions have mosquito larvicidal, adult knock down, adulticidal and repellent property. This is due to the bioactive compounds present in the essential oil and also due to the size-dependent property of nanoemulsions (Table 6).

3.3.1 Larvicidal Activity

Nanoemulsions of the plant essential oils such as *Artemisia dracunculus*, *Anethum graveolens*, *Azadirachta indica*, *Baccharis reticularia*, *Eucalyptus globulus*, *Lippia alba*, *Ocimum basilicum*, *Pterodon emarginatus*, *Ricinus communis*, *Rosmarinus officinalis* and *Vitex negundo*, etc., are known to have larvicidal property against various mosquito vectors such as *Ae. aegypti*, *An. culicifacies*, *An. stephensi* and *Cx. quinquefasciatus* (Anjali et al. 2012, Ghosh et al. 2013, Sugumar et al. 2014, Duarte et al. 2015, Balasubramani et al. 2017, Botas et al. 2017, Oliveira et al. 2017, Osanloo et al. 2017, 2018, Sogan et al. 2018, Sundararajan et al. 2018, Ferreira et al. 2019). Neem oil-in-water nanoemulsion with optimized droplet size of 31.03 nm size at a 1:3 (vol/vol) ratio of oil and surfactant demonstrated potent larvicidal activity against filariasis vector *Cx. quinquefasciatus*. The lethal concentration (LC_{50}) of the nanoemulsion against *Cx. quinquefasciatus* decreased when droplet size decreased from 251.43 nm to 31.03 nm with increase in surfactant concentration and the LC_{50} value for the ratio 1:3 nanoemulsion (31.03 nm of droplet size) is reported to be 11.75 ppm. This confirms the size dependent larvicidal activity of nanoemulsion, which explains that less amount of lower droplet size nanoemulsion is enough to kill the mosquito larva when compared to the larger droplet size nanoemulsion (Anjali et al. 2012). Rodrigues et al. (2014) developed a mosquito larvicidal nanoemulsion with Copaiba (*Copaifera duckei*) oleoresin. Many natural products of plant origin are potential insecticides, including diterpenes. One of the possible mechanisms for mosquito larvicidal activity is inhibition of acetylcholinesterase enzyme of the mosquitoes by these compounds (Sugumar et al. 2014).

3.3.2 Adulticidal Activity

Basil (*Ocimum sanctum*) essential oil based nanoemulsion has the ability to kill the adult mosquito of dengue and filariasis vector, i.e., *Ae. Aegypti* and *Cx. quinquefasciatus*. The Knock down (KD) and adulticidal bioassay was done following WHO guidelines. First, nanoemulsions of various dilutions were applied on to Whatman no. 1 filter papers and then twenty-five female mosquitoes (2–5 days old) were transferred to a plastic holding tube and exposed to the nanoemulsion.

Table 6: Plant essential oil based nanoemulsions for mosquito vector control.

Essential Oil	Plant Parts Used	Major Bioactive Components	Droplet Size (nm)	Target Mosquito	Activity	Reference
Cymbopogon nardus	–	Limonene, citronellal	150–220	*Ae. aegypti*	Repellent	Nuchuchua et al. 2009
Ocimum americanum	–	3-carene, caryophyllene				
Vetiver zizanioides	–	Vetiveric acid				
Cymbopogon nardus	–	D-limonene, citronellal	135	*Ae. aegypti*	Repellent	Sakulku et al. 2009
Azadirachta indica	–	–	31	*Cx. quinquefasciatus*	Larvicidal	Anjali et al. 2012
Ocimum basilicum	–	–	30	*Ae. aegypti*	Larvicidal	Ghosh et al. 2013
Eucalyptus globulus	–	Eucalyptol	20–40	*Cx. quinquefasciatus*	Larvicidal	Sugumar et al. 2014
Rosmarinus officinalis	Leaves	1,8-cineole (44.0%), camphor (16.1%), β-myrcene, α-pinene, verbenone, borneol, camphene	< 200	*Ae. aegypti*	Larvicidal	Duarte et al. 2015
Vitex negundo	Leaves	βterpinene, β-caryophyllene, hedycariol, 2R–acetoxymethyl-1,3,3-trimethyl-4t-(3-methyl2-buten-1-yl)-1t- cyclohexanol, eudesmol, nerolidol, α-humulene, γ-neoclovene	< 200	*Ae. aegypti*	Larvicidal	Balasubramani et al. 2017
Baccharis reticularia	Leaves	D-limonene, (*E*)-caryophyllene, bicyclogermacrene	90	*Ae. aegypti*	Larvicidal	Botas et al. 2017
Pterodon emarginatus	Fruits	β-caryophyllene, λ-cadinene, germacrene D, α-humulene	128	*Ae. aegypti*	Larvicidal	Oliveira et al. 2017
Artemisia dracunculus	–	P-Ally anisole, cis-ocimene, β-ocimene Y, limonene, 3-methoxy cinnamaldehyde	15	*An. stephensi*	Larvicidal	Osanloo et al. 2017
Ocimum sanctum	Whole plant	–	–	*Ae. aegypti*, *Cx. quinquefasciatus*	Knock down, adulticidal	Ramar et al. 2017
Anethum graveolens	–	p-Cymenealpha, α-phellandrene, Carvone, dill ether, cis-sabinol	11	*An. stephensi*	Larvicidal	Osanloo et al. 2018
Ricinus communis	Seeds	Ricinoleic acid, oleic acid, linoleic acid methyl esters	114	*An. culicifacies*	Larvicidal	Sogan et al. 2018
Ocimum basilicum	Leaves	Trans β-guaiene, α-cadinol, 9-Methoxybicyclo [6.1.0] nona-2, 4, 6-triene, phytol, eucalyptol	< 200	*Cx. quinquefasciatus*	Larvicidal	Sundararajan et al. 2018
Lippia alba	Leaves	Geranial, neral, limonene, elemol, γ-muurolene, bergamal	117	*Cx. quinquefasciatus*, *Ae. aegypti*	Larvicidal	Ferreira et al. 2019

Number of mosquitoes knocked down in the different interval tube was recorded and mosquitoes were transferred to a holding tube for 24 h of incubation to check for adulticidal effect. Basil oil nanoemulsion knocked down dengue and filariasis vector with KD_{50} and KD_{90} values of 7.01 mg/cm^2 and 29.94 mg/cm^2; 4.05 mg/cm^2 and 20.88 mg/cm^2 for *Ae. Aegypti* and *Cx. quinquefasciatus*, respectively. Basil oil nanoemulsion also has a mortal effect on adult *Ae. aegypti* and *Cx. quinquefasciatus* with LD_{50} (24 h) and LD_{90} (24 h) values of 28.60 mg/cm^2 and 69.82 mg/cm^2; 20.09 mg/cm^2 and 57.13 mg/cm^2 correspondingly (Ramar et al. 2017).

3.3.3 Repellent Activity

Mosquito repellent activity by nanoemulsion of citronella oil (*Cymbopogon nardus*), hairy basil oil (*Ocimum americanum*) and vetiver oil (*Vitiver zizanioides*) was evaluated using human-bait technique based on World Health Organization guidelines (WHO 2009) by Nuchuchua et al. (2009) and Sakulku et al. (2009). Nanoemulsion was applied on to the marked area of one forearm of human volunteer. The volunteer placed the test forearm in a mosquito cage containing non-blood fed female mosquitoes (3–5 days old) to check for the mosquito biting (i.e., probing or landing) and protection time was calculated. There was size-dependent mosquito repellency by the nanofomulations, which is justified by the fact that protection times against mosquito biting of the nanoemulsions prepared with high-pressure homogenization were significantly longer than the ones prepared without homogenization. Smaller size formulations have better repellency property, which is achieved by homogenization. Protection time of up to 4.7 h was observed with nanoemulsion with combination of essential oils, i.e., citronella oil, hairy basil oil and vetiver oil.

4. Chitosan Nanoparticle against Mosquitoes

Chitosan is a polymer extracted from the shrimp cells (*Penaues indicus*). Chitosan nanoparticles have high surface area to volume ratio, small particle size and compactness, which gives these nanoparticles their unique physico-chemical properties. These polymer-based nanoparticles are synthesized by cross-linking or physical interactions between the polymer chains by various bonds such as covalent bonds, electrostatic forces, hydrogen bonds or hydrophobic associations. These nanoparticles have drawn attention in biomedical applications due to non-toxicity, better biocompatibility and biodegradability. Ionic gelation method was used to synthesize chitosan nanoparticles. Anand et al. (2018) prepared anti-mosquito larval chitosan nanoparticles by ionic gelation method using sodium tripolyphosphate (TPP) as a reducing agent. The synthesized chitosan nanoparticles were spherical in shape and 8 nm in size as confirmed by High Resolution Transmission Electron Microscopy images, which exhibited larvicidal activity against third instars larvae of *Ae. aegypti* with a LC_{50} value of 66.42 mg/L and LC_{90} value of 92.58 mg/L suggesting the potential use of chitosan nanoparticles for mosquito vector control.

5. Nanoformulation of Organic Insect Repellent for Mosquito Vector Control

Water-dispersive nano-formulation of a poor water-soluble insect repellent diethylphenylacetamide (DEPA) using PEG polymerization and subsequent PIT emulsification was reported by Balaji et al. (2015). Nano-DEPA with mean hydrodynamic diameter of 153.74 nm was evaluated for larvicidal and cone bioassay against the Japanese encephalitis vector *Cx. tritaeniorhynchus* as per WHO guidelines. Nano-DEPA is more efficient mosquito control agent than bulk-DEPA with the median lethal concentration (48 h) value of 0.052 mg/L for Nano DEPA for 3rd instars of *Cx. tritaeniorhynchus* larvae, whereas the value is 0.416 mg/L for bulk-DEPA. The median knockdown value (60 min) for sucrose-fed female adult mosquitoes (2–3 days old) is 3.47% (v/v) and 5.37% (v/v) for nano-DEPA and bulk-DEPA, respectively. Further, the neurotoxic activity of bulk-DEPA and nano-DEPA treated mosquito larvae by acetylcholine esterase assay confirmed substantial neurotoxicity by nano-DEPA corroborated by the total protein content analysis. The acetylcholine esterase is one of the serine hydrolases, which is associated with the basement membrane around the cholinergic nerve terminals and breaks down the acetylcholine to terminate excitatory transmission.

6. Green Nanoparticles as Antimalarial Agents

Malaria is one of the mosquito vector borne diseases and is caused by the parasite *Plasmodium falciparum*. Green silver nanoparticles synthesized by using Ashoka (*Saraca asoca*) and Neem (*Azadirachta indica*) leaf extracts as reducing agent have been reported to exhibit antiplasmodial activity. The growth of *Plasmodium falciparum* parasite in human red blood cell culture is inhibited by these biosynthesized nanoparticles (Mishra et al. 2013). Larayetan et al. (2019) also reported strong antiplasmodial activity of silver nanoparticles, synthesized by *Callistemon citrinus* flower, leaf and seed extracts with the IC_{50} value of 2.99, 3.14 and 5.34 ppm, respectively. Biosynthesized TiO_2 nanoparticles using *Momordica charantia* leaf aqueous extract as bio-reductants also demonstrated antimalarial activity with the IC_{50} value of 53.42 ppm and 59.71 ppm for chloroquine-sensitive and chloroquine-resistant *Plasmodium falciparum* strains correspondingly. *In vitro* antiplasmodial assays confirmed that the bio-synthesized nanoparticles are more efficient than the plant extracts. Also, considering the increasing threat of resistance to Chloroquine by the *Plasmodium falciparum* strains, the biosynthesized nanoparticles are alternatives to the control and treatment of the malaria parasites (Gandhi et al. 2018).

7. Conclusion

Nanoformulations require a smaller quantity of bionanomaterials to develop a formulation, and the use of bio-reductants also makes these nanopesticides conducive and friendly to the environment. The removal of the volatile organic solvent from the pesticide formulation improves its bio-security property and makes it a "greener" strategy for the control of vectors of pathogenic diseases. A higher degree of delivery to the target of action, stability of formulation, water dispersion, cost effective and lower toxicity to non-target organisms make these bionanoformulations very efficient and highly eco-friendly. Based on the research done so far in this area, it can be concluded that the nanopesticides are an eco-safe efficient tool and can be applied easily for the control of vectors of diseases in humans and animals.

Acknowledgements

Authors thank to SERB, DST, Govt. of India (File No: PDF/2017/002767) for financial assistance and SoS in Life Science, Pt Ravishankar Shukla University, Raipur, India for providing necessary facilities.

References

Amerasan, D., T. Nataraj, K. Murugan, C. Panneerselvam, P. Madhiyazhagan, M. Nicoletti and G. Benelli. 2016. Myco-synthesis of silver nanoparticles using *Metarhizium anisopliae* against the rural malaria vector *Anopheles culicifacies* Giles (Diptera: Culicidae). J. Pest. Sci. 89: 249–256.

Anand, M., P. Sathyapriya, M. Maruthupandy and A.H. Beevi. 2018. Synthesis of chitosan nanoparticles by TPP and their potential mosquito larvicidal application. Frontiers in Laboratory Medicine 2: 72–78.

Angajala, G., R. Ramya and R. Subashini. 2014. *In-vitro* anti-inflammatory and mosquito larvicidal efficacy of nickel nanoparticles phytofabricated from aqueous leaf extracts of *Aegle marmelos* Correa. Acta Trop. 135: 19–26.

Anjali, C.H., Y. Sharma, A. Mukherjee and N. Chandrasekaran. 2012. Neem oil (*Azadirachta indica*) nanoemulsion—A potent larvicidal agent against *Culex quinquefasciatus*. Pest. Manag. Sci. 68: 158–163.

Azarudeen, R.M.S.T., M. Govindarajan, M.M. AlShebly, F.S. AlQahtani, A. Amsath, S. Senthilmurugan, P. Vijayan and G. Benelli. 2017. Size-controlled biofabrication of silver nanoparticles using the *Merremia emarginata* leaf extract: Toxicity on *Anopheles stephensi*, *Aedes aegypti* and *Culex quinquefasciatus* (Diptera: Culicidae) and non-target mosquito predators. J. Asia-Pac. Entomol. 20: 359–366.

Balaji, A.P.B., P. Mishra, R.S.S. Kumar, A. Mukherjee and N. Chandrasekaran. 2015. Nanoformulation of polyethylene glycol polymerised organic insect repellent by PIT emulsification method and its application for Japanese encephalitis vector control. Colloids Surf. B Biointerfaces 128: 370–378.

Balasubramani, G., R. Ramkumar, N. Krishnaveni, R. Sowmiya, P. Deepak, D. Arul and P. Perumal. 2015. GC–MS analysis of bioactive components and synthesis of gold nanoparticle using *Chloroxylon swietenia* DC leaf extract and its larvicidal activity. J. Photochem. Photobiol. B 148: 1–8.

Balasubramani, S., T. Rajendhiran, A.K. Moola and R.K.B. Diana. 2017. Development of nanoemulsion from *Vitex negundo* L. essential oil and their efficacy of antioxidant, antimicrobial and larvicidal activities (*Aedes aegypti* L.). Environ. Sci. Pollut. R. 24: 15125–15133.

Banu, A.N. and C. Balasubramanian. 2014a. Myco-synthesis of silver nanoparticles using *Beauveria bassiana* against dengue vector, *Aedes aegypti* (Diptera: Culicidae). Parasitol. Res. 113: 2869–2877.

Banu, A.N. and C. Balasubramanian. 2014b. Optimization and synthesis of silver nanoparticles using *Isaria fumosorosea* against human vector mosquitoes. Parasitol. Res. 113: 3843–3851.

Banu, A.N. and C. Balasubramanian. 2015. Extracellular synthesis of silver nanoparticles using *Bacillus megaterium* against malarial and dengue vector (Diptera: Culicidae). Parasitol. Res. 114: 4069–4079.

Banu, A.N., C. Balasubramanian and P.V. Moorthi. 2014. Biosynthesis of silver nanoparticles using *Bacillus thuringiensis* against dengue vector, *Aedes aegypti* (Diptera: Culicidae). Parasitol. Res. 113: 311–316.

Benelli, G. and M. Govindarajan. 2017. Green-synthesized mosquito oviposition attractants and ovicides: Towards a nanoparticle-based "lure and kill" approach? J. Clust. Sci. 28: 287–308.

Benelli, G., A. Caselli and A. Canale. 2017. Nanoparticles for mosquito control: Challenges and constraints. J. King Saud. Univ. Sci. 29: 424–435.

Bhuvaneswari, R., R.J. Xavier and M. Arumugam. 2016. Larvicidal property of green synthesized silver nanoparticles against vector mosquitoes (*Anopheles stephensi* and *Aedes aegypti*). J. King Saud. Univ. Sci. 28: 318–323.

Botas, G.D.S., R.A.S. Cruz, F.B. de Almeida, J.L. Duarte, R.S. Araújo, R.N.P. Souto, R. Ferreira, J.C.T. Carvalho, M.G. Santos, L. Rocha, V.L.P. Pereira and C.P. Fernandes. 2017. *Baccharis reticularia* DC and limonene nanoemulsions: Promising larvicidal agents for *Aedes aegypti* (Diptera: Culicidae) control. Molecules 22: 19901–14.

Chen, J.C., Z.H. Lin and X.X. Ma. 2003. Evidence of the production of silver nanoparticles via pretreatment of *Phoma* sp. 3.2883 with silver nitrate. Lett. Appl. Microbiol. 37: 105–108.

Duarte, J.L., J.R.R. Amado, A.E.M.F.M. Oliveira, R.A.S. Cruz, A.M. Ferreira, R.N.P. Souto, D.Q. Falcão, J.C.T. Carvalho and C.P. Fernandes. 2015. Evaluation of larvicidal activity of a nanoemulsion of *Rosmarinus officinalis* essential oil. Rev. Bras. Farmacogn. 25: 189–192.

Elango, G., S.M. Roopan, K.I. Dhamodaran, K. Elumalai, N.A. Al-Dhabi and M.V. Arasu. 2016a. Spectroscopic investigation of biosynthesized nickel nanoparticles and its larvicidal, pesticidal activities. J. Photochem. Photobiol. B 162: 162–167.

Elango, G., S.M. Roopan, N.A. Al-Dhabi, M.V. Arasu, K.I. Dhamodaran and K. Elumalai. 2016b. Coir mediated instant synthesis of Ni-Pd nanoparticles and its significance over larvicidal, pesticidal and ovicidal activities. J. Mol. Liq. 223: 1249–1255.

Elemike, E.E., D.C. Onwudiwe, A.C. Ekennia, R.C. Ehiri and N.J. Nnaji. 2017. Phytosynthesis of silver nanoparticles using aqueous leaf extracts of *Lippia citriodora*: Antimicrobial, larvicidal and photocatalytic evaluations. Mater. Sci. Eng. C 75: 980–989.

El-Sadawy, H.A., A.H. El Namaky, E.E. Hafez, B.A. Baiome, A.M. Ahmed, H.M. Ashry and T.H. Ayaad. 2018. Silver nanoparticles enhance the larvicidal toxicity of *Photorhabdus* and *Xenorhabdus* bacterial toxins: An approach to control the filarial vector, *Culex pipiens*. Trop. Biomed. 35: 392–407.

Elumalai, D., M. Hemavathi, C.V. Deepaa and P.K. Kaleena. 2017. Evaluation of phytosynthesised silver nanoparticles from leaf extracts of *Leucas aspera* and *Hyptis suaveolens* and their larvicidal activity against malaria, dengue and filariasis vectors. Parasite Epidemiol. Control 2: 15–26.

Elumalai, D., P.K. Kaleena, K. Ashok, A. Suresh and M. Hemavathi. 2016. Green synthesis of silver nanoparticle using *Achyranthes aspera* and its larvicidal activity against three major mosquito vectors. Engineering in Agriculture, Environment and Food 9: 1–8.

Ferreira, R.M., J.L. Duarte, R.A. Cruz, A.E. Oliveira, R.S. Araújo, J.C. Carvalho, R.H. Mourao, R.N. Souto and C.P. Fernandes. 2019. A herbal oil in water nano-emulsion prepared through an ecofriendly approach affects two tropical disease vectors. Rev. Bras. Farmacogn. (in press).

Gandhi, P.R., C. Jayaseelan, C. Kamaraj, S.R.R. Rajasree and R.R. Mary. 2018. *In vitro* antimalarial activity of synthesized TiO$_2$ nanoparticles using *Momordica charantia* leaf extract against *Plasmodium falciparum*. J. Appl. Biomed. 16: 378–386.

Gandhi, P.R., C. Jayaseelan, E. Vimalkumar and R.R. Mary. 2016. Larvicidal and pediculicidal activity of synthesized TiO$_2$ nanoparticles using *Vitex negundo* leaf extract against blood feeding parasites. J. Asia-Pac. Entomol. 19: 1089–1094.

Gandhi, P.R., C. Jayaseelan, R.R. Mary, D. Mathivanan and S.R. Suseem. 2017. Acaricidal, pediculicidal and larvicidal activity of synthesized ZnO nanoparticles using *Momordica charantia* leaf extract against blood feeding parasites. Exp. Parasitol. 181: 47–56.

Ghosh, V., A. Mukherjee and N. Chandrasekaran. 2013. Formulation and characterization of plant essential oil based nanoemulsion: Evaluation of its larvicidal activity against *Aedes aegypti*. Asian J. Chem. 25: S321–S323.

Ghramh, H.A., K.M. Al-Ghamdi, J.A. Mahyoub and E.H. Ibrahim. 2018. Chrysanthemum extract and extract prepared silver nanoparticles as biocides to control *Aedes aegypti* (L.), the vector of dengue fever. J. Asia-Pac. Entomol. 21: 205–210.

Gnanadesigan, M., M. Anand, S. Ravikumar, M. Maruthupandy, V. Vijayakumar, S. Selvam, M. Dhineshkumar and A.K. Kumaraguru. 2011. Biosynthesis of silver nanoparticles by using mangrove plant extract and their potential mosquito larvicidal property. Asian Pac. J. Trop. Med. 4: 799–803.

Govindarajan, M., M. Rajeswary, K. Veerakumar, U. Muthukumaran, S.L. Hoti and G. Benelli. 2016a. Green synthesis and characterization of silver nanoparticles fabricated using *Anisomeles indica*: Mosquitocidal potential against malaria, dengue and Japanese encephalitis vectors. Exp. Parasitol. 161: 40–47.

Govindarajan, M., M. Rajeswary, U. Muthukumaran, S.L. Hoti, H.F. Khater and G. Benelli. 2016b. Single-step biosynthesis and characterization of silver nanoparticles using *Zornia diphylla* leaves: A potent eco-friendly tool against malaria and arbovirus vectors. J. Photochem. Photobiol. B 161: 482–489.

Ishwarya, R., B. Vaseeharan, R. Anuradha, R. Rekha, M. Govindarajan, N.S. Alharbi, S. Kadaikunnan, J.M. Khaled and G. Benelli. 2017. Eco-friendly fabrication of Ag nanostructures using the seed extract of *Pedalium murex*, an ancient Indian medicinal plant: Histopathological effects on the Zika virus vector *Aedes aegypti* and inhibition of biofilm-forming pathogenic bacteria. J. Photochem. Photobiol. B 174: 133–143.

Ishwarya, R., B. Vaseeharan, S. Kalyani, B. Banumathi, M. Govindarajan, N.S. Alharbi, S. Kadaikunnan, M.N. Al-Anbr, J.M. Khaled and G. Benelli. 2018. Facile green synthesis of zinc oxide nanoparticles using *Ulva lactuca* seaweed extract and evaluation of their photocatalytic, antibiofilm and insecticidal activity. J. Photochem. Photobiol. B 178: 249–258.

Jinu, U., S. Rajakumaran, S. Senthil-Nathan, N. Geetha and P. Venkatachalam. 2018. Potential larvicidal activity of silver nanohybrids synthesized using leaf extracts of *Cleistanthus collinus* (Roxb.) Benth. ex Hook.f. and *Strychnos nux-vomica* L. nux-vomica against dengue, Chikungunya and Zika vectors. Physiol. Mol. Plant P. 101: 163–171.

Kalaimurugan, D., P. Sivasankar, K. Lavanya, M.S. Shivakumar and S. Venkatesan. 2019b. Antibacterial and larvicidal activity of *Fusarium proliferatum* (YNS2) whole cell biomass mediated copper nanoparticles. J. Clust. Sci. 30: 1071–1080.

Kalaimurugan, D., P. Vivekanandhan, P. Sivasankar, K. Durairaj, P. Senthilkumar, M.S. Shivakumar and S. Venkatesan. 2019a. Larvicidal activity of silver nanoparticles synthesized by *Pseudomonas fluorescens* YPS3 isolated from the Eastern Ghats of India. J. Clust. Sci. 30: 225–233.

Khader, S.Z.A., S.S.Z. Ahmed, J. Sathyan, M.R. Mahboob, K.P. Venkatesh and K. Ramesh. 2018. A comparative study on larvicidal potential of selected medicinal plants over green synthesized silver nano particles. Egyptian Journal of Basic and Applied Sciences 5: 54–62.

Kumar, D., G. Kumar, R. Das and V. Agrawal. 2018. Strong larvicidal potential of silver nanoparticles (AgNPs) synthesized using *Holarrhena antidysenterica* (L.) Wall. bark extract against malarial vector, *Anopheles stephensi* Liston. Process Saf. Environ. 116: 137–148.

Kumar, V.A., K. Ammani, R. Jobina, P. Subhaswaraj and B. Siddhardha. 2017. Photo-induced and phytomediated synthesis of silver nanoparticles using *Derris trifoliata* leaf extract and its larvicidal activity against *Aedes aegypti*. J. Photochem. Photobiol. B 171: 1–8.

Larayetan, R., M.O. Ojemaye, O.O. Okoh and A.I. Okoh. 2019. Silver nanoparticles mediated by *Callistemon citrinus* extracts and their antimalaria, antitrypanosoma and antibacterial efficacy. J. Mol. Liq. 273: 615–625.

Marimuthu, S., A.A. Rahuman, A.V. Kirthi, T. Santhoshkumar, C. Jayaseelan and G. Rajakumar. 2013. Eco-friendly microbial route to synthesize cobalt nanoparticles using *Bacillus thuringiensis* against malaria and dengue vectors. Parasitol. Res. 112: 4105–4112.

McClements, D.J. 2011. Edible nanoemulsions: Fabrication, properties, and functional performance. Soft Matter 7: 2297–2316.

Mishra, A., N.K. Kaushik, M. Sardar and D. Sahal. 2013. Evaluation of antiplasmodial activity of green synthesized silver nanoparticles. Colloids Surf. B 111: 713–718.

Murugan, K., M. Roni, C. Panneerselvam, A.l.T. Aziz, U. Suresh, R. Rajaganesh, R. Aruliah, J.A. Mahyoub, S. Trivedi, H. Rehman, H.A.N. Al-Aoh, S. Kumar, A. Higuchi, B. Vaseeharan, H. Wei, S. Senthil-Nathan, A. Canale and G. Benelli. 2018. *Sargassum wightii*-synthesized ZnO nanoparticles reduce the fitness and reproduction of the malaria vector *Anopheles stephensi* and cotton bollworm *Helicoverpa armigera*. Physiol. Mol. Plant P. 101: 202–213.

Murugan, K., C.M. Samidoss, C. Panneerselvam, A. Higuchi, M. Roni, U. Suresh, B. Chandramohan, J. Subramaniam, P. Madhiyazhagan, D. Dinesh, R. Rajaganesh, A.A. Alarfaj, M. Nicoletti, S. Kumar, H. Wei, A. Canale, H. Mehlhorn and G. Benelli. 2015b. Seaweed-synthesized silver nanoparticles: An eco-friendly tool in the fight against *Plasmodium falciparum* and its vector *Anopheles stephensi*? Parasitol. Res. 114: 4087–4097.

Murugan, K., G. Benelli, S. Ayyappan, D. Dinesh, C. Panneerselvam, M. Nicoletti, J.S. Hwang, P.M. Kumar, J. Subramaniam and U. Suresh. 2015a. Toxicity of seaweed-synthesized silver nanoparticles against the filariasis vector *Culex quinquefasciatus* and its impact on predation efficiency of the cyclopoid crustacean *Mesocyclops longisetus*. Parasitol. Res. 114: 2243–2253.

Murugan, K., J. Wei, M.S. Alsalhi, M. Nicoletti, M. Paulpandi, C.M. Samidoss, D. Dinesh, B. Chandramaohan, C. Panneerselvam, J. Subramaniam, C. Vadivalagan, H. Wei, P. Amuthavalli, A. Jaganathan, S. Devanesan, A. Higuchi, S. Kumar, A.T. Aziz, D. Nataraj, B. Vaseeharan, A. Canale and G. Benelli. 2017. Magnetic nanoparticles are highly toxic to chloroquine-resistant *Plasmodium falciparum*, dengue virus (DEN-2), and their mosquito vectors. Parasitol. Res. 116: 495–502.

Nalini, M., M. Lena, P. Sumathi and C. Sundaravadivelan. 2017. Effect of phyto-synthesized silver nanoparticles on developmental stages of malaria vector, *Anopheles stephensi* and dengue vector, *Aedes aegypti*. Egyptian J. Bas. Appl. Sci. 4: 212–218.

Nuchuchua, O., U. Sakulku, N. Uawongyart, S. Puttipipatkhachorn, A. Soottitantawat and U. Ruktanonchai. 2009. *In vitro* characterization and mosquito (*Aedes aegypti*) repellent activity of essential-oils-loaded nanoemulsions. AAPS Pharm. Sci. Tech. 10: 1234.

Oliveira, A.E.M.F.M., D.C. Bezerra, J.L. Duarte, R.A.S. Cruz, R.N.P. Souto, R.M.A. Ferreira, J. Nogueira, E.C. da Conceição, S. Leitão, H.R. Bizzo, P.E. Gama, J.C.T. Carvalho and C.P. Fernandes. 2017. Essential oil from *Pterodon emarginatus* as a promising natural raw material for larvicidal nanoemulsions against a tropical disease vector. Sustain. Chem. Pharm. 6: 1–9.

Osanloo, M., A. Amani, H. Sereshti, M.R. Abai, F. Esmaeili and M.M. Sedaghat. 2017. Preparation and optimization nanoemulsion of Tarragon (*Artemisia dracunculus*) essential oil as effective herbal larvicide against *Anopheles stephensi*. Ind. Crop Prod. 109: 214–219.

Osanloo, M., H. Sereshti, M.M. Sedaghat and A. Amani. 2018. Nanoemulsion of Dill essential oil as a green and potent larvicide against *Anopheles stephensi*. Environ. Sci. Pollut. Res. 25: 6466–6473.

Parthiban, E., N. Manivannan, R. Ramanibai and N. Mathivanan. 2019. Green synthesis of silver-nanoparticles from *Annona reticulata* leaves aqueous extract and its mosquito larvicidal and anti-microbial activity on human pathogens. Biotechnol. Rep. 21: e00297.

Prabakaran, K., C. Ragavendran and D. Natarajan. 2016. Mycosynthesis of silver nanoparticles from *Beauveria bassiana* and its larvicidal, antibacterial, and cytotoxic effect on human cervical cancer (HeLa) cells. RSC Adv. 6: 44972–44986.

Rajakumar, G. and A.A. Rahuman. 2011. Larvicidal activity of synthesized silver nanoparticles using *Eclipta prostrata* leaf extract against filariasis and malaria vectors. Acta Trop. 118: 196–203.

Rajasekharreddy, P. and P.U. Rani. 2014. Biofabrication of Ag nanoparticles using *Sterculia foetida* L. seed extract and their toxic potential against mosquito vectors and HeLa cancer cells. Mat. Sci. Eng. C 39: 203–212.

Raman, N., S. Sudharsan, V. Veerakumar, N. Pravin and K. Vithiya. 2012. *Pithecellobium dulce* mediated extra-cellular green synthesis of larvicidal silver nanoparticles. Spectrochim. Acta A, Mol. Biomol. Spectrosc. 96: 1031–1037.

Ramanibai, R. and K. Velayutham. 2015. Bioactive compound synthesis of Ag nanoparticles from leaves of *Melia azedarach* and its control for mosquito larvae. Res. Vet. Sci. 98: 82–88.

Ramanibai, R. and K. Velayutham. 2016. Synthesis of silver nanoparticles using 3, 5-di-t-butyl-4-hydroxyanisole from *Cynodon dactylon* against *Aedes aegypti* and *Culex quinquefasciatus*. J. Asia-Pac. Entomol. 19: 603–609.

Ramar, M., P. Manonmani, P. Arumugam, S.K. Kannam, R.R. Erusan, N. Baskaran and K. Murugan. 2017. Nano-insecticidal formulations from essential oil (*Ocimum sanctum*) and fabricated in filter paper on adult of *Aedes aegypti* and *Culex quinquefasciatus*. J. Entomol. Zool. Stud. 5: 1769–1774.

Rekha, R., M. Divya, M. Govindarajan, N.S. Alharbi, S. Kadaikunnan, J.M. Khaled, M.N. Al-Anbr, R. Pavela and B. Vaseeharan. 2019. Synthesis and characterization of crustin capped titanium dioxide nanoparticles: Photocatalytic, antibacterial, antifungal and insecticidal activities. J. Photochem. Photobiol. B, Biol. 111620.

Rodrigues, E.C.R., A.M. Ferreira, J.C.E. Vilhena, F.B. Almeida, R.A.S. Cruz, A.C. Florentino, R.N.P. Souto, J.C.T. Carvalho and C.P. Fernandes. 2014. Development of a larvicidal nanoemulsion with Copaiba (*Copaifera duckei*) oleoresin. Rev. Bras. Farmacogn. 24: 699–705.

Roopan, S.M., G. Madhumitha, A.A. Rahuman, C. Kamaraj, A. Bharathi and T.V. Surendra. 2013. Low-cost and eco-friendly phyto-synthesis of silver nanoparticles using *Cocos nucifera* coir extract and its larvicidal activity. Ind. Crop. Prod. 43: 631–635.

Sakulku, U., O. Nuchuchua, N. Uawongyart, S. Puttipipatkhachorn, A. Soottitantawat and U. Ruktanonchai. 2009. Characterization and mosquito repellent activity of citronella oil nanoemulsion. Int. J. Pharm. 372: 105–111.

Salunkhe, R.B., S.V. Patil, C.D. Patil and B.K. Salunke. 2011. Larvicidal potential of silver nanoparticles synthesized using fungus *Cochliobolus lunatus* against *Aedes aegypti* (Linnaeus, 1762) and *Anopheles stephensi* Liston (Diptera; Culicidae). Parasitol. Res. 109: 823–831.

Sanghi, R., P. Verma and S. Puri. 2011. Enzymatic formation of gold nanoparticles using Phanerochaete chrysosporium. Advances in Chemical Engineering and Science 1: 154.

Santhosh, S.B., C. Ragavendran and D. Natarajan. 2015. Spectral and HRTEM analyses of Annona muricata leaf extract mediated silver nanoparticles and its Larvicidal efficacy against three mosquito vectors *Anopheles stephensi*, *Culex quinquefasciatus*, and *Aedes aegypti*. J. Photochem. Photobiol. B, Biol. 153: 184–190.

Sap-Iam, N., C. Homklinchan, R. Larpudomlert, W. Warisnoicharoen, A. Sereemaspun and S.T. Dubas. 2010. UV irradiation-induced silver nanoparticles as mosquito larvicides. J. Appl. Sci. (Faisalabad) 10: 3132–3136.

Sastry, M., A.A. Absar, M.I. Khan and R. Kumar. 2003. Biosynthesis of metal nanoparticles using fungi and actinomycete. Current Science 85: 162–170.

Seetharaman, P.K., R. Chandrasekaran, S. Gnanasekar, G. Chandrakasan, M. Gupta, D.B. Manikandan and S. Sivaperumal. 2018. Antimicrobial and larvicidal activity of eco-friendly silver nanoparticles synthesized from endophytic fungi *Phomopsis liquidambaris*. Biocatal. Agric. Biotechnol. 16: 22–30.

Shanmugasundaram, T. and R. Balagurunathan. 2015. Mosquito larvicidal activity of silver nanoparticles synthesised using actinobacterium, *Streptomyces* sp. M25 against *Anopheles subpictus*, *Culex quinquefasciatus* and *Aedes aegypti*. J. Parasit. Dis. 39: 677–684.

Sharma, A., S. Kumar and P. Tripathi. 2019. A facile and rapid method for green synthesis of *Achyranthes aspera* stem extract-mediated silver nano-composites with cidal potential against *Aedes aegypti* L. Saudi J. Biol. Sci. 26: 698–708.

Silver, S. 2003. Bacterial silver resistance: molecular biology and uses and misuses of silver compounds. FEMS Microbiol. Rev. 27: 341–353.

Sogan, N., N. Kapoor, S. Kala, P.K. Patanjali, B.N. Nagpal, V. Kumar and V. Neena. 2018. Larvicidal activity of castor oil nanoemulsion against malaria vector *Anopheles culicifacies*. Int. J. Mosq. Res. 5: 01–06.

Soni, N. and R.C. Dhiman. 2017. Phytochemical, anti-oxidant, larvicidal, and antimicrobial activities of castor (*Ricinus communis*) synthesized silver nanoparticles. Chin. Herb. Med. 9: 289–294.

Soni, N. and S. Prakash. 2012a. Efficacy of fungus mediated silver and gold nanoparticles against *Aedes aegypti* larvae. Parasitol. Res. 110: 175–184.

Soni, N. and S. Prakash. 2012b. Synthesis of gold nanoparticles by the fungus *Aspergillus niger* and its efficacy against mosquito larvae. Rep. Parasitol. 2: 1–7.

Soni, N. and S. Prakash. 2012c. Entomopathogenic fungus generated nanoparticles for enhancement of efficacy in *Culex quinquefasciatus* and *Anopheles stephensi*. Asian Pac. J. Trop. Dis. 2: S356–S361.

Soni, N. and S. Prakash. 2013. Possible mosquito control by silver nanoparticles synthesized by soil fungus (*Aspergillus niger* 2587). Adv. Nanoparticles 2: 125–132.

Soni, N. and S. Prakash. 2014a. Green nanoparticles for mosquito control. The Scientific World Journal 2014: 496362.

Soni, N. and S. Prakash. 2014b. Microbial synthesis of spherical nanosilver and nanogold for mosquito control. Ann. Microbiol. 64: 1099–1111.

Soni, N. and S. Prakash. 2015. Antimicrobial and mosquitocidal activity of microbial synthesized silver nanoparticles. Parasitol. Res. 114: 1023–1030.

Sugumar, S., S.K. Clarke, M.J. Nirmala, B.K. Tyagi, A. Mukherjee and N. Chandrasekaran. 2014. Nanoemulsion of eucalyptus oil and its larvicidal activity against *Culex quinquefasciatus*. Bull. Entomol. Res. 104: 393–402.

Suman, T.Y., D. Elumalai, P.K. Kaleena and S.R. Rajasree. 2013. GC-MS analysis of bioactive components and synthesis of silver nanoparticle using *Ammannia baccifera* aerial extract and its larvicidal activity against malaria and filariasis vectors. Ind. Crop. Prod. 47: 239–245.

Sundararajan, B. and B.R. Kumari. 2017. Novel synthesis of gold nanoparticles using *Artemisia vulgaris* L. leaf extract and their efficacy of larvicidal activity against dengue fever vector *Aedes aegypti* L. J. Trace Elem. Med. Biol. 43: 187–196.

Sundararajan, B., A.K. Moola, K. Vivek and B.R. Kumari. 2018. Formulation of nanoemulsion from leaves essential oil of *Ocimum basilicum* L. and its antibacterial, antioxidant and larvicidal activities (*Culex quinquefasciatus*). Microb. Pathog. 125: 475–485.

Sundaravadivelan, C. and M.N. Padmanabhan. 2014. Effect of mycosynthesized silver nanoparticles from filtrate of *Trichoderma harzianum* against larvae and pupa of dengue vector *Aedes aegypti* L. Environ. Sci. Pollut. R. 21: 4624–4633.

Suresh, G., P.H. Gunasekar, D. Kokila, D. Prabhu, D. Dinesh, N. Ravichandran, B. Ramesh, A. Koodalingam and G.V. Shiva. 2014. Green synthesis of silver nanoparticles using *Delphinium denudatum* root extract exhibits antibacterial and mosquito larvicidal activities. Spectrochim. Acta A: Mol. Biomol. Spectrosc. 127: 61–66.

Suresh, U., K. Murugan, C. Panneerselvam, R. Rajaganesh, M. Roni, A.I.T. Aziz, H.A.N. Al-Aoh, S. Trivedi, H. Rehman, S. Kumar, A. Higuchi, A. Canale and G. Benelli. 2018. *Suaeda maritima*-based herbal coils and green nanoparticles as potential biopesticides against the dengue vector *Aedes aegypti* and the tobacco cutworm *Spodoptera litura*. Physiol. Mol. Plant Pathol. 10: 225–235.

Thammasittirong, A., K. Prigyai and S.N.R. Thammasittirong. 2017. Mosquitocidal potential of silver nanoparticles synthesized using local isolates of *Bacillus thuringiensis* subsp. *israelensis* and their synergistic effect with a commercial strain of *B. thuringiensis* subsp. *israelensis*. Acta Trop. 176: 91–97.

Udayabhanu, J., V. Kannan, M. Tiwari, G. Natesan, B. Giovanni and V. Perumal. 2018. Nanotitania crystals induced efficient photocatalytic color degradation, antimicrobial and larvicidal activity. J. Photochem. Photobiol. B, Biol. 178: 496–504.

Velayutham, K., A.A. Rahuman, G. Rajakumar, S.M. Roopan, G. Elango, C. Kamaraj, S. Marimuthu, T. Santhoshkumar, M. Iyappan and C. Siva. 2013. Larvicidal activity of green synthesized silver nanoparticles using bark aqueous extract of *Ficus racemosa* against *Culex quinquefasciatus* and *Culex gelidus*. Asian Pac. J. Trop. Med. 6: 95–101.

Vimala, R.T.V., G. Sathishkumar and S. Sivaramakrishnan. 2015. Optimization of reaction conditions to fabricate nano-silver using *Couroupita guianensis* Aubl. (leaf & fruit) and its enhanced larvicidal effect. Spectrochim. Acta A: Mol. Biomol. Spectrosc. 135: 110–115.

Vivekanandhan, P., S. Deepa, E.J. Kweka and M.S. Shivakumar. 2018. Toxicity of *Fusarium oxysporum*-VKFO-01 derived silver nanoparticles as potential inseciticide against three mosquito vector species (Diptera: Culicidae). J. Clust. Sci. 29: 1139–1149.

Yadav, R., H. Saini, D. Kumar, S. Pasi and V. Agrawal. 2019. Bioengineering of *Piper longum* L. extract mediated silver nanoparticles and their potential biomedical applications. Mat. Sci. Eng.: C 104: 109984.

Yan, L., S. Zhang, P. Chen, H. Liu, H. Yin and H. Li. 2012. Magnetotactic bacteria, magnetosomes and their application. Microbiol. Res. 167: 507–519.

Yazhiniprabha, M., B. Vaseeharan, A. Sonawane and A. Behera. 2019. *In vitro* and *in vivo* toxicity assessment of phytofabricated ZnO nanoparticles showing bacteriostatic effect and larvicidal efficacy against *Culex quinquefasciatus*. J. Photochem. Photobiol. B 192: 158–169.

Yazhiniprabha, M. and B. Vaseeharan. 2019. *In vitro* and *in vivo* toxicity assessment of selenium nanoparticles with significant larvicidal and bacteriostatic properties. Mater. Sci. Eng. C 103: 109763.

17

Nanobiotechnology and Chemotherapy

Utharkar S. Mahadeva Rao,[1] *Ashwini Ravi,*[2,*] *S. Vijayanand,*[2]
P.N. Sudha,[3] *T. Gomathi,*[3] *S. Pavithra,*[3] *S. Sugashini*[3] and *J. Hemapriya*[4]

1. Introduction

Cancer is the large group of diseases that results from abnormal proliferation of cells (Cooper 2000, WHO). Cancer can start in any organ and has the ability to go beyond their boundaries invading adjoining parts of the body and it may spread to other organs also. The process by which the cancer spreads to other parts of the body is called as metastasis. As of 2018, it is the second most leading cause of death globally and it was estimated to cause 9.6 million deaths (WHO). Globally, as of 2018, approximately 1.8 crore cases and 95 lakh deaths were estimated to occur because of cancer (Global fact sheet WHO). In India, as of 2018, approximately 11 lakh total cases and 7 lakh deaths were recorded because of cancer (India fact sheet, WHO).

Cancer is broadly classified into two types—malignant and benign. A benign tumour will be confined to the original location of onset and it never spreads to or invades other organs whereas the malignant tumour is capable of spreading and also invading other organs. Based on the type of body cells it originates from, they were classified into different types as carcinoma, sarcoma, lymphoma, leukaemia and myeloma. Carcinomas are those that occur in epithelial cells, sarcoma occurs in connective tissues whereas lymphoma occurs in immune system and those that affect the blood cells are called leukaemia. Cancer affecting the particular types of white blood cells is called as myeloma (Cooper 2000, DeBerardinis et al. 2008). In addition, based on the organ it is occurring, cancers have been named as lung cancer, breast cancer, stomach, colorectal, liver, cervical, oesophagus, ovarian cancer, larynx, thyroid, prostate, pancreas, lymphoma, leukaemia, etc. Among these, the most common cancer are lung cancer, colorectal, liver, stomach, breast, and Non – Hodgkin Lymphoma (Cancer, WHO; India fact sheet, WHO).

While most of the cancers are caused due to life habits, prevention is better way to stay away from cancer. There are many risk factors for cancer such as alcohol drinking, tobacco smoking, obesity, dietary factors, chronic infections, environmental and occupational risks (Cancer, WHO). In

[1] Professor, Universiti Sultan Zainal Abidin, Malaysia.
[2] Bioresource Technology Lab, Department of Biotechnology, Thiruvalluvar University, Serkkadu, Vellore, Tamil Nadu, India, 632115.
[3] Biomaterial Research Lab, Department of Chemistry, DKM College for Women (Autonomous), Vellore, Tamil Nadu, India, 632001.
[4] PG & Research Department of Microbiology, DKM College for Women (Autonomous), Vellore, Tamil Nadu, India, 632001.
* Corresponding author: vipni76@gmail.com

addition to these, other factors such as family history, age, life style and exposure to microorganisms such as bacteria and viruses can also be some reasons for acquiring cancer (Cancer, ATSDR). Some environmental and occupational compounds that cause cancer include asbestos, chromium, tars, silica, chemicals from leather industry, rubber industry, benzene, arsenic, chlorophenol, DDT, vinyl chloride, etc. (Cancer, ATSDR). While some of the chemicals that are carcinogenic are present in every day products such as 1,4-dioxane which are present in soaps and personal care products, tricholoroethylene is present in cleaners, 1-Bromopropane(n-propyl bromide, 1-BP) is present in degreasers and adhesives, and asbestos is present in building materials and or infrastructure (Singla 2020).

While cancer is caused due to many reasons as acquired lifestyle or by environmental issues, there are measures to prevent them before developing one. The possible preventive measures to control cancer is to avoid certain habits such as tobacco smoking, alcohol drinking, reducing exposure to carcinogens in environment and work place and especially when it is concerned with chronic diseases, care should be taken with proper food diets and healthy life style habits (Cancer, WHO; Cancer control, WHO). But when the cancer has already developed, the only possible way not to get the condition worse is early diagnosis. Like other diseases, cancer also shows symptoms such as weight loss, loss of appetite, weakness, headache, inability to swallow, etc. The symptoms vary with the type of cancer developed. During the onset of any one or few symptoms, diagnosis is made; cancer can be detected at the early stage which is Stage I or Stage II. After early diagnosis, treatment and palliative care can be provided to the patient. While treatments aim at curing disease and prolong life, palliative care involves meeting the needs of patient to get relieved from symptoms (Cancer control, WHO; Cancer, WHO).

Treatment involves mixture of therapies such as psychosocial support, radiotherapy and chemotherapy. WHO has been creating programmes and proposing agenda for cancer prevention and control. The 58th and 70th World Health Association (WHA) have proposed agenda to minimize the incidence of cancer throughout the world such as identifying the diagnosis methods used, tobacco use, about treatment methods, providing awareness about cancer to people, palliative care programmes, etc., for treating the cancer patients better, specially to prevent and control its incidence (58th and 70th WHA). The research on different types of cancer from its onset, mechanism to different kinds of treatment has been studied extensively by various researchers (Cancer research, Elsevier) around the globe for understanding them and to provide better treatment options.

The present chapter concentrates on green synthesized and microbial synthesized nanomaterials that are used for treating cancer.

2. Cancer Treatment Methods

There are different types of cancer treatment and the treatment given to every patient varies with the type of cancer the patient has. The different type of cancer treatments available are surgery, radiation therapy, chemotherapy, immunotherapy, targeted therapy, hormone therapy, stem cell transplant and precision medicine (NIH). Though there are wide varieties of treatments like mentioned, every method has been modified or advanced to treat cancer more effectively since it has the most difficulty to be completely removed from the system.

Surgery is the resection or operation of malignant or benign tumours from the body of infected patient. It is considered the most convenient treatment methods for predominant cancers and in early stage if operation is done, it may result in long remissions (Abbas and Rehman 2018, Damyanov et al. 2018). In most of the surgeries, the body is cut open and the tumour is removed and in some cases, laproscopy is done without making sizeable cuts and the cancer is removed carefully. Though this surgical technique is conventional, certain modification has also been made in this by using radiations as knife. Instead of making cuts and wounds, radiation has been used to resect the infected part. Different radiations such as Gamma Knife system, Linear accelerator system, proton beam therapy and stereotactic radiosurgery has been used as treatment methods for cancer (SEER Training Modules 2022).

3. Bionanomaterials in Cancer Treatment

Several biological entities have been used in the synthesis of nanoparticles which are used in cancer treatment. Gold nanoparticles synthesized from marine *Enterobacteria* sp. showed remarkable anticancer activity against HepG2 and A459 cell lines even at a low concentration of 100 µg concentration of nanoparticles (Rajeshkumar 2016). Similarly, silver nanoparticles synthesized from aqueous extract of cyanobacterium *Oscillatoria limnetica* showed better cytotoxic effect against human breast cell (MCF-7) and human colon cancer (HCT-116) with IC50 value of 6.147 µg/ml and 5.369 µg/ml, respectively (Hamouda et al. 2019). Nanospindles synthesized from bacterium along with MnOx was found to have effectively encapsulated cancer drug Doxorubicin and was found to have potential advantage in treating cancer cells (Bao et al. 2020).

In addition to microorganisms, plants have also been used in the synthesis of nanoparticles with anti-cancerous activity. Silver nanoparticles synthesized from aqueous extract of *Lonicera hypoglauca* flower caused cytotoxicity to MCF-7 cell lines by the induction of apoptosis (Jang et al. 2016). Similarly, silver nanoparticles synthesized from leaves of *Artemisia vulgaris* showed anti-cancer activities against Human epithelial cancer cell line (HeLa) and human breast cancer cell line (MCF-7) (Azarpazhooh et al. 2019). Silver nanoparticles synthesized from aqueous extract of Nepeta deflersiana plant showed anticancer ability against HeLa cells. The synthesized nanoparticles were found to activate the apoptotic and necrotic cycle in cancer cell lines causing its death (Al-Sheddi et al. 2018).

Similarly, silver nanoparticles synthesized from black peel pomegranate showed anti-cancer activity against tumour cell lines BT-20 and MCF-7 but showed no toxic effect on non-tumour cell lines L-929 (Khorrami et al. 2019). Silver nanoparticles synthesized using leaf extracts of *Cucumis prophetarum* (CpAg NP) showed anti proliferative effect on cancer cell lines A549, MDA-MB-231, HepG2, and MCF-7 and were found to have IC50 values of 105.8, 81.1, 94.2, and 65.6 µg/mL, respectively. In addition to these, the study also revealed that the nanoparticle synthesized was highly potential against MCF-7 than other cancer cell lines (Hemlata et al. 2020). Gold nanoparticles synthesized using Graviola, the fruit of *Annona muricata*, showed cytotoxicity against Hep2 liver cell line and the IC50 values for this cell line was found to be 10.94 µg/mL (Priya and Iyer 2020). Similarly, gold nanoparticles synthesized from *Scutellaria barbata* showed anticancer activity against pancreatic cancer line PANC-1 (Wang et al. 2019).

4. Nanobiomaterial Conjugates

Nanomaterial conjugates have been studied by various researchers around the world for cancer treatment and some of them are discussed as follows:

4.1 Protein Drug Conjugated Nanoparticles

The ability of nanoparticles to carry a diagnostic and therapeutic agent such as peptides, proteins and nucleic acids make them a better candidate for cancer therapy. These peptides have the ability to form tight nanoparticles coatings due to their small size and less disruption of the nanoparticles. In addition to these, it also helps in controlled surface loading, concurrent presentation of multiple targeted genes, high affinity and synergistic binding (Field et al. 2015). Various types of proteins such as silk fibroin (Wongpinyochit et al. 2015), collagen (Posadas et al. 2016), keratin (Zhi et al. 2015), albumin (Lohcharoenkal et al. 2014), gelatine (Bajpai and Choubey 2006), elastin (Herrero-Vanrell et al. 2005), gliadin (Lohcharoenkal et al. 2014), legumin (Lohcharoenkal et al. 2014), corn zein (Hurtado-López and Murdan 2006), soy protein (Teng et al. 2012), milk proteins (Lohcharoenkal et al. 2014), casein (Saralidze et al. 2010), haemoglobin (Paciello et al. 2016), fibrinogen (Saralidze et al. 2010), etc., have been used to create nanoparticles. These protein based nanoparticles are classified as viral like nanoparticles, non-viral like nanoparticles and *de novo* nanoparticles (Diaz et al. 2018).

The formulation techniques used in the synthesis of protein based nanoparticles include emulsification, desolvation, complex coacervation and electro spray method (Verma et al. 2018). These protein-based nanoparticles will be developed in such a way that it has three components for drug delivery, viz., interior cavity used in loading the therapeutic agent, the exterior surface which is functionalized with proteins or DNA markers to display cancer targeting ligands and the target site and interfaces between the protein subunits which can be engineered to allow the controlled release of drugs (Lee et al. 2016, Diaz et al. 2018). These entrapped molecules stay in the target site for longer periods releasing the therapeutic agent at certain intervals or can be triggered to release them by stimuli (Moein Moghimi 2006).

The advantage of using protein nanoparticles is that they are biocompatible, biodegradable, have low toxicity, are obtained from renewable source and have high drug binding capacity (Cheng et al. 2016). In addition to these, they have other advantages such as showing efficient activity at minimal dose, help in decreasing the resistance of drugs in the body, enhance the rate of dissolution and surface area embedded in nanoparticles and show less chemical reactions. Apart from cancer therapy, these protein based nanoparticles are also used in ocular therapy, as nasal drug and also oral drug for treating ailments like blood glucose, problems in GIT, etc. (Verma et al. 2018).

In a study by Oh et al. (2018), protein corona shield nanoparticles were developed by incorporating nanoparticles with supramolecularly pre coated recombinant fusion protein. The recombinant fusion protein comprises of Her2 binding affibody combined with glutathione S transferase. This protein shield corona nanoparticles were found to increase the tumour targeting ability, thereby enhancing the efficiency of cancer treatment. The ability of gold nanoparticles to be used as protein drug conjugate has been studied by Kalimuthu et al. (2018). In the study, the PEG coated AuNPs were studied for its efficiency to bind with protein conjugated drug. The protein conjugated drug used in the study was peptide P4 from A20 murine lymphoma cells and this peptide was conjugated with chlorambucil, mephalan or bendamustine to form protein drug conjugate. The synthesized Au Np was coated sequentially with citrate, PEG and the protein conjugated drug (AuNp-PEG-PDC). The study on effect of these protein conjugated drug nanoparticles showed that P4 with all three drugs killed cancer cell lines with an average time of 10.6 to 15.4 min and had a half life of 21.0–22.3 h. This study concluded that protein conjugate drug nanoparticles can be used as an efficient targeting system for treating cancer.

4.2 Liposomal Nanoparticles

Liposomes are the spherical vesicles consisting of lipid bilayer made up of phospholipids surrounded by an aqueous phase. The phospholipid bilayer of liposomes can be synthetic or naturally existing (Gubernator 2011). For the past few years, liposomes have been identified as a better carrier system for therapeutic agents because of their advantageous characteristics such as biocompatibility, biodegradability, low toxicity, provoking no immune response and ability to carry both hydrophilic and hydrophobic drugs (Voinea and Simionescu 2002). In addition to these, it also aids in controlled release of drugs at the target site (Strebhardt and Ullrich 2008). The fundamental cause of using liposome-based drug carrier system is its prolonged circulation time of the tissues with reduced toxicity to the healthy tissues around the target site (Sercombe et al. 2015). The liposomes are usually classified based on their size and number of phospholipid bilayers. Based on phospholipid bilayer membrane, they are categorized as unilamellar and multilamellar vesicles (Figure 1a

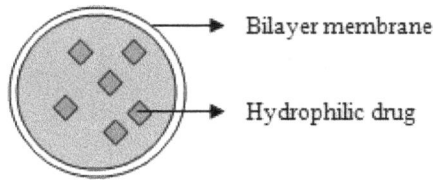

Figure 1a: Unilamellar liposomes (Alavi et al. 2017).

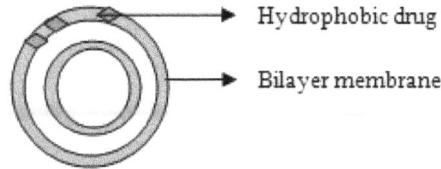

Figure 1b: Multilamellar liposomes (Alavi et al. 2017).

and 1b). The unilamellar vesicles are again classified as small unilamellar vesicles and large unilamellar vesicles. The major components of liposomes are phospholipids and cholesterol (Akbarzadeh et al. 2013, Patil and Jadhev 2014, Pattni et al. 2015). The structures of liposome play a vital role in drug carrier systems as they are used in encapsulating drugs of different size, shape and solubility in water (Luo et al. 2016).

Liposomes mediated cancer treatment follows passive and active targeting strategies. Usually, the pore size of the endothelial cells of tumour microvasculature is loose when compared to the tight structures that are found normally in capillaries. Therefore, in passive targeting strategies, liposome is prepared in such a way that it circulates in cancer cells and does not escape in capillaries, whereas in active targeting, variety of ligands are utilized to any antigens expressed by cancer cells (Zhang et al. 2011, Ravar et al. 2016). The ligands that are used for targeting cancer cells by liposome based drug carrier are Floate receptors (FRs), Epidermal growth factor receptor (EGFR), Vascular endothelial growth factor (VEGF), etc. (Deshpande et al. 2013). The loading of drugs in liposomes is of two types, based on the solubility-passive loading and active loading. The drugs that are hydrophilic are loaded by passive method whereas those which are weakly acidic and alkaline are loaded by active method (Pandey et al. 2016).

Hyaluronic acid–ceramide and egg phosphatidyl choline along with doxorubicin drug were found efficient against the human breast cancer (Park et al. 2014). The conjugated RNA A10 with PLA-block-PEG co polymers were found to successfully target prostate specific membrane antigen and also found to have increased drug delivery to prostate tumour tissue (Byme et al. 2008). In a study by Shmeeda et al. (2010), HER2 targeted PEGylated liposomal doxorubicin showed targeted delivery and greater accumulation of drugs in the tumour cells causing anti-cancer effect on the site. Similarly, Trastuzumab was conjugated to a maleimide PEG polymer to target HER2 in a multivalent fashion and the study identified the downregulation of Akt in cancer cell lines when compared with free or bivalent trastuzumab (Chiu et al. 2007). In this way, liposomal based drug carriers were found to be efficient in cancer therapy.

4.3 Polymeric Nanoparticles

The polymeric drug loaded nanoparticles have been considered as the promising drug carrier system for cancer treatment since they improve drug pharmacokinetics and also have the ability of permeation and retention effect to enhance the accumulation of drugs on the target site (Kapoor et al. 2015). In addition to these, they are also found to increase the half-life of encapsulated drug in plasma, have high loading capacity, efficient delivery of drugs to the intracellular region of targeted

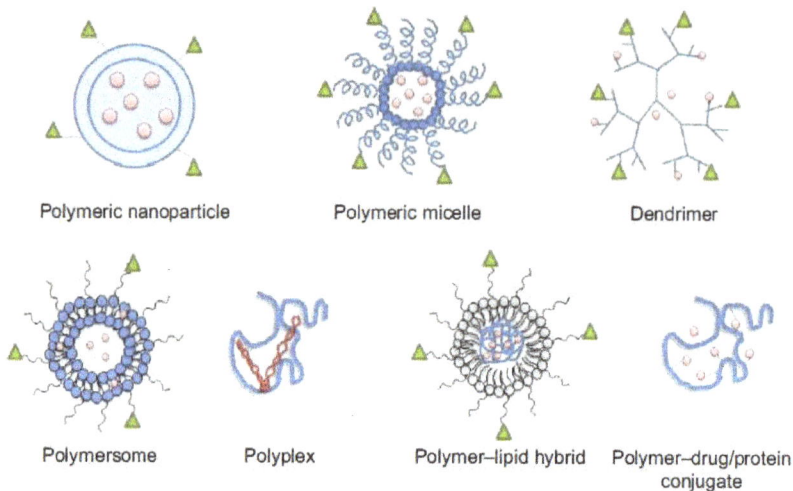

Figure 2: Types of polymeric nanoparticles (Prabhu et al. 2015).

cancer cells and above all, are biocompatible (Sah and Sah 2015). They also are capable of binding with targeting ligands making the delivery of drug on target site (Frohlich 2012).

The therapeutic agent in polymer based drug carrier systems can be either conjugated to the surface of the nanoparticle or encapsulated into the polymeric core. The polymeric nanoparticle platforms (Figure 2) can be categorized into various types as polymeric micelle, polymeric conjugate, dendrimer, polymer, some polyplex and polymer lipid hybrid system (Prabhu et al. 2015). For cancer therapy, polymers such as Poly(lactide-co-glycolide) (PLGA) (Italia et al. 2007), Polylactide (PLA) (Pandey et al. 2015), Polyglycolide, polycaprolactone (PCL) (Devalapally et al. 2008) and chitosan (Rao et al. 2015) were utilized. Several investigations on these polymer based drug carriers have shown promising results in treating cancer cells.

In a study by Amgoth and Dharmapuri (2016), polymeric nanoscale cuboid particles and spherical nanoporous capsules were developed with AB type block co polymer from L–alanine and polycaprolactone. The polymeric nanomaterials of both cuboid and spherical shapes were encapsulated with anticancer drug Imatinib and it has been observed from the study that spherical nanoporous capsules have the ability to kill leukaemia blood cancer cells K562. Ke et al. (2014) showed that the doxorubicin loaded PEG/PAC has higher anti-tumour activity and lower toxicity than the free doxorubicin drug. In this way, several works have been carried out and still performed to better understand the efficacy of polymeric nanoparticles in cancer therapy.

4.4 Dendrimer Nanoparticles

Dendrimers are a type of polymeric nanoparticles but have a classic structure which makes them unique. It consists of a central core and branches arising from them. They are classified into three types—polymers, hyperbranched polymers and brush polymers (Morikawa 2016). The structure of dendrimer is given in Figure 3.

Different types of dendrimers are synthesized based on the core that initiates the polymerization process as Poly(amidoamine) spherical dendrimers (PAMAM) (Esfand and Tomalia 2001), Poly (propylene amine) (PPI) (Kaur et al. 2016) and Poly L lysine (PLL) (Wu et al. 2013). Dendrimers can be synthesized by two routes, namely divergent method and convergent method (Tomalia et al. 1985, Hawker et al. 1990). In addition to these, other methods such as hypercore and branched monomer growth (Wooley et al. 1991), double exponential growth (Kawaguchi et al. 1995), lego chemistry (Maraval et al. 2003) and click chemistry (Wu et al. 2004) are also used. Dendrimers have

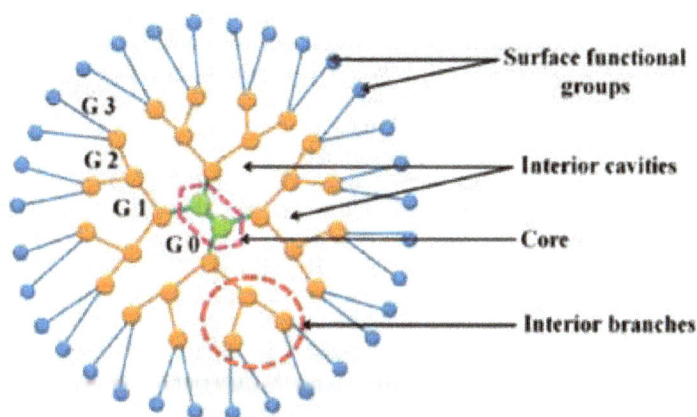

Figure 3: Dendrimers (Tomaalia et al. 2002).

several advantages such as soluble in water, biodegradable, biocompatible, minimal blood-protein binding, suitable carrier for drugs and genes, and controlled and sustainable drug release. In addition to these, they have also been used as co delivery system for simultaneous delivery of therapeutic agents (Abedi-Gaballu et al. 2018).

5. Conclusion

Cancer is still a treatable disease and not a curable disease. Though several treatment options such as radiotherapy, chemotherapy, etc., are available, the ability to kill the targeted cancer cells without affecting the healthy tissues surrounding them is still an unsolved issue. As a great breakthrough in such cases and to treat them efficiently, nanoparticles are the better options. Nanoparticles have been extensively used in various fields such as cosmetics, pharmaceutics, textiles, paints, food, etc. But their utilization in cancer treatment is inevitable and further investigations on them with the biological conjugates and drugs will efficiently improve the cancer treatment methods.

References

Abedi-Gaballu, F., G. Dehghan, M. Ghaffari, R. Yekta, S. Abbaspour-Ravasjani, B. Baradaran, J.E.N. Dolatabadi, and M.R. Hamblin. 2018. PAMAM dendrimers as efficient drug and gene delivery nanosystems for cancer therapy. App. Mater. Today 12: 177–190.

Akbarzadeh, A., R. Rezaei-Sadabady, S. Davaran, S.W. Joo, N. Zarghami, Y. Hanifehpour, M. Samiei, M. Kouhi and K. Nejati-Koshki. 2013. Liposome: Classification, preparation, and applications. Nanoscale Res. Lett. 8: 102.

Alavi, M., N. Karimi and M. Safaei. 2017. Application of various types of liposomes in drug delivery systems. Adv. Pharm. Bull. 7: 3.

Al-Sheddi, E.S., N.N. Farshori, M.M. Al-Oqail, S.M. Al-Massarani, Q. Saquib, R. Wahab, J. Musarrat, A.A. Al-Khedhairy and M.A. Siddiqui. 2018. Anticancer potential of green synthesized silver nanoparticles using extract of Nepeta deflersiana against human cervical cancer cells (HeLA). Bioinorg. Chem. Appl. 2018.

Amgoth, C. and G. Dharmapuri. 2016. Synthesis and characterization of polymeric nanoparticles and capsules as payload for anticancer drugs and nanomedicines. Mater. Today 3: 3833–3837.

Andrews, D.W., C.B. Scott, P.W. Sperduto, A.E. Flanders, L.E. Gaspar, M.C. Schell, M. Werner-Wasik, W. Demas, J. Ryu, J.P. Bahary and L. Souhami. 2004. Whole brain radiation therapy with or without stereotactic radiosurgery boost for patients with one to three brain metastases: Phase III results of the RTOG 9508 randomised trial. Lancet (London, England) 9422: 1665–1672.

Azarpazhooh, E., P. Sharayei, S. Zomorodi and H.S. Ramaswamy. 2019. Physicochemical and phytochemical characterization and storage stability of freeze-dried encapsulated pomegranate peel anthocyanin and *in vitro* evaluation of its antioxidant activity. Food Bioprocess Technology 12: 199–210.

Bajpai, A.K. and J. Choubey. 2006. Design of gelatin nanoparticles as swelling controlled delivery system for chloroquine phosphate. Journal of Materials Science. Materials in Medicine 17: 345–358.

Bao, Y.W., X.W. Hua, J. Zeng and F.G. Wu. 2020. Bacterial template synthesis of multifunctional nanospindles for glutathione detection and enhanced cancer-specific chemo-chemodynamic therapy. Res. 2020: 9301215.

Byrne, J.D., T. Betancourt and L. Brannon-Peppas. 2008. Active targeting schemes for nanoparticle systems in cancer therapeutics. Adv. Drug Delivery Rev. 60: 1615–1626.

Cancer—https://www.who.int/health-topics/cancer#tab=tab_1.

Cancer–Knowledge into action: https://www.who.int/cancer/modules/en/.

Cancer ATSDR https://www.atsdr.cdc.gov /emes/public/docs/Chemicals 20Cancer,%20and %20You% 20FS.pdf.

Cheng, D., X. Yong, T. Zhu, Y. Qiu, J.W.H. Zang, B. Zhu, B. Ma and J. Xie. 2016. Synthesis of protein nanoparticles for drug delivery. European J. BioMed. Res. 2: 8–11.

Chiu, G.N., L.A. Edwards, A.I. Kapanen, M.M. Malinen, W.H. Dragowska, C. Warburton, G.G. Chikh, K.Y. Fang, S. Tan, J. Sy and C. Tucker. 2007. Modulation of cancer cell survival pathways using multivalent liposomal therapeutic antibody constructs. Mol. Cancer Therap. 6: 844–855.

Cooper, G.M. 2000. The Cell: A Molecular Approach. 2nd edition. Sunderland (MA): Sinauer Associates.

Damyanov, C.A., I.K. Maslev, V.S. Pavlov and L. Avramov. 2018. Conventional treatment of cancer realities and problems. Ann Complementary and Comprehensive Medicine 1(1): 1–9.

DeBerardinis, R.J., J.J. Lum, G. Hatzivassiliou and C.B. Thompson, 2008. The biology of cancer: Metabolic reprogramming fuels cell growth and proliferation. Cell Metabolism 7: 11–20.

Deshpande, P.P., S. Biswas and V.P. Torchilin, 2013. Current trends in the use of liposomes for tumor targeting. Nanomed. 8: 1509–1528.

Devalapally, H., Z. Duan, M.V. Seiden and M.M. Amiji. 2008. Modulation of drug resistance in ovarian adenocarcinoma by enhancing intracellular ceramide using tamoxifen-loaded biodegradable polymeric nanoparticles. Clinical Cancer Research 14: 3193–3203.

Diaz, D., A. Care and A. Sunna. 2018. Bioengineering strategies for protein-based nanoparticles. Genes 7: 370.

Elsevier—Cancer Research: Current Trends and Future Perspectives—https://www.elsevier.com/?a=230374.

Esfand, R. and D.A. Tomalia. 2001. Poly (amidoamine)(PAMAM) dendrimers: From biomimicry to drug delivery and biomedical applications. Drug Discovery Today 6: 427–436.

Field, L.D., J.B. Delehanty, Y. Chen and I.L. Medintz. 2015. Peptides for specifically targeting nanoparticles to cellular organelles: Quo vadis? Accounts Chem. Res. 48: 1380–1390.

Fogh, S.E., D.W. Andrews, J. Glass, W. Curran, C. Glass, C. Champ, J.J. Evans, T. Hyslop, E. Pequignot, B. Downes and E. Comber. 2010. Hypo fractionated stereotactic radiation therapy: An effective therapy for recurrent high-grade gliomas. Journal of Clinical Oncology: Official J. American Soc. Clinical Oncology 28: 3048–3053.

Fröhlich, E. 2012. The role of surface charge in cellular uptake and cytotoxicity of medical nanoparticles. Int. J. Nanomed. 7: 5577.

Global fact sheet, WHO – https://www.who.int/cancer/country-profiles/Global_Cancer_Profile_2020.pdf.

Gubernator, J. 2011. Active methods of drug loading into liposomes: Recent strategies for stable drug entrapment and increased *in vivo* activity. Expert Opinion on Drug Delivery 8: 565–580.

Hamouda, R.A., M.H. Hussein, R.A. Abo-Elmagd and S.S. Bawazir. 2019. Synthesis and biological characterization of silver nanoparticles derived from the cyanobacterium Oscillatoria limnetica. Sci. Rep. 9: 13071.

Hassen-Khodja, R. 2004. Gamma knife and linear accelerator stereotactic radiosurgery.

Hawker, C.J. and J.M. Frechet. 1990. Preparation of polymers with controlled molecular architecture. A new convergent approach to dendritic macromolecules. J. American Chem. Soc. 112: 7638–764.

Hemlata, P.R.A.P., Meena Singh and K.K. Tejavath. 2020. Biosynthesis of silver nanoparticles using cucumis prophetarum aqueous leaf extract and their antibacterial and antiproliferative activity against cancer cell lines. ACS Omega 5: 5520–5528.

Herrero-Vanrell, R., A.C. Rincón, M. Alonso, V. Reboto, I.T. Molina-Martinez and J.C. Rodríguez-Cabello. 2005. Self-assembled particles of an elastin-like polymer as vehicles for controlled drug release. Journal of Controlled Release: Official Journal of the Controlled Release Society 102: 113–122.

Hurtado-López, P. and S. Murdan. 2006. Zein microspheres as drug/antigen carriers: A study of their degradation and erosion, in the presence and absence of enzymes. J. Microencapsulation 23: 303–314.

India fact sheet, WHO – who.int/cancer/country-profiles/IND_2020.pdf?ua=1.

Italia, J.L., D.K. Bhatt, V. Bhardwaj, K. Tikoo and M.R. kumar. 2007. PLGA nanoparticles for oral delivery of cyclosporine: Nephrotoxicity and pharmacokinetic studies in comparison to Sandimmune Neoral®. Journal of Controlled Release 119: 197–206.

Jang, S.J., I.J. Yang, C.O. Tettey, K.M. Kim and H.M. Shin. 2016. *In-vitro* anticancer activity of green synthesized silver nanoparticles on MCF-7 human breast cancer cells. Mater. Sci. Eng. C 68: 430–435.

Kalimuthu, K., B.C. Lubin, A. Bazylevich, G. Gellerman, O. Shpilberg, G. Luboshits and M.A. Firer. 2018. Gold nanoparticles stabilize peptide-drug-conjugates for sustained targeted drug delivery to cancer cells. J. Nanobiotechnol. 16: 34.

Kapoor, D.N., A. Bhatia, R. Kaur, R. Sharma, G. Kaur and S. Dhawan. 2015. PLGA: A unique polymer for drug delivery. Therapeutic Delivery 6: 41–58.

Kaur, D., K. Jain, N.K. Mehra, P. Kesharwani and N.K. Jain. 2016. A review on comparative study of PPI and PAMAM dendrimers. J. Nano Res. 18: 146.

Kawaguchi, T., K.L. Walker, C.L. Wilkins and J.S. Moore. 1995. Double exponential dendrimer growth. J. ACS 117: 2159–2165.

Ke, X.Y., V.W.L. Ng, S.J. Gao, Y.W. Tong, J.L. Hedrick and Y.Y. Yang. 2014. Co-delivery of thioridazine and doxorubicin using polymeric micelles for targeting both cancer cells and cancer stem cells. Biomaterials 35: 1096–1108.

Khorrami, S., A. Zarepour and A . Zarrabi. 2019. Green synthesis of silver nanoparticles at low temperature in a fast pace with unique DPPH radical scavenging and selective cytotoxicity against MCF-7 and BT-20 tumor cell lines. Biotechnol. Rep. 24: e00393.

Lee, I. and O. Sokolsky. 2010. Medical cyber physical systems. *In*: 2010 47th ACM/IEEE Design Automation Conference (DAC); IEEE.

Lee, E.J., N.K. Lee and I.S. Kim. 2016. Bioengineered protein-based nanocage for drug delivery. Adv. Drug Delivery Rev. 106: 157–171.

Lohcharoenkal, W., L. Wang, Y.C. Chen and Y. Rojanasakul. 2014. Protein nanoparticles as drug delivery carriers for cancer therapy. Biomed. Res. Int. 2014.

Luo, D., K.A. Carter, A. Razi, J. Geng, S. Shao, D. Giraldo, U. Sunar, J. Ortega and J.F. Lovell. 2016. Doxorubicin encapsulated in stealth liposomes conferred with light-triggered drug release. Biomaterials 75: 193–202.

Maraval, V., J. Pyzowski, A.M. Caminade and J.P. Majoral. 2003. "Lego" chemistry for the straightforward synthesis of dendrimers. J. Org. Chem. 68: 6043–6046.

Moein Moghimi, S. 2006. Recent developments in polymeric nanoparticle engineering and their applications in experimental and clinical oncology. Anti-Cancer Agents in Medicinal Chemistry (Formerly Current Medicinal Chemistry-Anti-Cancer Agents) 6: 553–561.

Morikawa, A. 2016. Comparison of properties among dendritic and hyperbranched poly (ether ether ketone)s and linear poly (ether ketone)s. Molecules 21: 219.

MR, K.P. and P.R. Iyer. 2020. Antiproliferative effects on tumor cells of the synthesized gold nanoparticles against Hep2 liver cancer cell line. Egypt. Liver J. 10: 1–12.

NIH—https://www.cancer.gov/about-cancer/treatment/types.

Oh, J.Y., H.S. Kim, L. Palanikumar, E.M. Go, B. Jana, S.A. Park, H.Y. Kim, K. Kim, J.K. Seo, S.K. Kwak and C. Kim. 2018. Cloaking nanoparticles with protein corona shield for targeted drug delivery. Nature Comm. 9: 1–9.

Paciello, A., G. Amalfitano, A. Garziano, F. Urciuolo and P.A. Netti. 2016. Hemoglobin-conjugated gelatin microsphere as a smart oxygen releasing biomaterial. Advanced Healthcare Mater. 5: 2655–2666.

Pandey, H., R. Rani and V. Agarwal. 2016. Liposome and their applications in cancer therapy. Brazilian Arch. Biol. Technol. 59.

Pandey, S.K., D.K. Patel, R. Thakur, D.P. Mishra, P. Maiti and C. Haldar. 2015. Anti-cancer evaluation of quercetin embedded PLA nanoparticles synthesized by emulsified nanoprecipitation. Int. J. Biol. Macromol. 75: 521–529.

Park, J.H., H.J. Cho, H.Y. Yoon, I.S. Yoon, S.H. Ko, J.S. Shim and J.H. Cho. 2014. Hyaluronic acid derivative-coated nanohybrid liposomes for cancer imaging and drug delivery. J. Controlled Release 174: 98–108.

Patil, Y.P. and S. Jadhav. 2014. Novel methods for liposome preparation. Chem. Phys. Lipids 177: 8–18.

Pattni, B.S., V.V. Chupin and V.P. Torchilin. 2015. New developments in liposomal drug delivery. Chem. Rev. 115: 10938–10966.

Posadas, I., S. Monteagudo and V. Cena. 2016. Nanoparticles for brain-specific drug and genetic material delivery, imaging and diagnosis. Nanomed. 11: 833–849.

Prabhu, R.H., V.B. Patravale and M.D. Joshi. 2015. Polymeric nanoparticles for targeted treatment in oncology: Current insights. Int. J. Nanomed. 10: 1001.

Rajeshkumar, S. 2016. Anticancer activity of eco-friendly gold nanoparticles against lung and liver cancer cells. J. Gen. Eng. Biotechnol. 14: 195–202.

Rao, W., H. Wang, J. Han, S. Zhao, J. Dumbleton, P. Agarwal, W. Zhang, G. Zhao and J. Yu. 2015. Chitosan-decorated doxorubicin-encapsulated nanoparticle targets and eliminates tumor reinitiating cancer stem-like cells. ACS Nano 9: 5725–5740.

Ravar, F., E.M. Saadat, P. Gholami, M. Dehghankelishadi, S. Mahdavi Azami and F.A. Dorkoosh. 2016. Hyaluronic acid-coated liposomes for targeted delivery of paclitaxel, *in-vitro* characterization and *in-vivo* evaluation. J. Controlled Release 229: 10–22.

Sah, E. and H. Sah. 2015. Recent trends in preparation of poly (lactide-co-glycolide) nanoparticles by mixing polymeric organic solution with antisolvent. J. Nanomater. 2015.

Saralidze, K.L., H. Koole and M.L. Knetsch. 2010. Polymeric microspheres for medical applications. Materials 3: 3537–3564.

SEER Training Modules, Radiation Therapy. U.S. National Institutes of Health, National Cancer Institute. 17th October, 2022 <https://training.seer.cancer.gov/>.

Sercombe, L., T. Veerati, F. Moheimani, S.Y. Wu, A.K. Sood and S. Hua. 2015. Advances and challenges of liposome assisted drug delivery. Front. Pharm. 6: 286.

Shmeeda, H., Y. Amitay, J. Gorin, D. Tzemach, L. Mak, J. Ogorka, S. Kumar, A. Zhang and A. Gabizon. 2010. Delivery of zoledronic acid encapsulated in folate-targeted liposome results in potent *in vitro* cytotoxic activity on tumor cells. J. Controlled Release 146: 76–83.

Singla, V. 2020. Carcinogens in products: Inadequate protections raise cancer risks. Trends in Cancer 6: 619–622.

Strebhardt, K. and A. Ullrich. 2008. Paul Ehrlich's magic bullet concept: 100 years of progress. Nature Rev. Cancer 8: 473–480.

Teng, Z., Y. Luo and Q. Wang. 2012. Nanoparticles synthesized from soy protein: Preparation, characterization, and application for nutraceutical encapsulation. J. Agri. Food Chem. 60: 2712–2720.

Tomalia, D.A. and J.M. Fréchet. 2002. Discovery of dendrimers and dendritic polymers: A brief historical perspective. J. Poly. Sci. A Poly. Chem. 40: 2719–2728.

Tomalia, D.A., H. Baker, J. Dewald, M. Hall, G. Kallos, S. Martin, J. Roeck, J. Ryder and P. Smith. 1985. A new class of polymers: Starburst-dendritic macromolecules. Polymer J. 17: 117–132.

Verma, D., N. Gulati, S. Kaul, S. Mukherjee and U. Nagaich. 2018. Protein based nanostructures for drug delivery. J. Pharm. 2018.

Voinea, M. and M. Simionescu. 2002. Designing of 'intelligent' liposomes for efficient delivery of drugs. J. Cell Mol. Med. 6: 465–474.

Wang, L., J. Xu, Y. Yan, H. Liu, T. Karunakaran and F. Li. 2019. Green synthesis of gold nanoparticles from Scutellaria barbata and its anticancer activity in pancreatic cancer cell (PANC-1). Artificial Cells Nanomed. Biotechnol. 47: 1617–1627.

Wongpinyochit, T., P. Uhlmann, A.J. Urquhart and F.P. Seib. 2015. PEGylated silk nanoparticles for anticancer drug delivery. Biomacromol. 16: 3712–3722.

Wooley, K.L., C.J. Hawker and J.M.J. Fréchet. 1991. Hyperbranched macromolecules via a novel double-stage convergent growth approach. Journal of the American Chemical Society 113: 4252–4261.

Wu, J., W. Huang and Z. He. 2013. Dendrimers as carriers for siRNA delivery and gene silencing: A review. The Sci. World J. 630654.

Wu, P., A.K. Feldman, A.K. Nugent, C.J. Hawker, A. Scheel, B. Voit, J. Pyun, J. Fréchet, K.B. Sharpless and V.V. Fokin. 2004. Efficiency and fidelity in a click-chemistry route to triazole dendrimers by the copper(i)-catalyzed ligation of azides and alkynes. Angewandte Chemie (International ed. in English) 43: 3928–3932.

Zhang, L., Y. Jiang, Y. Zheng, Y. Zeng, Z. Yang, G. Huang, D. Liu, M. Gao, X. Shen, G. Wu and X. Yan. 2011. Selective killing of Burkitt's lymphoma cells by mBAFF-targeted delivery of PinX1. Leukemia 25: 331–340.

Zhi, X., Y. Wang, P. Li, J. Yuan and J. Shen. 2015. Preparation of keratin/chlorhexidine complex nanoparticles for long-term and dual stimuli-responsive release. RSC Adv. 5: 82334–82341.

WHA, 58th—https://apps.who.int/iris/bitstream/handle/10665/20372/WHA58_22en.pdf?sequence=1&isAllowed=y.

WHA, 70th—https://apps.who.int/iris/bitstream/handle/10665/275676/A70_R12-en.pdf?sequence=1&isAllowed=y.

Part III
Environmental Applications

18

Industrial Wastewater Sources and Treatment Strategies by using Bionanomaterials

K. Vijayalakshmi,[1],* J. Annie Kamala Florence[2] and Prabhakarn Arunachalam[3]

1. Introduction

Water is very essential for all the life forms to exist on this planet. Due to the development of industrialization process, population explosion, rapid urbanization, and energy utilization, the generated waste from domestic and industrial sources has rendered many water bodies unwholesome and hazardous to man and other living resources (Amuda and Ibrahim 2006, Kumari et al. 2019). Human activities create wastewater that can be catastrophic to the environment. Rivers and groundwater tables contaminated by wastewater render the water resource unusable and therefore it is imperative that before the wastewater is released into the environment, it should be treated. Mostly, the chemical substances that are nonbiodegradable were considered as emerging pollutants in wastewater and this will persist in the environment, bioaccumulate through the food web and pose a risk of causing adverse effects not only to human health but also to the environment, microflora, etc. (Larramendy and Soloneski 2015, Kumar et al. 2015a, Sidhu et al. 2019). Recent researches reported that around 34 billion gallons of wastewater was generated everyday and in order to prevent the threat to the environment and the human health, the wastewater pollutants were reduced to less than maximum permissible limits by various treatment procedures. Depending upon the degree and type of wastewater, the nature of the treatment and the engineering scale of the plant can be decided.

Advances in the field of nanotechnology help in resolving challenges in wastewater treatment and assist in providing clean water. In recent decades, treating the industrial wastewater with nanomaterials is the most important and widespread one. Nanomaterials are fabricated with features, such as high aspect ratio, reactivity, tunable pore volume, and electrostatic, hydrophilic, and hydrophobic interactions, which are useful in adsorption, catalysis, sensoring, and optoelectronics (Das et al. 2014).

[1] Department Chemistry, Madras Christian College, East Tambaram, Chennai, Tamil Nadu, India – 632001.
[2] Department of Chemistry, Voorhees College, Vellore, Tamil Nadu, India.
[3] Electrochemistry Science Research Chair (ESRC), Chemistry Department, College of Science, King Saud University, Riyadh 11451, Saudi Arabia.
* Corresponding author: kumarhandins@rediffmail.com

2. Wastewater

Across the globe, the wastewaters are extensively generated on daily basis (Rathour et al. 2019) from domestic and industrial sources, which pose several challenges such as water crisis and environmental deterioration. The water quality is a result of natural phenomena and the act of human beings. Any water that can be contaminated by human use is termed as the wastewater, which mainly comprises of a combination of domestic, industrial, commercial or agricultural activities. A complex matrix of significant concentrations of solids (total solids: 350–1200 mg/L), dissolved and particulate matter (chemical oxygen demand: 250–1000 mg/L), microorganisms (upto 109 number/mL) (Warwick et al. 2013), nutrients, heavy metals, micropollutants, organic and inorganic matter in varying degrees of concentration were found to be the major components present in the wastewater.

The effluent refers to the sewage or liquid waste that is discharged into the waterbodies, either from direct sources or from treatment plants, which are mainly referred to as the wastewater, and this volume of wastewater has increased over two and a half times that the quantity generated two decades ago. The aqueous discard which results from substances having been dissolved or suspended in water typically during the use of water in an industrial manufacturing process or the cleaning activities that take place along with that process is termed as industrial wastewater and depending on the type of industry, the produced industrial wastewater has very variable quality and volume.

The sources of wastewater are mainly classified into two categories—domestic sewage and non-domestic sewage. The wastewaters generated by home dwellings, public restrooms, hotels, restaurants, resorts, schools, places of worship, hospitals and sport stadium were included in the domestic sewage whereas the non-domestic sewage represents the wastewater generated from floods, runoff, cleaning centers and agricultural facilities.

3. Sources of Industrial Wastewater

Generally, the industrial wastewater mainly includes organic matter, inorganics (sodium, potassium, calcium, magnesium, copper, lead, nickel, and zinc), pathogens, and nutrients (most notably nitrogen and phosphorus). The industrial wastewaters are highly biodegradable or not at all and may or may not contain compounds recalcitrant to treatment. The effluents which are discharged from the battery manufacturing industries, electric power plants, food industries, iron and steel industries, mines and quarries, organic chemicals manufacturing, petroleum refining and petrochemicals, pulp and paper industries, textile mills, industrial oil contamination, wood preserving, etc., pollute the environment to a greater extent and these are considered as major sources of industrial wastewater.

Chromium, cobalt, copper, cyanide, iron, lead, manganese, cadmium, mercury, nickel, oil and grease, phenols, ammonium, sulphides, silver and zinc (Battery manufacturing USEPA 2017) were found to be the major pollutants generated at the battery manufacturing plants, textile mills, fossil fuel power stations, petroleum refining, petrochemicals, and paper and pulp industries. In addition, the wastewater streams from the electric power stations include flue-gas desulfurization, fly ash, bottom ash and the contamination of waste streams from iron and steel industries include gasification products such as benzene, naphthalene, anthracene, cyanide, ammonia, phenols, cresols together with a range of more complex organic compounds known collectively as polycyclic aromatic hydrocarbons (PAH) (Wastewater Characteristics EPA 2002). The specific pollutants such as benzene, chloroform, napthalene, phenols, toluene and vinyl chloride were found to be discharged from the organic chemical manufacturers and depending on the types of products manufactured such as bulk organic chemicals, resins, pesticides, plastics, or synthetic fibers, these discharging pollutants will vary widely from plant to plant.

4. Kinds and Impacts of Wastewater Components

4.1 pH

Regardless of whether treatment is physical/chemical or biological, pH has a direct influence on wastewater treatability. pH extremely affects the solubility of metals, salts and organic chemicals. At neutral to slightly alkaline pH-7 to 8, the bacteria and other organisms play an effective role in wastewater treatment. The rate of change of pH depends on the chemical reaction times which are directly associated with the tank volume, amount of mixing and all other aspects of the treatment procedure. The pH adjustment requires residence or contact time during the wastewater treatment for the pH to change as needed.

4.2 Temperature

Temperature of water is a very important critical parameter to monitor for any biological wastewater system and this is because of its effect on aquatic life, reaction rates and chemical reactions. The normal range of temperature in wastewater ranges between 10 degree celsius to 20 degree celsius. The wastewater treatment system efficiency is affected by the extremes in temperature. The release of effluent from wastewater treatment plants can impact water bodies by altering water temperatures.

4.3 Total Solids, Suspended Solids, Filterable Solids, Settleable Solids

The discharge of untreated wastewater into the aquatic environment can lead to the development of sludge deposits and anaerobic conditions. Total solids represent the combination of dissolved solids, suspended solids and settleable solids in water. Dissolved solids consist of calcium, chlorides, nitrate, phosphorous, iron, sulphur and other ion particles that will pass through a filter with pores of around 2 microns in size. Fraction of total solids in any wastewater that can be settled gravitationally can be classified into organic (volatile) and inorganic (fixed) fractions. The small solid particles which remain as colloid in the suspension in water due to the motion of water are referred to as the suspended solids and the particles which are retained by filtering a wastewater sample are also termed as suspended solids.

Clay particles, slit, plankton algae, fine organic debris and other particulate matter represent the suspended solids in wastewater. High concentration of suspended solids can serve as carriers of toxins which readily cling to the suspended particles. Particles which settle out of the wastewater sample during one hour settling test using an Inhoff cone were termed as settleable solids.

4.4 Nutrient Salts (Nitrogen, Phosphorus, Sulphur)

The valuable nutrient salts present in the wastewater were found to be nitrogen, phosphorous and potassium. Wastewater treatment systems that are commonly used typically manage to reduce BOD and nitrogen effectively but often are not as effective at reducing phosphorous to acceptable levels. When these nutrients are available in unwanted excessive amounts, they can fuel rapid plant growth.

On a dry weight basis, about 1 to 6 percent nitrogen is typically found to be present in treated sludges (Metclf and Eddy 1991) and this nitrogen in treated sludge occurs in both organic and plant available inorganic forms. By contrast, sludges typically contain 0.8 to 6.1 percent phosphorous and it is present as organic and inorganic forms in sludges.

The concentration of nutrients in wastewater depends upon the water supply, the quality of the wastewater and type and degree of wastewater treatment. Along with carbon, both nitrogen and phosphorous are essential nutrients for growth. In all the living organisms, the nitrogen is an essential nutrient and this acts as one of the main constituents (Barnes and Bliss 1983, Rahmat et al. 2011). About 10–40 mg of nitrogen and from few mg to 30 mg of phosphorous per liter is present in the conventionally treated municipal wastewaters (Asano et al. 1985). Sulphur present within an anaerobic digestion wastewater system is readily converted to an odourless sulphate.

4.5 Hazardous Substance

Hazardous substances in wastewater include a variety of chemical compounds like heavy metals, pharmaceuticals, perfluorinated compounds, etc. Certain typical problems such as large water volumes and small fluctuating concentration of the substances were encountered when certain amounts of hazardous substances existing in the natural water were analysed. The hazardous waste with a chemical composition or other properties make it capable of causing illness, death, or some other harm to humans and other life forms when mismanaged or released into the environment.

4.6 Corrosion-inducing Substances

The metallic materials used in the construction of an effluent treatment were mainly affected by the corrosion process. The corrosion of metal occurs due to the set of oxidation and reduction reactions. Certain chemicals present in the liquid media, e.g., Cl_2 and SO_4^{2-}, may also be responsible for the other reduction reactions. Linhardt investigated the corrosion of stainless steel pipes in process water distribution system of a leather plant where MIC having manganese oxidising bacteria are responsible for the attack (Linhardt 2005). Ram and his coworkers reported about the fact that in secondary stage mill effluent, certain anions, viz., SO_4^{2-}, PO_4^{3-}, NO_2^- and NO_3^- impart inhibition, whereas Cl-effluent and chlorophenols enhance the corrosion of mill effluent (Ram et al. 2015b).

4.7 Cleaning Agents, Disinfectants, and Lubricants

If the wastewater is excessively diluted with washing water or is highly concentrated such as undiluted blood or milks, certain problems can arise and in such case, the presence of cleaning agents, disinfectants, pesticides or antibiotics can have detrimental impacts on treatment processes. Physical, chemical or biological processes were utilized to remove, inactivate, reduce or destroy pathogenic microorganisms by using cleaning agents and disinfectants. Chlorine is one of the most practical and widely used disinfectants for wastewater which is found to be effective against wide spectrum of pathogenic organisms. Chlorite compounds in solid or liquid form were considered to be the common chlorine containing disinfection products.

The effectiveness of disinfectant can be measured by evaluating the chlorine residue which remains in the wastewater effluent. The chlorine residue can prolong disinfection even after initial treatment. The disinfection can be enhanced by initiating a reaction between free chlorine and ammonia nitrogen by proper mixing.

Permanent, automatic lubrication protects the metal surfaces against corrosion, reduction and wear. The maintainence cost of wastewater and sewage treatment plants has been significantly reduced by the use of lubricators and at the same time, these lubricants also increase the lifespan of the mechanical parts. The environmental performance can be improved by the use of waterbased lubricants replacing oil based lubricants. The change of lubricant enabled a sludge that can be easily filtered and handled.

5. Wastewater Treatment Strategies

The removal of contaminants from the wastewater or sewage was carried out by wastewater treatment process. Toxic, other harmful materials as well as the components that are non biodegradable can reduce the efficiency of many wastewater treatment operations. Wastewater treatment is mainly composed of three major pretreatment strategies. They are primary, secondary and tertiary treatment processes.

5.1 Primary Treatment

The wastewater, with large debris and grit removed, is directed to primary treatment operations. The flow velocity of wastewater is reduced to afford hydraulic detention times of between 2 and 4 hours during primary treatment and In this case Initial separation occurs, with 40 to 50% of the heavier settleable solids forming a raw or primary sludge on the bottom of the settling tanks and lighter materials to float to the tank's surface. This treatment process allows the material suspended in the wastewater to either float to the surface or sink to the bottom and followed by that, the small and large particles present within the wastewater, with a typical volatile solids content of 75%, are collected by sedimentation-filtration process and discharged to other process operations for further treatment by sedimentation and filtration process. Primary treatment alone has been unable to meet many communities' demand for higher water quality over the years and hence in order to meet the requirements, secondary treatment has been introduced.

5.2 Secondary Treatment

Common biodegradable contaminants were reduced to the safe levels by the secondary wastewater treatment process. By making use of the bacteria in sewage, about 83 percent of the organic matter has been removed in secondary stage of treatment. In secondary wastewater treatment through various aerobic biological process such as biofiltration, aeration and oxidation ponds, the biological content of wastewater has been substantially degraded. The additional sediment from the wastewater is removed by making use of biofiltration which uses sand filters, contact filters or trickling filters. By the introduction of air into the wastewater, the oxygen saturation is increased by the aeration process and since this can last for upto 30 hours, it is found to be very effective.

5.3 Tertiary Treatment

The removal of pathogens, inorganic compounds, and substances such as nitrogen and phosphorus from the wastewater was carried out by utilizing the tertiary treatment process. Followed by the secondary treatment, the tertiary treatment of effluent was done by involving a series of additional steps. This treatment process was done mainly to raise the quality of water to ensure it for domestic, industrial and drinking purposes. Total dissolved solids, trace heavy metals, bacteria, viruses and other dissolved contaminants were found to be greatly reduced by the tertiary RO wastewater recycling process along with the proper pretreatment in the primary and secondary stages. Another efficient tertiary treatment option is the reverse osmosis (RO) (Kimura et al. 2004), which mainly removes a variety of pollutants, but it also has certain limitations related to cost and significant water rejects.

6. Industrial Wastewater Treatment

Industrial wastewater treatment represents the processes used for treating wastewater that is produced by industries as an undesirable by-product. One such industrial treatment is the brine treatment which mainly involves the removal of dissolved Na ions from the salt in waste stream. In order to remove other contaminants, such as metals and other ions that give water hardness such as calcium (Ca), additional treatments may be done and this will depend on the origin of certain brine wastewater. Electrical consumption, chemical usage, or physical footprint can be reduced by the optimization of brine treatment systems. Brine treatment technologies might include the membrane filtration processes, such as reverse osmosis, ion exchange processes such as electrodialysis or weak acid cation exchange, or evaporation processes. But due to the potential for fouling caused by hardness salts or organic contaminants, or damage to the reverse osmosis membranes from hydrocarbons, the reverse osmosis process may not be viable for brine treatment.

Highest purity effluent, even distillate-quality was produced from the evaporation process as they enable the highest degree of concentration, as high as solid salt and these evaporation processes are also more tolerant of organics, hydrocarbons, or hardness salts. The reduction of certain solids such as waste product, organic materials and sand is often a goal of industrial wastewater treatment and some common ways to reduce solids include primary sedimentation (clarification), Dissolved Air Flotation or (DAF), belt filtration (microscreening), and drum screening.

The discharge of wastewaters from large-scale industries such as oil refineries, petrochemical plants, chemical plants and natural gas processing plants commonly contain gross amounts of oil and suspended solids. The oil present in industrial waste water may be free light oil, heavy oil, which tends to sink, and emulsified oil, often referred to as soluble oil. API oil-water separator is a device which is designed to separate the oil and suspended solids from their wastewater effluents. A small specific gravity difference exists between suspended solids and the wastewater and due to this the suspended solids settles to the bottom of the separator as a sediment layer, the oil rises to top of the separator and the cleansed wastewater is the middle layer between the oil layer and the solids (Beychok and Milton 1967).

By making use of the trickling filters or activated sludge process, the biodegradable organic material of plant or animal origin from the wastewaters can be treated. Activated sludge is a biochemical process in which the organic pollutants from the sewage and industrial wastewater can be biologically oxidized using air (or oxygen) and microorganisms and as a result, the waste sludge (or floc) containing the oxidized material can be produced. The trickling filter process is among the oldest and most well characterized treatment technologies, which involve adsorption of organic compounds in the wastewater by the microbial slime layer. The slime layer requires the oxygen, the dissolved air which was furnished from the diffusion of wastewater over the media for the biochemical oxidation of the organic compounds releases carbon dioxide gas, water and other oxidized end products.

Certain methods such as distillation, adsorption, ozonation, vitrification, incineration, chemical immobilisation or landfill disposal were utilized to remove the generated waste like the synthetic organic materials including solvents, paints, pharmaceuticals, pesticides, and products from coke production. Under controlled conditions, the acids and alkalis can usually be neutralised. Within the wastewater, the dissolved organics can be *incinerated* by the advanced oxidation process. In recent years, certain smart capsules provide a possible means for the remediation of contaminated water. There are three types of capsule under investigation: alginate-based capsules, carbon nanotubes, and polymer swelling capsules.

Generally, the water treatment methods had been very limited but in recent years, with ever expanding human knowledge and advancing technologies, various environment friendly and effective treatment technologies were utilized to treat wastewater (Ali et al. 2011, Saleh et al. 2012, Gupta and Nayak 2012). Utilization of organic or, in some instances, inorganic substances, for food were included in the biological methods which rely upon mainly the living organisms. Also, major portions of contamination were removed by the biological treatment method which serves as secondary treatment stage. The amount of organic compounds in wastewater is determined by both BOD and COD.

Without necessarily changing their chemical structures, certain physical processes such as sedimentation, floatation, filtering, stripping, ion exchange, adsorption and other processes were used to remove dissolved and non dissolved substances from wastewater. The formation of insoluble gas followed by stripping, chemical precipitation, chemical oxidation and reduction or other chemical reaction that involve exchanging or sharing of electrons between the atoms were included as chemical methods utilized for wastewater treatment process. Nanobiomaterials have strong adsorption capacities and reactivity due to their small sizes, and large specific surface areas and hence, in particular, the application of nanobiomaterials in water and wastewater treatment has drawn wide attention in recent decades (Lu et al. 2016). Many reports available in literature showed that the nanomaterials served a leading role in order to remove different pollutants from waste waters (Zhang et al. 2019). Some of the nanomaterials utilized as adsorbents were the doped metal oxides, doped carbon nanotubes (single and double walled), nanosorbents, etc.

6.1 Coagulation and Flocculation

Coagulation and flocculation are usually followed by sedimentation, filtration and disinfection, in the primary stage, succeeded by chlorination. Before it is finally distributed to the consumers, this method is used worldwide for water treatment and in order to make the water as fit for use by the consumers, various types of coagulants such as inorganic coagulants, synthetic polymers and biological coagulants were used in typical water treatment processes (Binti Saharudin and Nithyanandam 2013).

Charge neutralization, adsorption, precipitative coagulation, bridge formation (related to the high molecular weight of biomolecules) and electrostatic patch were cited as the mechanisms of coagulation/flocculation which involve the removal of dissolved and particulate contaminants using naturally occurring coagulants (Yin 2010). Muruganandha and his coworkers studied about the efficacy of organic coagulants such as aloevera, moringa oleifera and cactus (*O. ficus-indica*) in treating the industrial effluent. Results revealed that among the various organic coagulants, the *M. Oleifera* seeds gave the highest removal of turbidity and COD for the given samples of waste water (Muruganandham et al. 2017).

Coagulation of the wastewater with a natural coagulant such as by-product chitosan has been carried out by Nguyen Van Nhi Tran and his coworkers. Results reveal that the coagulation by chitosan removed up to 99.4% of turbidity at pH = 10.6 with a chitosan dose of 86.4 mg·L^{-1} (Tran et al. 2020). A popular option for upgrading existing treatment facility to incorporate removal of selected micro pollutants is the enhanced coagulation process which offers several advantages especially in terms of cost effectiveness, simplicity in design, and ease of operation. More than 95% percent removal was attained at 5 mg Fe/L using Fe(VI) oxidation-coagulation in case of pharmaceutical compounds such as diclofenac (Lee et al. 2009).

The performances of bioflocculants in the removal of suspended solids, heavy metals and other organic pollutants from various types of wastewater were commented by Huiru Li and his coworkers. Certain microbial flocculants were produced from gram positive bacteria (*Rhodococcus erythropolis, Nocardia calcarea, Stink bugs bug nocardia, Corynebacterium*, etc.), gram negative bacteria (*Alcaligenes latus*, etc.), and other microorganisms (*Agrobacterium, Dematium, Pseudo single cell fungus, Eritrea's bacillus genus, acinetobacter, Soy sauce aspergillus, Paecilomyces*, etc.) (Porwal et al. 2015, Aljuboori et al. 2014). The observed results show that among them, the flocculant produced by *Paecilomycin* has good flocculation effect on food wastewater, coal slurry wastewater, and textile wastewater and the flocculant produced by *Rhodococcus erythropolis* has good flocculation effect on fine coal wastewater, papermaking wastewater, and activated sludge.

Nkosinathi Goodman Dlamini and his coworkers reported about the optimization and application of bioflocculant passivated copper nanoparticles in the wastewater treatment. Results showed that the synthesized copper nanoparticles showed an excellent flocculation property at a low concentration of 0.2 mg/mL and they flocculate independent of cation, are thermostable and work at weak acidic, neutral, and alkaline pH (Dlamini et al. 2019).

The application of hematite (α-Fe$_2$O$_3$) nanoparticles in coagulation and flocculation processes of River Nile Rosetta branch surface water has been studied by Ahmed A Almarasy and his coworkers. Percentage turbidity removal as high as 93.8% has been achieved by making use of the hematite nanoparticles as a coagulant. Also in addition, aside from turbidity removal upon using hematite nanoparticles in focculation, by performing the reaction with Fe^{3+} ions capping hematite nanoparticles, other sharp reductions in concentrations of phosphate ion contaminant can occur (Almarasy et al. 2019).

6.2 Membrane Filtration and Reverse Osmosis

Over the past two decades, the use of membranes for water and wastewater treatment has been promoted to a larger extent due to increased water scarcity and severe regulations in industrialized countries. As a result, especially in industries, the membrane separation technologies like microfiltration (MF), ultrafiltration (UF), nanofiltration (NF), and reverse osmosis (RO) have been developed as the main contributors to the resolution of water-related problems (Alireza Zirehpour and Ahmad Rahimpour 2016).

The microfiltration of oily waste water involving a study of flux decline and feed types has been investigated by Sutrisna and her coworkers. Both ultrafiltration and microfiltration have been used to separate oil from water and it was evident from the obtained reports that among them, microfiltration with small pore size has been considered as a promising technique to treat oily waste water (Sutrisna et al. 2019).

One of the promising technologies which is utilized for treating wastewater and as well as for reusing water from wastewater is the nanofiltration technology. More and more attention of usage of nanofiltation membrane for wastewater treatment is mainly due to the reduced energy consumption compared to reverse osmosis (RO) and higher rejection compared to ultrafiltration process (Mulyanti and Susanto 2018). Investigation was done on the effect of photooxidation on nanofiltration membrane fouling during wastewater treatment from the confectionery industry by Anna Marszałek and Ewa Puszczało. Results showed that the membrane contamination is minimized by the use of photooxidation methods as pretreatment. It was found that the use of UV reduced the phenomenon of fouling of nanofiltration membranes and the NF process resulted in a total decrease in absorbance, 99% TOC removal, and 98% color removal and that provided the required quality of treated wastewater which can be reused in industrial applications (Marszałek and Puszczało 2020) .

RO is a form of membrane separation which is highly capable of separating dissolved solids, bacteria, viruses and other selected dissolved substances by using a pressure to force a solution through a membrane that retains the solute and allows the pure solvent to pass to the other side. A novel technique termed as reverse osmosis is developed and field tested to "hide" concentrate in dewatered sludge (digested sludge after biological treatment and dewatering) that is withdrawn from the system as a sludge moisture. The treatment of oil refinery wastewater with reverse osmosis membranes and its separation into product quality water and sludge withdrawn from the system has been evaluated. Results showed that the maximum recovery value is determined that corresponds to the maximum permitted by water regulations concentration value of oil and other impurities in RO product water (Pervov et al. 2018).

Novel nanocomposites, films, hybrids, matrixes and membranes were fabricated via constitutional morphological alterations to owe substantial water flux/permeability and salt rejection sought in desalination. Investigation on the use of a highly stable and electrochemically active membrane made solely of CNTs in wastewater treatment process showed that it finds significant applications in both chemical as well as biological wastewater treatment (James et al. 2016).

6.3 Adsorption

Adsorption is a wastewater purification technique for removing a wide range of compounds from industrial wastewater. In terms of initial cost, flexibility, ease of operation and effectiveness towards wide range of pollutants, adsorption is preferred over other techniques of wastewater treatment (Han et al. 2008). The adsorption effect of adsorbent was evaluated by the important index termed as adsorption capacity. Various low-cost alternative adsorbents from agricultural solid waste, industrial solid waste, agricultural by-products and biomass are used in wastewater treatment.

Study on the enhanced adsorption of methylene blue (MB) dye ion onto the activated carbon (AC) modified by three surfactants in aqueous solution was researched by Yu Khuang and his coworkers. Anionic surfactants—sodium lauryl sulfate (SLS) and sodium dodecyl sulfonate (SDS)—and cationic surfactant—hexadecyl trimethyl ammonium bromide (CTAB)—were used for the modification of AC and the observed results showed that AC modified by anionic surfactant was effective for the adsorption of MB dye in both modeling water and real water (Khuang et al. 2020).

In recent years, researchers revealed that, because of high specific surface area, high porosity and active surface, the nano-adsorbents from organic or inorganic sources have high affinity for adsorption and they are highly effective in isolating wide range of contaminants of varying molecular sizes (Grace et al. 2016). Without releasing any toxic payload, the nanoadsorbents work efficiently and rapidly, and regeneration is another significant characteristic of nano-adsorbents (Pacheco et al. 2006).

In the bygone decade, the global issue of water pollution and treatment has been tackled by the introduction of numerous adsorbents and among them the NPs have emerged as an excellent option. Owing to their extremely high specific surface area and reduced surface imperfections, unique chemical, physical, optical, electrical, magnetic and biological characteristics, the nanoparticles act as excellent adsorbents and these are effective towards a huge range of pollutants (Qu et al. 2013).

Usually nanoadsorbents possesses some unique properties like small size, catalytic potential, high reactivity, large surface area, ease of separation, and large number of active sites for interaction with different contaminants and hence these are selected as ideal adsorbent materials for the removal of inorganic and organic pollutants from wastewater (Ali 2012).

Carbon-based nanoadsorbents such as carbon nanotubes (CNTs) (Yang and Xing 2010), surface oxidized CNTs (Vukovic et al. 2010), metal-based nanoadsorbents such as iron oxide, titanium dioxide, zinc oxide, and alumina, magnetic nanosorbents such as maghemite (γ-Fe_2O_3), hematite (α-Fe_2O_3), and spinel ferrites ($M^{2+}Fe_2O_4$, where M^{2+}: Fe^{2+}, Cd^{2+}, Cu^{2+}, Ni^{2+}, Co^{2+}, Mn^{2+}, Zn^{2+}, Mg^{2+}) were employed for removing the toxic materials from the contaminated wastewater.

In order to enhance furthermore the adsorption capacity, recent researchers proposed the synthesis of hybrid nanocellulose based renewable material in combination with several other nanomaterials as adsorbents to remove the toxic contaminants from wastewater since it has a combination of high surface area with high material strength (Voisin et al. 2017).

Ge and his coworkers synthesized Fe_3O_4 magnetic nanoparticles (Fe_3O_4 MNPs) modified with 3-aminopropyltriethoxysilane (APS) and copolymers of acrylic acid (AA) and crotonic acid (CA) as polymer shells (Fe_3O_4 @APS@AAco-CA MNPs). By the use of this polymer shell, the interparticle aggregation has been prevented and also the dispersion stability of the nanostructures has been improved. Results showed that the polymer modified MNPs successfully removed heavy metal ions, such as Cd^{2+}, Zn^{2+}, Pb^{2+}, and Cu^{2+} from aqueous solution with high maximum adsorption capacity at pH 5.5 and can be reusable in at least four cycles (Ge et al. 2012).

7. Future Perspectives for Research

In order to meet the requirements for health, water and resource protection, specific technologies for wastewater treatment have to be further developed and implemented in the future. The focus of future developments is on the compliance with hygienic requirements, protection against eutrophication and climate protection in terms of minimizing the emission of greenhouse gases. Future research addressing the safety of these systems and their economics and scalability is expected to help overcome current limitations and allow the use of numerous nanobiomaterials along with superior technologies to revolutionize the treatment of wastewater.

8. Conclusion

Production of wastewater from industrialized basins should be managed well and this has received a great deal of scientific, technical, and regulatory attention. Nanotechnology provides the opportunity; unique properties of nanoparticles are ideal candidate for developing rapid water-treatment technology. The development of various researches has been aimed at improving the safety and health of humans and environmental fauna. The engineered materials provide a noninvasive tool for wastewater treatment and the knowledge gaps in the area of risk management and risk assessment of engineered technologies are closed by the coordination of research.

Acknowledgement

The authors are grateful to the authorities of Madras Christian College, East Tambaram, Chennai for the support. Thanks are also to the editor for the given opportunity to review such an innovating field.

References

Ali, I. 2012. New generation adsorbents for water treatment. Chem. Rev. 112: 5073–5091.

Ali, I., T.A. Khan and M. Asim. 2011.Removal of arsenic from water by electro coagulation and electrodialysis techniques. Sep. Purif. Rev. 40: 25–42.

Alireza Zirehpour and Ahmad Rahimpour. 2016. Membranes for wastewater treatment. pp. 159–208. *In*: Visakh, P.M. and Olga Nazarenko (eds.). Nanostructured Polymer Membranes. Scrivener Publishing LLC.

Aljuboori, A.H., Y. Uemura, N.B. Osman and S. Yusup. 2014. Production of a bioflocculant from Aspergillus niger using palm oil mill effluent as carbon source. Bioresour. Technol. 171: 66–70.

Almarasy Ahmed, A., Saleh A. Azim and El-Zeiny M. Ebeid. 2019. The application of hematite (α-Fe2O3) nanoparticles in coagulation and focculation processes of River Nile Rosetta branch surface water. SN Applied Sciences 1: 6.

Amuda, O.S. and A.O. Ibrahim. 2006. Industrial wastewater treatment using natural material as adsorbent. Afr. J. Biotechnol. 5: 1483–1487.

Anna Marszałek and Ewa Puszczało. 2020. Effect of photooxidation on nanofiltration membrane fouling during wastewater treatment from the confectionery industry. Water 12: 793, 1–13.

Asano, T., R.G. Smith and G. Tchobanoglous. 1985. Municipal wastewater: Treatment and reclaimed water characteristics. *In*: Pettygrove, G.S. and T. Asano (eds.). Irrigation with Reclaimed Municipal Wastewater—A Guidance Manual, Lewis Publishers, Inc. Chelsea, Michigan.

Barnes, D. and P.J. Bliss. 1983. Biological Control of Nitrogen in Wastewater Treatment. E. Spon. London and New York.

Battery Manufacturing Effluent Guideline. 2017. Washington, D.C.:.U.S. Environmental Protection Agency (EPA).

Beychok and R. Milton. 1967. Aqueous Wastes from Petroleum and Petrochemical Plants (1st ed.). John Wiley & Sons. 67019834.

Binti Saharudin, N.F.A and R. Nithyanandam. 2013. Wastewater Treatment by using Natural Coagulant. 2nd Eureka 4: 213–217.

Christopher Warwick and Antonio Guerreiro. 2013. Sensing and analysis of soluble phosphates in environmental samples: A review. Biosensors and Bioelectronics 41: 15, 1–11.

Das, R., M.E. Ali, S.B.A. Hamid, S. Ramakrishna and Z.Z. Chowdhury. 2014. Carbon nanotube membranes for water purification: A bright future in water desalination. Desalination 336: 97–109.

Dlamini Nkosinathi Goodman, Albertus Kotze Basson and Viswanadha Srirama Rajasekhar Pullabhotla. 2019. Optimization and application of bioflocculant passivated copper nanoparticles in the wastewater treatment. Int. J. Environ. Res. Public Health 16: 2185.

Ge, F., M.M. Le, H. Ye and B.X. Zhao. 2012. Effective removal of heavy metal ions Cd^{2+}, Zn^{2+}, Pb^{2+}, Cu^{2+} from aqueous solution by polymer-modified magnetic nanoparticles. J. Hazard. Mater. 211-212: 366–372.

Grace, A.N., C. Santhosh, V. Velmurugan, G. Jacob, S.K. Jeong and A. Bhatnagar. 2016. Role of nanomaterials in water treatment applications: A review. Chem. Eng. J. 306: 1116–1137.

Gupta, V.K. and A. Nayak. 2012. Cadmium removal and recovery from aqueous solutions by novel adsorbents prepared from orange peel and Fe2O3 nanoparticles. Chem. Eng. J. 180: 81–90.

Haijiao Lu, Jingkang Wang, Marco Stoller, Ting Wang, Ying Bao and Hongxun Hao. 2016. An overview of nanomaterials for water and wastewater treatment. Advances in Material Science and Engineering 2016: 1–10.

Han, R., D. Ding, Y. Xu, W. Zou, Y. Wang, Y. Li and L. Zou. 2008. Use of rice husk for the adsorption of congo red from aqueous solution in column mode. Bioresour. Technol. 99: 2938–2946.

James, S.A. and Z. Zhou. 2016. Electrochemical carbon nanotube filters for water and wastewater treatment. Nanotechnol. Rev. 5: 41–50.

Khuang Yu, Xiaoping Zhang and Shaoqi Zhou. 2020. Adsorption of methylene blue in water onto activated carbon by surfactant modification. Water 12: 587, 1–19.

Kimura, K., S. Toshima, G. Amy and Y. Watanabe. 2004. Rejection of neutral endocrine disrupting compounds (EDCs) and pharmaceutical active compounds (PhACs) by RO membranes. J. Memb. Sci. 245: 71–78.

Kumar, V., S. Singh, N. Kashyap, S. Singla, P. Bhadrecha and P. Kaur. 2015a. Bioremediation of heavy metals by employing resistant microbial isolates from agricultural soil irrigated with industrial waste water. Orient. J. Chem. 31: 357–361.

Larramendy, M.L. and S. Soloneski. 2015. Emerging pollutants in the environment: Current and further implications. InTechOpen.

Lee, Y., S.G. Zimmermann, A.T. Kieu and U. Von Gunten. 2009. Ferrate (Fe(VI)) application for municipal wastewater treatment: A novel process for simultaneous micropollutant oxidation and phosphate removal. Environ. Sci. Technol. 43: 3831–3838.

Li, R., L. Zhang and P. Wang. 2015. Rational design of nanomaterials for water treatment. Nanoscale 7: 17167–17194.

Metcalf and Eddy, Incorporated. 1991. Wastewater Engineering: Treatment Disposal and Reuse. New York: McGraw-Hill.

Mulyanti, R. and H. Susanto. 2018. Wastewater treatment by nanofiltration membranes. IOP Conf. Series: Earth and Environmental Science 142: 012017.

Muruganandam, L., M.P. Saravana Kumar, Amarjit Jena, Sudiv Gulla and Bhagesh Godhwani. 2017. Treatment of waste water by coagulation and flocculation using biomaterials. IOP Conf. Series: Materials Science and Engineering 263: 032006.

Pacheco, S., M. Medina, F. Valencia and J. Tapia. 2006. Removal of inorganic mercury from polluted water using structured nanoparticles. J. Environ. Eng. 132: 342–349.

Paul Linhardt. 2005. Corrosion at Welds of Stainless Steel Pipes, Influenced by Manganese Oxidizers, NACE Conference Papers (NACE International) Corrosion. 3–7 April, Houston, Texas.

Pervov Alexei, Konstantin Tikhonov and Nikolay Makisha. 2018. Application of reverse osmosis techniques to treat and reuse biologically treated wastewater. IOP Conf. Series: Mater. Sci. Eng. 365: 062026.

Porwal, H.J., A.V. Mane and S.G. Velhal. 2015. Biodegradation of dairy effluent by using microbial isolates obtained from activated sludge. Water Resour. Ind. 9: 1–15.

Priya Kumari, Masood Alam and Weqar Ahmed Siddiqi. 2019. Usage of nanoparticles as adsorbents for waste water treatment: An emerging trend. Sustainable Materials and Technologies 22: e00128.

Qu, X.L., J. Brame, Q. Li and J.J.P. Alvarez. 2013. Nanotechnology for a safe and sustainable water supply: Enabling integrated water treatment and reuse. Acc. Chem. Res. 46: 834–843.

Rahmat, Muhammad Ridwan Fahmi and Azner Abidin. 2011. Characteristic of colour and COD removal of azo dye by advanced oxidation process and biological treatment. International Conference on Biotechnology and Environment Management 18(2011).

Ram, C., C. Sharma and A.K. Singh. 2015b. Corrosion investigations on secondary stage paper mill effluent. Anti-Corros. Meth. Mater. 62: 327–333.

Rohit Rathour, Vidhi Kalola, Jenny Johnson, Kunal Jain, Datta Madamwar and Chirayu Desai. 2019. Treatment of various types of wastewaters using microbial fuel cell systems, microbial electrochemical technology, sustainable platform for fuels, chemicals and remediation. pp. 665–692. *In*: Venkata Mohan, Sunita Varjani and Ashok Pandey (eds.). Biomass, Biofuels and Biochemicals, Elsevier.

Saleh, T.A. and V.K. Gupta. 2012. Column with CNT/magnesium oxide composite for lead (II) removal from water. Environ. Sci. Pollut. Res. 19: 1224–1228.

Sidhu, G.K., S. Singh, V. Kumar, S. Datta, D. Singh and J. Singh. 2019. Toxicity, monitoring and biodegradation of organophosphate pesticides: A review. Crit. Rev. Environ. Sci. Technol. https://doi. org/10.1007/s00128990044.

Sutrisna, P.D., J. Candrawan and W.W. Tangguh. 2019. Microfiltration of oily waste water: A study of flux decline and feed types. IOP Conf. Series: Materials Science and Engineering 543: 012079.

Tran Nguyen Van Nhi, Qiming Jimmy Yu, Tan Phong Nguyen and San-Lang Wang. 2020. Coagulation of chitin production wastewater from shrimp scraps with by-product chitosan and chemical coagulants. Polymers 12: 607, 1–17.

Voisin, H., L. Bergström, P. Liu and A.P. Mathew. 2017. Nanocellulose-based materials for water purification. Nanomaterials 7(3): 57.

Vukovic, G.D., A.D. Marinkovic, M. Colic, M.D. Ristic, R. Aleksic, A.A. PericGrujic and P.S. Uskokovic. 2010. Removal of cadmium from aqueous solutions by oxidized and ethylenediamine-functionalized multi-walled carbon nanotubes. Chem. Eng. J. 157: 238–248.

Wang, Z., L.C. Ciacchi and G. Wei. 2018. Recent advances in nanoporous membranes for water purification: Review. Nanomaterials 8(2): 65.

Wastewater Characterization Development Document for final effluent limitations guidelines and standards for the iron and steel manufacturing point source category (Report). EPA. 2002. pp. 7-1ff.

Yang, K. and B. Xing. 2010. Adsorption of organic compounds by carbon nanomaterials in aqueous phase: Polanyi theory and its application. Chem. Rev. 110: 5989–6008.

Yin, C.Y. 2010. Emerging usage of plant-based coagulants for water and wastewater treatment. Process Biochemistry 45: 1437–1444.

Zhang, Y.H., C.Z. Hu, F. Liu, Y. Yuan, H. Wu and A. Li. 2019. Effects of ionic strength on removal of toxic pollutants from aqueous media with multifarious adsorbents: A review. Sci. Total Environ. 646: 265–279.

19

Applications of Metallic Nanomaterials for the Desalination of Water
Perspectives and Limitations

Anjum Rabab, Heena Tabassum and *Iffat Zareen Ahmad**

1. Introduction

In the modern era, water desalination is increasing day by day to provide pure drinking water which is free from various impurities. Scarcity in water affects one out of three people in every continent of the world (UNICEF and WHO 2006). This situation is getting worse as the need of water is increasing with the population growth, climatic changes and increase in industrial as well as household uses (Shannon 2010). Desalination process is applied on water coming from seas, brackish groundwater, oceans, etc. The word "desalination" basically refers to desalting of water, but at the same time, it also removes chemical and microbial impurities (Darwish 2007). Desalination process aims to remove access salt and minerals from saline water to convert it into fresh drinking water. Various publications and laboratories also have discussed the importance of pure water not only for people's health but also for some energy and security purpose (Shannon et al. 2008, Torcellini et al. 2003).

In the past few decades, nanomaterials have been applied successfully in different fields of medicine, biology, sensing, etc. They are tiny in size, and their properties are different from various conventional materials. Multiple applications of nanomaterials for the treatment of water are widely used because of their high mobility, reactivity, and adsorption capacity. This chapter aims to throw light on various technologies of water desalination, their benefits, and limitations, using nanomaterials. These technologies should be eco-friendly, cheaper, and more reliable. The use of the appropriate technology depends upon the type of water we are feeding and the quality of water required (Yeston et al. 2006).

Various existing water desalination methods: There exists a vast variety of water desalination methods. Here are some of the conventional methods. Basically, there are two main commercial desalination methods: thermal processes and membrane processes (Figure 1).

Department of Bioengineering, Integral University, Dasauli, Kursi Road, Lucknow-226026, Uttar Pradesh, India.
* Corresponding author: iffat@iul.ac.in

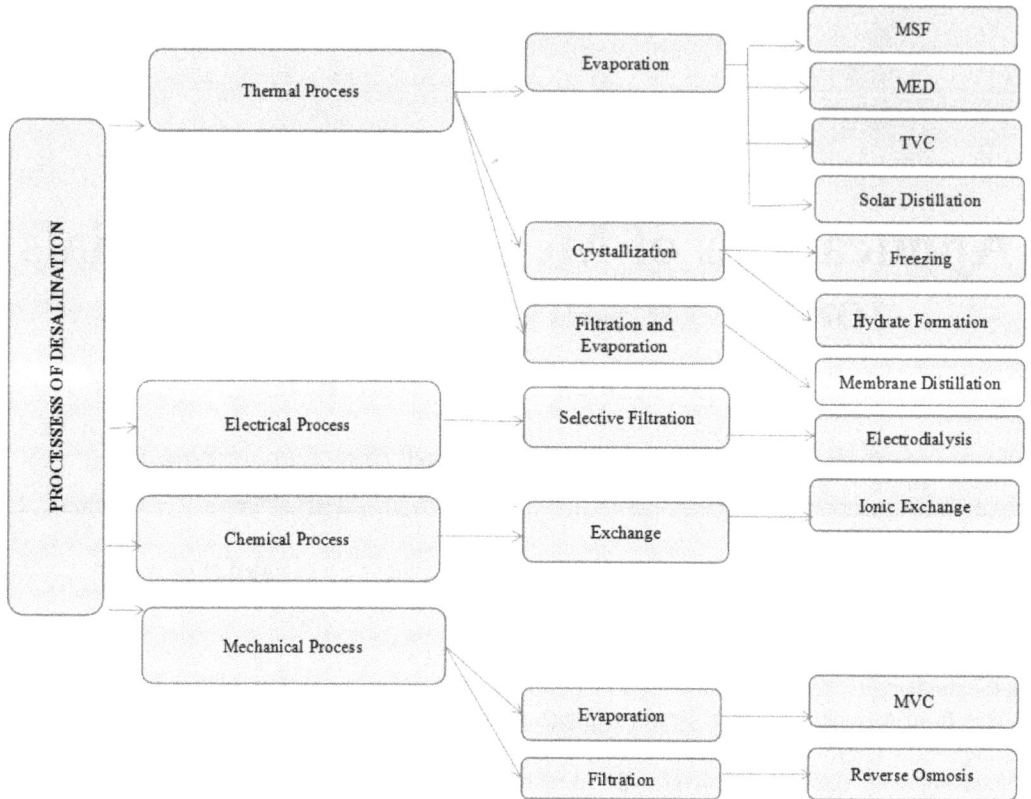

Figure 1: The schematic representation of the desalination processes.

(1) Thermal Desalination Process: This process is based on evaporation and condensation methods. In this process, saline water is boiled or heated until it produces water, and then the water is condensed to produce pure water (Darwish and El Dessouky 1996). But generally, this method is not used for brackish water desalination as the cost of this method is extremely high. This method is used to remove chemical impurities that are present in the water, and if there are some petroleum impurities in water, then all the volatiles cannot be removed due to the use of only one theoretical plate in each stage. Standard thermal desalination processes include (Kim et al. 2010):

(a) Evaporation

- Multi-Stage Flash Distillation (MSF)
- Multiple Effect Distillation (MED)
- Vapour Compression Evaporation (TVC)
- Solar Water Desalination (Jackson et al. 1965)

(b) Crystallization

- Freezing
- Hydrate Formation

(c) Filtration and Evaporation

- Membrane Distillation (Hille 1992, Laksminarayanaiah 1969)

(2) Electrical Process: This process is more straightforward than other methods of desalination and dramatically consumes much less energy than conventional methods. A small electric field is created which effectively removes salts from brackish water. This method overcomes the problem of using a membrane for removal of salt particles from water at a micro scale (Service 2006). The most widely used way in the electrical process is:

- Electrodialysis (Allison 2013)

(3) Chemical Process: Chemicals are used in desalination to prevent membrane fouling and scaling. It includes two main types of chemicals: the first type comprises coagulants, chlorination agents, de-chlorination agents, flocculants and biocides, etc. The second type comprises strong chemicals which effectively remove membrane fouling. The most widely used method of a chemical process is:

- Ion-Exchange (Forgacs et al. 1972)

(4) Mechanical Process: Brackish water is transported into the plant which undergoes a screening process involving an automated filtration process to remove coarse and fine particles. Water is heated, and fresh water is drawn out via an outlet. The main methods involved in the mechanical process are:

- Mechanical Vaporisation Process (MVC)
- Reverse Osmosis (RO) (Fritzmann et al. 2007)

Role of nanotechnology in desalination: Nano is derived from a Greek word, which means "dwarf" (Shannon 2010, Kemery et al. 1998). To develop antibacterial agents that can effectively destroy pathogens from water is a significant issue. Today various types of processes are available in the market for water purification, like boiling, filtration, distillation, sedimentation, and oxidation, etc. In the past few decades, nanotechnology has played a vital role in this field. It provides better water purification techniques using different types of nanoparticles/nanomaterial. Nanomaterials are usually a single unit having a size between 1–100 nanometres (Kemery et al. 1998). Water desalination by nanotechnology uses nanoscopic particles like carbon nanotubes, etc., for filtration (Asatekin et al. 2006). Nanosensors based on titanium oxide and palladium play a vital role in the detection of impurities in water samples. The contaminants which nanotechnology seizes depend upon the types of technique used and the stage of water purification (Spiegler 1962). It is used for the removal of sediments, charged particles, bacteria, and other harmful pathogens. Using this technology, one can trace the toxic elements like oil and arsenic and remove them as well.

One of the main advantages of nanotechnology over conventional methods is that less pressure is used to pass water through the filter (Bhattacharya and Ghosh 2004); this makes this technology more efficient. They also have a large surface area as compared to the conventional method. It can remove negatively charged impurities like viruses, bacteria, inorganic and organic colloids faster. Conventional methods are very time consuming whereas these techniques are simple and easy to use. The other most important advantage of this technology is bacterial biofilm possibility which is built not only on the surface but also on the centre inside the carrier, place where bacterial toxic effects are more protected from the environment. This demonstrates that substrate penetration exists in this technology and oxygen of microbes is also possible.

Positive effects of the use of nanotechnology: Nanotechnology has both negative and positive effects. Some of its positive effects are:

1. Hybrid automobiles, an alternative energy method, decrease the price by novel developments.
2. The use of raw materials will become more efficient. Catalysts increase the chemical reaction.
3. By the use of advanced nanotechnology, environmental monitoring and protection is now more accurate and secure.
4. They have improved ability for the detection and elimination of various pollutions.

5. From the atmosphere, they remove greenhouse gases and other harmful pollutants.

6. They decrease the use of large industrial plants for various purposes.

Negative effects of nanotechnology: Some of the disadvantages of this technology are:

1. They have lower recovery and lower recycling rates.

2. High energy demand for synthesizing nanoparticles.

3. There is lack of trained engineers and experienced workers.

4. To find out the toxicity of the material is a critical factor.

5. To predict the impacts of environment, life cycle risk assessment is a major factor.

6. To characterize the particle, there is a lack of information and methods which makes the existing technology more difficult.

2. Different Types of Metallic Nanomaterials used for Desalination

In this section, we will discuss about various metallic nanoparticles that are used for water purification. They are of immense interest owing to their unusual physical and chemical properties. They show an extraordinary UV-VIS absorption band which is not present in the bulk metal spectra. This band can be attributed to the collective excitation of conduction electrons when the size of metal particles is less than the mean free path of the electron in the metal. This phenomenon is commonly known as Localized Surface Plasmon Resonance (Figure 2). All the nanoparticles have properties like adsorption, high reactivity and catalysis. Here we will discuss zero valent metal nanomaterials, metal oxide nanomaterials and carbon nanotubes and carbon composites. Nanomaterials give us different new ways for treating water impurities without the formation of DBP. They have a wide range of new improved properties due to the increase in surface area.

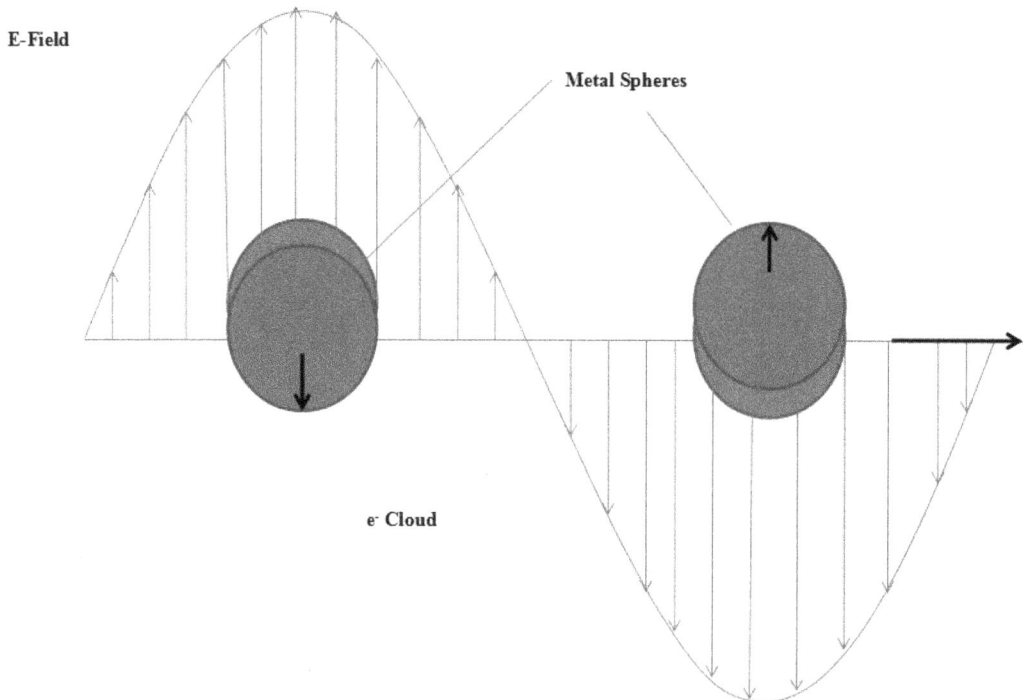

Figure 2: Surface Plasmon absorption phenomenon. The electric field of the incident light's electric field induces a polarization of the free surface electrons with respect to the heavier ionic core. Free surface electron's dipolar coherent oscillation is known as the plasmon ban.

2.1 Silver Nanomaterial

These particles have unique optical, thermal and electrical characteristics. These particles have very strong antibacterial properties against variety of microorganisms like fungi, bacteria, viruses and pathogens because they are highly toxic to microbes. Due to their toxicity, they are widely used for water purification. Because they have electrical conductivity, they are also useful in scientific domain. These particles have the ability to damage the cell membrane and cause the death of the cell. Upon coming in contact with the bacteria, free radicals are generated. Ag particles can react with the sulphur and phosphorous present in the DNA of the cell and can destroy it. When they come in contact with water, their direct application may cause problems due to their aggregation in aqueous media, hence their efficiency decreases when they are used for long term (Veerapaneni et al. 2007). They are cost effective, although their mechanism is still not clear. Emergency water treatment can be done by filtration through whatman filter paper with the silver nanomaterial deposition.

Properties

(a) These particles have extraordinary efficiency to absorb and scatter light.

(b) Due to their optical properties and by analysing spectral properties in solution, a lot of information regarding the physical state of the particle can be obtained.

(c) In a solution, molecules are associated with their surface to form a double layer of charge to stabilize them and to prevent aggregation.

(d) They are used in biosensors and in quantitative detection.

(e) They are used in cosmetics, appliances, dressing material, paints, footwear industries, etc.

(f) For transparent conductors, they provide coatings to conductors.

(g) They are ideal for various technologies like biomedical, antimicrobial and optical applications, etc.

2.1.1 Recent Trends in Desalination using Silver Nanoparticles

Owing to its high affinity towards iodine species, silver nanoparticles (Shim et al. 2018) are immobilized on a cellulose acetate membrane (Ag-CAM) and biogenic silver nanoparticles containing the radiation-resistant bacterium *Deinococcus radiodurans* (Ag-DR) were developed and investigated for desalination in removing radioactive iodine from water. A simple filtration using Ag-CAM (approximately 1.5 mL/s) provided an excellent removal efficiency (> 99%) as well as iodide anion-selectivity. The results were visualized by using single-photon emission computed tomography (SPECT) scanning.

In a recent study, using a facile hydrothermal reaction an antibacterial silver/reduced graphene oxide (Ag/rGO) hydrogel consisting of controlled porous rGO network and well-dispersed Ag nanoparticle was synthesized. Scanning electron microscopy, transmission electron microscope, X-ray diffraction, mercury porosimetry, and Fourier transform IR spectroscopy were used for characterisation. Ag nanoparticles find good structural support due to the 3D structure of rGO network. Experiments showed that the Ag/rGO hydrogel exhibits good efficacy against *Escherichia coli*. The study indicated that bacteria cells are inactivated due to damage of cell membrane-induced by silver nanoparticles and rGO nanosheets when they flow through Ag/rGO hydrogel.

Biofouling is known to cause major problem in the application of membrane technology in water and wastewater treatment. In a recent study, different amounts of biogenic silver nanoparticles (bio-Ag0) were embedded in polyethersulfone (PES) membranes, via phase-inversion method. The effects of the bio-Ag0 content on the structure of the membrane and its filtration performance were studied. It was concluded that silver-containing nanostructures were uniformly distributed on membrane surface. Bio-Ag0 incorporation slightly increased the hydrophilicity of the PES membrane and increased the permeate flux. The antibacterial and anti-biofouling properties were tested with pure cultures of *Escherichia coli* and *Pseudomonas aeruginosa*. The bio-Ag0/PES composite membranes, even with the lowest concentration of biogenic silver (140 mg bio-Ag0 m−2), exhibited both excellent antibacterial activity, and prevention of bacterial attachment to the membrane surface.

2.2 Iron Particles

Iron particles have many key properties including adsorption, precipitation and oxidation in the presence of oxygen. They are low in cost. It works as an excellent reducing agent as compared to many redox labile impurities. In anaerobic conditions, it can be oxidized by water and hence generates Fe^{2+} and H_2, both of which are forceful reducing agents for contaminants (U.S. Department of Energy 2006). They are successfully applied for the removal of various impurities like dyes, phenols, metals, phosphates, nitrates, radio elements, etc. They have also been used for soil remediation. From the degradation system, they also have aggregation, oxidation, and separation difficulty. Doping with other metals can enhance their performance. Aggregation can be prevented by surface coating and conjugation process (Zelmanov and Semiat 2015).

From the past few years, iron oxides are extensively used for the removal of heavy metals because of their simplicity and availability. The size of iron oxide is very tiny; therefore, their separation and recovery from polluted water is a significant concern. Iron oxides can be used as a sorbent for the removal of heavy metals in the water. Their adsorption capacity is much higher than any other nanoparticle (Qiao and Aluru 2004).

Properties

(a) They are also useful for treating ground impurities containing pesticides, chlorinated compounds, etc.

(b) Particles made of ferro/ferromagnetic material may exhibit different form of magnetism known as super-paramagnetism.

(c) One of the largest applications of these particles is to use them as magnetic recording media.

(d) Various magnetic tapes and camcorder have high capacity with the help of these nanoparticles.

(e) Recent methods for producing iron nanoparticles completely rely on dispersing them into mercury.

(f) Iron oxide particles have wide applications because of their magnetic properties.

2.2.1 Recent Trend in Desalination using Iron Particles

In a recent study, phosphate removal from water solution using iron (Fe^{+3}) oxide/hydroxide nanoparticles-based agglomerates (Ag-Fe) suspension was investigated for efficiency and cost-effectiveness. The influence of inorganic ions (HCO^{-3}, Cl$^-$, and SO_4^{-2}) on the adsorption process was reported. The adsorption properties of the adsorbent were investigated in solutions containing different inorganic ions concentrations at different pH levels. A strong concentration effect of the used AggFe, HCO_3 and pH level of the purified water solution on removal efficiency was shown. This technique achieved a residual phosphate concentration of less than 0.02 mg/l as P, which is acceptable by water quality regulations, and at least 95–98% regeneration efficiency of the phosphate with the proposed adsorbent (Zelmanov and Semiat 2011).

Thermoresponsive magnetic nanoparticles (MNPs) may be used as a draw solute to extract pure water from brackish or seawater via forward osmosis (FO). A distinct advantage is the efficient regeneration of the draw solute and the recovery of water via heat-facilitated magnetic separation. However, the osmotic pressure attained by this type of draw solution is too low to counteract that of seawater (Gong et al. 2007, Qiao and Aluru 2004). In this study, FO draw solution based on multifunctional Fe_3O_4 nanoparticles grafted with copolymer poly (sodium styrene-4-sulfonate)-co-poly (N-isopropylacrylamide) (PSSS-PNIPAM) was being designed. The resulting regenerable draw solution showed high osmotic pressure for seawater desalination. This was enabled by three essential functional components being integrated within the nanostructure (Kemery et al. 1998): (i) a Fe_3O_4 core which allows magnetic separation of the nanoparticles from the solvent, (ii) a thermoresponsive polymer, PNIPAM, which enables reversible clustering of particles for improved magnetic capturing at a temperature above its low critical solution temperature (LCST), and (iii) a polyelectrolyte, PSSS, which provides an osmotic pressure that is well above that of seawater (Zhao et al. 2013).

2.3 Zinc Nanoparticles

The impurities degradation rate of zinc is faster than iron as it is a stronger reluctant. It is mainly used in sunscreens and paint industries as it can absorb UV lights, and plays an important role in rubber, pharmacy and food-related industries. Zinc oxide particles are good antibacterial agents as they are stable under harsh processing situations. Its antimicrobial activity causes the release of oxygen from its surface which can cause damage to microorganisms. They are able to damage cell membrane and can penetrate intracellular contents. Its decreased size increases its antimicrobial property. These nanoparticles are basically an inorganic compound which is white in colour and are soluble in water (Ben-Sasson et al. 2016).

The municipal and industrial water generally contains oils, proteins, carbohydrates, pollutants, pathogens and other impurities (Qiao et al. 2006). These impurities can be purified in three steps. Their nanostructure forms multifunctional membrane which is very efficient in removing water impurities by developing photodegradation activity under light. They absorb water and after some time, the pure water can be collected in the bottom of the container.

Properties

(a) These nanoparticles' biomedical uses stem from their anticancer properties, drug delivery, and bioimaging activity.

(b) These nanoparticles are listed as a safe substance.

(c) They have high surface area and high activity to block various pathogens.

(d) These nanoparticles are also used in the treatment of cancer.

(e) These are low cost and low toxic nanoparticles that are strongly used in the biomedical research.

(f) Zinc is commonly found in all parts of the body. They play an important role in body metabolism; hence, these nanoparticles can be easily absorbed by body.

2.3.1 Trends in Desalination using Zinc Nanoparticles

Zinc oxide (ZnO) nanoparticles are extensively used for the synthesis of various materials. The nanoparticles, when they enter into the environment, drastically affect aquatic life. Their antibacterial properties deter the biological treatment process of wastewater treatment plants. In a recent study, the effectiveness of ultrafiltration (UF) membranes (Asatekin et al. 2007) for the removal of ZnO nanoparticles was studied. ZnO nanoparticles were separated in the presence of natural organic matter (NOM) and alkalinity using a commercial membrane. Membrane flux and retention were studied for different concentrations of ZnO (1 mg/L, 10 mg/L, and 100 mg/L). Membranes were studied using a scanning electron microscope (SEM), energy-dispersive X-ray (EDX), and atomic force microscopy (AFM). At higher concentrations (> 10 mg/L), ZnO nanoparticles tend to aggregate and increase in size, resulting in 95 to 98% retention (Mehta et al. 2016).

In another study, zinc oxide (ZnO) grafted electrodes, and micro/nanomaterials (nanoparticles, nanorods, micro sheets and microspheres) on activated carbon cloth (ACC) for water desalination by capacitive deionization (CDI) were fabricated. ZnO micro/nanomaterials were grown hydrothermally on ACC and were used as electrodes in a flow cell for brackish water desalination. Structures of ZnO on ACC surfaces were found to effectively remove salt from salinated water. Experiments of desalination were carried out using 100 ppm sodium chloride (NaCl) solution employing a flow rate of 2 ml/min in a parallel plate configuration with an electrode area of ~ 8.4 cm² under an applied potential of 1.2 V (DC). Enhanced salt removal efficiency of 22% for ZnO micro sheet grafted ACC electrodes as well as ZnO nanorod grafted ACC electrodes was achieved (Myint et al. 2014).

2.4 Titanium Dioxide Nanoparticle

To remove impurities from water, photocatalytic degradation technology has been applied since the last few years (Shahid et al. 2015). This technology is based on light and catalyst, i.e., in the presence of light and catalyst, water impurities are oxidized successfully into low molecular weight interlude product and it is transformed into carbon dioxide, water and other anions. In this technology, most of the photocatalysts are metal oxides, whereby the use of titanium dioxide has significantly increased during the past few decades. Because of its sublimate photocatalytic activity, cheaper prices, and chemical/biological stability, it is one of the most remarkable photo catalysts till date (Choi et al. 2007). These nanoparticles are suitable for the removal of different types of impurities including chlorine, organic compounds, cyanide, pesticides, phenol, hydrocarbons, phenol, heavy metals, etc. Although its production process is very complicated, to recover this nanoparticle from treated water is very difficult, particularly when it is used in suspension (Yacou et al. 2015).

The band structure of TiO_2 is composed of valence band and conduction band and there is a band gap between these two bands. Study shows that more than 30k refractory compounds can be removed by titanium dioxide under ultraviolet irradiation. It has a large specific surface area and high UV absorption capacity. Photolytic method plays a very major role in water desalination because it can remove almost all type of organic impurities from water. The water produced by this method is fully decontaminated and does not contain any secondary pollutants (Isha et al. 2015, Al Mayyahi et al. 2018).

Properties

(a) Titanium oxide has a high band gap and is non-toxic in nature.
(b) They are chemically stable, having catalytic effect and are low in price.
(c) They are used in lead, chromium, mercury-containing, nitrogen, oily impure water treatment.
(d) This nanoparticle is an active ingredient in sunscreen.
(e) Its production process is continuous and is suitable for industrial applications.

2.4.1 Recent Trends in Desalination using TiO_2 Nanoparticles

Photocatalysis is a suitable technique for the desalination and purification of water. Thin layers of TiO_2 deposited on three various substrates are being prepared by a sol-gel process along with the employment of a dip-coating technique for application. Prepared layers were characterized by XRD, SEM, AFM, UV–vis, and Raman spectroscopy. Photo-electrochemical properties were determined by exploiting amperometry and linear voltammetry and to obtain prepared TiO_2 photocatalyst's photoinduced properties. Photocatalytic decomposition efficiency of individual compounds was evaluated. Furthermore, individual compounds' resistance to the photocatalytic decomposition process was evaluated together with the possibility of formation of intermediates (Kumar et al. 2017).

In another study, TiO_2 nanoparticles (Shahid et al. 2015) were inserted into the polyamide layer of thin film composite membrane. Before interfacial polymerization with m-phenylenediamine-aqueous solution, the nanoparticles were dispersed in a trimesoyl chloride-hexane solution. In another study, TiO_2 nanoparticles were inserted into the polyamide layer of thin film composite membrane. Before interfacial polymerization with m-phenylenediamine-aqueous solution, the nanoparticles were dispersed in a trimesoyl chloride-hexane solution. AFM, SEM, water flux, salt rejection, and fouling resistance evaluation was done. The results indicated that TiO_2 has the potential of improving the membrane performance (Hille 1992). Water flux exhibited an increment from 40 to 65 L/m² h when the concentration of NPs was increased from 0 to 0.1 wt. %, while NaCl rejection was found to be > 96%. The modified membrane showed higher resistance against organic fouling and robust antibacterial efficiency (Al Mayyahi 2018).

Recently, magnetised TiO_2 has been produced to offer a solution for the photocatalyst separation problem. Magnetised TiO_2 photocatalysts have been extensively studied but the lack of articles discussing the water purification processes is still slowing any advance in this field. In general, the effectiveness of organic removal by magnetised TiO_2 is lower as compared to single phase TiO_2. The future prospect of this field is emphasized in order to develop a novel, cheap and efficient magnetised TiO_2 photocatalyst, which promises high organic removal properties.

In another study, TiO_2 membranes dip-coated on commercial α-Al2O3 tubes were used for water desalination. TiO_2 thin films (~ 400 nm) were synthesised from titanium isopropoxide and were exposed to a gentle thermal treatment. The TiO_2 membranes were evaluated in a perforation setup for saline, seawater and brine concentrations at a temperature of 75°C. Water fluxes for brackish water were found to be as high as 10.5 kg m^{-2} h^{-1}, values of up to 4.0 and 6.0 kg m^{-2} h^{-1} were found for the highest brine concentration. Salt rejection was very high (> 99%) for all conditions, thereby demonstrating the feasibility of mesoporous inorganic TiO_2 structures for desalination (Yacou et al. 2015).

In another study, using a simple sol-gel route, employing self-assembled surfactant molecules as pore directing agents along with acetic acid-based sol-gel route, TiO_2 thin films and TiO_2/Al_2O_3 composite membranes with simultaneous photocatalytic, disinfection, separation, and anti-biofouling properties have been fabricated. The highly porous TiO_2 nanomaterial exhibited high specific surface area and porosity, narrow pore size distribution, homogeneity without cracks and pinholes, active anatase crystal phase, and small size. These TiO_2 materials were highly efficient in water purification (Choi et al. 2007).

In a recent study, hybrid TiO_2 catalysts, which is TiO_2 and biomass ash mixture, was synthesized via wet impregnation and calcined at a temperature of 800°C. Scanning Electron Microscopy (SEM) and Braunauer-Emmett-Teller (BET) was used to characterize the photocatalyst. A photoreactor was used for investigation, which was equipped with UV light and operated for up to six hours. The catalyst to seawater weight ratio of 1:400 stirring speed of 600 rpm was maintained. The salt concentration, turbidity and pH of the water were determined later on. The result indicated that the salt concentration gradually decreased for more than 25% as the contact time was increased up to six hours (Isha et al. 2015).

In another study, the conversation of wastewater into pure water by desalination effect was evaluated. Solar energy was used as the primary source of energy in this method. Three types of investigations were carried out for the setup: (1) wastewater was poured into the chamber, (2) black stones were placed over the bottom of the plate, (3) paraffin wax used as a phase change material. A mixture of titanium oxide and paraffin wax was poured into a copper tube and was placed over the surface plate. Solar energy stored by the paraffin wax liberates its stored heat. The absorbed heat energy cannot escape as the double glass solar was fully insulated by Polyurethane (PU) foam. The brackish water is being converted into pure, potable drinking water by utilising solar energy (Pandey et al. 2021).

2.5 Carbon Nanotubes

Carbon nanotubes belong to the class of nanomaterial having 2D hexagonal carbon lattice, joined in a direction forming a hollow cylinder. These have great capacity to absorb variety of impurities, and fast kinetics. These nanoparticles have large specific surface area and are selective towards aromatics. They are mostly combined with other metals to mechanical and electrical properties as well as adsorption. In some carbon tubes, there is a zigzag path that goes all around the path. By combining the two allotropes of carbon, i.e., carbon and fullerene, a new material carbon nanobuds can be created (Gethard et al. 2010).

They can be used as a gatekeeper for the removal of multiple pollutants from water. Basically, they can divided into two categories: (a) single-walled carbon nanotubes, (b) multiple-walled carbon nanotubes. Both these nanotubes can be used directly and indirectly for water desalination. These nanotube membranes are reusable, scalable and less complex. They are eco-friendly and simple to use (Holt et al. 2006, Majumder et al. 2005).

Properties

(a) The strength of these materials is very high.
(b) They have very useful absorption and Raman spectroscopic properties.
(c) They are used in the designing of number of bicycle components.
(d) Performance depends upon fabrication and processing methods.
(e) They often decrease water flux and membrane water rejection capacity.
(f) Carbon nanotubes have the ability to replace RO and UF membranes without any consumption of energy.

2.5.1 Recent Trends in Desalination using Carbon Nanotubes

Traditional desalination methods (Glueckanf 1966) are operationally intensive and less energy efficient whereas adsorption-based techniques are easy and simple to use for water purification, but their capacity for salt removal is limited. In a recent study, plasma-modified ultralong carbon nanotubes exhibited ultrahigh specific adsorption capacity for salt (exceeding 400% by weight), which is higher than that found in the current activated carbon-based water treatment systems. The adsorption capacity in ultra-long carbon nanotube-based membranes is exploited to remove salt, as well as organic and metal contaminants (Yang et al. 2013). These ultra-long carbon nanotube-based membranes give a promising potential for next-generation rechargeable, potable water purification systems with superior properties of desalination, disinfection and filtration properties.

In another study, multi-wall carbon nanotubes (Pu et al. 2004) and silica nanocomposite (Qiao and Aluru 2005b) (CNT/SiO_2) were being synthesised. It was observed by scanning electron microscopy, energy dispersive X-ray spectroscopy, Fourier transform infrared spectroscopy and high-resolution transmission electron microscopy, etc. The resulting $MWCNT/SiO_2$ gave promising adsorption performance (~ 95%) over silica nanoparticles (Qiao and Aluru 2005a) (~ 50%) and CNTs (~ 45%). The work also highlighted the recyclability of the nanocomposite with high efficiencies and supported its potential for environmental applications. It is anticipated that the results bear marvellous potential in the sorption domain and can be effectively used for purification of water (Saleh et al. 2016).

A novel approach was used for the synthesis of a silver doped-CNT membrane, with the sole objective of utilizing the anti-toxic properties of CNTs and silver. Binders were not used for the synthesis of this membrane; instead, silver particles themselves served as a binding material for the CNTs when sintered at high temperature. Firstly, CNTs were impregnated with different loadings of silver (1, 10 and 20 wt.%) via a wet chemistry technique. Impregnated CNTs were then subjected to a pressure of 200 MPa and sintered at 800°C for 3 h in order to form a compact disk (membrane)

(Shukla et al. 2019). The materials were characterized by scanning electron microscopy, energy-dispersive X-ray spectroscopy, X-ray diffraction and thermogravimetric analysis, the membranes were characterized by measuring the porosity, contact angle, diametrical compression test, pure water permeate flux and antibacterial properties. The biofouling membrane affinity was studied using *Escherichia coli* (*E. coli*). The produced membrane exhibited a high water permeate flux and also showed strong antibacterial properties. These membranes would be highly advantageous when used in a continuous filtration system for the effective removal of different water contaminants via desalination, adsorption and sieving.

It was demonstrated in another study that when CNTs were immobilised in the pores of a hydrophobic membrane, it positively altered the water-membrane interaction to promote water permeability across the membrane while preventing the permeation of liquid into the pores of the immobilised membrane. The nanotube incorporation led to an increment of 1.85 and 15 times in flux and salt reduction for a salt concentration of $34\,000$ mg L^{-1} and at 80°C, respectively.

2.6 Copper Nanoparticles

A block D element of the periodic table, copper is a ductile material having high thermal and electrical conductivity. Its nanoparticles are round and they appear as brown or black powder. It is too soft to be used in several applications and hence it has to amalgamate with other metals to form various alloys like that of brass (copper-zinc alloy). Copper nanoparticles are highly flammable and are known to be toxic for aquatic organisms (Ben-Sasson et al. 2016).

Properties

(a) It acts as an antibiotic, anti-microbial and anti-fungal agent.
(b) It is used as a diet supplement.
(c) It is used to produce high strength metals and alloys.
(d) It is used as an efficient catalyst in several chemical reactions and for the synthesis of methanol and glycol.
(e) It is used as sintering agent and in capacitor materials.
(f) It can also be used in conductive inks and pastes in place of expensive and expensive noble metals in different printed electronics, displays, etc.
(g) It is used as a lubricant additive for various nanometals.

2.6.1 Recent Trends in Desalination using Copper Nanoparticles

In recent years, copper based nanoparticles have been used as a potential anti-fouling agent in various membrane processes owing to their strong antibacterial activity and low cost. Biocidal nanoparticles are coated on the surface of a thin-film composite RO membrane. It is being characterised using SEM imaging and X-Ray photoelectron spectroscopy. It exhibited surface properties similar to that of pristine membrane. This study demonstrated the reduction in biofouling along with antibacterial activity against *E. Coli*.

In another study, copper nanoparticles exhibited a facile and time efficient process for salt permeation. Copper nanoparticles modified by polyethylenenimine along with high density of positive charges play a very important role in tuning of hydrophilicity and salt permeation (82% Na_2SO_4, 98% NaCl) (Zhu et al. 2016).

3. Advanced Water Desalination Techniques

The thought of fresh water generation from salty water (Lange 2010, Glueckanf 1966) is not new, but with time new advanced technologies with variety of properties are emerging every day (Miller

2003, Buros 2000, Trieb 2007). These technologies emerged on the basis of new impurities found in water. In the past decades, only filtration and chlorination were enough for the desalination (Darwish 2007) in municipal water treatment (Hasson and Semiat 2006, Zhou and Tol 2005, Reid 1966, Siemens 2010, Pilat 2001, Fritzmann et al. 2007), but from the last 20 years there is a drastic change in the water desalination approaches as now there are new impurities found in water where only chlorination and filtration are not able to purify the water (Electrochemical Capacitors for water desalination 2009, Allison 2013, Darwish and El-Dessouky 1996). Here are some of the new emerging technologies for water desalination:

3.1 Microfluidic and Nanofluidic Concentration Polarization Technique

Concentration polarization occurs when there is increase or decrease in the concentration of a component in the boundary layer which is close to the membrane surface (Prakash et al. 2009). In this system, multiple fluidic elements are linked in a series to identify substance present in a fluid. It is the process that rises in the junction of microfluidic and nanofluidic channels (Prakash et al. 2008, Daiguji et al. 2004, 2005) when an electric field is applied there. The depletion layer forms in this junction where charged particles repel each other. Larger nonporous membrane is more efficient in terms of energy used for water purification (Kim et al. 2010, Karmik et al. 2005).

3.2 Advance Membrane

Membrane technology is widely used technology for water purification (Lakshminarayanaiah 1969). This technology is not only used in industries but also in municipal wastewater treatment. One of the advanced membrane technique consists of number of carbon nanotubes. The water flows effortlessly by carbon nanotubes. Membranes are able to transform impure water into pure water which is fit for drinking (Hille 1992). But they have some disadvantages also, i.e., using membrane techniques for water purification requires more energy as compared to other techniques (Pedley 1983). But these membranes are able to remove salts and organic compounds from water as these impurities cannot pass through the membrane which results in fouling. Low-pressure membranes are very powerful for particulate contaminants while high-pressure membranes are very dominant for the effective removal of both organic and inorganic impurities.

3.3 Forward Osmosis

Forward osmosis is one of the examples of membrane process and is the sustainable alternative to reverse osmosis (Gray et al. 2006), FO has many benefits as there is a low hydraulic pressure that is required for this process. This process requires low consumption of energy and hence it is cheaper in cost and is economically and technically feasible (McCutcheon and Elimelech 2007). FO has many applications such as generation of power, water desalination, and in food processing also (Holloway et al. 2007). It is suitable for the rejection of moderate to high organic impurities. There is relatively low membrane fouling in FO as compared to pressure driven processes (Yip et al. 2010). This method is able to allow high water flux and the rejection rate of dissolved contaminants is also very high. The major difference between RO and FO is that the RO requires energy-intensive hydraulic pressure for their operation while FO requires only osmotic pressure for their operation (Cath et al. 2006).

3.4 Humidification and Dehumidification Water Desalination

HDH desalination is a thermal water purification technology based on nature's rain cycle. The main components used in this technology are (1) humidifier, (2) dehumidifier, and (3) heater. In the humidifier component, air is humidified directly in contact with salty water whereas in dehumidifier, moist air is in indirect contact with cold salty water. This process causes water vapour to condense and produce fresh drinking water (Ettouney 2005). This model is based on first law of thermodynamics, which is balance of mass and energy for combined system. This technology is ideal for small scale water desalination industries. This technology is very simple and requires low maintenance. There are very low water vapours in the carrier gas and there is low energy recovery in this method. This process is simple but requires more research and development to improve its efficiency and performance (Bourouni et al. 2001, Enezi et al. 2006).

3.5 Microbial Desalination Cells

It is one of the new technologies that remove the salinity of water in one solution (cathode) and generate the electric power from organic method and bacteria in another solution (anode). In this method, salinity requires large volume of anode solution than saline water. The most important parameter for this technology is rate of desalination which completely depends upon the concentration of salt in the water. Desalination performance is also affected by anode and cathode chambers. The electrons travel from anode to cathode through wire which connects the two electrodes (Cao et al. 2009). The performance of this technology can be increased by adding multiple ion exchange membranes between anode and cathode membranes. Another major concern in this method is the imbalance of pH between the anode and cathode when there is more than one chamber between the electrodes (Laoui et al. 2015).

4. Conclusion

As we know, 70% of the Earth is surrounded by water and it is required in every aspect of life. Advanced water technologies are needed to ensure high quality of drinking water, for the removal of pollutants, etc. In this chapter, we have discussed about the role of nanotechnology in desalination of water, the use of various nanomaterials, and their properties for water purification. One of the significant advantages of using nanomaterials is that they can integrate all their properties, which results in a multifunctional system.

Further research and studies are still needed to deal with the challenges and limitations of nanomaterials. Only a few nanomaterials are used commercially till date. Further improvement is required to improve its economic efficiency. Some nanomaterials are very toxic and put a negative impact on the environment and human health, and further studies are needed to reduce their toxicity.

References

Al Mayyahi, A. 2018. TiO_2 polyamide thin film nanocomposite reverses osmosis membrane for water desalination. Membranes 8(3): 66.

Allison, R.P. 2013. Electrodialysis treatment of surface and waste waters. Technical paper, GE Water and Process Technologies Paper TP1032EN. (http://www.gewater.com /pdf/Technical%20Papers_Cust /Americas/English/ TP103EN.pdf).

Asatekin, A., Seokate Kang, Menachem Elimelech and Anne M. Mayes. 2007. Anti-fouling ultrafiltration membranes containing polyacrylonitrile-graft-poly (ethlene oxide) comb copolymer additives. Journal of Membrane Science 298: 136–146.

Asatekin, A., Seokate Kang, Menachem Elimelech, Anne M. Mayes and Eberhard Morgenroth. 2006. Antifouling nanofiltration membranes for membrane bioreactors from self-assembling graft copolymers. Journal of Membrane Science 285: 81–89.

Ben-Sasson Moshe, Xinglin Lu, Siamak Nejati, Humberto Jaramillo and Menachem Elimelech. 2016. *In situ* surface functionalization of reverse osmosis membranes with biocidal copper nanoparticles. Desalination 388(2016): 1–8.

Bhattacharya, A. and P. Ghosh. 2004. Nanofiltration and reverse osmosis membranes: Theory and application in separation of electrolytes. Reviews in Chemical Engineering 20: 111–173.

Bourouni, K., M.T. Chaibi and L. Tadrist. 2001. Water desalination by humidification and dehumidification of air: State of art. Desalination 137: 167–176.

Buros, O.K. 2000. The ABCs of desalting. International Desalination Association. http://www.idadesal.org/pdf/ABCs.pdf.

Cao, X., Xia Huang, Peng Liang, Kang Xiao and Yingjun Zhou. 2009. A new method for water desalination using microbial desalination cells. Environmental Science Technology 43(18): 7148–7152.

Cath, T.Y., A.E. Childress and M. Elimelech. 2006. Forward osmosis: Principles, applications, and recent developments. Journal of Membrane Science 281: 70–87.

Choi, H., E. Stathatos and D.D. Dionysiou. 2007. Photocatalytic TiO_2 films and membranes for the development of efficient wastewater treatment and reuse systems. Desalination 202(1-3): 199–206.

Daiguji, H., P. Yang and A. Majumdar. 2004. Ion transport in nanofluidic channels. Nano Letters 4(1): 137–142.

Daiguji, H., Y. Oka and K. Shirono. 2005. Nanofluidic diode and bipolar transistor. Nano Letters 5(11): 2274–2280.

Darwish, M.A. 2007. Desalting: Fuel energy cost in Kuwait in view of $75/barrel oil price. Desalination 208: 306–320.

Darwish, M.A. and H. El-Dessouky. 1996. The heat recovery thermal vapour-compression desalting system: A comparison with other thermal desalination processes. Applied Thermal Engineering 16: 523–527.

Enezi, G.L., E. Hisham and N. Fawzy. 2006. Desalination process low temperature humidification dehumidification. Energy Conversion and Management 47: 470–484.

Ettouney, H. 2005. Design and analysis of humidification dehumidification desalination processes. Desalination 183: 341–352.

Forgacs, C., N. Ishibashi, J. Leibovitz, J. Sinkovic and K.S. Spiegler. 1972. Polarization at ion-exchange membranes in electrodialysis. Desalination 10: 181–214.

Fritzmann, C., J. Lowenberg, T. Wintgens and T. Melin. 2007. State-of-the-art reverse osmosis desalination. Desalination 216: 1–76.

Gethard Ken, Ornthida Sae-Khow and Somenath Mitra. 2010. Water desalination using carbon-nanotube-enhanced membrane distillation. ACS Applied Materials and Interfaces 3(2): 110–114.

Glueckauf, E. 1966. Seawater desalination—in perspective. Nature 211: 1227–1230.

Gong, X., Jingyuan Li, Hangjun Lu, Rongzheng Wan, Jichen Li, Jun Hu and Haiping Fang. 2007. A charge-driven molecular water pump. Nature Nanotechnology 2: 709–712.

Gray, G.T., R.L. McCutcheon and M. Elimelech. 2006. Internal concentration polarization in forward osmosis: Role of membrane orientation. Desalination 197: 1–8.

Hasson, D. and R. Semiat. 2006. Scale control in saline and wastewater desalination. Israel Journal of Chemistry 46: 97–104.

Hille, B. 1992. Ionic Channels of Excitable Membranes. Sinauer Associates, Sunderland, MA.

Holloway, R.W., Amy E. Childress, Keith E. Dennet and Tzahi Y. Cath. 2007. Forward osmosis for concentration of anaerobic digester concentrate. Water Research 41: 4005–4014.

Holt, J.K., Hyung Gyu Park, Yinmin Wang, Michael Stadermann, Alexander B. Artyukhin, Costas P. Grigoropoulos, Aleksandr Noy and Olgica Bakajin. 2006. Fast mass transport through sub-2nm carbon nanotubes. Science 213: 1034–1037.

Innovations. Electrochemical capacitors for water desalination, January 22, 2009. (http://www.innovationsreport.com/html/reports/energy_engineering/electrochemical_capacitor_water_desalination_125885.html).

Isha, R. and N.H. Abd Majid. 2015. A potential hybrid TiO2 in photocatalytic seawater desalination. In Advanced Materials Research. Trans Tech Publications Ltd. (Vol. 1113, pp. 3–8).

Jackson, R.D. and C.H.M. Van Bavel. 1965. Solar distillation of water from soil and plant materials: A simple desert survival technique. Science 149: 1377–1378.

Karnik, R., Rong Fan, Min Yue, Deyu Li, Peidong Yang and Arun Majumdar. 2005. Electrostatic control of ions and molecules in nanofluidic transistors. Nano Letters 5(5): 943–948.

Kemery, P.J., J.K. Steehler and P.W. Bohn. 1998. Electric field mediated transport in nanometer diameter channels. Langmuir 14(10): 2884–2889.

Kim, S.J., S.H. Ko, K.H. Kang and J. Han. 2010. Direct seawater desalination by ion concentration polarization. Nature Nanotechnology 5: 297–301.

Kumar, M. Ravi, M. Sridhar, S. Madhan Kumar and C. Vignesh Vasanth. 2017. Experimental investigation of solar water desalination with phase change material and TiO. Imperial. J. Interdisc. Res. (IJIR) 1128–1134.

Laksminarayanaiah, N. 1969. Transport Phenomena in Membranes. Academic Press, New York.

Lange, K.E. 2010. Get the salt out. National Geographic 217(4): 32–35.

Laoui Tahar, Adnan M. Al-Amer, Amjad B. Khalil, Aamir Abbas, Marwan Khraisheh and Muataz Ali Atieh. 2015. Novel anti-microbial membrane for desalination pretreatment: A silver nanoparticle-doped carbon nanotube membrane. Desalination 376: 82–93.

Majumder, M., Nitin Chopra, Rodney Andrews and Bruce J. Hinds. 2005. Experimental observation of enhanced liquid flow through aligned carbon nanotube membranes. Nature 438: 44–45.

McCutcheon, J.R. and M. Elimelech. 2007. Modelling water flux on forward osmosis: Implications for improved membrane design. AIChE Journal 53: 1736–1744.

Mehta, N., S. Basu and A. Kumar. 2016. Separation of zinc oxide nanoparticles in water stream by membrane filtration. Journal of Water Reuse and Desalination 6(1): 148–155.

Miller, J.E. 2003. Review of water resources and desalination technologies. Sandia National Laboratories, Livermore, CA.

Myint, Myo Tay Zar, Salim H. Al-Harthi and Joydeep Dutta. 2014. Brackish water desalination by capacitive deionization using zinc oxide micro/nanostructures grafted on activated carbon cloth electrodes. Desalination 344: 236–242.

Pandey, A.K., R.R. Kumar, B. Kalidasan, I.A. Laghari, M. Samykano, R. Kothari and V.V. Tyagi. 2021. Utilization of solar energy for wastewater treatment: Challenges and progressive research trends. Journal of Environmental Management 297: 113300.

Pedley, T.J. 1983. Calculation of unstirred layer thickness in membrane transport experiments: A survey. Quarterly Review of Biophysics 16: 115–150.

Pilat, B. 2001. Practice of water desalination by electrodialysis. Desalination 139: 385–392.

Prakash, S., A. Piruska, E.N. Gatimu and P.W. Born. 2008. Development of a hydrogel-bridged nanofluidic system for water desalting. Paper Presented at 233rd National Meeting and Exposition. American Chemical Society, Chicago, 2008.

Prakash, S., M.B. Karacor and S. Banerjee. 2009. Surface modification in microsystems and nanosystems. Surface Science Reports 64: 233–254.

Pu, Q., Jongsin Yun, Henryk Temkin and Shaorong Liu. 2004. Ion-enrichment and ion-depletion effect of nanochannel structures. Nano Letters 4(6): 1099–1103.

Qiao, R. and N.R. Aluru. 2004. Charge inversion and flow reversal in a nanochannel electroosmotic flow. Physical Review Letters 92(19): 198301-1–198301-4.

Qiao, R. and N.R. Aluru. 2005a. Atomistic simulation of KCl transport in charged silicon nanochannels: Interfacial effects. Colloids and Surfaces A 267: 103–109.

Qiao, R. and N.R. Aluru. 2005b. Surface charge induced asymetric electrokinetic transport in confined silicon nanochannels. Applied Physics Letters 86: 143105–143107.

Qiao, R., J.G. Georgiadis and N.R. Aluru. 2006. Differential ion transport induced electroosmosis and internal recirculation in heterogeneous osmosis membranes. Nano Letters 6(5): 995–999.

Reid, C.E. 1966. Principles of reverse osmosis. pp. 1–14. *In*: U. Merten (ed.). Desalination by Reverse Osmosis, MIT Press, Cambridge, MA.

Saleh Tawfik, A. 2016. Nanocomposite of carbon nanotubes/silica nanoparticles and their use for adsorption of Pb (II): from surface properties to sorption mechanism. Desalination and Water Treatment 57(23): 10730–10744.

Service, R.F. 2006. Desalination freshens up. Science 313: 1088–1090.

Shahid Mohammad, Andrew McDonagh, Jong Ho Kim and Ho Kyong Shon. 2015. Magnetised titanium dioxide (TiO2) for water purification: Preparation, characterisation and application. Desalination and Water Treatment 54(4-5): 979–1002.

Shannon, M.A. 2010. Fresh for less. Nature Nanotechnology 5: 248–250.

Shannon, M.A., Paul W. Bohn, Menachem Elimelech, John G. Georiadis, Benito J. Marinas and Anne M. Mayes. 2008. Science and technology for water purification in the coming decades. Nature 452: 301–310.

Shim, Ha, Jung Yang, Sun-Wook Jeong, Chang Lee, Lee Song, Sajid Mushtaq, Dae Choi, Yong Choi and Jongho Jeon. 2018. Silver nanomaterial-immobilized desalination systems for efficient removal of radioactive iodine species in water. Nanomaterials 8(9): 660.

Shukla, A.K., J. Alam, M.A. Ansari, M. Alhoshan, M. Alam and A. Kaushik. 2019. Selective ion removal and antibacterial activity of silver-doped multi-walled carbon nanotube/polyphenylsulfone nanocomposite membranes. Materials Chemistry and Physics 233: 102–112.

Siemens. 2010. What will the future of salt water desalination look like? Case study. 2010. (http://aunz.siemens.com/PicFuture/Documents/WaterCasestudy_desalination.pdf).

Spiegler, K.S. 1962, Salt Water Purification, New York: John Wiley & Sons.

Torcellini, P., N. Long and R. Judkoff. 2003. Consumptive water use for U.S. power production. National Renewable Energy Laboratory, Golden, CO.

Trieb, F. 2007. Concentrating solar power for water desalination, G.A.C. (DLR) and I.o.T. Thermodynamics (eds.). Federal Ministry for the Environment, Nature Conservation and Nuclear Energy, Stuttgart, Germany.

U.S. Department of Energy. 2006. Energy demands on water resources. Sandia National Laboratory, Livermore, CA, 1–80.

UNICEF and WHO. 2006. Water for Life: Making it Happen. United Nations, Geneva.

Veerapaneni, S., Bruce Long, Scott Freeman and Rick Bond. 2007. Reducing energy consumption for seawater desalinations. Journal of the American Water Works Association 99: 95–106.

Yacou, C., S. Smart and J.C.D. da Costa. 2015. Mesoporous TiO_2 based membranes for water desalination and brine processing. Separation and Purification Technology 147: 166–171.

Yang, Hui Ying, Zhao Jun Han, Siu Fung Yu, Kin Leong Pey, Kostya Ostrikov and Rohit Karnik. 2013. Carbon nanotube membranes with ultrahigh specific adsorption capacity for water desalination and purification. Nature Communications 4: 2220.

Yeston, J., Robert Coontz, Jesse Smith and Caroline Ash. 2006. A thirsty world. Science 313: 1067.

Yip, N.Y., A. Tiraferri, W.A. Phillip, Jessica D. Schiffman and Menachem Elimelech. 2010. High performance thin-film composite forward osmosis membrane. Environmental Science Technology 44: 3812–3818.

Zelmanov, G. and R. Semiat. 2011. Phosphate removal from water and recovery using iron (Fe^{3+}) oxide/hydroxide nanoparticles-based agglomerates suspension (AggFe) as adsorbent. Environ. Eng. Manage. J. 10: 1923–1933.

Zelmanov Grigori and Raphael Semiat. 2015. The influence of competitive inorganic ions on phosphate removal from water by adsorption on iron (Fe^{3+}) oxide/hydroxide nanoparticles-based agglomerates. Journal of Water Process Engineering 5: 143.

Zhao, Q., N. Chen, D. Zhao and X. Lu. 2013. Thermoresponsive magnetic nanoparticles for seawater desalination. ACS Applied Materials & Interfaces 5(21): 11453–11461.

Zhou, Y. and R.S.J. Tol. 2005. Evaluating the costs of desalination and water transport. Water Resources Research 41(W03003): 1–10.

Zhu Junyong, Adam Uliana, Jing Wang, Shushan Yuan, Jian Li, Miaomiao Tian and Kenneth Simoens. 2016. Elevated salt transport of antimicrobial loose nanofiltration membranes enabled by copper nanoparticles via fast bioinspired deposition. Journal of Materials Chemistry A 4(34): 13211–13222.

20

Nanobiomaterials in Bioremediation

Ashwini Ravi,[1] *J. Hemapriya,*[2] *P.N. Sudha,*[3] *S. Revathi,*[3]
R. Sasirekha[3] and *S. Vijayanand*[1,*]

1. Introduction

The production of man-made chemicals has increased from one million to 400 million between the years 1930 and 2000 (Toxic chemicals - WWF). The global production of chemicals as of 2018 was found to be 3,347 billion euros. This global production has a 2.5% increase when comparing with previous year 2017, which was found to be 3,266 billion euros. China stands in the first place followed by Europe and USA, whereas India takes the 6th position in global chemical production (Cefic facts and Figures 2020). In India, chemical manufacturing industries are considered one of the main capital generating industry and an intrinsic component for the developing industry. In 2017–2018, 1.69% of total GVA has been provided by chemicals and chemical product sector which is an increase from 1.07% during 2016–2017 (National accounts statistics 2017). The production of total petrochemicals in India increased from 13,448 MT during the year 2014–15 to 16,269 MT during the year 2018–19. There has been considerable increase in all sorts of chemicals such as organic and inorganic chemicals, pesticides, dyes and pigments, polymers, detergents and performance plastics (Annual Report 2019–20).

The chemicals produced by these industries have a wide range of applications from baby products to cosmetics. In a seminar entitled "Taking it all in: Environmental toxins and your health", Skerret 2016 quoted Rachel Carson's words from Silent Spring which says "There was once a town in the heart of America where all life seemed to be in harmony with its surroundings. But it quickly built to a sustained, meticulously reported account of the toll that widespread aerial spraying of DDT, dieldrin, aldrin, chlordane, heptachlor, and other synthetic pesticides was taking on birds, raccoons, fish, bees, and even the supposed beneficiaries of spraying — humans". The book was published in September 27, 1962. 60 decades later, we are still talking about environmental pollution with respect to chemicals and the production rate has increased 400 folds in the last century. These chemicals have been categorized based on their persistence and its effect on the environment as very persistent (vP) and Very bioaccumulative chemicals, Endocrine disruptors and those chemicals that cause cancer, reproductive problems and damages DNA (Toxic chemicals - WWF 7th EAP).

---

[1] Bioresource Technology Lab, Department of Biotechnology, Thiruvalluvar University, Serkkadu, Vellore, Tamil Nadu, India, 632115.
[2] PG & Research Department of Microbiology, DKM College for Women (Autonomous), Vellore, Tamil Nadu, India, 632001.
[3] Biomaterial Research Lab, Department of Chemistry, DKM College for Women (Autonomous), Vellore, Tamil Nadu, India, 632001.
*Corresponding author: vipni76@gmail.com

Though each type mentioned above has detrimental effects on the environment, intensive care is given to the Very persistent chemicals since they have unforeseeable and unimaginable effects due their log stay in the environment. But the detection of chemicals in the environment is still uncertain as the biodegradation half life of only 220 chemicals out of 95,000 industrially manufactured chemicals have been identified and only 1000 chemicals have the data on bio accumulation (UNEP 2013). Also, only minimum amount of data have been made publicly available for 14% of all chemicals. This unavailability of information makes it very difficult to design a basic safety assessment (Toxic chemicals - WWF). A study relating to bio monitoring the population in United States during 2019 showed presence of 75 toxic chemicals in peoples' blood and urine (The Fourth Report 2009). Similarly, a study conducted in workers of different occupations such as shoe polishing, thinner handling, painting and in related areas, furniture polishing, textile dyeing, printing press, petrol bunks and dry porting showed the presence of several types of Volatile Organic Compounds in their blood. Also, the amount of these VOCs increased with increase in working hours, work experience and disease (Yaqub et al. 2020). On 13th July 2015, a study on breast milk in Sirsa, Haryana, India showed the presence of pesticides in mother's milk. The study reported that "[t]he presence of such residues in mother's milk in Sirsa was pegged at 0.12 milligrams per kg, a figure about hundred times the estimates of the World Health Organization" (India Today 2013).

The better way to harness these detrimental effects of chemical to the environment is to remediate them. Several remediation methods are available to protect the environment from toxic chemicals but bionanoremediation is the more advanced method than other procedures. The present chapter deals with the nanobioremediation of such toxic chemicals in the environment.

2. Bioremediation

A brochure by Department of Nuclear Energy, International Atomic Energy Agency, Austria states: "Taking care of the environment today is a sustainable act for the generations of tomorrow. Avoiding the need for excessive remediation programmes after the end of operations is a fundamental aspect of life cycle thinking of any nuclear facility or industry handling radioactive material" (IAEA). Though this statement is given for radioactive materials, the first half of it is applicable to all contaminants in the environment including radioactive ones. Like stated, sustainability plays a vital role for well being of the future generations. Keeping in mind that the present generations should not leave behind an empty vessel for future generations, saving some for the future will not only help them thrive well but also not to experience adverse effects of what we make today. The present generation has already faced a lot improvements from transportation to communication through technology and this will continue. All of these technologies are possible only through several manufacturing industries. But this manufacturing industry, despite producing easy life, creates an environmental hazard. The industries contaminate the soil, water and air as wastes, effluents and smoke, respectively. In addition to these, the increased usage of vehicles, pesticides/fertilizers, etc., has also caused enough contamination in the environment. The contamination caused has already showed its effects on the environment such as increased temperature, soil pollution, etc. In order to stop these detrimental effects and also to protect the environment from further damage, remediation is important.

According to Japan Remediation agreement (2003), remediation can be defined as "any cleanup, response, removal, remedial, corrective or other action to clean up, detoxify, decontaminate, treat, contain, prevent, cure, mitigate or otherwise remedy any release of any hazardous substance; any action to comply with any environmental law or governmental approval; and any inspection, investigation (including subsurface investigations), study, monitoring, assessment, sampling and testing (including soil and/or groundwater sampling activities), laboratory or other analysis, or evaluation relating to any hazardous substances or to anything referred to herein". There are several remediation techniques and the technique to be used will be based on the type of pollution as soil, water or air (Hamby 1996).

The remediation techniques used for soil pollution includes various physical and chemical methods like thermal soil remediation, encapsulation, air sparging (Pavel and Gravilescu 2008), pyrometallurgical techniques (Lombi et al. 1998), hydrometallurgical techniques (Lombi et al. 1998), leaching (Braunig et al. 2019), chelating (Lombi et al. 2008), complexing (Lombi et al. 1998), amendments (Kabata-Pendias and Pendias 1992), chemical oxidation (Rybnikova et al. 2017), reduction (Peterson and Girling 1981), neutralization (Adriano 1986), vitrification (Ballesteros et al. 2017), etc. As far as water pollution is concerned, various techniques such as air stripping, activated carbon filtration, chemical oxidation (SDWF), precipitation and coagulation (EPA 2000), ion exchange (Alexandratos 2009), reverse osmosis (Bhattacharya and Gosh 2004, Pawlak et al. 200, Natraj et al. 2009), photcatalytic method have been utilized to clear pollutants of the water body (Ibhadon and Fitzpatrick 2013), etc.

While several physical and chemical methods were performed to remediate contaminated soil and water, bioremediation presenst various advantages. The method can be performed *in situ* and *ex situ* and has various advantages such as low cost, onsite treatment, permanent elimination of waste from the contaminated site, minimizing the disturbing of site, and it can be applied to more widespread and diluted contaminants in case of *in situ* operations; in case of *ex situ* operation, it is a cost effective, fast process, can be controlled and can remediate various pollutants (Prokop et al. 2000, Iwamoto and Nasu 2001). Bioremediation involves two important biological entities: microbes and plants. The remediation employing microbes is called as microbial remediation and those which employ plants are called as phytoremediation. Apart from these, photocatalysis by using natural pigments is also performed. During the recent years, nanoparticles synthesized from natural entities either plant or microbe are also utilized for bioremediation of hazardous compounds.

3. Biological Synthesis of Nanoparticles

The nanoparticles were synthesized by using various physical methods such as grilling, milling, thermal ablation (Shedbalker et al. 2014) and chemical methods such as sol gel synthesis, chemical reduction, irradiation reduction, etc. (Arshad 2017). These two methods fall into two approaches of nanoparticles synthesis called "top down" approach and "bottom up" approach. However, these methods have disadvantages such as high cost, inefficient, complicated procedures and generation of hazardous and toxic wastes (Li 2011). The generation of hazardous waste being the major concern to protect environment and to prevent from further build up of these wastes, biological synthesis has been concentrated throughout the world.

Green synthesis or biological synthesis is an environmental friendly approach of nanoparticles synthesis without causing harm to the environment. It includes the use of plants and microorganisms in synthesizing nanoparticles in a controlled environment to produce desired size, shape and characteristics (Singh et al. 2016) (Figure 1). Adding to these, it also has advantages such as low cost, low energy requirement and easy process function (Dhuper et al. 2012).

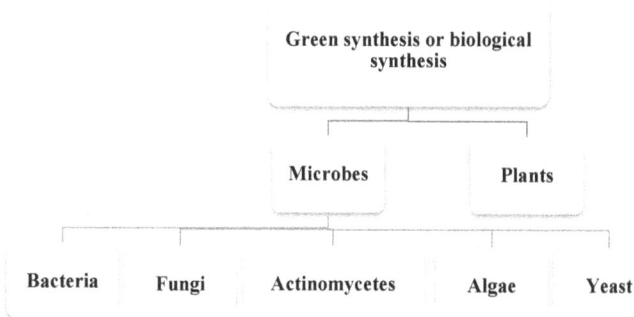

Figure 1: Green synthesis of nanoparticles.

3.1 Synthesis of Nanoparticles by Microbes

Microbial synthesis of nanoparticles includes the use of wide variety of microbes such as bacteria, fungi, actinomycetes, algae and yeast for synthesis of nanoparticles. The various microbes that are used for the synthesis of different nanoparticles are given in Table 1 below:

Table 1: Synthesis of nanoparicles by various microbial community.

S. No.	Name of the Microorganism	Nanoparticle Synthesized	Size	Reference
Bacteria:				
1.	*E. coli*	Au	20	(Du et al. 2007)
		Ag	50,	(Gurunathan et al. 2009,
			5–10	Baltazar-Encarnación et al. 2019)
		CdTe	2–3.2	(Bao et al. 2010)
		Cu	10–40	(Singh et al. 2010)
2.	*Klebsilla pneumoniae*	Ag	5–32	(Ahmad et al. 2007,
			28.2–122	Shahverdi et al. 2007)
		Au	10–15	(Prema et al. 2016)
3.	*Pseudomonas aeruginosa*	Au	15–30	(Husseiny et al. 2007)
		Se	–	(Yadav et al. 2008)
4.	*Bacillus cereus*	Ag	4–5	(Babu and Gunasekaran 2009)
5.	*Enterobacter* sp.	Hg	2–5	(Sinha and Khare 2011)
6.	*Bacillus licheniformis*	Ag	4–20	(Schulter et al. 2014)
7.	*Lactobacillus* sp.	Ti	40–60	(Prasad et al. 2007)
8.	*Lactobacillus fermentum*	Ag	15–40	(Sintubin et al. 2009)
9.	*Enterococcus faecalis*	Se	29–195	(Shoeibi et al. 2017)
10.	*Lactococcus lactis*	Ag	5–50	(Viorica et al. 2017)
		Se	38–152	(Xu et al. 2019)
11.	*Thermomonospora* sp.	Au	520 nm	(Ahmad et al. 2003b)
12.	*Pseudomonas stutzuri*	Ag	–	(Joerger et al. 2000)
13.	*Lactobacillus fermentum*	Ag	11.2 ± 0.9	(De Gusseme et al. 2010)
14.	*M. magnetotacticum*	Fe_3S_4	47.1	(Philipse and Maas 2002)
15.	*M. gryphiswaldense*	Iron oxide	35–50	(Manucci et al. 2014)
16.	*Desulfovibrio desulfuricans*	Pd	–	(Yong et al. 2002)
17.	*Serratia* sp.	Cu	10–30	(Hasan et al. 2008)
18.	*Enterobacter cloacae*	Se	–	(Losi and Frankenberger 1997)
19.	*Pseudomonas alkaphila*	Se	50–500	(Zhang et al. 2011)
20.	*Sulfurospyrillum barnesii*	Te	< 50	(Baesaman et al. 2007)
21.	*Bacillus brevis*	Ag	41–68	(Saravanan et al. 2018)
22.	*Shewanella loihica*	Cu	10–16	(Lv et al. 2018)
23.	*Halomonas salina*	Au	30–100	(Shah et al. 2017)
24.	*Micrococcus yummanensis*	Au	53.8	(Jafari et al. 2018)
25.	*Pseudomonas deceptionensis*	Ag	10–30	(Jo et al. 2016)
26.	*Bacillus methylotropicus*	Ag	10–30	(Makarov et al. 2014)
27.	*Bacillus amiloliquifaciens*	CdS	3–4	(Mukherjee et al. 2014)
28.	*Bacillus pumilus*	Ag	77–92	(Singh et al. 2015)
29.	*Bacillus* sp.	Ag	7–31	(Deljou and Goudarzi 2016)

Table 1 contd. ...

...Table 1 contd.

S. No.	Name of the Microorganism	Nanoparticle Synthesized	Size	Reference
30.	*Geobacillus* sp.	Au	5–50	(Correa-Llantén et al. 2013)
31.	*Salmonella typhirium*	Ag	50–150	(Ghorbani 2013)
32.	*Lactobacillus sporogenes*	ZnO	5–15	(Scott et al. 2008)
33.	*Lactobacillus plantarum*	Ag	19	(Sintubin et al. 2014)
34.	*Bacillus thuringiensis*	Ag	15	(Jain et al. 2010)
35.	*Bacillus megaterium*	Ag	6–50	(Dahikar and Bhutada 2013)
36.	*Pseudomonas denitrificans*	Au	15–20	(Mewada et al. 2012)
37.	*Pseudomonas fluorescens*	Au	50–70	(Rajasree and Suman 2012)
38.	*Marinobacter pelagius*	Au	< 10	(Sharma et al. 2012)
39.	*Aeromonas* sp.	Ag	6.4	(Rai et al. 2009)
40.	*Thermoanaerobacter ethanolicus*	Co, Ni	–	(Rai et al. 2009)
41.	*Stentrophonas maltophila*	Ag		(Oves et al. 2013)
42.	*Pseudomonas putida*	Se Ag	100–500 6–16	(Avendano et al. 2016, Zhang et al. 2019)
43.	*Exiguobacterium*	Ag	4.4	(Tamboli and Lee 2013)
44.	*Cinetobacter*	Au	10–12	(Wadhwani et al. 2018)
45.	*Halococcus sulfodinae*	Te	10–180	(Presentato et al. 2016)
46.	*Citrobacer freundii*	Se	580	(Wang et al. 2017)
47.	*Alkaligens faecalis*	Ag	30–50	(Divya et al. 2019)
48.	*Altermonas macleodii*	Ag	70	(Mehta et al. 2014)
49.	*Halomonas elongate*	ZnO	18.11	(Taran et al. 2018)
50.	*Lactobacillus johnsonii*	ZnO	4–7	(Al Zaharni et al. 2018)
51.	*Aeromonas hydrophila*	ZnO	57.72	(Jyaseelan et al. 2013)
Fungi:				
52.	*Penicillium*	Au	< 135	(Sadowski et al. 2008)
53.	*Hormoconis resinae*	Ag	20–80	(Varshney et al. 2009)
54.	*Fusarium oxysporum*	Ag BaTiO$_3$ ZrO$_2$ CdS Au Pt Si Ti	5–15 4–5 3–11 5–20 22 5–30 9.8 ± 0.2 10.2 ± 0.1	(Ahmad et al. 2003a) (Bansal et al. 2006) (Lefevre et al. 2010) (Ahmad et al. 2002) (Thakker et al. 2013) (Syed and Ahmad 2012) (Bansal et al. 2005) (Bansal et al. 2005)
55.	*Aspergillus fumigates*	Ag	5–25	(Bhainsa and D Souza 2006)
56.	*Neurospora crassa*	Pt	17–76	(Castro Longoria et al. 2012)
57.	*Aspergillus niger*	ZnO	53–69	(Kalpana et al. 2018)
58.	*Aspergillus terreus*	Ag	16–57	(Singh and Vidyasagar 2018)
59.	*Candida glabrata*	Ag	2–15	(Jalal et al. 2018)
60.	*Penicillium polonicum*	Ag	10–15	(Neethu et al. 2018)
61.	*Trichoderma harzianum*	Au	32–44	(Tripathi et al. 2018)
62.	*Aspergillus nidulans*	CoO	20.29	(Vijayanandan et al. 2018)

Table 1 contd. ...

...Table 1 contd.

S. No.	Name of the Microorganism	Nanoparticle Synthesized	Size	Reference
63.	*Cladosporium* sp.	Ag	24	(Popli et al. 2018)
64.	*Penicillium chrysogenum*	Pt	5–40	(Subramaniyan et al. 2018)
65.	*Rhizopus stolonifer*	Ag	2.86	(Abdel Rahim et al. 2017)
66.	*Penicillium fellutanum*	Ag	5–25	(Kathiresan et al. 2009)
67.	*Aspergillus flavus*	Ag	17 ± 5.9	(Jain et al. 2011)
68.	*Coriolus versicolor*	Ag	15–35	(Deniz et al. 2019)
69.	*Alternaria alternate*	ZnO	45–150	(Sarkar et al. 2014)
70.	*Aspergillus aeneus*	ZnO	100–140	(Jain et al. 2013)
71.	*Candida albicans*	ZnO	25	(Shamsuzzaman et al. 2017)
72.	*Fusarium* sp.	ZnO	> 100	(Velmurugan et al. 2010)
73.	*Rhizopus oryzae*	Ag	5–65	(Das et al. 2012)
Yeast:				
74.	*Yarrowia lipolytica*	Au	15	(Mithila et al. 2009)
75.	*Rhodosporadium diobovatum*	Lead	2–5	(Seshadri et al. 2011)
76.	*Pichia kudriavzevii*	ZnO	10–61	(Moghaddam et al. 2017)
77.	*Pichia fermentans*	Ag, ZnO	NA	(Chauhan et al. 2015)
Algae:				
78.	*Plectonema boryanum*	Au Pt	10–6000 < 30	(Lengke et al. 2006a) (Lengke et al. 2006a)
79.	*Chlorella salina*	Ag	53.1–71.9	(Merin et al. 2010)
80.	*Spirulina platensis*	Ag	7–16	(Govindraju et al. 2008)
81.	*Oscillatoria willei*	Ag	100–200	(Mubarak Ali et al. 2011)
82.	*Microcoleus* sp.	Ag	40–80	(Sudha et al. 2013)
83.	*Lyngbya sp*	Ag	40–80	(Sudha et al. 2013)
84.	*Sargassum tenerrimum*	Au	35	(Ramakrishna et al. 2016)
85.	*Sargassum wightii*	ZrO_2	18	(Kumaresan et al. 2018)
86.	*Chlorella vulgaris*	Pd	5–20	(Arsiya et al. 2017)
87.	*Nostoc* sp.	Ag	51–100	(Sonker et al. 2017)
88.	*Galaxaura elongate*	Au	3.85–77	(Abdel-Raouf et al. 2017)
89.	*Nostoc linckia*	Ag	5–60	(Vanlalveni et al. 2018)
90.	*Oscillatoria limnetica*	Ag	3.30–17.97	(Hamouda et al. 2019)
Actinomycetes:				
91.	*Streptomyces cyaneus*	Ag	12.63	(El-Batal et al. 2015)
92.	*Sreptomyces* sp.	Ag	5	(Karthik et al. 2014)
93.	*Rhodococcus* sp.	Au Ag	5–15 10	(Zheng et al. 2010) (Otari et al. 2010)
94.	*Sreptomyces griseoruber*	Au	5–50	(Ranjitha and Rai 2017)
95.	*Streptomyces capillispiralis*	Cu	3.6–5.9	(Saad et al. 2018)
Virus:				
96.	*Tobacco Mosaic Virus*	Co and Ni	50	(Shenton et al. 1999)
97.	M13 bacteriophage	TiO2	20–40	(Chen et al. 2015)

3.2 Synthesis of Nanoparticles by Plants

Phytonanotechnology or Green synthesis of nanoparticles deals with the synthesis of nanoparticles by plant materials. Like microorganisms, synthesis of nanoparticles by plants also has several advantages such as biocompatibility, scalability and medical applicability. It also has other advantages such as low cost, is ecofriendly, stable and rapid. Various plants that were used to synthesize nanoparticles are given in Table 2.

Table 2: Nanoparticles synthesized by various plants.

S. No.	Name of the Microorganism	Nanoparticle Synthesized	Size nm	Reference
1.	*Shorea tumbuggaia*	Ag	40	(Venkateswaralu et al. 2010)
2.	*Ocimum sanctum*	Pt Pt Ag	23 2 14	(Soundarrajan et al. 2012, Prabhu and Gajendran 2017) (Jain and Mehata 2017)
3.	*Cardiospermum halicacabum*	Ag	422–447	(Shekawat et al. 2013)
4.	*Pogostemon benghalensis*	Ag	–	(Gogoi 2013)
5.	*Fenugreek*	Se	50–150	(Ramamurthy et al. 2013)
6.	*Psidium guajava*	Ag	100–500	(Sriram and Pandidurai 2014)
7.	*Mentha*	Au	34	(Jafarizad et al. 2015)
8.	*Pelargonium*	Au	33.8	(Jafarizad et al. 2015)
9.	*Clitoria ternatea*	Ag	20	(Krithiga et al. 2015)
10.	*Solanum nigrum*	Ag	28	(Krithiga et al. 2015)
11.	*Hibiscus sabdariffa*	ZnO	12–250	(Bala et al. 2015)
12.	*Acer pentapomicum*	Au	19–24	(Khan et al. 2018b)
13.	*Bauhinia acuminate*	Ag	–	(Antony et al. 2016)
14.	*Biophytum sensitivum*	Ag	–	(Antony et al. 2016)
15.	*Azima tetracantha*	Au	540	(Hariharan et al. 2016)
16.	*Quercus glauca*	Pt	5–15	(Karthik et al. 2016)
17.	*Polyscias scutellaria*	Au	5–20	(Yulizar et al. 2017)
18.	*Passiflora caerulea*	ZnO	70	(Santhoskumar et al. 2017)
19.	*Alpinia katsumadai*	Ag	12.6	(He et al. 2017)
20.	*Nigella arvensis*	Au	3–37	(Chahardoli et al. 2018)
21.	*Galenia africana*	Au	15–25	(Elbgaory et al. 2017)
22.	*Hypoxis hemerocallidea*	Au	15–25	(Elbagory et al. 2017)
23.	*Parthenium hysterophorous*	ZnO	16–45	(Data et al. 2017)
24.	*Broccoli*	Se	50–150	(Kapur et al. 2017)
25.	*Curcuma neilgherrensis*	ZnO	1.56–5.08	(Parthasarathy et al. 2017)
26.	*Coriandrum sativum*	Ag	6.45	(Khan et al. 2018a)
27.	*Wedelia trilobata*	Au	10 – 50	(Dey et al. 2018)
28.	*Ziziphus zizyphus*	Au	51.8 ± 0.8	(Aljabali et al. 2018)
29.	*Muntingia calabura* L.	Au	36.93, 78.2	(Wahab et al. 2018)
30.	*Gloriosa superb*	Pt Pd	49.65 ± 1.99 36.26 ± 0.91	(Rokade et al. 2018)
31.	*Corchorus olitorius*	Au	37–50	(Ismail et al. 2018)
32.	*Hawthorn fruit*	Se	113	(Cui et al. 2018)

Table 2 contd. ...

...Table 2 contd.

S. No.	Name of the Microorganism	Nanoparticle Synthesized	Size nm	Reference
33.	*Origanum vulgare*	Ag	12	(Shaik et al. 2018)
34.	*Anogeissus latifolia*	Pd	4.8 ± 1.6	(Kora and Rastogi 2018)
35.	*Impatiens balsamina*	Ag	12–20	(Aritonang et al. 2019)
36.	*Lantana camara*	Ag Au	3.2–12 30–50	(Aritonang et al. 2019, Sane et al. 2013)
37.	*Aloe vera*	Se	121	(Fardsadegh and Malmiri 2019)
38.	*Mimosa tenuiflora*	Au	20–200	(Rodriguez Leon et al. 2019)
39.	*Phyllanthus emblica*	Ag	39	(Masum et al. 2019)
40.	*Averrhoa blimbi* L.	Ag	29	(Sagadevan et al. 2019)
41.	*Plum*	Ag	47	(Sagadevan et al. 2019)
42.	*Anthriscus sylvestris*	Au	18	(Ghahremanzadeh et al. 2019)
43.	*Ferula gummosa*	Au	30–32	(Ghahremanzadeh et al. 2019)
44.	*Achillea eriohora*	Au	56	(Ghahremanzadeh et al. 2019)
45.	*Coffea arabica*	Au	14–500	(Keijok et al. 2019)
46.	*Dates*	Pt	1.3–2.6	(Al-dadadi 2019)
47.	*Sargassum muticum*	ZnO	30–57	(Azizi et al. 2014)
48.	*Cynara scolymus*	Ag	98.47 ± 2.04	(Erdogan et al. 2019)
49.	*Cucumis prophetarum*	Ag	30–50	(Hemlata et al. 2020)
50.	*Lentinan extracts*	Pd	2.35–3.32	(Han et al. 2019)
51.	*Solanum torvum*	ZnO	34–40	(Ezealisiji et al. 2019)
52.	*Cassia auriculata*	Se	10–20	(Anu et al. 2020)
53.	*Macrotyloma uniflorum*	Au	14–17	(Aromal et al. 2012)
54.	*Crysophyllum cainito*	Pd	59.18–169.24	(Majumdar et al. 2017)
55.	*Azadirachta indica*	Ag Pt	34 5–50	(Ahmed et al. 2015, Thirumurugan et al. 2016)
56.	*Deverra tortuosa*	ZnO	9.26–31.18	(Selim et al. 2020)
57.	*Jatropa gossypifolia*	Pt	20	(Jeypaul et al. 2018)
58.	*Jatropa glandulifera*	Pt	100	(Jeypaul et al. 2018)
59.	*Diospyros montana*	Se	4–16	(Kokila et al. 2017)
60.	*Moringa oleifera*	Ag	9 & 11	(Moodley et al. 2018)
61.	*Nerium oleander*	Au	2–10	(Tahir et al. 2015)
62.	*Albizia lebbeck*	ZnO	66.25	(Umar et al. 2019)
63.	*Diopyros kaki*	Pt	2–12	(Song et al. 2010)
64.	*Salvia spinosa*	Ag	5.13	(Pirtarighat et al. 2018)
65.	*Camellia sinesis*	ZnO	853	(Shah et al. 2015)
66.	*Betel*	Au	26–40	(Sane et al. 2013)
67.	*Bitter gaurd*	Au	44–63	(Sane et al. 2013)
68.	*Euphorbia granulate*	Pd	25–35	(Nazrollahzadeh et al. 2015)
69.	*Pimpinella tirupatiensis*	Pd	12.25	(Narassiah et al. 2017)
70.	*Matricaria chamomilla* L.	ZnO	51.2 ± 3.2	(Ogunyemi et al. 2019)

Table 2 contd. ...

...Table 2 contd.

S. No.	Name of the Microorganism	Nanoparticle Synthesized	Size nm	Reference
71.	*Olea europaea*	ZnO	41 ± 2	(Ogunyemi et al. 2019)
72.	*Lycopersicum esculentum*	ZnO0	51.6 ± 3.6	(Ogunyemi et al. 2019)
73.	*Eucalyptus globulus*	ZnO	11.6	(Reddy et al. 2017)
74.	*Ginkgo bloba*	Ag	10–16	(Ren et al. 2016)
75.	*Banana*	Pd	50	(Banker et al. 2010)
76.	*Lawsonia inermis*	Fe	21	(Naseem and Farrukh 2015)
77.	*Gardenia jasminoides*	Fe	32	(Naseem and Farrukh 2015)
78.	*Mango peel*	Fe	–	(Desalgn et al, 2019)

4. Nanoparticles in Bioremediation

Nanoparticles found their application in vast fields of advancements such as cosmetics, medicine, paints, food products, preservatives, etc. The application of nanoparticles in remediation of xenobiotics is also increasing every day. But during the past few decades, green synthesis or microbial synthesis of these nanoparticles has made its way in bioremediation since these biologically synthesized nanoparticles will be benign to environment and also can be produced at a lower cost. The nanoparticles that are widely used in bioremediation are discussed as follows.

4.1 Iron Nanoparticles

Zero valent iron nanoparticles (nZVI) have been used in the remediation of both contaminated soil and ground water (Chekli et al. 2016). These zero valent iron nanoparticles can be used for both on site applications and *in situ* applications (Grieger et al. 2010). In water treatment, zero valent iron nanoparticles have a metallic iron core and a thin oxide surface which helps them in adsorbing or reducing the contaminants causing its removal and transportation (Mueller 2010), whereas in remediation of soil, zero valent iron nanoparticles have been used as reductant (Machado et al. 2017).

     Zero valent nanoparticles have been successfully used in the remediation of various contaminants in both soil and water such as chromium (Kouhiyan Afzal et al. 2017, Galdames et al. 2017), chromate-copper-arsenate (Chekli et al. 201), arsenic (Galdames et al. 2017), pesticides (Simkovic et al. 2015), etc. These zero valent nanoparticles have been used along with or modifications with other compounds for better remediation of the contaminated sites. Su et al. (2014) demonstrated the modification of zero valent iron nanoparticles by sulphide (S-nZVI) and its efficiency in removal of cadmium from contaminated area. It has been found from the study that S-nZVI removes cadmium from the contaminated area and the iron nanoparticle persisted without any changes throughout the experiment which was conducted for two months (Su et al. 2014). Another study by (Khorsandi et al. 2017) showed the modification of nZVI with starch in the effective removal of boron from contaminated site.

     Madhavi et al. (2013) showed that nZVI when used along with farmyard manure (FMY) was found to effectively remove chromium from the contaminated soil and also enhanced the tolerance of plant *Brassica juncea* (Madhavi et al. 2013). Similar study was performed by Pillai and Kottekottil (2016), where nZVI has been tested for its efficiency to remove an organochlorine pesticide endosulfan in the presence of three plants, viz., *Alpinia calcarata*, *Ocimum sanctum* and *Cymbopogon citratus*. Though the efficiency of the plants to remove endosulfan decreased overtime, very less endosulfan concentration was found in all the plants, which was attributed to nZVI. The zero valent iron nanoparticles caused the removal of pesticide resulting in minimal concentration of the same in all the three plant species (Pillai and Kottekottil 2016). nZVI, in addition to be used with modifications, is also used in nano-hybrid systems. A nano-hybrid system produced from yeast *Candida* sp. SMNO4 and nZVI has efficiently degraded a drug Cefdinir from contaminated water, proving that nZVI can be utilized in treating pharmaceutical wastewater (Selvi and Das 2016).

Despite using the physical or chemical mediated synthesis of nZVI by top down approach or bottom up approach, green synthesis or microbial synthesis has also been done. Zero valent iron nanoparticles synthesized from plant showed better activity in removing metals from contaminated site. nZVI synthesized from *Emblica officinalis* leaf extract efficiently removed lead from aqueous solution within 24 hrs (Kumar et al. 2015). In addition, nZVI synthesized from *Eucalyptus globulus*, *Mangifera indica*, *Syzygium cumini* and *Psidium guajava* was found to be effective in oxidizing arsenic to arsenate, thus reducing the toxicity of arsenic (Rana et al. 2018). Green synthesized nZVI from *Shirazi thyme* leaf and pistachio green hulls was efficient in removal of phosphorous from contaminated water resources.

Apart from its utilization in removing metals from contaminated site by nZVI, it is also employed in removal of other toxins such as dyes and other xenobiotics. Zero valent iron nanoparticles synthesized from *Vaccinium floribundum* removes petroleum hydrocarbons effectively from the contaminated soil (Murgueitio et al. 2018). nZVI synthesized from mulberry leaves was also found to be effective in removal of dye methyl orange from contaminated water (Lim et al. 2018). In addition, the dye methylene blue was also removed by nZVI synthesized from mulberry leaves (Lim et al. 2018) and *Calotrois gigantean* (Sravanthi et al. 2018). The study by Sravanthi et al. (2018) also shows the removal of aniline by green synthesized nZVI. Comparative study on green synthesized nZVI from black tea along with activated carbon and nZVI was performed by (Karam et al. 2020) in which it was confirmed that green synthesized iron nanoparticles were more efficient than the activated carbon and nZVI in the removal of colour removal from textile wastewater (Karam et al. 2020). Apart from using green synthesized nZVI as such, encapsulation of nZVI was also performed for the effective use of bioremediation. Anu and Vijay (2016) showed the encapsulation of green synthesized nZVI from *Camellia sinensis* and the effective degradation of dyes methy red and methy orange (Anu and Vijay 2016).

Zero valent iron nanoparticles can be considered as an effective nanoparticle for environmental bioremediation. They have been used in the bioremediation of soil and water contaminated by various xenobiotics such as dyes, metals, pharmaceutical products, etc. In addition to these, a study on green synthesized nZVI from *Spinacia oleracea* showed the reduction of BOD and COD in a municipal waste from 56 mg/L to 22 mg/L and 405 mg/L to 85 mg/L, respectively (Turakhia et al. 2018). The study confirms that nZVI can be used efficiently in bioremediation of waste water. Several studies in green synthesized nZVI is still performed onsite and in laboratories to explore their efficacy in bioremediation.

4.2 Uranium Nanoparticles

Uranium is a heavy naturally occurring material present in Earth's crust and especially in rocks. It is an alpha emission radiation and a radioactive material which, when exposed, causes radiotoxicity and chemotoxicity (Gavrilescu 2009). The concentration of uranium is increasing in the environment due to industrial and mining activities. In addition, the leaching of these uranium containing waste materials contaminates not only soil but also the ground water (Lloyd and Renshaw 2005, Choy et al. 2006, Gavrilescu 2009). The removal of uranium from contaminated water by physical or chemical methods leaves secondary waste streams. Therefore, microbial bioremediation was concentrated for the effective removal of uranium from contaminated sites (Wu et al. 2006, Appukuttan et al. 2006).

The mobility of uranium in the environment is determined by its species produced from oxidation or reduction. Under oxidative conditions, UO_2 is produced whereas under reducing conditions U(VI) is produced (Choppin et al. 2002). The predominant species of any uranium ore deposit was considered to be crystalline uraninite which is formed by the abiotic reduction of hexavalent uranium U(VI). But recently, the predominant species in uranium ore deposits was found to be non crystalline uranium (VI) which is formed by the biogenic action of microbes (Battacharya et al. 2017). Microbes interact by variety of mechanisms with uranium (Acharya 2015). Therefore, bioremediation of uranium was considered more successful than any other methods. Uranium bioremediation is done by many microbial species such as *Actinobacteria* sp. (Li et al. 2018), *Synechoccus elongates* (Acharya et al. 2009), etc. *Ex situ* bioremediation of uranium contaminated sites by *Acidithiobacillus ferroxidans* strain showed that microbes can remove uranium effectively (Romero et al. 2016). Now, geoscientists are more interested in uraninite for successful remediation of uranium. This uraninite molecule being in size of nano scale can effectively interact with U(VI) present in the environment and aid in their removal (Bargar et al. 2008). Several works are still performed in this type of bioremediation to clear uranium from the site of contamination.

4.3 Gold Nanoparticles

Gold nanoparticles, with their biocompatibility due to better conjugation with biomolecules along with their low or non-cytotoxicity, were used widely in the field of medicine such as diagnostics, therapeutics, *in vivo* imaging, antimicrobial agents, etc. (Jeong et al. 2014, Connor et al. 2005, Qin et al. 2014, Nayan et al. 2017). Gold nanoparticles synthesized by fungus *Candida albicans* have been used in the detection of liver cancer (Chauhan et al. 2011), those produced from *Mangifera indica* have been used as antibacterial, anticancer and antiangiogenic agent (Vimalraj et al. 2018), gold nanoparticles synthesized from *Brassica olerace* have antimicrobial activity against human pathogenic bacteria *Staphylococcus aureus*, *Klebsilla pneumonia* and human pathogenic fungi *Aspergillus flavus*, *Aspergillus niger* and *Candida albicans* (Piruthiviraj et al. 2016), etc. Apart from these, they are also used in electronics and chemistry due to their remarkable opto-electric properties (Sharma et al. 2012).

The catalytic activity of gold nanoparticles has created a sole place for it in bioremediation (Maye et al. 2003). Green synthesis or microbial synthesis of gold nanoparticles and its application in bioremediation adds further advantage in using gold nanoparticles. Gold nanoparticles synthesized from *Mangifera indica* (Nayan et al. 2017), *Avicennia marina* (Nabikhan et al. 2017) and *Breynia rhamnoides* (Gangula et al. 2011) degrade 4 Nitrophenol. Similarly, the synthesized gold nanoparticles from *Avicennia marina* also detoxify 4 Aminophenol (Gangula et al. 2011). Apart from these, gold nanoparticles synthesized from *Lawsonia inermis* are also efficient in degrading pesticides such as DDT (Abd El-Aziz et al. 2018). In addition to the degradation of chemicals and pesticides, they are also efficient in degrading dye compounds. A study by Tripathi et al. (2017) showed the effectiveness of gold nanoparticles synthesized from *Trichoderma harzianum* in degrading methylene blue dye.

4.4 Silver Nanoparticles

Silver nanoparticles have been exploited in medicine due to their natural antimicrobial activities. Along with that, they are also explored in bioremediation of various xenobiotics especially dye compounds. Green synthesized silver nanoparticle by *Viburnum opulus* was effectual in degrading dyes such as carmoisine, brilliant blue and tartrazine (David and Moldovan 2020). Methylene blue, which is predominantly used in several laboratory applications, was degraded by silver nanoparticles synthesized from various biological entities such as pomegranate peel (Joshi et al. 2018), *Morinda tinctoria* (Vanaja et al. 2014), *Gmelina arborea* (Saha et al. 2017), *Gymnema sylvestre* (Kumar et al.

2019), etc. In addition to methylene blue, another laboratory dye predominantly used is malachite green. It is used for staining microbes. Green synthesized silver nanoparticle by Aegle marmelos degrades this laboratory dye malachite green efficiently (Femila et al. 2014).

Bhakya et al. (2015) showed degradation of several dye compounds such as methyl violet, saffranin, eosin methylene blue and methyl orange by silver nanoparticles synthesized from *Helicteres isora*. Similarly, the azo dye Congo red was degraded by silver nanoparticles synthesized from *Bacillus pumillus* (Modi et al. 2015). In addition to degradation of dye compounds, silver nanoparticles also degrade pesticides and other chemicals. Deka and Sinha (2015) showed the degrading ability of pesticide chlorpyrifos by silver nanoparticles synthesized from fungal isolate *Penicillium pinophilum* and those synthesized from *Aloe barbedensis* and *Azadirachta indica* have better decontaminating ability of naphthalene by absorbtion (Abbas et al. 2020). Though silver nanoparticles are exploited in various fields, it is still explored for advancements in bioremediation.

4.5 Other Nanoparticles

In addition to the above discussed nanoparticles, other nanoparticles such as copper, palladium, cadmium and zinc were also explored for their role in bioremediation. For instance, copper nanoparticles synthesized from fish scales waste of *Labeo rohita* were used in the degradation of methylene blue. The synthesized nanoparticles were found to efficiently degrade the dye with a degradation rate of 96% (Sinha and Ahmaruzzaman 2015). Similarly, zinc oxide nanoparticles synthesized from culture filtrates of *Aspergillus niger* were found to degrade Bismarck brown dye (Kalpana et al. 2018). Cadmium sulphate nanoparticles synthesized from *Pseudomonas aeruginosa* JP-11 were found to remove Cd(II) by adsorption (Raj et al. 2016). Palladium nanoparticles synthesized from fruit extract catalytically removes Nitroarenes (Nasrollahzadeh et al. 2016). The calcium alginate nanoparticles synthesized from honey were found to effectively remove Cr(VI) by sorption (Geetha et al. 2015). In this way, other nanoparticles have also been explored in bioremediation of toxic chemicals.

5. Nanobioremediation

Nanoparticle has already been utilized in several industries for wide range of applications. During recent times, they have also been studied for its role in biodegradation. Biodegradation of biosynthesized nonmaterial is discussed as follows:

5.1 Solid Waste Remediation

Solid wastes are the disposals from everyday activities such as household wastes, food wastes, vegetable waste, etc. The disposal of these wastes is a significant problem both in urban and rural areas. The processes that involve collection of solid waste for its disposal is called as solid waste management and the means of solid waste disposal is the main concern of it (Ashish et al. 2014, Abdel-Shafy and Mansour 2018). Though methods such as landfills, incineration, composting, etc., are widely followed, a better non disadvantageous method is very much essential. Therefore, recycling solid wastes in any form have been concentrated as a method to dispose them from environment. In this process, nanoparticles were found to have a substantial role.

Leachate is the main disadvantage of landfills and a study on leachate treatment by nanoparticles was conducted by (Hu et al. 2016). He used the bacterium *Phanerochaete chrysosporium* and loaded them with nitrogen doped TiO_2 nanoparticles in treating leachate with low biodegradability ratio and succeeded in removing heavy metals and also observed the reduction in percentage of organic compounds. In addition to using nanoparticles as direct source for degrading municipal waste, it was also utilized as catalyst. Pure iron is used as a catalyst in enhancing the degradation of food wastes, which is usually carried out in anaerobic digester. This digester generates methane which can be utilized as energy source and also the slurry generated can be utilized as organic fertilizer. This anerobic digestion process is improved by the addition of iron. (Sreekanth and Sahu 2015) used iron nanoparticles instead of bulk iron nanoparticles in improvising the digestion further since nanoparticles are always advantageous than the bulk particles. With the addition of iron nanoparticles, the digestion which will take place usually for 3 months was completed in 2 months after its addition.

Despite its application in degrading solid waste, nanoparticles are majorly studied for recycling them. Several works were carried out in synthesizing nanoparticles from solid waste and utilizing them for better further applications making multiple advantages. Solid municipal waste containing agricultural sugar cane bagasse, corn residues, scrap tire chips, postconsumer polyethylene, polyethylene terephthalate was first treated by pyrolysis and then utilized for the production of carbon nanotubes (Zhuo et al. 2012). Instead of using mixture of solid waste, any one solid waste from industry or households was collected and studied for synthesis of nanoparticles. In that way, carbon nanoparticles were synthesized from rapeseed oil cake which is a solid waste of oil extraction. The synthesized carbon nanoparticles were also reported to exhibit better antibacterial activity by disrupting their cell wall and it was said that it can be utilized in water treatment plants and also in food processing plants (Purkayastha et al. 2014).

Phenolic extracts of bilberry waste and spent coffee grounds were utilized in the production of silver nanoparticles (Baiocco et al. 2016). Tea waste/Fe_3O_4 was synthesized from tea waste by (Fan et al. 2017). The study showed the ability of TW//Fe_3O_4 to reduce Cr(VI) to Cr(III) and was said to be a better option to adsorb heavy metals since it was found to have better adsorbing ability. Egg shells from the campus restaurant were utilized by Habte et al. (2019) to synthesize calcium oxide nanoparticles and proved to be a promising source of calcium nanoparticles. Similarly, rice husk, bamboo leaf, walnut shell, coconut shell, wheat straw, etc., have also been studied for their role in synthesizing nanoparticles (Zamani et al. 2019).

5.2 Heavy Metal Remediation

Heavy metals are highly toxic to the environment and they pose a serious threat to the living community since they cause various disorders. Since they are highly soluble, they easily find their way into the food chain (Devatha et al. 2016). Several methods such as electrochemical techniques, chemical precipitation, ion exchange, membrane filtration and adsorption have been used to remove heavy metals from water bodies, especially sewage or waste water but these methods pose various disadvantages such as high cost, use of chemicals, etc. Therefore, other methods that nullify the use of these chemicals and those which are cost effective were explored (Abo-State et al. 2015, Singh et al. 2018). Bioremediation is such technique which utilizes the naturally available components of environment in removing heavy metals. During the past few years, bioremediation was synergistically performed with nanotechnology in removing heavy metals.

Heavy metals such as cadmium, nickel, lead, chromium, etc., have been efficiently removed from the water bodies by various green synthesized nanoparticles. Multicomponent nanoparticles produced from extract of *Vaccinium floribundum* Kunth by using ferric chloride and sodium sulphate were found to be efficient in removing copper and zinc from water and the nanoparticles were also found to be efficient in immobilizing heavy metals from contaminated soil (Abril et al. 2018). Copper was also removed by iron oxide nanoparticles synthesized from marine algae *Enteromorpha* sp. by means of adsorption (Ercan et al. 2019). The removal of copper ions was also studied by Shittu and Ihebunna (2017). They showed the synthesis of silver nanoparticles from leaf extract of Piliostigma thonningi and its efficacy in removal of copper ions, which is 82.1% with a contact time of 60 mins. In addition to copper ions, they also demonstrated the removal of magnesium, iron and lead by the synthesized silver nanoparticles, which showed maximum removal of 71.8%, 96.9% and 97.89%, respectively with the contact time of 60 mins.

Another heavy metal that is of primary concern and is widely spread in the environment is lead. Lead is widely used in electroplating, mining, dying, metal finishing, etc., and it causes serious illness in human beings when consuming the lead contaminated water such as renal failure, nausea, convulsions, cancer, coma and affects the metabolism and intelligence (Halim et al. 2003, Rao et al. 2010, Lingamdinne et al. 2016, 2017). Therefore, lead is kept in top tier of heavy metals and its removal by nanoparticles was studied by various researchers. Titanium dioxide nanoparticles synthesized from leaf extract of *Syzygium cumini* showed 82.53% removal of lead. The synthesized nanoparticles in addition to removing lead also reduced the Chemicals Oxygen Demand (COD) to 75.5% (Sethy et al. 2020). Lingamdinne et al. (2019) showed the synthesis of iron nanoparticles from tangerine peel extract (T-Fe_2O_3). With 0.6 g of synthesized T-Fe_2O_3 nanoparticles, 99% lead was removed. Lead was also removed (93% removal) by iron nanoparticles synthesized from leaf extract of *Trigonella foenum graecum*.

Cadmium is another harmful heavy metal usually produced as a result of volcaic eruption but most of the modern world cadmium is due to the extraction of metal itself, because of its use as fungicide and from landfills that are poorly maintained (from glatstein). It causes various health effects in plants, animals and humans. (Yazdani et al. 2018) showed the removal of cadmium (82.2%) effectively by palladium nanoparticles synthesized from brown alga Dictyota indica. Hassan et al. (2018) reported 63.5% removal of cadmium by silver nanoparticles synthesized from marine *Pseudomonas* sp. strain H64. In addition to cadmium, he also reported the removal of other metals such as nickel, copper and lead whose removal was comparatively low than cadmium. Cadmium was also removed by iron nanoparticles synthesized from bacteria *Escherichia coli*. The highest removal of 96.9% was achieved at pH 5.5 and contact time 95.8 mins. Removal of cadmium by CdS fluorescent nanoparticles synthesized from Antartic *Pseudomonas* strain was reported by (Glatstein et al. 2018). In addition to Cd, the nanoparticles were also found to be efficient in the removal of arsenic. In a study by Al-Qahtani (2017), the zero valent silver nanoparticles synthesized from *Benjamina* leaf extract removed almost 85% of cadmium from contaminated water.

In addition to utilizing nanoparticles for removal of heavy metals, they are also used for identifying them. (Maiti et al. 2015) demonstrated the two methods for detecting mercury and copper ions by silver nanoparticles synthesized from juice of *Citrullus lanatus*. Initially, the copper ions were detected by changes in absorbance due to the complex formation of metal ion. Then the silver nanoparticles were further modified with 3-mercapto-1, 2, propanediol for the detection of mercury ions. This proves that nanoparticles have wide range of application in the field of bioremediation.

5.3 Hydrocarbon Degradation

Hydrocarbons are the most problematic pollutants in the environment. Hydrocarbons in the form of oil spills and PET bottles have major difficulty in removing from environment. Therefore, they are always treated by bioremediation. From the discovery of Pseudomonas putida by Anand Chakraborthy till date, many microorganisms have been explored for removal of oil spills as well as plastics. During the recent years, nanoparticles have also been used along with microbes and plants for their removal. Polychlorinated biphenyls (PCBs) are chlorinated hydrocarbons used in capacitors and transformers and also as plasticizers. The degradation of these PCBs by palladium nanoparticles synthesized by *Desulfovibrio desulfuricans* was reported by (Baxter-plant et al. 2004). Similarly, palladium nanoparticles synthesized by *Shewanella oneidensis* were also found to efficiently remove PCBs by dehalogenation (De Windt et al. 2005).

Polyethylene is a derivative of hydrocarbon and has lesser proneness to degradation. Polyethylene are of different types based on density such as Ultra high molecular weight polyethylene (UHMWP), High density polyethylene (HDPE), Medium density polyethylene (MDPE), Low density polyethylene (LDPE) and Very low density polyethylene (VLDPE). In a study by (Jayaprakash and Palemalli 2019), both LDPE and HDPE were found to be removed efficiently by silver nanoparticles synthesized from *Aspergillus oryzae*. The removal efficiency was found to be 64.5% for LDPE and 44.4% for HDPE. Instead of degrading polyethylene products, utilizing them for synthesis of nanoparticles was also performed. In a study by (Al Essawy et al. 2016), carbon nanoparticles were synthesized from polyethylene terephthalate (PET). With this kind of recycling and reusing approach, the pollutant can be removed and a useful product can also be produced making them more advantageous than biodegradation.

Oil spills pose a great threat to the organisms of ocean. The consistent oil spills in ocean during transport seriously affect the marine organisms. Therefore, removing them from oceans is a great deal. To perform this, several researchers have been concentrating on microorganisms, plants and nanoparticles. Crude oil or its forms such as petrol, diesel, lubricants or their components such as phenanthrene were studied for degradation by biosynthesized nanoparticles. $CoFe_2O_4$ synthesized from bark extracts of *Aesculus hippocastanum* has the efficiency to remove 78.5% crude oil from contaminated water (Muhy and Duman 2018). A patent on degrading oil spills by magnetic nanoparticles by the extracts from aerial parts of plant *Anthemis pseudocotula* was filed in 2017 by (Abdullah et al. 2018) and the patent for that was issued on February 2018. Despite using plants or microbes for synthesizing nanoparticles for degrading hydrocarbons, the natural magnetic iron ore was utilized for producing ZnO/CuO/iron nanoparticles and with the help of it, Polycyclic aromatic hydrocarbons were removed by adsorption (Sajadi et al. 2018).

5.4 Degradation of Pesticides

Pesticides are those that protect plant from various insects. Since green revolution, using fertilizers and pesticides or any other chemicals that protects plant have been widely used. But the effect of these chemical compounds on human and also their persistent stay in the environment brought concern among environmentalists. Therefore, removing these pesticides has been studied widely. Biosynthetic iron nanoparticles from yeast extract efficiently degrade the pesticide dichlorovas by liberating phosphate ions from them. The compound itself is broken down into chlorine, carbon, phosphate and hydrogen as separate entities making them harmless (Mehrotra et al. 2017). Dichloro Diphenyl Trichloroethane (DDT) is the most hazardous pesticide that has been banned by several governments around the globe because of its toxic effects to human and animals. This hazardous chemical was degraded to 77.4% with 20 mg/L biosynthesized gold nanoparticles from Lawsonia inermis with a contact time of 72 hrs which is highly efficient with respect to aforesaid persistent pollutant (Abdel aziz et al. 2018).

Chlorpyrifos, another commonly used pesticide, was removed efficiently by silver nanoparticles synthesized from fungal secretes of Penicillium pinophilum. The coefficient of determination for degradation of chlorpyrifos was found to be 0.949 at pH 3, 0.973 at pH 6, and 0.974 at pH 9 (Deka and Sinha 2015). Methyl parathion is again a commonly used pesticide known for its interference in cholinesterase activity that affects nervous system in insects. It not only affects insects but also humans and animals when consumed. Such a harmful pesticide was degraded efficiently by gold nanoparticle synthesized from β-D glucan of *Tricholoma crassum* (Berk.) (Pattanayak et al. 2018). In addition to biosynthesized nanoparticles, nanocomposites also play a role in biodegradation of pesticides. Chitosan silver nnaocomposites remove the pesticide Atrazine from water filtration system (Saifuddin et al. 2011).

Despite using biosynthesized nanoparticles as biodegrading component, they can also be used as natural pesticides. (Vadlapudi and Amanchy 2017) reported the efficiency of biogenic silver nanoparticles synthesized from leaf extract of *Myriostachya wightiana* against the pest that spoils food in storage such as *Tribolium castaneum* (Flour beetle), *Rhyzopertha dominica* (F.) (Lesser grain borer) and *Sitophilus oryzae* L. (Rice weevil). They also described the efficiency of synthesized silver nanoparticles against plant pathogens such as *Xanthomonas campestris* and *Ralstonia solanacearum.*

5.5 Ground Water Remediation

Ground water is the source of wide variety of pollutants since any chemical like pesticide, fertilizers, pharmaceutical products or plastics when contaminates soil automatically contaminates the ground water. Therefore, treating ground water minimizes the travel of such harmful substances and reach several biotic communities of the environment. To perform such remediation, zero valent iron nanoparticles are widely used. These zero valent nanoparticles find its application in removing several contaminants such as arsenic (Kanel et al. 2005), nitrates (Vodiyanitskii and Mineev 2015), hexavalent chromium (Li et al. 2019), and vinyl chloride and reduces total solids, and suspended solids from ground water (Wei et al. 2010). The iron nanoparticles on stabilizing with natural gum Carboxy methyl cellulose (CMC) were found to play much efficient role in ground water remediation (Liu et al. 2015).

5.6 Uranium Remediation

Uranium is the widely known radionuclide and its removal has been studied extensively. The bioreduction of uranium by *Shewanella oneidensis* (Lovely et al. 1991), *Geobacter*, *Desulfovibrio desulfuricans* and *Desulfovibrio vulgaris*, etc., can reduce uranium (Lovely and Philips 1992a, b). But the use of nanoparticles in uranium removal was highly challenging. (Shao et al. 2016) reported the immobilization of uranium in FeS stabilized by gelatine and CMC. Recent research findings reveal that uraninite, which is produced by the microbial action on uranium bulk particles, can be efficiently used in removing uranium from the environment.

5.7 Dye Degradation

Dye degradation is one of the extensively studied processes by many researchers since it is the most polluting component of textile industries which is one of the largest industries. Textile effluents on their disposal not only contaminate water but produce colours due to the presence of dyes in them. Also, these dyes pose serious health effects to all organisms in an ecosystem. Silver nanoparticles are the most extensively studied among all nanoparticles and they were known for their antimicrobial activity. In addition, they were also used for biodegradation of several pollutants and one of them was dyes. Silver nanoparticles synthesized from leaf extract of *Aegle marmelos* degrade malachite green, which is the dye commonly used in laboratory experiments (Femila et al. 2014). Another dye used in laboratories, methylene blue, was degraded by silver nanoparticles synthesized from *Polygonum hydropiper* (Bonnia et al. 2016) and *Morinda tinctoria* (Vanaja et al. 2014). Those which were synthesized from fruit extract of *Viburnum Opulus* L. degrade organic dye such as tartrazine, carmoisine and brilliant blue (David and Moldovan 2020). Silver nanoparticles synthesized using *Hypnea musciformis* (Selvam and Sivakumar 2015) and *Gymnema sylvestre* (Pinagle et al. 2018) degrade dye methyl orange efficiently. Eosin Y, which is a water soluble organic dye, was found to be degraded by silver nanoparticles synthesized from *Trigonella foenum graecum* (Vidhu and Philip 2014). In addition to these, silver nanoparticles synthesized from *Plumbago zeylanica* showed its efficiency in degrading dyes such as Eosin Y, phenol red, methylene blue and methyl red (Roy and Bharadvaja 2019). Silver nanoparticles, in addition to degrading dyes, also degrade the component required for preparation. P nitrophenol is one such component degraded by silver nanoparticles synthesized from *Bacillus amyloliquefaciens* (Samuel et al. 2020).

Copper nanoparticles were also found to play significant role in dye degradation. (Batool et al. 2019) described the degradation kinetics of azo dye congo red by copper nanoparticles synthesized from *Aloe vera*. Similarly, copper nanoparticles synthesized from *Drypetes sepiaria* degrade the aforesaid congo red dye efficiently (Narassaiah et al. 2017a). Copper nanoparticles synthesized from leaf extract of *Achillea millefolium* have greater potency in degrading dye methylene blue (Rabiee et al. 2020). The congo red dye and methyl blue is also degraded by gold nanoparticles synthesized from *Bacillus marisflavi* (Nadaf and Kanase 2016). Congo red was also degraded by Fe_3O_4 synthesized by extracts of *Cnicus Benedictus* (Ruiz-Baltazaar et al. 2020). Malachite green is also degraded by magnetic nanoparticles synthesized from tragacanth gum (Taghavi et al. 2019) and zinc oxide nanoparticles synthesized from *Garcinia mangostana* (Aminuzzaman et al. 2018). Gold nanoparticles synthesized from cotton ball peels were found to degrade methyl orange and methylene blue (Narassaiah et al. 2018). In addition to these various kinds of nanoparticles, selenium nanoparticles synthesized from *Erwinia herbicola* were found to degrade dyes such as methylene blue, methyl orange and eichrome black T (Srivastava and Mukhopadhyay 2014).

6. Conclusion

Nanoparticles were already exploited in various fields such as cosmetics, pharmaceutical industry, medicine, food, etc. They are still explored in various fields and one among them is biodegradation. Several nanoparticles synthesized from biomaterials were found to degrade several pollutants. With further advancements, the nanoparticles can still be explored for better biodegradation and degrade less degrading components.

References

Abbas, S., S. Nasreen, A. Haroon and M.A. Ashraf. 2020. Synthesis of silver and copper nanoparticles from plants and application as adsorbents for naphthalene decontamination. Saudi J. Biol. Sci. 27(4): 1016–1023.

Abd El-Aziz, A.R., M.R. Al-Othman and M.A. Mahmoud. 2018. Degradation of DDT by gold nanoparticles synthesised using *Lawsonia inermis* for environmental safety. Biotechnol. Biotechnol. Equip. 32(5): 1174–1182.

Abdel Rahim, K., S.Y. Mahmoud, A.M. Ali, K.S. Almaary, A.E. Mustafa and S.M. Husseiny. 2017. Extracellular biosynthesis of silver nanoparticles using *Rhizopus stolonifer*. Saudi J. Biol. Sci. 24(1): 208–216.

Abdel-Raouf, N., N.M. Al-Enazi and I.B. Ibraheem. 2017. Green biosynthesis of gold nanoparticles using Galaxaura elongata and characterization of their antibacterial activity. Arab. J. Chem. 10: S3029–S3039.

Abdel-Shafy, H.I. and M.S.M. Mansour. 2018. Solid waste issue: Sources, composition, disposal, recycling, and valorization, Egypt. J. Pet. 27: 1275–1290.

Abdullah, M.M.S., A.M. Atta, H.A. Al-Lohedan, H.Z. Alkhathlan, M. Khan and A.O. Ezzat. 2018. Biosynthesized magnetic metal nanoparticle for oil spill, Patent No: US 9,901,903 B1, Filed date: Sep 14, 2017, Patent date: Feb 27, 2018.

Abo-State, M.A.M. and A.M. Partila. 2015. Microbial production of silver nano particles by *Pseudomonas aeruginosa* cell free extract. J. Eco. Heal. Env. 3: 91–98.

Abril, M., H. Ruiz and L. Cumbal. 2018. Biosynthesis of multicomponent nanoparticles with extract of mortiño (*Vaccinium floribundum Kunth*) berry: Application on heavy metals removal from water and immobilization in soils. J Nanotechnol. Article id 9504807,10.1155/2018/9504807.

Acharya, C., D. Joseph and S.K. Apte. 2009. Uranium sequestration by a marine cyanobacterium, *Synechococcus elongatus* strain BDU/75042. Bioresour. Technol. 100: 2176–2181.

Acharya, C. 2015. Microbial bioremediation of uranium: An overview. BARC Newslett. March–April, 27–30.

Adriano, D.C. 1986. Trace Elements in the Terrestrial Environment. Springer-Verlag, New York.

Ahmad, A., P. Mukherjee, D. Mandal, S. Senapati, M.I. Khan, R. Kumar and M. Sastry. 2002. Enzyme mediated extracellular synthesis of CdS nanoparticles by the fungus, Fusarium oxysporum. ACS 124(41): 12108–12109.

Ahmad, A., P. Mukherjee, S. Senapati, D. Mandal, M.I. Khan, R. Kumar and M. Sastry. 2003a. Extracellular biosynthesis of silver nanoparticles using the fungus Fusarium oxysporum. Colloids Surf. B. 28(4): 313–318.

Ahmad, A., S. Senapati, M.I. Khan, R. Kumar and M. Sastry. 2003b. Extracellular biosynthesis of monodisperse gold nanoparticles by a novel extremophilic actinomycete, *Thermomonospora* sp. Langmuir 19(8): 3550–3553.

Ahmad, R.S., F. Ali, R.S. Hamid and M. Sara. 2007. Synthesis and effect of silver nanoparticles on the antibacterial activity of different antibiotics against *Staphylococcus aureus* and *Escherichia coli*. Nanomed. Nanotechnol. 3: 168–171.

Ahmed, S., M. Saifullah Ahmad, B.L. Swami and S. Ikram. 2016. Green synthesis of silver nanoparticles using *Azadirachta indica* aqueous leaf extract. J. Radiat. Res. Appl. Sci. 9(1): 1–7.

Al-Qahtani, K.M. 2017. Cadmium removal from aqueous solution by green synthesis zero valent silver nanoparticles with Benjamina leaves extract. Egypt. J. Aquat. Res. 43: 269–274.

Alexandratos, S.D. 2009. Ion-exchange resins: A retrospective from industrial and engineering chemistry research. Ind. Eng. Chem. Res. 48(1): 388–398.

Aljabali, A.A.A., Y. Akkam, M.S. Al Zoubi, K.M. Al-Batayneh, B. Al-Trad, O. Abo Alrob, A.M. Alkilany, M. Benamara and D.J. Evans. 2018. Synthesis of gold nanoparticles using leaf extract of *Ziziphus zizyphus* and their antimicrobial activity. Nanomater. 8(3): 174.

Al-Radadi, N.S. 2019. Green synthesis of platinum nanoparticles using Saudi's dates extract and their usage on the cancer cell treatment. Arab. J Chem. 12(3): 330–349.

Al-Zahrani, H., A. El-Waseif and D. El-Ghwas. 2018. Biosynthesis and evaluation of TiO_2 and ZnO nanoparticles from *in vitro* stimulation of *Lactobacillus johnsonii*. J. Innov. Pharm. Biol. Sci. 5(1): 16–20.

Aminuzzaman, M., L.P. Ying, W.S. Goh and A. Watanabe. 2018. Green synthesis of zinc oxide nanoparticles using aqueous extract of Garcinia mangostana fruit pericarp and their photocatalytic activity. Bullet. Mater. Sci. 41(2): 50.

Annual report (2019-2020) Government of India, Ministry of Chemicals and Fertilizers, Department of Chemicals and Petrochemicals.

Antony, E., M. Sathiavelu and S. Arunachalam. 2016. Synthesis of silver nanoparticles from the medicinal plant bauhinia acuminata and biophytum sensitivumâ€"a comparative study of its biological activities with plant extract. Int. J. Appl. Pharm. 9(1): 22–29.

Anu, K., S. Devanesan, R. Prasanth, M. AlSalhi, S. Ajithkumar and G. Singaravelu. 2020. Biogenesis of selenium nanoparticles and their anti-leukemia activity. J. King Saud. Univ. Sci. 32. 10.1016/j.jksus.2020.04.018.

Anu, Y. and M.D. Vijay. 2016. *Camellia sinensis* mediated synthesis of iron nanoparticles and its encapsulation for decolorization of dyes. BCAIJ 10(1): 020–029.

Appukuttan, D., A.S. Rao and S.K. Apte. 2006. Engineering of *Deinococcus radiodurans* R1 for bioprecipitation of uranium from dilute nuclear waste. Appl. Environ. Microbiol. 72(12): 7873–7878.

Aritonang, H.F., H. Koleangan and A.D. Wuntu. 2019. Synthesis of silver nanoparticles using aqueous extract of medicinal plants' (*Impatiens balsamina* and *Lantana camara*) fresh leaves and analysis of antimicrobial activity. Int. J. Microbial. 2019: 8642303. https://doi.org/10.1155/2019/8642303.

Aromal, S.A., V.K. Vidhu and D. Philip. 2012. Green synthesis of well-dispersed gold nanoparticles using Macrotyloma uniflorum. Spectrochim. Acta A Mol. Biomol. Spectrosc. 85(1): 99–104.

Arpita, R. and B. Navneeta. 2019. Silver nanoparticle synthesis from *Plumbago zeylanica* and its dye degradation activity. Bioinspir. Biomim. Nanobiomater. 8(2): 130–140.

Arshad, A. 2017. Bacterial synthesis and applications of nanoparticles. Nano. Sci. Nano. Technol. 11(2): 119.

Arsiya, F., M.H. Sayadi and S. Sobhani. 2017. Green synthesis of palladium nanoparticles using Chlorella vulgaris. Mater. Lett. 186: 113–115.

Ashmi, M., G. Oza, P.Dr. Sunil and S. Madhuri. 2012. Extracellular synthesis of gold nanoparticles using Pseudomonas denitrificans and comprehending its stability. J. Microbial. Biotechnol. Res. 2(4): 493–499.

Avendaño, R., N. Chaves, P. Fuentes, E. Sánchez, J.I. Jiménez and M. Chavarría. 2016. Production of selenium nanoparticles in *Pseudomonas putida* KT2440. Sci. Rep. 6: 37155. https://doi.org/10.1038/srep37155.

Azizi, S., M.B. Ahmad, F. Namvar and R. Mohamad. 2014. Green biosynthesis and characterization of zinc oxide nanoparticles using brown marine macroalga Sargassum muticum aqueous extract. Mater. Lett. 116: 275–277.

Babu, M.M.G. and P. Gunasekaran. 2009. Production and structural characterization of crystalline silver nanoparticles from *Bacillus cereus* isolate. Colloids Surfaces B 74(1): 191–195.

Baesman, S.M., T.D. Bullen, J. Dewald, D. Zhang, S. Curran, F.S. Islam, T.J. Beveridge and R.S. Oremland. 2007. Formation of tellurium nanocrystals during anaerobic growth of bacteria that use the oxyanions as respiratory electron acceptors. Appl. Environ. Microbiol. 73(7): 2135–2143.

Baiocco, D., R. Lavecchia, S. Natali and A. Zuorro. 2016. Production of metal nanoparticles by agro-industrial wastes: A green opportunity for nanotechnology. Chem. Eng. Trans. 47(6): 67–72.

Bala, N., S. Saha, M. Chakraborty, M. Maiti, S. Das, R. Basu and P. Nandy. 2015. Green synthesis of zinc oxide nanoparticles using *Hibiscus subdariffa* leaf extract: Effect of temperature on synthesis, anti-bacterial and anti-diabetic activity. RSC Adv. 5: 4993–5003.

Ballesteros, S., J.M. Rincón, B. Rincón-Mora and M.M. Jordán. 2017. Vitrification of urban soil contamination by hexavalent chromium. J. Geochem. Explor. 174: 132–139.

Baltazar-Encarnación, E., C.E. Escárcega-González, X.G. Vasto-Anzaldo, M.E. Cantú-Cárdenas and J.R. Morones-Ramírez. 2019. Silver nanoparticles synthesized through green methods using Escherichia coli Top 10 (Ec-Ts) growth culture medium exhibit antimicrobial properties against nongrowing bacterial strains. J. Nanomat. Article ID 4637325: 1–8. https://doi.org/10.1155/2019/4637325.

Bankar, A., B. Joshi, A.R. Kumar and S. Zinjarde. 2010. Banana peel extract mediated novel route for the synthesis of palladium nanoparticles. Mater. Lett. 64: 1951–1953.

Bansal, V., D. Rautaray, A. Bharde, K. Ahire, A. Sanyal, A. Ahmad and M. Sastry. 2005. Fungus-mediated biosynthesis of silica and titania particles. J. Mater. Chem. 15(26): 2583–2589.

Bansal, V., P. Poddar, A. Ahmad and M. Sastry. 2006. Room-temperature biosynthesis of ferroelectric barium titanate nanoparticles. Journal of the American Chemical Society 128(36): 11958–11963.

Bao, H., Z. Lu, X. Cui, Y. Qiao, J. Guo, J. Anderson and C. Li. 2010. Extracellular microbial synthesis of biocompatible CdTe quantum dots. Acta Biomater. 6(9): 3534–3541.

Bargar, J.R., R. Bernier-Latmani, D.E. Giammar and B.M. Tebo. 2008. Biogenic uraninite nanoparticles and their importance for uranium remediation. Elements 4(6): 407–412.

Batool, M., M.Z. Qureshi, F. Hashmi, N. Mehboob and A.S. Shah. 2019. Congo red azo dye removal and study of its kinetics by aloe vera mediated copper oxide nanoparticles. Indones. J. Chem. 19(3): 626–637.

Baxter-Plant, V.S., I.P. Mikheenko, M. Robson, S.J. Harrad and L.E. Macaskie. 2004. Dehalogenation of chlorinated aromatic compounds using a hybrid bioinorganic catalyst on cells of Desulfovibrio desulfuricans. Biotechnol. Lett. 26(24): 1885–1890.

Bazylizinki, D.A., B.R. Heywood and S. Mann. 1993. Fe3O4 and Fe3S4 in a bacterium. Nature 366: 218.

Bhainsa, K.C. and S.F. D'Souza. 2006. Extracellular biosynthesis of silver nanoparticles using the fungus *Aspergillus fumigatus*. Colloid Surf. B Biointerfaces 47(2): 160–164.

Bhakya, S., S. Muthukrishnan, M. Sukumaran, M. Muthukumar, S.T. Kumar and M.V. Rao. 2015. Catalytic degradation of organic dyes using synthesized silver nanoparticles: A green approach. J. Biorem. Biodegred. 6(5): 1–9.

Bhattacharya, A. and P. Ghosh. 2004. Nanofiltration and reverse osmosis membranes: Theory and application in separation of electrolytes. Rev. Chem. Eng. 20(1-2): 111–173.

Bhattacharyya, A., K.M. Campbell, S.D. Kelly, Y. Roebbert, S. Weyer, R. Bernier-Latmani and T. Borch. 2017. Biogenic non-crystalline U(IV) revealed as major component in uranium ore deposits. Nat. Comm. 8(1): 15538, 1–8. https://doi.org/10.1038/ncomms15538.

Bonnia, N., M. Kamaruddin, M. Nawawi, S. Ratim, H. Azlina and E. Ali. 2016. Green biosynthesis of silver nanoparticles using 'Polygonum hydropiper' and study its catalytic degradation of methylene blue. Procedia Chem. 19. Doi: 10.1016/j.proche.2016.03.058.

Bräunig, J., C. Baduel, C.M. Barnes and J.F. Mueller. 2019. Leaching and bioavailability of selected perfluoroalkyl acids (PFAAs) from soil contaminated by firefighting activities. Sci. Total Environ. 646: 471–479.

Castro-Longoria, E., S.D. Moreno-Velázquez, A.R. Vilchis-Nestor, E. Arenas-Berumen and M. Avalos-Borja. 2012. Production of platinum nanoparticles and nanoaggregates using *Neurospora crassa*. J. Microbial. Biotechnol. 22(7): 1000–1004.

Cefic Facts and Figures of the European Chemical Industry, 2020. https://www.francechimie.fr/media/52b/the-european-chemical-industry-facts-and-figures-2020.pdf.

Chahardoli, A., N. Karimi, F. Sadeghi and A. Fattahi. 2018. Green approach for synthesis of gold nanoparticles from Nigella arvensis leaf extract and evaluation of their antibacterial, antioxidant, cytotoxicity and catalytic activities. Artif. Cells Nanomed. Biotechnol. 46(3): 579–588.

Chauhan, A., S. Zubair, S. Tufail, A. Sherwani, M. Sajid, C.R. Suri, A. Azam and M. Owais. 2011. Fungus-mediated biological synthesis of gold nanoparticles: Potential in detection of liver cancer. International Journal of Nanomedicine 6: 2305–2319.

Chauhan, R., A. Reddy and J. Abraham. 2015. Biosynthesis of silver and zinc oxide nanoparticles using Pichia fermentans JA2 and their antimicrobial property. Appl. Nanosci. 5(1): 63–71.

Chekli, L., G. Brunetti, E.R. Marzouk, A. Maoz-Shen, E. Smith, R. Naidu, H.K. Shon, E. Lombi and E. Donner. 2016. Evaluating the mobility of polymer-stabilised zero-valent iron nanoparticles and their potential to co-transport contaminants in intact soil cores. Environ. Pollut. 216: 636–645.

Chen, P.Y., X. Dang, M.T. Klug, N.M.D. Courchesne, J. Qi, M.N. Hyder, A.M. Belcher and P.T. Hammond. 2015. M13 virus-enabled synthesis of titanium dioxide nanowires for tunable mesoporous semiconducting networks. Chem. Mater. 27(5): 1531–1540.

Choppin, G., J.O. Liljenzin and J. Rydberg. 2002. Behavior of radionuclides in the environment. Radiochemistry Nuclear Chemistry (Third edition), Butterworth- Heinemann, London, 653–685.

Choy, C.C., G.P. Korfiatis and X. Meng. 2006. Removal of depleted uranium from contaminated soils. J. Hazard. Mater. 136: 53–60.

Connor, E.E., J. Mwamuka, A. Gole, C.J. Murphy and M.D. Wyatt. 2005. Gold nanoparticles are taken up by human cells but do not cause acute cytotoxicity. Small 1(3): 325–327.

Correa-Llantén, D.N., S.A. Mu-oz-Ibacache, M.E. Castro, P.A. Mu-oz and J.M. Blamey. 2013. Gold nanoparticles synthesized by Geobacillus sp. strain ID17 a thermophilic bacterium isolated from Deception Island. Antarctica. Microb. Cell Fact. 12(75): 1–6.

Cui, D., T. Liang, L. Sun, L. Meng, C. Yang, L. Wang, T. Liang and Q. Li. 2018. Green synthesis of selenium nanoparticles with extract of hawthorn fruit induced HepG2 cells apoptosis. Pharm. Biol. 56(1): 528–534.

Das, S.K., C. Dickinson, F. Lafir, D.F. Brougham and E. Marsili. 2012. Synthesis, characterization and catalytic activity of gold nanoparticles biosynthesized with Rhizopus oryzae protein extract. Green Chem. 14(5): 1322–1334.

Datta, A., C. Patra, H. Bharadwaj, S. Kaur, N. Dimri and R. Khajuria. 2017. Green synthesis of zinc oxide nanoparticles using *Parthenium hysterophorus* leaf extract and evaluation of their antibacterial properties. J. Biotechnol. Biomater. 7(3): 1000271. 10.4172/2155-952X.1000271.

David, L. and B. Moldovan. 2020. Green synthesis of biogenic silver nanoparticles for efficient catalytic removal of harmful organic dyes. Nanomater. 10(2): 202.

De Windt, W., P. Aelterman and W. Verstraete. 2005. Bioreductive deposition of palladium (0) nanoparticles on Shewanella oneidensis with catalytic activity towards reductive dechlorination of polychlorinated biphenyls. Environ. Microbiol. 7(3): 314–325. https://doi.org/10.1111/j.1462-2920.2005.00696.x.

De-Gusseme, B., L. Sintubin, L. Baert, E. Thibo, T. Hennebel, G. Vermeulen, M. Uyttendaele, W. Verstraete and N. Boon. 2010. Biogenic silver for disinfection of water contaminated with viruses. Appl. Environ. Microbiol. 76(4): 1082–1087.

Deka, A. and S. Sinha. 2015. Mycogenic silver nanoparticle biosynthesis and its pesticide degradation potentials. International Journal of Technology Enhancements and Emerging Engineering Research 3: 108–113.

Deljou, A. and S. Goudarzi. 2016. Green extracellular synthesis of the silver nanoparticles using thermophilic *Bacillus* sp. AZ1 and its antimicrobial activity against several human pathogenetic bacteria. Iran J. Biotechnol. 14(2): 25–32.

Deniz, F., A. Adıgüzel and M. Mazmanci. 2019. Turkish journal of engineering the biosynthesis of silver nanoparticles by cytoplasmic fluid of coriolus versicolor. Turk. J Eng. 3(2): 92–96. 10.31127/tuje.429072.

Deplanche, K., I. Caldelari, I.P. Mikheenko, F. Sargent and L.E. Macaskie. 2010. Involvement of hydrogenases in the formation of highly catalytic Pd(0) nanoparticles by bioreduction of Pd(II) using *Escherichia coli* mutant strains. Microbiol. 156(9): 2630–2640.

Devatha, C.P., A.K. Thalla and S.Y. Katte. 2016. Green synthesis of iron nanoparticles using different leaf extracts for treatment of domestic waste water. J. Clean. Prod. 139: 1425–1435.

Dey, A., A. Yogamoorthi and S. Sundarapandian. 2018. Green synthesis of gold nanoparticles and evaluation of its cytotoxic property against colon cancer cell line. RJLBPCS 4(6): 1–17. 10.26479/2018.0406.01.

Dhuper, S., D. Panda and P.L. Nayak. 2012. Green synthesis and characterization of zero valent iron nanoparticles from the leaf extract of *Mangifera indica*. Nano. Trends: J. Nanotech. App. 13(2): 16–22.

Divya, M., G. Seghalkiran, S. Hassan and J. Selvin. 2019. Biogenic synthesis and effect of silver nanoparticles (AgNPs) to combat catheterrelated urinary tract infections. Biocatal. Agric. Biotechnol. 18: 101037.

Du, L.W., H. Jiang, X.H. Liu and E.K. Wang. 2007. Biosynthesis of gold nanoparticles assisted by Escherichia coli DH5a and its application on direct electrochemistry of hemoglobin. Electrochem. Commun. 9: 1165–1170.

El Essawy, A., A.H. Konsowa, M. Elnouby and H.A. Farag. 2016. A novel one step synthesis for carbon based nanomaterials from polyethyleneterephthalate (PET) bottles waste. J. Air Waste Manag. Assoc. DOI: 10.1080/10962247.2016.1242517.

Elbagory, A.M., M. Meyer and A.A. Hussein. 2017. Green synthesis of gold nanoparticles from South African plant extracts for the treatment of skin infection Wounds. J. Nanomed. Nanotechnol. 8(4). DOI: 10.4172/2157-7439-C1-049.

El-Batal, A.I., S.S. Mona and M.S.S. Al-Tamie. 2015. Biosynthesis of gold nanoparticles using marine streptomyces cyaneus and their antimicrobial, antioxidant and antitumor (*in vitro*) activities. J. Chem. Pharm. Res. 7: 1020–1036.

Environmental Protection Agency US 2000. Wastewater technology sheet: chemical precipitation. United State Environmental Protection, 832-F-00-018.

Ercan, G., D. Uzunoğlu, M. Ergüt and A. Özer. 2019. Biosynthesis and characterization of iron oxide nanoparticles from *Enteromorpha* spp. extract: Determination of adsorbent properties for copper (II) ions. Int. Adv. Res. Eng. J. 3(1): 65–74.

Erdogan, O., M. Abbak, G.M. Demirbolat, F. Birtekocak, M. Aksel, S. Pasa and O. Cevik. 2019. Green synthesis of silver nanoparticles via Cynara scolymus leaf extracts: The characterization, anticancer potential with photodynamic therapy in MCF7 cells. PloS One 14(6): e0216496.

Ezealisiji, K.M., X. Siwe-Noundou, B. Maduelosi, N. Nwachukwu and R.W.M. Krause. 2019. Green synthesis of zinc oxide nanoparticles using *Solanum torvum* (L.) leaf extract and evaluation of the toxicological profile of the ZnO nanoparticles-hydrogel composite in Wistar albino rats. Int. Nano Lett. 9(2): 99–107.

Fan, S., Y. Wang, Y. Li, J. Tang, Z. Wang, J. Tang, X. Li and K. Hu. 2017. Facile synthesis of tea waste/Fe$_3$O$_4$ nanoparticle composite for hexavalent chromium removal from aqueous solution. RSC Adv. 7(13): 7576–7590.

Fardsadegh, B. and H. Jafarizadeh-Malmiri. 2019. Aloe vera leaf extract mediated green synthesis of selenium nanoparticles and assessment of their *in vitro* antimicrobial activity against spoilage fungi and pathogenic bacteria strains. Green Process. Synth. 8(1): 399–407.

Femila, E.M.E., R. Srimathi and D. Charumathi. 2014. Removal of malachite green using silver nanoparticles via adsorption and catalytic degradation. Int. J. Pharm. Pharm. Sci. 6(8): 579–583.

Galdames, A., A. Mendoza, M. Orueta, I.S.D.S. García, M. Sánchez, I. Virto and J.L. Vilas. 2017. Development of new remediation technologies for contaminated soils based on the application of zero-valent iron nanoparticles and bioremediation with compost. Res. Efficient Technol. 3: 166–176.

Gangula, A., R. Podila, L. Karanam, C. Janardhana and A.M. Rao. 2011. Catalytic reduction of 4-nitrophenol using biogenic gold and silver nanoparticles derived from *Breynia rhamnoides*. Langmuir 27(24): 15268–15274.

Gavrilescu, M., L. Vasile and I. Cretescu. 2009. Characterization and remediation of soils contaminated with uranium. J. Hazard. Mater. 163: 475–510.

Geetha, P., M.S. Latha, S.S. Pillai, B. Deepa, K.S. Kumar and M. Koshy. 2015. Green synthesis and characterization of alginate nanoparticles and its role as a biosorbent for Cr (VI) ions. J. Mol. Struct. 1105: 54–60.

Ghahremanzadeh, R., F.Y. Samadi and M. Yousefi. 2019. Green synthesis of gold nanoparticles using three medicinal plant extracts as efficient reducing agents. IJCCE 38(1): 1–10.

Ghorbani, H.R. 2013. Biosynthesis of silver nanoparticles using *Salmonella typhirium*. J. Nanostruct. Chem. 3: 29.

Glatstein, D.A., N. Bruna, C. Gallardo-Benavente, D. Bravo, M.E. Carro Perez, F.M. Francisca and J.M. Pérez-Donoso. 2018. Arsenic and cadmium bioremediation by antarctic bacteria capable of biosynthesizing CdS fluorescent nanoparticles. J. Environ. Eng. 144(3): 04017107.

Gogoi, S. 2013. Green synthesis of silver nanoparticles from leaves extract of ethnomedicinal plants—Pogostemon benghalensis (B) O. Ktz. Adv. Appl. Sci. Res. 4(4): 274–278.

Goldenman, G. 2017. The strategy for a non-toxic environment of the 7th Environment Action Programme Sub-study d–Very persistent chemicals, Milieu Ltd.

Govindraju, K., S.K. Basha, V.G. Kumar and G. Singaravelu. 2008. Silver, gold and bimetallic nanoparticles production using single-cell protein (*Spirulina platensis*) Geitler. J. Mater. Sci. 43: 5115–5122.

Grieger, K.D., A. Fjordboge, N.B. Hartmann, E. Eriksson, P.L. Bjerg and A. Baun. 2010. Environmental benefits and risks of zero-valent iron nanoparticles (nZVI) for *in situ* remediation: Risk mitigation or trade-off. J. Contam. Hydrol. 118(3-4): 165–83.

Gurunathan, S., K. Kalishwaralal, R. Vaidyanathan, D. Venkataraman, S.R.K. Pandian, J. Muniyandi, N. Hariharan and S.H. Eom. 2009. Biosynthesis, purification and characterization of silver nanoparticles using Escherichia coli. Colloids Surf B Biointerfaces 74(1): 328–335.

Habte, L., N. Shiferaw, D. Mulatu, T. Thenepalli, R. Chilakala and J.W. Ahn. 2019. Synthesis of nano-calcium oxide from waste eggshell by sol-gel method. Sustainability 11(11): 3196. Doi: 10.3390/su11113196.

Hai, V., P. Poddar, A. Ahmad and M. Sastry. 2006. Room-temperature biosynthesis of ferroelectric barium titanate nanoparticles. J. Am. Chem. Soc. 128(36): 11958–11963.

Halim, A.S.H., A.M.A. Shehata and M.F. Shahat. 2003. Removal of lead ions from industrial waste water by different types of natural materials. Water Res. 37: 1678–1683.

Hamby, D.M. 1996. Site remediation techniques supporting environmental restoration activities—A review. Sci. Total Environ. 191: 203–224.

Hamouda, R.A., M.H. Hussein, R.A. Abo-elmagd and S.S. Bawazir. 2019. Synthesis and biological characterization of silver nanoparticles derived from the cyanobacterium *Oscillatoria limnetica*. Sci. Rep. 9(1): 1–17.

Han, Z., L. Dong, J. Zhang, T. Cui, S. Chen, G. Ma, X. Guo and L. Wang. 2019. Green synthesis of palladium nanoparticles using lentinan for catalytic activity and biological applications. RSC Adv. 9(65): 38265–38270.

Hariharan, A., T.N. Begum, M.H. Ilyas, H.S. Jahangir, P. Kumpati, S. Mathew, A. Govindaraju and I. Qadri. 2016. Synthesis of plant mediated gold nanoparticles using *Azima tetracantha Lam*. Leaves extract and evaluation of their antimicrobial activities. Pharm. J. 8(5): 507–512.

Hasan, S., S. Singh, R.Y. Parikh, M.S. Dharne, M.S. Patole, B.L.V. Prasad and Y.S. Shouche. 2008. Bacterial synthesis of copper/copper oxide nanoparticles. J. Nanosci. Nanotechnol. 8(6): 3191–3196.

Hassan, S.E.D., S.S. Salem, A. Fouda, M.A. Awad, M.S. El-Gamal and A.M. Abdo. 2018. New approach for antimicrobial activity and bio-control of various pathogens by biosynthesized copper nanoparticles using endophytic actinomycetes. J. Radait. Res. Appl. Sci. 11(3): 262–270.

Hassan, S.W. and H.H. El-latif. 2018. Characterization and applications of the biosynthesized silver nanoparticles by marine Pseudomonas sp. H64. J. Pure Appl. Microbiol. 12(3): 1289–1299.

He, Y., F. Wei, Z. Ma, H. Zhang, Q.R. Yang, B. Yao, Z. Huang, J. Li, C. Zeng and Q. Zhang. 2017. Green synthesis of silver nanoparticles using seed extract of *Alpinia katsumadai*, and their antioxidant, cytotoxicity, and antibacterial activities. RSC Adv. 39842–39851.

Hemlata, Meena, P.R., A.P. Singh and K.K. Tejavath. 2020. Biosynthesis of silver nanoparticles using *Cucumis prophetarum* aqueous leaf extract and their antibacterial and antiproliferative activity against cancer cell lines. ACS Omega 5(10): 5520–5528.

Hu, L., G. Zeng, G. Chen, H. Dong, Y. Liu, J. Wan and Z. Yu. 2016. Treatment of landfill leachate using immobilized *Phanerochaete chrysosporium* loaded with nitrogen doped TiO_2 nanoparticles. J. Hazard. Mater. 301: 106–118.

Husseiny, M.I., M.A. El-Aziz, Y. Badr and M.A. Mahmoud. 2007. Biosynthesis of gold nanoparticles using *Pseudomonas aeruginosa*. Spectrochim. Acta A 67(3-4): 1003–1006.

Ibhadon, A.O. and P. Fitzpatrick. 2013. Hetergeneous photocatalysts; recent advances and applications. Catalysts 3: 189–218.

Ismail, E.H., A. Saqer, E. Assirey, A. Naqvi and R.M. Okasha. 2018. Successful green synthesis of gold nanoparticles using a *Corchorus olitorius* extract and their antiproliferative effect in cancer cells. Int. J Mol. Sci. 19(9): 2612.

Iwamoto, T. and M. Nasu. 2001. Current bioremediation practice and perspective. J. Biosci. Bioeng. 92: 1.

Jafari, M., F. Rokhbakhsh-Zamin, M. Shakibaie, M.H. Moshafi, A. Ameri, H.R. Rahimi and H. Forootanfar. 2018. Cytotoxic and antibacterial activities of biologically synthesized gold nanoparticles assisted by *Micrococcus yunnanensis* strain J2. Biocatal. Agric. Biotechnol. 15: 245–253.

Jafarizad, A., K. Safaee, S. Gharibian, Y. Omidi and D. Ekinci. 2015. Biosynthesis and *in-vitro* study of gold nanoparticles using mentha and pelargonium extracts. Procedia Mater. Sci. 11: 224–230.

Jain, D., S. Kachhwaha, R. Jain, G. Srivatsava and S.L. Kothari. 2010. Novel microbial route to synthesize silver nanoparticles using spore crystal mixture of Bacillus thuringiensis. Indian J. Exp. Biol. 48: 1152–1156.

Jain, N., A. Bhargava, S. Majumdar, J. Tarafdar and J. Panwar. 2011. Extracellular biosynthesis and characterization of silver nanoparticles using *Aspergillus flavus* NJP08: A mechanism perspective. Nanoscale 3(2): 635–641.

Jain, N., A. Bhargava, J.C. Tarafdar, S.K. Singh and J. Panwar. 2013. A biomimetic approach towards synthesis of zinc oxide nanoparticles. Appl. Microbiol. Biotechnol. 97(2): 859–869.

Jain, S. and M.S. Mehata. 2017. Medicinal plant leaf extract and pure flavonoid mediated green synthesis of silver nanoparticles and their enhanced antibacterial property. Sci. Rep. 7: 15867.

Jalal, M., M.A. Ansari, M.A. Alzohairy, S.G. Ali, H.M. Khan, A. Almatroudi and K. Raees. 2018. Biosynthesis of silver nanoparticles from oropharyngeal *Candida glabrata* isolates and their antimicrobial activity against clinical strains of bacteria and fungi. Nanomater. 8(8): 586.

Jayaprakash, V. and U.M.D. Palempalli. 2019. Studying the effect of biosilver nanoparticles on polyethylene degradation. Appl. Nanosci. 9: 491–504.

Jayaseelan, C., R. Ramkumar, A.A. Rahuman and P. Perumal. 2013. Green synthesis of gold nanoparticles using seed aqueous extract of *Abelmoschus esculentus* and its antifungal activity. Ind. Crop. Prod. 45: 423–429.

Jeong, E.H., G. Jung, C. Am Hong and H. Lee. 2014. Gold nanoparticle (AuNP)-based drug delivery and molecular imaging for biomedical applications. Arch. Pharm. Res. 37(1): 53–59.

Jeyapaul, U., M.J. Kala, A.J. Bosco, P. Piruthiviraj and M. Easuraja. 2018. An eco-friendly approach for synthesis of platinum nanoparticles using leaf extracts of *Jatropa gossypifolia* and *Jatropa glandulifera* and its antibacterial activity. Orient. J Chem. 34(2): 783–790.

Jo, J.H., P. Singh, Y.J. Kim, C. Wang, R. Mathiyalagan, C.G. Jin and D.C. Yang. 2016. *Pseudomonas deceptionensis* DC5-mediated synthesis of extracellular silver nanoparticles. Artificial Cells Nanomed. Biotechnol. 44(6): 1576–1581.

Joerger, R., T. Klaus and C.G. Granquist. 2000. Biologically produced silver-carbon composite materials for optically thin film coatings. Adv. Mater. 12(6): 407–409.

Joshi, S.J., S.J. Geetha, S. Al-Mamari and A. Al-Azkawi. 2018. Green synthesis of silver nanoparticles using pomegranate peel extracts and its application in photocatalytic degradation of methylene blue. Jundishapur J. Nat. Pharm. Prod. 13(3): e67846.

Kabata-Pendias, A. and H. Pendias. 1992. Trace Elements in Soils and Plants. CRC Press, Boca Raton, FL, USA.

Kalpana, V., B.A.S. Kataru, N. Sravani, T. Vigneshwari, A. Panneerselvam and V.D. Rajeswari. 2018. Biosynthesis of zinc oxide nanoparticles using culture filtrates of Aspergillus niger: Antimicrobial textiles and dye degradation studies. OpenNano 3: 48–55.

Kanel, S.R., B. Manning, L. Charlet and H. Choi. 2005. Removal of arsenic (III) from groundwater by nanoscale zero-valent iron. Environ. Sci. Technol. 39(5): 1291–1298.

Kapur, M., K. Soni and K. Kohli. 2017. Green synthesis of selenium nanoparticles from broccoli, characterization, application and toxicity. Adv. Tech. Biol. Med. 5(1): 1000198.

Karam, A., K. Zaher and A.S. Mahmoud. 2020. Comparative studies of using nano zerovalent iron, activated carbon, and green synthesized nano zerovalent iron for textile wastewater color removal using artificial intelligence, regression analysis, adsorption isotherm, and kinetic studies. Air Soil Water Res. 13: 1178622120908273.

Karthik, L., G. Kumar, A.V. Kirthi, A.A. Rahuman and K.V. Bhaskara Rao. 2014. Streptomyces sp. LK3 mediated synthesis of silver nanoparticles and its biomedical application. Bioprocess Biosyst. Eng. 37(2): 261–267.

Karthik, R., R. Sasikumar, S.M. Chen, M. Govindasamy, J.K. Kumar and V. Muthura. 2016. Green synthesis of platinum nanoparticles using *Quercus Glauca* extract and its electrochemical oxidation of hydrazine in water samples. Int. J. Electrochem. Sci. 11: 8245–8255.

Kathiresan, K., S. Manivannan, M.A. Nabeel and B. Dhivya. 2009. Studies on silver nanoparticles synthesized by a marine fungus, *Penicillium fellutanum* isolated from coastal mangrove sediment. Colloid Surf. B, Biointerfaces 71(1): 133–137.

Keijok, W., R. Pereira, L. Alvarez, A. Prado, A.R. Da Silva, J. Ribeiro, J. Oliveira and M. Guimarães. 2019. Controlled biosynthesis of gold nanoparticles with Coffea arabica using factorial design. Sci. Rep. 9: 16019.

Khan, M.Z., F.K. Tareq, M.A. Hossen and M.N.A.M. Roki. 2018a. Green synthesis and characterization of silver nanoparticles using Coriandrum sativum leaf extract. J. Eng. Sci. Technol. 13(1): 158–166.

Khan, S., J. Bakht and F. Syed. 2018b. Green synthesis of gold nanoparticles using acer pentapomicum leaves extract its characterization, antibacterial, antifungal and antioxidant bioassay. Dig. J. Nanomater. Biostructures 13(2): 579–589.

Khorsandi, H., A.A. Aghapour, A. Azarnioush, S. Nemati, H.R. Khalkhali and S. Karimzadeh. 2017. An analysis of boron removal from water using modifed zero-valent iron nanoparticles. Desalin. Water Treat., 1-6. Doi: 10.5004/dwt.2017.20505.

Kokila, K., N. Elavarasan and V. Sujatha. 2017. *Diospyros montana* leaf extract-mediated synthesis of selenium nanoparticles and their biological applications. New J. Chem. 41(15): 7481–7490.

Kora, A.J. and L. Rastogi. 2018. Green synthesis of palladium nanoparticles using gum ghatti (Anogeissus latifolia) and its application as an antioxidant and catalyst. Arab. J Chem. 11(7): 1097–1106.

Kouhiyan Afzal, M.T., A. Farrokhian Firouzi and M. Taghavi. 2017. Synthesis of bare and four different polymer-stabilized zero-valent iron nanoparticles and their efficiency on hexavalent chromium removal from aqueous solutions. J. Water Environ. Nanotechnol. 2(4): 278–289.

Krithiga, N., A. Rajalakshmi and A. Jayachitra. 2015. Green synthesis of silver nanoparticles using leaf extracts of *Clitoria ternatea* and *Solanum nigrum* and study of its antibacterial effect against common nosocomial pathogens. Journal of Nanoscience 2015(2015): 1–8.

Kumar, R., N. Singh and S.N. Pandey. 2015. Potential of green synthesized zero-valent iron nanoparticles for remediation of lead-contaminated water. Int. J. Environ. Sci. Technol. 12: 3943–3950.

Kumar, M.S., N. Supraja and E. David. 2019. Photocatalytic degradation of methylene blue using silver nanoparticles synthesized from Gymnema sylvestre and antimicrobial assay. Nov. Res. Sci. 2(2): 1–7. DOI: 10.31031/NRS.2019.2.000532.

Kumaresan, M., K. Vijai Anand, K. Govindaraju, S. Tamilselvan and V. Ganesh Kumar. 2018. Seaweed *Sargassum wightii* mediated preparation of zirconia (ZrO_2) nanoparticles and their antibacterial activity against gram positive and gram negative bacteria. Microb. Pathogen. 124: 311–315.

Lefèvre, C., F. Abreu, U. Lins and D. Bazylinski. 2010. Nonmagnetotactic multicellular prokaryotes from low-saline, nonmarine aquatic environments and their unusual negative phototactic behavior. Applied and Environmental Microbiology 76(10): 3220–3227.

Lengke, M.F., B. Ravel, M.E. Fleet, G. Wanger, R.A. Gordon and G. Southam. 2006a. Mechanisms of gold bioaccumulation by filamentous cyanobacteria from gold (III)−chloride complex. Environ. Sci. Technol. 40(20): 6304–6309.

Lengke, M.F., M.E. Fleet and G. Southam. 2006b. Synthesis of platinum nanoparticles by reaction of filamentous cyanobacteria with platinum(IV)-chloride complex. Langmuir 22(17): 7318–7323.

Li, L., T. Zeng and S. Xie. 2018. Uranium (VI) bioremediation by Acinetobacter sp. USCB2 isolated from uranium tailings area. IOP Conf. Ser. Earth Environ. Sci. 170: 052043.

Li, X., H. Xu, Z. Chen and C. Guofang. 2011. Biosynthesis of nanoparticles by microorganisms and their applications. J. Nanomater. Article ID 270974: 1–16.

Li, Z., S. Xu, G. Xiao, L. Qian and Y. Song. 2019. Removal of hexavalent chromium from groundwater using sodium alginate dispersed nano zero-valent iron. J. Environ. Manag. 244: 33–39.

Lingamdinne, L.P., J.R. Koduru and R. Rao Karri. 2019. Green synthesis of iron oxide nanoparticles for lead removal from aqueous solutions. Key Eng. Mater. 805: 122–127.

Lingamdinne, L.P., J.R. Koduru, H. Roh, Y.L. Choi, Y.Y. Chang and J.K. Yang. 2016. Adsorption removal of Co(II) from waste-water using graphene oxide. Hydrometallurgy 165: 90–96.

Lingamdinne, L.P., Y.Y. Chang, J.K. Yang, J. Singh, E.H. Choi, M. Shiratani, J.R. Koduru and P. Attri. 2017. Biogenic reductive preparation of magnetic inverse spinel iron oxide nanoparticles for the adsorption removal of heavy metals. Chem. Eng. J. 307: 74–84.

Liu, W., S. Tian, X. Zhao, W. Xie, Y. Gong and D. Zhao. 2015. Application of stabilized nanoparticles for *in situ* remediation of metal-contaminated soil and groundwater: A critical review. Curr. Pollut. Rep. 1(4): 280–291.

Lloyd, J.R. and J.C. Renshaw. 2005. Microbial transformations of radionuclides: Fundamental mechanisms and biogeochemical implications. Metal Ions Biol. Syst. 44: 205–240.

Lombi, E., W.W. Wenzel and D.C. Adriano. 1998. Soil contamination, risk reduction and remediation. Land Contamination and Reclamation 6(4): 183–197.

Losi, M. and W.T. Frankenberger. 1997. Reduction of selenium by *Enterobacter cloacae* SLD1a-1: Isolation and growth of bacteria and its expulsion of selenium particles. Appl. Environ. Microbiol. 63: 3079–3084.

Lovley, D.R. and E.J. Phillips. 1992a. Reduction of uranium by Desulfovibrio desulfuricans. Appl. Environ. Microbiol. 58: 850–856.

Lovley, D.R. and E.J.P. Phillips. 1992b. Bioremediation of uranium contamination with enzymatic uranium reduction. Environ. Sci. Technol. 26: 2228–2234.

Lovley, D.R., E.J.P. Phillips, Y.A. Gorby and E.R. Landa. 1991. Microbial reduction of uranium. Nature 350: 413–416.

Lv, Q., B. Zhang, X. Xing, Y. Zhao, R. Cai, W. Wang and Q. Gu. 2018. Biosynthesis of copper nanoparticles using Shewanella loihica PV-4 with antibacterial activity: Novel approach and mechanisms investigation. J. Hazard. Mater. 347: 141–149.

Machado, S., J.G. Pacheco, H.P.A. Nouws, J.T. Albergaria and C. Delerue-Matos. 2017. Green zero-valent iron nanoparticles for the degradation of amoxicillin. International J. Environ. Sci. Technol. 14(5): 1109–1118.

Madhavi, V., T.N.V.K.V. Prasad, V.B.R. Ambavaram and G. Madhavi. 2013. Plant growth promoting potential of nano-bioremediation under cr (VI) stress. Int. J. Nanotechnol. Appl. 3(3): 1–10.

Maiti, S., G. Barman and J.K. Laha. 2015. Detection of heavy metals (Cu^{+2}, Hg^{+2}) by biosynthesized silver nanoparticles. Appl. Nanosci. 6(4): 529–538.

Majumdar, R., S. Tantayanon and B. Bag. 2017. Synthesis of palladium nanoparticles with leaf extract of *Chrysophyllum cainito* (Star apple) and their applications as efficient catalyst for C–C coupling and reduction reactions. Int. Nano Lett. 7(4): 267–274.

Makarov, V., A. Love, O. Sinitsyna, S. Makarova, I. Yaminsky, M. Taliansky and N. Kalinina. 2014. Green nanotechnologies: Synthesis of metal nanoparticles using plants. Acta Nat. 6(1): 35–44.

Mannucci, S., L. Ghin, G. Conti, S. Tambalo, A. Lascialfari, T. Orlando, D. Benati, P. Bernardi, N. Betterle, R. Bassi, P. Marzola and A. Sbarbati. 2014. Magnetic nanoparticles from Magnetospirillum gryphiswaldense increase the efficacy of thermotherapy in a model of colon carcinoma. PloS One 9(10): e108959.

Masum, M., M.M. Siddiqa, K.A. Ali, Y. Zhang, Y. Abdallah, E. Ibrahim, W. Qiu, C. Yan and B. Li. 2019. Biogenic synthesis of silver nanoparticles using *Phyllanthus emblica* fruit extract and its inhibitory action against the pathogen *Acidovorax oryzae* strain RS-2 of rice bacterial brown stripe. Front. Microbiol. 10: 820.

Maye, M.M., J. Luo, L. Han, N.N. Kariuki and C.J. Zhong. 2003. Synthesis, processing, assembly and activation of core-shell structured gold nanoparticle catalysts. Gold Bull. 36: 75–82.

Mehrotra, N., R.M. Tripathi, F. Zafar and M.P. Singh. 2017. Catalytic degradation of dichlorvos using biosynthesized zero valent iron nanoparticles. IEEE Transac. NanoBiosci. 16(4): 280–286.

Mehta, A., C. Sidhu, A.K. Pinnaka and R.A. Choudhury. 2014. Extracellular polysaccharide production by a novel osmotolerant marine strain of *Alteromonas macleodii* and its application towards biomineralization of silver. Plos One 9(6): e98798.

Merin, D.D., S. Prakash and B.V. Bhimba. 2010. Antibacterial screening of silver nanoparticles synthesized by marinemicro algae. Asian Pac. J. Trop. Med. 3(10): 797–799.

Mishra, A.R., S.A. Mishra and A.V. Tiwari. 2014. Solid waste management—case study. IJART 2(1): 396–399.

Mithila, A., J. Swanand, R.K. Ameeta, Z. Smita and K. Sulabha. 2009. Biosynthesis of gold nanoparticles by the tropical marine yeast Yarrowia lipolytica NCIM 3589. Mater. Lett. 63(15): 1231–1234.

Modi, S., B. Pathak and M.H. Fulekar. 2015. Microbial synthesized silver nanoparticles for decolorization and biodegradation of azo dye compound. J. Environ. Nanotechnol. 4(2): 37–46.

Moghaddam, A.B., M. Moniri, S. Azizi, R.A. Rahim, A.B. Ariff, W.Z. Saad, F. Namvar and M. Navaderi. 2017. Biosynthesis of ZnO nanoparticles by a new pichia kudriavzevii yeast strain and evaluation of their antimicrobial and antioxidant activities. Molecules 22(6): 872.

Moodley, J.S., S.B.N. Krishna, K. Pillay and P. Govender. 2018. Green synthesis of silver nanoparticles from Moringa oleifera leaf extracts and its antimicrobial potential. Adv. Nat. Sci. Nanosci. Nanotechnol. 9(1): 015011.

Mubarak-Ali, D., M. Sasikala, M. Gunasekaran and N. Thajuddin. 2011. Biosynthesis and characterization of silver nanoparticles using marine cyanobacterium, *Oscillatoria willei* ntdm01. Dig. J. Nanomater. Biostruct. 6(2): 385–390.

Mueller, N.C., J. Braun, J. Bruns, M. Černík, P. Rissing, D. Rickerby and B. Nowack. 2012. Application of nanoscale zero valent iron (NZVI) for groundwater remediation in Europe. Environ. Sci. Pollut. Res. 19(2): 550–558.

Muhy, H.M. and F. Duman. 2018. Biosynthesis, characterization and removal efficiency for petroleum leakage of the Cofe2o4 nanoparticles. Al-Mustansiriyah J. Sci. 29(3 ICSSSA 2018 Conference Issue): 21–28.

Mukherjee, S., D. Chowdhury, R. Kotcherlakota, S.B.V. Patra, M.P. Bhadra, B. Sreedhar and C.R. Patra. 2014. Potential theranostics application of bio-synthesized silver nanoparticles (4-in-1 system). Theranostics 4(3): 316–335.

Murgueitio, E., L. Cumbal, M. Abril, A. Izquierdo, A. Debut and O. Tinoco. 2018. Green synthesis of iron nanoparticles: Application on the removal of petroleum oil from contaminated water and soils. J. Nanotechnol. 2018. Article ID 4184769, 8 pages, https://doi.org/10.1155/2018/4184769.

Nabikhan, A., S. Rathinam and K. Kandasamy. 2017. Biogenic gold nanoparticles for reduction of 4-nitrophenol to 4-aminophenol: An eco-friendly bioremediation. IET Nanobiotechnol. 12(4): 479–483.

Nadaf, N. and S. Kanase. 2016. Biosynthesis of gold nanoparticles by Bacillus marisflavi and its potential in catalytic dye degradation. Arab. J. Chem. 12. 10.1016/j.arabjc.2016.09.020.

Narasaiah, P., B. Mandal and N. Sarada. 2017. Biosynthesis of copper oxide nanoparticles from drypetes sepiaria leaf extract and their catalytic activity to dye degradation. IOP Conf. Ser. Mater. Sci. Eng. 263: 022012. 10.1088/1757-899X/263/2/022012.

Narasaiah, P., B.K. Mandal and S.N. Chakravarthula. 2017. Synthesis of gold nanoparticles by cotton peels aqueous extract and their catalytic efficiency for the degradation of dyes and antioxidant activity. IET Nanobiotechnol. 12(2): 156–165. Doi: 10.1049/iet-nbt.2017.0039.

Narasaiah, P., B.K. Mandal and N.C. Sarada. 2017. Green synthesis of Pd NPs from Pimpinella tirupatiensis plant extract and their application in photocatalytic activity dye degradation. In IOP Conf. Ser. Mater. Sci. Eng. 263: 022013.10.1088/1757-899X/263/2/022013.

Naseem, T. and M.A. Farrukh. 2015. Antibacterial activity of green synthesis of iron nanoparticles using Lawsonia inermis and Gardenia jasminoides leaves extract. J. Chem. 2015.

Nasrollahzadeh, M. and S.M. Sajadi. 2016. Pd nanoparticles synthesized *in situ* with the use of Euphorbia granulate leaf extract: Catalytic properties of the resulting particles. J. Colloid Interface Sci. 462: 243–251.

Nasrollahzadeh, M., S.M. Sajadi, A. Rostami-Vartooni, M. Alizadeh and M. Bagherzadeh. 2016. Green synthesis of the Pd nanoparticles supported on reduced graphene oxide using barberry fruit extract and its application as a recyclable and heterogeneous catalyst for the reduction of nitroarenes. J. Colloid Inter. Sci. 466: 360–368.

Nataraj, S.K., K.M. Hosamani and T.M. Aminabhavi. 2009. Nanofiltration and reverse osmosis thin film composite membrane module for the removal of dye and salts from the simulated mixtures. Desalination 249(1): 12–17.

Nayan, V., S.K. Onteru and D. Singh. 2017. *Mangifera indica* flower extract mediated biogenic green gold nanoparticles: Efficient nanocatalyst for reduction of 4-nitrophenol. Environ. Prog. Sustainable Energy 37: 283–294.

Neethu, S., S.J. Midhun, M.A. Sunil, S. Soumya, E.K. Radhakrishnan and M. Jyothis. 2018. Efficient visible light induced synthesis of silver nanoparticles by Penicillium polonicum ARA 10 isolated from Chetomorpha antennina and its antibacterial efficacy against Salmonella enterica serovar Typhimurium. J. Photochem. Photobiol. B 180: 175–185.

Ogunyemi, S.O., Y. Abdallah, M. Zhang, H. Fouad, X. Hong, E. Ibrahim, M.M.I. Masum, A. Hossain, J. Mo and B. Li. 2019. Green synthesis of zinc oxide nanoparticles using different plant extracts and their antibacterial activity against *Xanthomonas oryzae* pv. oryzae. Artif. Cells Nanomed. Biotechnol. 47(1): 341–352.

Otari, S.V., R. Patil, N. Nadaf, S.J. Ghosh and S. Pawar. 2012. Green biosynthesis of silver nanoparticles from an actinobacteria Rhodococcus sp. Mater. Lett. 72: 92–94.

Oves, M., M.S. Khan, A. Zaidi, A.S. Ahmed, F. Ahmed, E. Ahmad, A. Sherwani, M. Owais and A. Azam. 2013. Antibacterial and cytotoxic efficacy of extracellular silver nanoparticles biofabricated from chromium reducing novel OS4 strain of Stenotrophomonas maltophilia. PLoS One 8(3): e59140. Doi: 10.1371/journal. pone.0059140. Epub 2013 Mar 21. PMID: 23555625; PMCID: PMC3605433.

Parthasarathy, G., S. Manickam, M. Venkatachalam and V.K. Evanjelene. 2017. Biological synthesis of zinc oxide nanoparticles from leaf extract of *Curcuma neilgherrensis* wight. Int. J. Mater. Sci. 12: 73–86.

Pattanayak, S., S. Chakraborty, S. Biswas, D. Chattopadhyay and M. Chakraborty. 2018. Degradation of methyl parathion, a common pesticide and fluorescence quenching of Rhodamine B, a carcinogen using β-D glucan stabilized gold nanoparticles. J. Saudi Chem. Soc. 22(8): 937–948.

Pavel, L.V. and M. Gavrilescu. 2008. Overview of *ex situ* decontamination techniques for soil cleanup. Environ. Eng. Manag. J. 7(6): 815–834.

Pawlak, Z., S. Zak and L. Zablocki. 2005. Removal of hazardous metals from groundwater by reverse osmosis. Pollut. J. Environ. Stud. 15(4): 579–583.

Peterson, P.J. and C.A. Girling. 1981. Other trace metals. pp. 222–229. *In*: Lepp, N.W. (ed.). Effect of Heavy Metal Pollution on Plants. Effects of Trace Metals on Plant Function. Applied Science Publishers, London, 1.

Philipse, A.P. and D. Maas. 2002. Magnetic colloids from magnetotactic bacteria: Chain formation and colloidal stability. Langmuir 18(25): 9977–9984.

Pillai, H. and J. Kottekottil. 2016. Nano-phytotechnological remediation of endosulfan using zero valent iron nanoparticles. J. Environ. Prot. 7: 734–744.

Pingale, S.S., S.V. Rupanar and M. Chaskar. 2018. Plant-mediated biosynthesis of silver nanoparticles from *Gymnema sylvestre* and their use in phtodegradation of Methyl orange dye. J. Water Environ. Nanotechnol. 3(2): 106–115.

Pirtarighat, S., M. Ghannadnia and S. Baghshahi. 2019. Green synthesis of silver nanoparticles using the plant extract of *Salvia spinosa* grown *in vitro* and their antibacterial activity assessment. J. Nanostruct. Chem. 9: 1–9.

Piruthiviraj, P., A. Margret and P.P. Krishnamurthy. 2016. Gold nanoparticles synthesized by *Brassica oleracea* (Broccoli) acting as antimicrobial agents against human pathogenic bacteria and fungi. Appl. Nanosci. 6: 467–473.

Popli, D., V. Anil, A.B. Subramanyam, M.N. Namratha, V.R. Ranjitha, S.N. Rao, R.V. Rai and M. Govindappa. 2018. Endophyte fungi, Cladosporium species-mediated synthesis of silver nanoparticles possessing *in vitro* antioxidant, anti-diabetic and anti-Alzheimer activity. Artificial Cells, Nanomed. Biotechnol. 46(sup1): 676–683.

Prabhu, N. and T. Gajendran. 2017. Green synthesis of noble metal of platinum nanoparticles from *Ocimum sanctum* (Tulsi) plant-extracts. IOSR J. Biotechnol. Biochem. 3(1): 107–112.

Prasad, K., A.K. Jha and A. Kulkarni. 2007. Lactobacillus assisted synthesis of titanium nanoparticles. Nanoscale Res. Lett. 2(5): 248–250.

Prema, P., P. Iniya and G. Immanuel. 2016. Microbial mediated synthesis, characterization, antibacterial and synergistic effect of gold nanoparticles using *Klebsiella pneumoniae* (MTCC-4030). RSC Adv. 6: 4601–4607.

Presentato, A., E. Piacenza, M. Anikovskiy, M. Cappelletti, D. Zannoni and R.J. Turner. 2016. *Rhodococcus aetherivorans* BCP1 as cell factory for the production of intracellular tellurium nanorods under aerobic conditions. Microb. Cell Fact. 15(1): 204.

Prokop, G., M. Schamann and I. Edelgaard. 2000. Management of contaminated sites in western Europe. European Environment Agency, Copenhagen.

Purkayastha, D.M., A.K. Manhar, M. Mandal and C.L. Mahanta. 2014. Industrial waste-derived nanoparticles and microspheres can be potent antimicrobial and functional ingredients. J. Appl. Chem. 2014. DOI: 10.1155/2014/171427.

Qin, J., C. Peng, B. Zhao, K. Ye, F. Yuan, Z. Peng, X. Yang, L. Huang, M. Jiang, Q. Zhao, G. Tang and X. Lu. 2014. Noninvasive detection of macrophages in atherosclerotic lesions by computed tomography enhanced with PEGylated gold nanoparticles. Int. J. Nanomed. 9: 5575–5590.

Rabiee, N., M. Bagherzadeh, M. Kiani, A.M. Ghadiri, F. Etessamifar, A.H. Jaberizadeh and A. Shakeri. 2020. Biosynthesis of copper oxide nanoparticles with potential biomedical applications. Int. J Nanomed. 15: 3983–3999.

Rai, M., A. Yadav and A. Gade. 2009. Silver nanoparticles as a new generation of antimicrobials. Biotech. Adv. 27(1): 76–83.

Raj, R., K. Dalei, J. Chakraborty and S. Das. 2016. Extracellular polymeric substances of a marine bacterium mediated synthesis of CdS nanoparticles for removal of cadmium from aqueous solution. J. Colloid Inter. Sci. 462: 166–175.

Rajasree, S.R.R. and T.Y. Suman. 2012. Extracellular biosynthesis of gold nanoparticles using a gram negative bacterium Pseudomonas fluorescens. Asian Pac. J. Trop. Dis. 2: S796–799.

Ramakrishna, M., D. Rajesh Babu, R. Gengan, S. Chandra and G. Nageswara Rao. 2016. Green synthesis of gold nanoparticles using marine algae and evaluation of their catalytic activity. J. Nanostruct. Chem. 6(1): 1–13.

Ramamurthy, C.H., K.S. Sampath, P. Arunkumar, M.S. Kumar, V. Sujatha, K. Premkumar and C. Thirunavukkarasu. 2013. Green synthesis and characterization of selenium nanoparticles and its augmented cytotoxicity with doxorubicin on cancer cells. Bioprocess Biosyst. Eng. 36(8): 1131–1139.

Rana, A., N. Kumari, M. Tyagi and S. Jagadevan. 2018. Leaf-extract mediated zero-valent iron for oxidation of Arsenic (III): Preparation, characterization and kinetics. Chem. Eng. J. 347: 91–100.

Ranjitha, V.R. and V.R. Rai. 2017. Actinomycetes mediated synthesis of gold nanoparticles from the culture supernatant of *Streptomyces griseoruber* with special reference to catalytic activity. 3 Biotech 7(5): 299.

Rao, K.R., T. Srinivasan and C. Venkateswarlu. 2010. Mathematical and kinetic modeling of biofilm reactor based on ant colony optimization. Process Biochem. 45(6): 961–972.

Ren, Y.Y., H. Yang, T. Wang and C. Wang. 2016. Green synthesis and antimicrobial activity of monodisperse silver nanoparticles synthesized using Ginkgo Biloba leaf extract. Phys. Lett. A 380(45): 3773–3777.

Rodríguez-León, E., B.E. Rodríguez-Vázquez, A. Martínez-Higuera, C. Rodríguez-Beas, E. Larios-Rodríguez, R.E. Navarro, R. López-Esparza and R.A. Iñiguez-Palomares. 2019. Synthesis of gold nanoparticles using *Mimosa tenuiflora* extract, assessments of cytotoxicity, cellular uptake, and catalysis. Nanoscale Res. Lett. 14(1): 334.

Rokade, S.S., K.A. Joshi, K. Mahajan, S. Patil, G. Tomar, D.S. Dubal, V.S. Parihar, R. Kitture, J.R. Bellare and S. Ghosh. 2018. *Gloriosa superba* mediated synthesis of platinum and palladium nanoparticles for induction of apoptosis in breast cancer. Bioinorg. Chem. Appl. 2018: 4924186.

Romero-González, M., B.C. Nwaobi, J.M. Hufton and D.J. Gilmour. 2016. *Ex-situ* bioremediation of U(VI) from contaminated mine water using Acidithiobacillus ferrooxidans strains. Front. Environ. Sci. 4: 39.

Ruíz-Baltazar, Á.D.J., S.Y. Reyes-López, D. Larrañaga-Ordáz, N. Méndez-Lozano, M.A. Zamora-Antuñano and R. Perez. 2020. Magnetic nanoparticles of Fe3O4 biosynthesized by cnicus benedictus extract: Photocatalytic study of organic dye degradation and antibacterial behavior. Preprints, 2020070223 (Doi: 10.20944/preprints202007.0223.v1).

Rybnikova, V., N. Singhal and K. Hanna. 2017. Remediation of an aged PCP-contaminated soil by chemical oxidation under flow-through conditions. Chem. Eng. J. 314: 202–211.

Sadowski, Z., I.H. Maliszewska, B. Grochowalska, I. Polowczyk and T. Kozlecki. 2008. Synthesis of silver nanoparticles using microorganisms. Mater. Sci. Poland. 26(2): 419–424.

Sagadevan, S., S. Vennila, P. Singh, J. Lett, M. Johan, A. Marlinda, B. Muthiah and M. Lakshmipathy. 2019. Facile synthesis of silver nanoparticles using Averrhoa bilimbi L. and Plum extracts and investigation on the synergistic bioactivity using *in vitro* models. Green Process. Synth. 8(1): 873–884.

Saha, J., A. Begum, A. Mukherjee and S. Kumar. 2017. A novel green synthesis of silver nanoparticles and their catalytic action in reduction of Methylene Blue dye. Sustain. Environ. Res. 27(5): 245–250.

Saifuddin, N., C.Y. Nian, L.W. Zhan and K.X. Ning. 2011. Chitosan-silver nanoparticles composite as point-of-use drinking water filtration system for household to remove pesticides in water. Asian J. Biochem. 6(2): 142–159.

Sajadi, S.M., K. Kolo, M. Pirouei, S.A. Mahmud, J.A. Ali and S.M. Hamad. 2018. Natural iron ore as a novel substrate for the biosynthesis of bioactive-stable ZnO@ CuO@ iron ore NCs: A magnetically recyclable and reusable superior nanocatalyst for the degradation of organic dyes, reduction of Cr (vi) and adsorption of crude oil aromatic compounds, including PAHs. RSC Adv. 8(62): 35557–35570.

Samadhan, D. and S. Bhutada. 2013. Biosynthesis of silver nanoparticles using bacillus megaterium and their antibacterial potential. IJADD 3(1): 13–19.

Samuel, M., S. Jose, S. Ethiraj, T. Mathimani and A. Pugazhendhi. 2020. Biosynthesized silver nanoparticles using Bacillus amyloliquefaciens; Application for cytotoxicity effect on A549 cell line and photocatalytic degradation of p-nitrophenol. J. Photochem. Photobiol. B, Biol. 202: 111642.

Sane, N., B. Hungund and N. Ayachit. 2013. Biosynthesis and characterization of gold nanoparticles using plant extracts. ICANMEET, 295–299. Doi: 10.1109/ICANMEET.2013.6609296.

Santhoshkumar, J., V. Kumar and R. Shanmugam. 2017. Synthesis of zinc oxide nanoparticles using plant leaf extract against urinary tract infection pathogen. Resource-Efficient Technologies 3(4): 459–465.

Saravanan, M., S.K. Barik, D. MubarakAli, P. Prakash and A. Pugazhendhi. 2018. Synthesis of silver nanoparticles from Bacillus brevis (NCIM 2533) and their antibacterial activity against pathogenic bacteria. Microb. Pathog. 116: 221–226.

Sarkar, J., M. Ghosh, A. Mukherjee, D. Chattopadhyay and K. Acharya. 2014. Biosynthesis and safety evaluation of ZnO nanoparticles. Bioprocess Biosyst. Eng. 37(2): 165–171.

Schlüter, M., T. Hentzel, C. Suarez, M. Koch, W.G. Lorenz, L. Böhm, R.A. Düring, K.A. Koinig and M. Bunge. 2014. Synthesis of novel palladium(0) nanocatalysts by microorganisms from heavy-metal-influenced high-alpine sites for dehalogenation of polychlorinated dioxins. Chemosphere 117: 462–470.

Scott, R.S., S.R. Frame, P.E. Ross, S.E. Loveless and G.L. Kennedy. 2008. Inhalation toxicity of 1,3-propanediol in the rat. Inhal. Toxicol. 17(9): 487–493.

SDWF - https://www.safewater.org/fact-sheets-1/2017/1/23/cleaning-up-after-pollution.

Selim, Y.A., M.A. Azb, I. Ragab and H.M.M. Abd El-Azim. 2020. Green synthesis of zinc oxide nanoparticles using aqueous extract of *Deverra tortuosa* and their cytotoxic activities. Sci. Rep. 10(1): 3445.

Selvam, G.G. and K. Sivakumar. 2015. Phycosynthesis of silver nanoparticles and photocatalytic degradation of methyl orange dye using silver (Ag) nanoparticles synthesized from *Hypnea musciformis* (Wulfen) JV Lamouroux. Appl. Nanosci. 5(5): 617–622.

Selvi, A. and N. Das. 2016. Nano-bio hybrid system for enhanced degradation of cefdinir using Candida sp. SMN04 coated with zero-valent iron nanoparticles. J. App. Pharm. Sci. 6(9): 009–017.

Seshadri, S., K. Saranya and M. Kowshik. 2011. Green synthesis of lead sulfide nanoparticles by the lead resistant marine yeast, *Rhodosporidium diobovatum*. Biotechnol. Progress 27(5): 1464–1469.

Sethy, N., Z. Arif, P. Mishra and P. Kumar. 2020. Green synthesis of TiO2 nanoparticles from Syzygium cumini extract for photo-catalytic removal of lead (Pb) in explosive industrial wastewater. Green Process. Synth. 9(1): 171–181.

Shah, R., G. Oza, S. Pandey and M. Sharon. 2017. Biogenic fabrication of gold nanoparticles using Halomonas salina. J. Microbiol. Biotechnol. 2: 485–492.

Shah, R.K., F. Boruah and N. Parween. 2015. Synthesis and characterization of ZnO nanoparticles using leaf extract of *Camellia sinesis* and evaluation of their antimicrobial efficacy. Int. J Curr. Microbiol. App. Sci. 4(8): 444–450.

Shahverdi, A.R., S. Minaeian, H.R. Shahverdi, H. Jamalifar and A.A. Nohi. 2007. Rapid synthesis of silver nanoparticles using culture supernatants of Enterobacteria: A novel biological approach. Process Biochem. 42(5): 919–923.

Shaik, M.R., M. Khan, M. Kuniyil, A. Al-Warthan, H.Z. Alkhathlan, M.R.H. Siddiqui, J.P. Shaik, A. Ahamed, A. Mahmood, M. Khan and S.F. Adil. 2018. Plant-extract-assisted green synthesis of silver nanoparticles using *Origanum vulgare* L. extract and their microbicidal activities. Sustainability 10(4): 913.

Shamsuzzaman, Mashrai, A., H. Khanam and R.N. Aljawfi. 2017. Biological synthesis of ZnO nanoparticles using C. albicans and studying their catalytic performance in the synthesis of steroidal pyrazolines. Arab. J. Chem. 10(2): 2013. Doi: 10.1016/j.arabjc.2013.05.004.

Shao, D., X. Ren, J. Wen, S. Hu, J. Xiong, T. Jiang, X. Wang and X. Wang. 2016. Immobilization of uranium by biomaterial stabilized FeS nanoparticles: Effects of stabilizer and enrichment mechanism. J. Hazard. Mater. 302: 1–9.

Sharma, N., A.K. Pinnaka, M. Raje, F.N.U. Ashish, M.S. Bhattacharyya and A.R. Choudhury. 2012. Exploitation of marine bacteria for production of gold nanoparticles. Microb. Cell Fact. 11(1): 86.

Shedbalkar, U., R. Singh, S. Wadhwani, S. Gaidhani and B.A. Chopade. 2014. Microbial synthesis of gold nanoparticles: Current status and future prospects. Adv. Colloid Interface Sci. 209: 40–48. Doi: 10.1016/j. cis.2013.12.011.

Shekhawat, M.S., M. Manokari, N.M. Kannan, J. Revathi and R.P. Latha. 2013. Synthesis of silver nanoparticles using *Cardiospermum halicacabum* L. leaf extract and their characterization. Int. J. Phytopharm. 2(5): 15–20.

Shenton, W., T. Douglas, M. Young, G. Stubbs and S. Mann. 1999. Inorganic-organic nanotube composites from template mineralization of tobacco mosaic virus. Adv. Mater. 11(3): 253–256.

Shittu, K.O. and O. Ihebunna. 2017. Purification of simulated waste water using green synthesized silver nanoparticles of *Piliostigma thonningii* aqueous leave extract. Adv. Nat. Sci. Nanosci. Nanotechnol. 8(4): 045003.

Shoeibi, S. and M. Mashreghi. 2017. Biosynthesis of selenium nanoparticles using Enterococcus faecalis and evaluation of their antibacterial activities. Journal of Trace Elements in Medicine and Biology: Organ of the Society for Minerals and Trace Elements (GMS) 39: 135–139.

Šimkovič, K., J. Derco and M. Valičková. 2015. Removal of selected pesticides by nano zero-valent iron. Acta Chim. Slov. 8(2): 152–155.

Sin, N.L., M.N. Wei, K.L. Jit and X.C. Hui. 2018. Performance of mulberry leaves mediated green synthesis zero-valent iron nanoparticles in dye removal. Int. J Eng. Technol. 7(3.36): 113–117.

Singh, P., Y.J. Kim, D. Zhang and D.C. Yang. 2016. Biological synthesis of nanoparticles from plants and microorganisms. Trends Biotech. 34. 10.1016/j.tibtech.2016.02.006.

Singh, P.S. and G. Vidyasagar. 2018. Biosynthesis of antibacterial silver nano-particles from Aspergillus terreus. World News Nat. Sci. 16: 117–124.

Singh, R., U.U. Shedbalkar, S.A. Wadhwani and B.A. Chopade. 2015. Bacteriagenic silver nanoparticles: Synthesis, mechanism, and applications. Appl. Microbial. Biotechnol. 99(11): 4579–4593.

Singh, R., J. Vora, S.B. Nadhe, S.A. Wadhwani, U.U. Shedbalkar and B.A. Chopade. 2018. Antibacterial activities of bacteriagenic silver nanoparticles against nosocomial *Acinetobacter baumannii*. J. Nanosci. Nanotechnol. 18: 3806–3815.

Singh, V., R. Patil, A. Anand, P. Milani and W. Gade. 2010. Biological synthesis of copper oxide nanoparticles using *Escherichia coli*. Curr. Nanosci. 6: 365–369.

Sinha, A. and S.K. Khare. 2011. Mercury bioaccumulation and simultaneous nanoparticle synthesis by *Enterobacter sp.* cells. Biores. Technol. 102: 4281–4284.

Sinha, T. and M. Ahmaruzzaman. 2015. Green synthesis of copper nanoparticles for the efficient removal (degradation) of dye from aqueous phase. Environ. Sci. Pollut. Res. Int. 22(24): 20092–20100.

Sintubin, L., W. De Windt, J. Dick, J. Mast, D. van der Ha, W. Verstraete and N. Boon. 2009. Lactic acid bacteria as reducing and capping agent for the fast and efficient production of silver nanoparticles. Appl. Microbiol. Biotechnol. 84(4): 741–749.

Siripireddy, B. and B.K. Mandal. 2017. Facile green synthesis of zinc oxide nanoparticles by Eucalyptus globulus and their photocatalytic and antioxidant activity. Adv. Powder Technol. 28(3): 785–797.

Skerret, P.J. 2016. Silent Spring at 50: Connecting human and environmental health, taking it all. In: Environmental toxins and your health, Longwood seminars, The Joseph B. Martin Conference Center, The New Research Building, Harvard Medical School, Boston.

Song, J.Y., E.Y. Kwon and B.S. Kim. 2010. Biological synthesis of platinum nanoparticles using Diopyros kaki leaf extract. Bioprocess Biosyst. Eng. 33(1): 159–164.

Sonker, A.S., J. Pathak, V. Kannaujiya and R. Sinha. 2017. Characterization and *in vitro* antitumor, antibacterial and antifungal activities of green synthesized silver nanoparticles using cell extract of Nostoc sp. strain HKAR-2. Can. J. Biotechnol. 1(1): 26–37.

Soundarrajan, C., A. Sankari, P. Dhandapani, S. Maruthamuthu, S. Ravichandran, G. Sozhan and N. Palaniswamy. 2012. Rapid biological synthesis of platinum nanoparticles using Ocimum sanctum for water electrolysis applications. Bioprocess Biosyst. Eng. 35(5): 827–833.

Sravanthi, K., D. Ayodhya and P. Yadgiri Swamy. 2018. Green synthesis, characterization of biomaterial-supported zero-valent iron nanoparticles for contaminated water treatment. J. Anal. Sci. Technol. 9: 3. https://doi.org/10.1186/s40543-017-0134-9.

Sreekanth, M. and D. Sahu. 2015. Effect of iron oxide nanoparticle in bio digestion of a portable food-waste digester. J. Chem. Pharm. Res. 7(9): 353–359.

Sriram, T. and V. Pandidurai. 2014. Synthesis of silver nanoparticles from leaf extract of Psidium guajava and its antibacterial activity against pathogens. IJCMAS 3(3): 146–152.

Srivastava, N. and M. Mukhopadhyay. 2014. Biosynthesis of SnO_2 nanoparticles using bacterium *Erwinia herbicola* and their photocatalytic activity for degradation of dyes. Ind. Eng. Chem. Res. 53(36): 13971–13979.

Su, Y., A.S. Adeleye, A.A. Keller, Y. Huang, C. Dai, X. Zhou and Y. Zhang. 2015. Magnetic sulfide-modified nanoscale zerovalent iron (S-nZVI) for dissolved metal ion removal. Water Res. 74: 47–57.

Subramaniyan, S.A., S. Sheet, M. Vinothkannan, D.J. Yoo, Y.S. Lee, S.A. Belal and K.S. Shim. 2018. One-pot facile synthesis of Pt nanoparticles using cultural filtrate of microgravity simulated grown *P. chrysogenum* and their activity on bacteria and cancer cells. J. Nanosci. Nanotechnol. 18(5): 3110–3125.

Sudha, S.S., K. Rajamanickam and J. Rengaramanujam. 2013. Microalgae mediated synthesis of silver nanoparticles and their antibacterial activity against pathogenic bacteria. Indian J. Exp. Biol. 52: 393–399.

Syed, A. and A. Ahmad. 2012. Extracellular biosynthesis of platinum nanoparticles using the fungus Fusarium oxysporum. Colloid Surf. B Biointerfaces 97: 27–31.

Taghavi Fardood, S., F. Moradnia, M. Mostafaei, Z. Afshari, V. Faramarzi and S. Ganjkhanlu. 2019. Biosynthesis of $MgFe_2O_4$ magnetic nanoparticles and its application in photo-degradation of malachite green dye and kinetic study. Nanochem. Res. 4(1): 86–93.

Tahir, K., S. Nazir, B. Li, A.U. Khan, Z.U.H. Khan, P.Y. Gong, S.U. Khan and A. Ahmad. 2015. Nerium oleander leaves extract mediated synthesis of gold nanoparticles and its antioxidant activity. Mater. Lett. 156: 198–201.

Tamboli, D.P. and D.S. Lee. 2013. Mechanistic antimicrobial approach of extracellularly synthesized silver nanoparticles against gram positive and gram negative bacteria. J. Hazard. Mater. 260: 878–884.

Taran, M., M. Rad and M. Alavi. 2018. Biosynthesis of TiO_2 and ZnO nanoparticles by Halomonas elongata IBRC-M 10214 in different conditions of medium. BioImpacts 8: 81–89.

Thakker, J.N., P. Dalwadi and P.C. Dhandhukia. 2013. Biosynthesis of gold nanoparticles using *Fusarium oxysporum* f. sp. cubense JT1, a plant pathogenic fungus. ISRN Biotechnol. 515091. https://doi.org/10.5402/2013/515091.

The Fourth Report. 2009. Fourth National Report on Human Exposure to Environmental Chemicals, Department of Health and Human Services, Centers for Disease Control and Prevention. National Center for Environmental Health USA. Available from: URL: http://www.cdc.gov/exposurereport/fourthreport.pdf.

Thirumurugan, A., P. Aswitha, C. Kiruthika, S. Nagarajan and A.G. Christy. 2016. Green synthesis of platinum nanoparticles using *Azadirachta indica*—An eco-friendly approach. Mater. Lett. 170: 175–178.

Toxic chemicals—World wide fund for nature https://wwf.panda.org/knowledge_hub/teacher_resources/webfieldtrips/toxics/.

Tripathi, R.M., B.R. Shrivastav and A. Shrivastav. 2018. Antibacterial and catalytic activity of biogenic gold nanoparticles synthesised by *Trichoderma harzianum*. IET Nanobiotechnol. 12(4): 509–513.

Turakhia, B., P. Turakhia and S. Shah. 2018. Green synthesis of zero valent iron nanoparticles from *Spinacia oleracea* (spinach) and its application in waste water treatment. J. Adv. Res. Appl. Sci. 5(1): 46–51.

Umar, H., D. Kavaz and N. Rizaner. 2018. Biosynthesis of zinc oxide nanoparticles using *Albizia lebbeck* stem bark, and evaluation of its antimicrobial, antioxidant, and cytotoxic activities on human breast cancer cell lines. Int. J. Nanomed. 14: 87–100.

UNEP. 2013. https://wedocs.unep.org/handle/20.500.11822/8607.

Vadlapudi, V. and R. Amanchy. 2017. Phytofabrication of silver nanoparticles using Myriostachya wightiana as a novel bioresource, and evaluation of their biological activities. Braz. Arch. Biol. Technol. 60: e17160329.

Vanaja, M., K. Paulkumar, M. Baburaja, S. Rajeshkumar, G. Gnanajobitha, C. Malarkodi, M. Sivakavinesan and G. Annadurai. 2014. Degradation of methylene blue using biologically synthesized silver nanoparticles. Bioinorg. Chem. Appl., 2014. Article 742346. 10.1155/2014/742346.

Vanlalveni, C., K. Rajkumari, A. Biswas, P.P. Adhikari, R. Lalfakzuala and L. Rokhum. 2018. Green synthesis of silver nanoparticles using nostoc linckia and its antimicrobial activity: A novel biological approach. Bionanosci. 8(2): 624–631.

Varshney, R., A.N. Mishra, S. Bhadauria and M.S. Gaur. 2009. A novel microbial route to synthesize silver nanoparticles using fungus *Hormoconis resinae*. Dig. J. Nanomater. Bios. 4(2): 349–355.

Velmurugan, P., J. Shim, Y. You, S. Choi, S. Kamala-Kannan, K.J. Lee, H.J. Kim and B.T. Oh. 2010. Removal of zinc by live, dead, and dried biomass of Fusarium spp. isolated from the abandoned-metal mine in South Korea and its perspective of producing nanocrystals. J. Hazard. Mater. 182(1-3): 317–324.

Venkateswarlu, P., S. Ankanna, T.N.V.K.V. Prasad, K. Elumalai, P.C. Nagajyothi and N. Savithramma. 2010. Green synthesis of silver nanoparticles using Shorea tumbuggaia stem bark. Int. J. Drug. Dev. Res. 2(4): 720–723.

Vidhu, V.K. and D. Philip. 2014. Catalytic degradation of organic dyes using biosynthesized silver nanoparticles. Micron (Oxford, England: 1993) 56: 54–62.

Vijayanandan, A.S. and R.M. Balakrishnan. 2018. Biosynthesis of cobalt oxide nanoparticles using endophytic fungus *Aspergillus nidulans*. J. Environ. Manage. 218: 442–450.

Vimalraj, S., T. Ashokkumar and S. Saravanan. 2018. Biogenic gold nanoparticles synthesis mediated by *Mangifera indica* seed aqueous extracts exhibits antibacterial, anticancer and anti-angiogenic properties. Biomed. Pharmacotherapy 105: 440–448.

Viorica, R.P., P. Pawel, M. Kinga, Z. Michal, R. Katarzyna and B. Boguslaw. 2017. *Lactococcus lactis* as a safe and inexpensive source of bioactive silver composites. Appl. Microbiol. Biotechnol. 101(19): 7141–7153.

Vodyanitskii, Y.N. and V.G. Mineev. 2015. Degradation of nitrates with the participation of Fe(II) and Fe(0) in groundwater: A review. Eurasian Soil Sc. 48: 139–147.

Wadhwani, S.A., U.U. Shedbalkar, R. Singh and B.A. Chopade. 2018. Biosynthesis of gold and selenium nanoparticles by purified protein from Acinetobacter sp. SW 30. Enzyme Microb. Technol. 111: 81–86.

Wahab, W., A. Karim, A. Asmawati and I.W. Sutapa. 2018. Bio-synthesis of gold nanoparticles through bioreduction using the aqueous extract of *Muntingia calabura* L. Leaves. Orient. J. Chem. 34(1): 401–409.

Wang, X., D. Zhang, X. Pan, D.J. Lee, F.A. Al-Misned, M.G. Mortuza and G.M. Gadd. 2017. Aerobic and anaerobic biosynthesis of nano-selenium for remediation of mercury contaminated soil. Chemosphere 170: 266–273.

Wei, Y.T., S.C. Wu, C.M. Chou, C.H. Che, S.M. Tsai and H.L. Lien. 2010. Influence of nanoscale zero-valent iron on geochemical properties of groundwater and vinyl chloride degradation: A field case study. Water Res. 44(1): 131–140.

Wu, W.M., J. Carley, T. Gentry, M.A. Ginder-Vogel, M. Fienen, T. Mehlhorn, H. Yan, S. Caroll, M.N. Pace, J. Nyman, J. Luo, M.E. Gentile, M.W. Fields, R.F. Hickey, B. Gu, D. Watson, O.A. Cirpka, J. Zhou, S. Fendorf, P.K. Kitanidis, P.M. Jardine and C.S. Criddle. 2006. Pilot-scale *in situ* bioremediation of uranium in a highly contaminated aquifer. 2. Reduction of U(VI) and geochemical control of U(VI) bioavailability. Environ. Sci. Technol. 40(12): 3986–3995.

Xu, C., Q. Lei, M. Li, Y. Shuqi, G. Yu, D. Xina, Z. Baohua and R. Alexandra. 2019. Biosynthesis of polysaccharides-capped selenium nanoparticles using *Lactococcus lactis* NZ9000 and their antioxidant and anti-inflammatory activities. Front. Microbiol. 10, Article 1632: 1–12. DOI: 10.3389/fmicb.2019.01632.

Yadav, V., N. Sharma, R. Prakash, K.K. Raina, L.M. Bharadwaj and N. Tejo-Prakash. 2008. Generation of selenium containing nano-structures by soil bacterium, *Pseudomonas aeruginosa*. Biotechnol. 7: 299–304.

Yaqub, G., A. Hamid, N. Khan, S. Ishfaq, A. Banzir and T. Javed. 2020. Biomonitoring of workers exposed to volatile organic compounds associated with different occupations by headspace GC-FID. J. Chem. Article ID 6956402, 1–8, https://doi.org/10.1155/ 2020/6956402.

Yazdani, A., M. Sayadi and A. Heidari. 2018. Green biosynthesis of palladium oxide nanoparticles using *Dictyota indica* seaweed and its application for adsorption. J. Water Environ. Nanotechnol. 3(4): 337–347.

Yong, P., J.P.G. Farr, I.R. Harris and L.E. Macaskie. 2002. Palladium recovery by immobilized cells of *Desulfovibrio desulfuricans* using hydrogen as the electron donor in a novel electrobioreactor. Biotechnol. Lett. 24(3): 205–212.

Yulizar, Y., T. Utari, H.A. Ariyanta and D. Maulina. 2017. Green method for synthesis of gold nanoparticles using *Polyscias scutellaria* leaf extract under UV light and their catalytic activity to reduce methylene blue. J. Nanomater. 1–6. 10.1155/2017/3079636.

Zamani, A., A. Marjani and Z. Mousavi. 2019. Agricultural waste biomass-assisted nanostructures: Synthesis and application. Green Process. Synth. 8(1): 421–429.

Zhang, P., Z. Wang, L. Liu, L.H. Klausen, Y. Wang, J.L. Mi and M. Dong. 2019. Modulation the electronic property of 2D monolayer MoS_2 by amino acid. Appl. Mater. Today 14: 151–158.

Zhang, W., Z. Chen, H. Liu, L. Zhang, P. Gao and D. Li. 2011. Biosynthesis and structural characteristics of selenium nanoparticles by *Pseudomonas alcaliphila*. Colloid. Surfaces B: Biointerfaces 88(1): 196–201.

Zheng, D., C. Hu, T. Gan, X. Dang and S. Hu. 2010. Preparation and application of a novel vanillin sensor based on biosynthesis of Au–Ag alloy nanoparticles. Sensor Actuat. B Chem. 148(1): 247–252.

Zhuo, C., J. Alves, J. Tenório and Y. Levendis. 2012. Synthesis of carbon nanomaterials through up-cycling agricultural and municipal solid wastes. Ind. Eng. Chem. Res. 51(7): 922–2930.

21

Bio-Nanotechnology in Agriculture
New Opportunities and Future Prospects

W.L. Isuru Wijesekara,[1] *S. Gokila,*[2] *T. Gomathi,*[2] *Supriya Prasad,*[3]
M. Deepa[3] *and P.N. Sudha*[3,*]

1. Introduction

The population of the world, which is currently 7.6 billion, will increase to 8.6 billion in 2030, 9.8 billion in 2050, and 11.2 billion in 2100. (Chattopadhyay and Patel 2014). This suggests that in order to assure food security, new systems for energy, water, and food will be required. On the other hand, increasing food production necessitates the use of natural resources, land, water, and energy (Amil Usmani et al. 2017). Therefore, it will be necessary for scientific research to produce new paradigms and techniques in the very near future in order to address a wide range of very complicated problems. Here are a few instances: (i) How will we feed our kids? (ii) How can we boost agricultural yields while also lessening the impact of agriculture on the environment? (iii) How do plants contribute to the ecological services that civilization depends on, such as photosynthesis, nitrogen fixation, and the organic matter cycle? (iv) Can global agricultural systems adapt to the changing climate? (Mukhopadhyay 2014).

The traditional agricultural inputs (land, water, energy, fertilisers, and pesticides) are used inefficiently, and a significant portion of the plant protection products sprayed annually are lost or are not available to the target (Huang et al. 2015). Additionally, agriculture (the raising of crops, maintaining animals, and deforestation) contributes significantly to greenhouse gas emissions, accounting for around 24% of the annual global total (Prasad et al. 2017). Another important challenge facing the primary industry is waste creation. Approximately 90 million tons of agricultural waste are produced annually in Europe (Prasad et al. 2014). The European Commission has named nanotechnology as one of its six "Key Enabling Technologies" that supports the transition to a greener economy by fostering sustainable competitiveness and growth in a variety of industrial applications (Bhattacharyya et al. 2016). We must respond to the following query before we can further explore the potential advantages of using nanoscience in agriculture. Why and how are Engineered Nano Materials (ENMs) and nanotechnology expected to address the aforementioned problems.

---

[1] Department of Food Science & Technology, Faculty of Applied Sciences, University of Sri Jayewardenepura, Gangodawila, Nugegoda, Sri Lanka. Email: isuruw@sci.sjp.ac.lk

[2] Biomaterials Research Lab, Department of Chemistry, D.K.M. College for Women, Vellore, Tamil Nadu, India.

[3] Department of Chemistry, Muthurangam Govt. Arts College, Vellore, Tamil Nadu, India.

* Corresponding author: drparsu8@gmail.com

Recent scientific literature, which reveals significant opportunities for nanoscience and nanotechnology to improve sustainability of agri-food system provides specific answers. It is evident from a quantitative standpoint that interest in this field of study increased dramatically between the end of the twentieth century and the start of the twenty-first century when considering the expansion of scientific literature on nanotechnology (Patil et al. 2016).

The distinctive physicochemical characteristics of nanomaterials, such as their large surface area, catalytic reactivity, size, and form, have the potential to open new paradigms and bring new approaches in agriculture. Such novel paradigms call for novel terminology. Although no precise definition of the phrase "agri-nanotechniques" was given, it has been used to refer to nanosystems employed for the delivery of nutrient components to crops. The word "phytonanotechnology" was used to denote the use of nanotechnology in a wide sense to plant production systems or, more broadly, to plant science (Kah and Hofmann 2014). Although the use of nanotechnology in agriculture is still in its infancy, it is quite likely that new terminology will be created in the future to describe more precise technical advancements.

There is a high demand in the agricultural sectors of the modern era for quick, dependable, and affordable solutions for the detection, monitoring, and diagnosis of biological host molecules . Green nanotechnology refers to the development of nanoparticles from plant systems because the use of chemically created nanomaterials is now seen as hazardous in nature.

Green nanotechnology is a safe method that uses less energy, produces less waste, and emits fewer greenhouse gases. Since these products are produced using renewable resources, the environmental impact of these operations is minimal (Kookana et al. 2014). Green nanotechnology has made great progress, and nanomaterials are environmentally sustainable. The usage of green nanotechnology to perform its duties has become more popular during the past ten years. How green nanotechnology will be environmentally sustainable in the future is still unclear. The reduction of these dangers is necessary to advance green nanotechnology solutions (Abdel-Aziz et al. 2016).

Without the use of agrochemicals like pesticides, fertilisers, etc., sustainable production and efficiency in modern agriculture are unthinkable. However, every agrochemical has certain possible drawbacks, such as water contamination or food product residues that endanger human and environmental health. As a result, careful management and control of inputs may be able to minimise these risks (Patra and Baek 2017). In order to revolutionise agricultural methods, decrease and/or eliminate the impact of modern agriculture on the environment, and improve both the quality and quantity of crops, a high-tech agricultural system that uses engineered smart nanotools may be developed (Kalpana and Rajeswari 2017).

Nanotechnology is present and plays a crucial role in the development of biosensors, making it an ideal field for utilising many of its strengths. Due to the unique characteristics of nanomaterials, biosensors' sensitivity and performance could be considerably increased in their applications; nevertheless, numerous novel signal transduction technologies may also be implemented in biosensors (Duhan et al. 2017). Furthermore, the use of nanomaterials enables the miniaturisation of numerous biosensors into small, intelligent, and compact devices like nanosensors and other nanosystems, which are crucial in biochemical analysis. Additionally, it aids in the quick detection of mycotoxins found in a variety of foods. Therefore, as an alternative to conventional technologies, nanotechnology can not only reduce uncertainty but also coordinate the management strategies of agricultural production (Kumar et al. 2014). Numerous agro-nanotech developments offer quick technological answers for the issues that plague contemporary industrial agriculture. The uses of nanotechnology in agriculture that may assure the sustainability of agriculture and the environment are summarised in the current review.

2. Bio-Nanotechnology and Agricultural Development

Through the regulation of nutrients, bio-nanotechnology plays a significant role in agriculture. It can also help enhance plant disease resistance, find mycotoxins in food, monitor water quality, and use pesticides for crops to grow sustainably (Jampilek et al. 2015). Applications that are pertinent to agriculture and food technology include those involving nanotubes, fullerenes, and biosensors with controlled delivery systems. The application of nanotechnology is made possible by a number of potential advantages, including improvements in food quality and safety, less agricultural effort, increased absorption of nanoscale nutrients from the soil, etc. Nanotechnology in agriculture seeks to reduce the use of dangerous herbicides, reduce nutrient loss during fertilisation, and increase output through improved pest and nutrient management (Wu et al. 2017). Due to the use and manufacturing of potentially dangerous chemicals, chemically synthesised nanomaterials are seen as toxic by nature; in their stead, nanomaterials made from plant materials are regarded as green and safe and the technique is known as nanotechnology or bio nanotechnology (Dey et al. 2017).

2.1 Nano-pesticides

Bio-pesticides currently have a unique position in controlling target illnesses of pest and insect origin, in contrast to the synthetic pesticides that are currently available on the market. The use of engineered nanomaterials or bio-nanotechnology is a cutting-edge technology in the field of biopesticides. It is common knowledge that insects and other pests are the main destructors of agriculture and its byproducts. Because of their higher solubility, selectivity, permeability, and stability, as well as other qualities, nano-pesticides may play a significant role in the control of pests, insects, and host infections (Dimkpa 2014). Thus, it is essential to create non-toxic, eco-friendly, and effective nano-pesticide delivery methods to increase agriculture productivity. This also helps to lessen the harmful environmental effects on ecosystems (Singh et al. 2016). Due to the electrostatic interaction of metal nanoparticles with bacterial cell membranes and their accumulation in cytoplasm, metal nanoparticles demonstrate good anti-pathogenic, antibacterial, and anti-fungal actions. In agriculture, microorganisms play a significant role in sustaining the ecosystem, crop productivity, and soil health. Therefore, knowing that agricultural plants free of any harmful nanocomposite from ecotoxicological perspectives is essential to understanding the unrivalled potential for increased production of agricultural crops (Marzbani et al. 2015).

2.2 Nano-fertilizers

In the recent years, nano-fertilizers have been on the market on their own, but agricultural fertilisers in particular are still not produced. Zinc, silica, iron, titanium dioxide, gold nano-rods, core shell QDs, and other materials may be found in nano-fertilizers (Afrouzi et al. 2015). Studies on the absorption and toxicity of different metal oxide nanoparticles have been carried out often in an effort to boost crop productivity. By guaranteeing that the plants absorb the nutrients as effectively as possible, carbon nanotubes and other nanoparticles, such as those of silver, zinc oxide, and other materials, can significantly improve plant growth (Prema and Kandasamy 2017). However, a number of variables, such as plant species susceptibility and other parameters, such as the concentration, composition, size, and chemical properties of nanomaterials, are crucial to its performance. Ion beam microscopy, Raman chemical imaging spectroscopy, transmission electron microscopy, and confocal laser scanning microscopy are other techniques used to confirm the uptake and intracellular destiny of nanoparticles. In the presence of proteins and cell medium, the size, degree of aggregation, and zeta potential of metal oxide NPs are investigated (Jeon 2016). Smart agriculture, which connects to other ecosystems, is a way to achieve significance of long-term improvement in the expression of environment in the twenty-first century. Thus, using nanoscale particles has many advantages over conventional methods.

2.3 Nano-technologies in Food Industry

Due to the anti-pathogenic qualities and improved properties of nanoparticles, bio-nanotechnology can improve consumables at the nanoscale while resourcing bioactive components in edibles (Ghaani et al. 2016a). Nanoscale biomaterials can assist in purifying systems for better food quality as well as pathogen detection. Target delivery, paper materials, nano-encapsulation, control and nutraceuticals delivery, intelligent packaging, nano-coating of plastics, and nano-additives are only a few of the significant topics discussed (Ghaani et al. 2016b). Currently, some nutrients, mostly vitamins, are targeted into the bloodstream and encapsulated. Nanoparticles were added to some foods and beverages without changing the flavour or appearance. Ice cream and spreads contain nanoparticle emulsions, which can enhance the consistency and texture of the ice cream (Mlalila et al. 2016). Researchers are still working to develop smart food packaging materials that can provide information about foods that are packed. In order to monitor the oxidation of food, various packing materials now incorporate "nano-sensors." Nano-sensors embedded in such packaging materials alert users to food contamination when oxidation in products like milk and meat results in a colour change (Pan 2015).

In summary, nanoparticles are effective against a variety of pathogens that cause food borne disease due to their broad-spectrum antipathogenic capabilities. According to Annamalai and Nallamuthu (2015), metal nanoparticles have a few prototypes for their antimicrobial mechanism of action, including oxidative stress and cell damage, metal ion release, or non-oxidative mechanisms. This property of metal nanoparticles is very beneficial for extending the shelf life of foods. By efficiently enmeshing nanoparticles in the materials used to make food storage packaging, pathogenic growth on stored meals has been reduced. Thus, nanotechnology is a process that looks to the future and serves as a form of agricultural biosecurity.

3. Future Perspectives of Bio-Nanotechnology

The newest and next technology is bio-nanotechnology, which has highly special qualities in agriculture through increased crop yield and output using nanopesticides, nanofertilizers, etc. Precision farming methods and abilities, healthy components chosen with care, enhanced food texture and quality, packaging and labelling, etc. are all related to agriculture. In the near future, agricultural research may need to pay more attention to several particular areas.

> ➤ New environmentally friendly and secure methods of delivering specialised foods, plant nutrients, etc. Potential medicinal applications for these systems also exist.

> ➤ The sensors-based bio-nanotechnology plays an effective function in the food business, as well as in the control of insects and pests.

> ➤ It is important to thoroughly examine the increased features of nanomaterials, including their size, structures, surface chemistry, dose delivery, exposure to the environment, immunological response, accumulation in eco-systems, retention period, etc.

At least two key areas of the management of primary production can be improved upon by research in order to satisfy future demands: (1) greater crop yield and production rate, (2) improved resource usage effectiveness, and (3) decreased waste generation.

3.1 Increased Production Rate and Crop Yield

CUtilizing plant breeding, fertilisers, and plant-protection products has increased crop yields (Chari et al. 2017). Agriculture productivity growth has decreased since the Green Revolution of the 1960s–1970s, and we are currently in need of a second revolution in agricultural technology (Pandey 2018). However, it is likely that considerable improvements in crop output will result from increasing the efficiency of the photosynthetic process rather than increasing the doses of conventional agronomic elements.

The basis for plant photosynthesis is food security. In cool, humid regions, 85% of plant species are most prevalent and effective at photosynthesis. They include crops like cotton, sugar beets, tobacco, soybeans, and cereal grains including wheat, rice, barley, and oats. The radiant energy from solar light can be transformed by photosynthetic organisms into chemical energy, which is then stored in sugars. The procedure combined biochemical activities like NADPH and ATP with biophysical processes like electron transport and photosynthetically active radiation (PAR) absorption. There are some areas that need improvement in photosynthesis (Gomathi et al. 2019).

Reconsider the biophysical mechanisms of photosynthesis by taking a step back. More specifically, we consider solar radiation as the energy source that drives the process. Between 400 and 700 nm in the solar spectrum, visible light, which makes up 43% of solar light, roughly correlates to radiation that is photosynthetically active (PAR). Chlorophyll-a and chlorophyll-b, which are photosynthetic pigments, absorb photons as permitted by their absorption spectrum when sunlight strikes the leaf surface, supplying energy to the metabolic mechanism of photosynthesis (Mout et al. 2017).

3.2 Increase in Efficiency of Resource Utilization

Fertilization plays a significant role in modern agriculture since optimal crop nutrition is a crucial component of food security. The amount of macronutrients (N, P, K, S, Ca, and Mg) and micronutrients (B, Fe, Mn, Cu, Zn, Mo, and Cl) applied to agricultural areas has a significant impact on crop productivity (Rossi et al. 2019). According to a conservative estimate based on an analysis of the findings of several long-term field studies on crop productivity, commercial fertiliser inputs may be responsible for 30% to 50% of crop yield (Du et al. 2019).

The effectiveness with which plants utilise the available mineral nutrients is known as nutrient utilisation efficiency (NUE). Due to the physical and chemical features of the soil, leaching, gaseous losses, and fertiliser characteristics, NUE of agricultural plants is lower than 50% in all agroecosystems (Wang et al. 2019). As an illustration, consider the situation with urea $[CO(NH_2)_2]$, one of the most significant N-fertilizers (46% N by weight). Nevertheless, the byproducts created in soil as a result of urea degradation and the soil enzymes urease, volatilization, and hydrolysis (Tirani et al. 2019). Large amounts of nitrogen are lost if ammonia cannot be easily absorbed by plant roots.

We have a 2-fold consequence because between 1950 and 2000, fertiliser consumption grew almost 20 times for N and 7 times for P, respectively (Disfani et al. 2017). On the one hand, the reduced fertiliser dose efficiency suggests that production costs are rising in order to sustain high production. There is a chance that we will cause environmental pollution. Micronutrients perform crucial physiological functions in plant metabolism by acting as activators of particular enzymes, despite typically being present in plants in concentrations below 100 ppm. Numerous micronutrients aid in or activate plants' defence mechanisms against ailments or abiotic stress (Iqbal et al. 2019). Furthermore, plants are the sources of these necessary components for both animals and people. Crop productivity and food's nutritional content are restricted by soil micronutrient deficits or a lack of micronutrient availability in the soil.

The most common method of micronutrient application for crops is soil application. Under unfavorable conditions (neutral to alkaline soil pH), microelements frequently precipitate and become less bioavailable (Pallavi Mehta et al. 2016). It has been reported that the fertilizer-micronutrient use efficiency by crops is lower than 5% (Rizwan et al. 2019). To overcome the soil limiting factors, a second strategy widely used to provide micronutrients to crops is via leaf treatments. However, the roots are where plants largely take up nutrients. Micronutrient absorption by leaves is constrained, and basipetal flow (phloem) does not carry them from the leaves to the roots (Malandrakis et al. 2019).

The most effective fertilisation management strategies are those that help farmers meet the three key goals of sustainable agriculture: production, profitability, and environmental protection. One of the primary cornerstones of this vision is improving NUE in agricultural production (Wang et al. 2016). Given that it has demonstrated the viability of the so-called "smart fertiliser," nanotechnology can significantly contribute to the improvement of agricultural sustainability. In other words, nutrients are transported by nanostructures, which also enable their controlled release.

The minimising of losses and the release of nutrients are significantly influenced by the design of smart fertilisers. In the field, these materials are irrigated onto crops or sprayed onto plant canopies. Creating and testing carriers that enable the regulated release of nitrogen on a schedule that may be synced with the physiological requirements of crops is the scientific challenge. Studies on the interplay between nanoparticles and biota still produce inconsistent findings. This also happens when studying nanofertilizers.

3.3 Reduction of Waste Production

Growing understanding of the significance of sustainability is evident, especially in light of the expanding worldwide population (Shenashen et al. 2017). The establishment of a circular economy based on resource regeneration is closely related to this problem. Waste reduction is one of the principles of a circular economy. Agriculture-related waste is described by the Organization for Economic Cooperation and Development (OECD) as "waste produced as a result of various agricultural operations including manure and other wastes from farms, poultry houses and slaughterhouses; harvest waste; fertilizer run-off from fields; pesticides that enter into water, air or soils; and salt and silt drained from fields" (Vanti et al. 2019). Remainders and wastes account for a sizeable share of lost agri-food production (Giannousi et al. 2013). It will therefore be crucial to investigate cutting-edge technologies that can present fresh chances to accomplish complete sustainability. Nanotechnology is thought to be able to make a big contribution in this field as well (Imada et al. 2016). The advancement of cutting-edge techniques for the exploitation and valorization of agricultural wastes and byproducts is a vital contribution of nanotechnology to strengthening the fundamental ideas of the circular economy.

Since it makes up the majority of plant tissues, cellulose is the most prevalent biopolymer on the planet. The lignocellulosic substance found in wood, which is the most significant industrial source of cellulose, is where cellulose is found most frequently. Other materials with cellulose content include agricultural waste, water plants, grasses, and other plant matter (Imada et al. 2016). According to estimates, photosynthesis worldwide produces 1011–1012 tonnes of cellulose annually (Li et al. 2018).

Micro and macrofibrils serve as the building blocks of cellulose fiber's hierarchical structure in plant tissues. Depending on the source of the cellulose, elementary fibrils (nanofibers) with a diameter of 3–35 nm make up microfibrils (Davarpanah et al. 2016).

Nanocellulose is a novel bio-based nanomaterial with great optical characteristics, high strength, and specific surface area that has gained a lot of attention recently. In the field of nanocomposites, nanocellulose can be isolated and chemically manipulated for a variety of purposes (Huang et al. 2017). Numerous agricultural crops and waste products, including soy hulls and wheat straw, sugar beet pulp, potato pulp and rutabaga, are already considered as raw materials for innovative, economically viable techniques of producing nanocellulose (Kale and Gawade 2016).

4. Conclusion

Bio-nanotechnology is a young field with a lot of potential to change agriculture and the food sector. A smart understanding of bionanotechnology is crucial for agriculture. Although there is a wealth of knowledge regarding specific nanomaterials, many nanoparticles still have very low levels of toxicity. These nanoparticles are rarely utilised to evaluate dangers and their effects on human health due to a lack of information. There are many opportunities and possibilities in agriculture as basic understanding of how nanomaterials, the fundamental components of bio-nanotechnology, interact with the cells and their biological effects develops.

We have discussed several recently established concepts in this chapter that relate to potential contributions of nanotechnologies to the primary sector. Some concepts are currently strongly projected into the future, if not entirely visionary. While some other theories are highly specific, some of them actually have some preliminary experimental data. As a result, we can feel pretty positive about the future.

It is necessary to solve a number of issues related to the practical elements of using nanomaterials in agriculture. How will nanopesticides, nanoherbicides, or fertilisers be used in actual field settings? Which safety standards need to be taken into account? Which tools or machinery will be employed? Are these the same tools or machinery that are used for bulk materials? What standards of workplace safety should there be? The authorities will need to lay down regulations for these and many other elements. Clearly, the industries have high expectations at this point.

In conclusion, extensive fundamental understanding of the destiny of nanomaterials in the agro-environment is still required for the use of nanomaterials in agriculture. The linkages between agriculture and nanotechnology are, however, more developed and, at the same time, quite promising when it comes to the valorization of waste materials. It is appropriate to state once more that nanotechnologies are undergoing a turbulent evolution. This implies that programmes now in development will soon be surpassed by innovations that address different problems in the area of sustainable agriculture. This notion is nothing more than the inspiration behind the expansion of human knowledge and the bolstering of technological applications.

References

Abdel-Aziz, H.M., M.N. Hasaneen and A.M. Omer. 2016. Nano chitosan-NPK fertilizer enhances the growth and productivity of wheat plants grown in sandy soil. Spanish J. Agri. Res. 14: 0902.

Afrouzi, Y.M., P. Marzbani and A. Omidvar. 2015. The effect of moisture content on the retention and distribution of nano-titanium dioxide in the wood. Maderas Ciencia y Tecnología 17: 385–390.

Amil Usmani, M., I.H. Khan, A.S. Bhat, R. Pillai, N.K. Ahmad, M. Mohamad Haafiz and M. Oves. 2017. Current trend in the application of nanoparticles for waste water treatment and purification: A review. Curr. Org. Synth. 14: 206–226.

Annamalai, J. and T. Nallamuthu. 2015. Characterization of biosynthesized gold nanoparticles from aqueous extract of Chlorella vulgaris and their antipathogenic properties. Applied Nanosci. 5: 603–607.

Bhattacharyya, A., P. Duraisamy, M. Govindarajan, A.A. Buhroo and R. Prasad. 2016. Nano-biofungicides: Emerging trend in insect pest control. Adv. Applic. Through Fungal Nanobiotechnol. 1: 307–319.

Chari, N., L. Felix, M. Davoodbasha, A.S. Ali and T. Nooruddin. 2017. *In vitro* and *in vivo* antibiofilm effect of copper nanoparticles against aquaculture pathogens. Biocatalysis Agri. Biotechnol. 10: 336–341.

Chattopadhyay, D. and B. Patel. 2014. Nano metal particles: Synthesis, characterization and application to textiles. Manuf. Nanostructures 2: 184–215.

Davarpanah, S., A. Tehranifar, G. Davarynejad, J. Abadia and R. Khorasani. 2016. Effects of foliar applications of zinc and boron nano-fertilizers on pomegranate (*Punica granatum* cv. Ardestani) fruit yield and quality. Sci. Hortic. 210: 57–64.

Dey, N., D. Bhagat, D. Cherukaraveedu and S. Bhattacharya. 2017. Utilization of red-light-emitting CdTe nanoparticles for the trace-level detection of harmful herbicides in adulterated food and agricultural crops. Chem. J. 12: 76–85.

Dimkpa, C.O. 2014. Can nanotechnology deliver the promised benefits without negatively impacting soil microbial life? J. Basic Microbiol. 54: 889–904.

Disfani, M.N., A. Mikha, M.Z. Kassaee and A. Maghari. 2017. Effects of nano Fe/SiO$_2$ fertilizers on germination and growth of barley and maize. Arch. Agron. Soil Sci. 63: 817–826.

Du, W., J. Yang, Q. Peng, X. Liang and H. Mao. 2019. Comparison study of zinc nanoparticles and zinc sulphate on wheat growth: From toxicity and zinc biofortification. Chemosphere 227: 109–116.

Duhan, J.S., R. Kumar, N. Kumar, P. Kaur, K. Nehra and S. Duhan. 2017. Nanotechnology: The new perspective in precision agriculture. Biotechnol. Rep. 15: 11–23.

Ghaani, M., N. Nasirizadeh, S.A.Y. Ardakani, F.Z. Mehrjardi, M. Scampicchio and S. Farris. 2016a. Development of an electrochemical nanosensor for the determination of gallic acid in food. Analytical Methods 8: 1103–1110.

Ghaani, M., C.A. Cozzolino, G. Castelli and S. Farris. 2016b. An overview of the intelligent packaging technologies in the food sector. Trends Food Sci. Technol. 51: 1–11.

Giannousi, K., I. Avraamides and C. Dendrinou-Samara. 2013. Synthesis, characterization and evaluation of copper-based nanoparticles as agrochemicals against *Phytophthora infestans*. RCS Adv. 3: 21743–21752.

Gomathi, T., K. Rajeshwari, V. Kanchana, P.N. Sudha and K. Parthasarathy. 2019. Impact of nanoparticle shape, size, and properties of the sustainable nanocomposites. *In*: Inamuddin, Thomas S., R. Kumar Mishra and A.M. Asiri (eds.). Sustainable Polymer Composites and Nanocomposites. Springer International Publishing; Cham, Switzerland.

Huang, M., Z. Wang, L. Luo, S. Wang, X. Hui, G. He, H. Cao, X. Ma, T. Huang, Y. Zhao and C. Diao. 2017. Soil testing at harvest to enhance productivity and reduce nitrate residues in dryland wheat production. Field Crops Res. 212: 153–164.

Huang, S., L. Wang, L. Liu, Y. Hou and L. Li. 2015. Nanotechnology in agriculture, livestock, and aquaculture in China. A review. Agronomy Sustainable Develop. 35: 369–400.

Imada, K., S. Sakai, H. Kajihara and S. Tanaka. 2016. Magnesium oxide nanoparticles induce systemic resistance in tomato against bacterial wilt disease. Plant Pathol. 65: 551–560.

Iqbal, M., N.I. Raja, M. Hussain, M. Ejaz and F. Yasmeen. 2019. Effect of silver nanoparticles on growth of wheat under heat stress. IJST A Sci. 43: 387–395.

Jampilek, J., K. Zaruba, M. Oravec, M. Kunes, P. Babula, P. Ulbrich, I. Brezaniova, R. Opatrilova, J. Triska and P. Suchy. 2015. Preparation of silica nanoparticles loaded with nootropics and their *in vivo* permeation through blood-brain barrier. BioMed. Res. Int. 2: 1–2.

Jeon, C.S. 2016. Surface functionalization of bioanalytical applications: Virus decorated gold microshells and modified synaptic cell adhesion molecules.

Kah, M. and T. Hofmann. 2014. Nanopesticide research: Current trends and future priorities. Environ. Int. 63: 224–235.

Kale, A.P. and S.N. Gawade. 2016. Studies on nanoparticle induced nutrient use efficiency of fertilizer and crop productivity. Green Chem. Technol. Lett. 2: 88–92.

Kalpana, V. and V.D. Rajeswari. 2017. Biosynthesis of metal and metal oxide nanoparticles for food packaging and preservation: A green expertise 1: 293–316.

Kookana, R.S., A.B. Boxall, P.T. Reeves, R. Ashauer, S. Beulke, Q. Chaudhry, G. Cornelis, T.F. Fernandes, J. Gan, M. Kah and I. Lynch. 2014. Nanopesticides: Guiding principles for regulatory evaluation of environmental risks. J. Agri. Food Chem. 62: 4227–4240.

Kumar, S., G. Bhanjana, A. Sharma, M. Sidhu and N. Dilbaghi. 2014. Synthesis, characterization and on field evaluation of pesticide loaded sodium alginate nanoparticles. Carbohydrate Polymers 101: 1061–1067.

Li, C., J. Li, Y. Li and G. Fu. 2018. Cultivation techniques and nutrient management strategies to improve productivity of rain-fed maize in semi-arid regions. Agric. Water Manag. 210: 149–157.

Malandrakis, A.A., N. Kavroulakis and C.V. Chrysikopoulos. 2019. Use of copper, silver and zinc nanoparticles against foliar and soil-borne plant pathogens. Sci. Total Environ. 670: 292–299.

Marzbani, P., Y.M. Afrouzi and A. Omidvar. 2015. The effect of nano-zinc oxide on particleboard decay resistance. Maderas Ciencia y Tecnología 17: 63–68.

Mlalila, N., D.M. Kadam, H. Swai and A. Hilonga. 2016. Transformation of food packaging from passive to innovative via nanotechnology: Concepts and critiques. J. Food Sci. Technol. 53: 3395–3407.

Mout, R., M. Ray, G. Yesilbag Tonga, Y. Lee, W. Tay, T. Sasaki and K. Rotello. 2017. Direct cytosolic delivery of CRISPR/Cas9-ribonucleoprotein for efficient gene editing. ACS Nano. 11: 2452–2458.

Mukhopadhyay, S.S. 2014. Nanotechnology in agriculture: Prospects and constraints. Nanotech. Sci. Appl. 7: 63.

Pallavi Mehta, C.M., R. Srivastava, S. Arora and A.K. Sharma. 2016. Impact assessment of silver nanoparticles on plant growth and soil bacterial diversity. 3 Biotech 6: 254.

Pan, Z. 2015. Food decontamination using nanomaterials. MOJ Food Process Technol. 1: 00011.

Pandey, G. 2018. Challenges and future prospects of agri-nanotechnology for sustainable agriculture in India. Environ. Technol. Innov. 11: 299–307.

Patil, S.S., U.U. Shedbalkar, A. Truskewycz, B.A. Chopade and A.S. Ball. 2016. Nanoparticles for environmental clean-up: A review of potential risks and emerging solutions. Environ. Technol. Innovation 5: 10–21.

Patra, J.K. and K.H. Baek. 2017. Antibacterial activity and synergistic antibacterial potential of biosynthesized silver nanoparticles against food borne pathogenic bacteria along with its anticandidal and antioxidant effects. Front. Microbiol. 8: 1–2.

Prasad, R., A. Bhattacharyya and Q.D. Nguyen. 2017. Nanotechnology in sustainable agriculture: Recent developments, challenges, and perspectives. Front. Microbiol. 8: 1–2.

Prasad, R., V. Kumar and K.S. Prasad. 2014. Nanotechnology in sustainable agriculture: Present concerns and future aspects. African J. Biotech. 13: 705–713.

Prema, R.S. and S. Kandasamy. 2017. Synthesis and characterization of zinc oxide and iron oxide nano particles using Sesbania Grandiflora leaf extract as reducing agent. J. Nanosci. 1: 1–7.

Rizwan, M., S. Ali, M.Z. Ur Rehma, S. Malik, M. Adrees, M.F. Qayyum, S.A. Alamri, M.N. Alyemeni and P. Ahmad. 2019. Effect of foliar applications of silicon and titanium dioxide nanoparticles on growth, oxidative stress, and cadmium accumulation by rice (*Oryza sativa*). Acta Physiol. Plant 41: 3567.

Rossi, L., L.N. Fedenia, H. Sharifan, X. Ma and L. Lombardini. 2019. Effects of foliar application of zinc sulfate and zinc nanoparticles in coffee (*Coffea arabica* L.) plants. Plant Physiol. Biochem. 135: 160–166.

Shenashen, M., A. Derbalah, A. Hamza, A. Mohamed and S. El Safty. 2017. Antifungal activity of fabricated mesoporous alumina nanoparticles against root rot disease of tomato caused by Fusarium oxysporium. Pest Manag. Sci. 73: 1121–1126.

Singh, P., Y.J. Kim, D. Zhang and D.C. Yang. 2016. Biological synthesis of nanoparticles from plants and microorganisms. Trends Biotechnol. 34: 588–599.

Tirani, M.M., M.M. Haghjou and A. Ismaili. 2019. Hydroponic grown tobacco plants respond to zinc oxide nanoparticles and bulk exposures by morphological, physiological and anatomical adjustments. Funct. Plant Biol. 46: 360–375.

Vanti, G.L., V.B. Nargund, K.N. Basavesha, R. Vanarchi, M. Kurjogi, S.I. Mulla, S. Tubaki and R.R. Patil. 2019. Synthesis of *Gossypium hirsutum*-derived silver nanoparticles and their antibacterial efficacy against plant pathogens. Appl. Organomet. Chem. 33: e4630.

Wang, S.H., F.Y. Wang, S.C. Gao and X.G. Wang. 2016. Heavy metal accumulation in different rice cultivars as influenced by foliar application of nano-silicon. Water Air Soil Pollut. 227–228.

Wang, Y., Y. Lin, Y. Xu, Y. Yin, H. Guo and W. Du. 2019. Divergence in response of lettuce (var. ramosa Hort.) to copper oxide nanoparticles/microparticles as potential agricultural fertilizer. Environ. Pollut. Bioavailab. 3: 80–84.

Wu, S., D. Li, J. Wang, Y. Zhao, S. Dong and X. Wang. 2017. Gold nanoparticles dissolution based colorimetric method for highly sensitive detection of organophosphate pesticides. Sensors and Actuators B: Chemical 238: 427–433.

Index

Editors

Dr. Thandapani Gomathi, Ph.D., is presently working as an Assistant Professor in the Department of Chemistry, at DKM College for Women (Autonomous), Affiliated to Thiruvalluvar University, Vellore, Tamilnadu, India. She received her M.Sc., M.Phil., and Ph.D. degrees in Chemistry from Thiruvalluvar University in 2006, 2008 and 2015. She was awarded the University Ninth rank in M.Sc. Chemistry from Thiruvalluvar University, Vellore (2006). Her major research interests are in the field of drug designing and drug delivery, nanotechnology, polymer chemistry, environmental chemistry, biopolymers for biomedical applications. She is highly enthusiastic about pursuing research and developmental activities and research is her passion. Furthermore, she expended her research fields up to the development of nano drug delivery system. To date, she has authored around 105 research papers and 15 book chapters.

Dr. P.N. Sudha is presently working as a Principal at DKM College for Women (Autonomous), affiliated to Thiruvalluvar University, Vellore, Tamilnadu, India. She obtained her Doctoral degree in Chemistry–Biology Interdisciplinary Science from Madras University. She is a fellow of International Congress of Chemistry and Environment, and National Environmentalists Association. Her name was included in *Marquis Who's Who in the World* (2011). She has been listed in the worlds top 2% scientists consecutively for the past three years. Her research interests are polymer chemistry, environmental chemistry, bioremediation, and photochemistry. She has authored 11 books and 13 book chapters. She is a life member of Asian Chitin Chitosan Society. She is a co editor and reviewer of international journals. Her immense experience in research is the key asset for the development of research among the young women researchers in rural areas. She expanded her research fields up to the use to biomaterials such as chitin and chitosan for biomedical application. She has undertaken projects funded by UGC, DST, TNSCST, and MHRD. She has guided 40 PhD scholars and 50 MPhil scholars. She has published 190 research articles in journals mainly on the wastewater treatment and biomedical applications of chitin and chitosan.

Dr. Sabu Thomas, PhD is the Founder Director and Professor of the International and Interuniversity Centre for Nanoscience and Nanotechnology and also full professor of Polymer Science and Engineering at the School of Chemical Sciences of Mahatma Gandhi University, Kottayam, Kerala, India. He is an outstanding leader with sustained international acclaims for his work. Dr. Thomas's ground breaking inventions in polymer nanocomposites, polymer blends, green bionanotechnological and nano-biomedical sciences, have made transformative differences in the development of new materials for automotive, space, housing and biomedical fields. In collaboration with India's premier tyre company, Apollo Tyres, Professor Thomas's group invented new high performance barrier rubber nanocomposite membranes for inner tubes and inner liners for tyres. Professor Thomas has received a number of national and international awards which include: Fellowship of the Royal Society of Chemistry, London, Distinguished Professorship from Josef Stefan Institute, Slovenia, MRSI medal, Nano Tech Medal, CRSI medal, Distinguished Faculty Award, and Sukumar Maithy Award for the best polymer researcher in the country. He is in the list of most productive researchers in India and holds a position of No. 5. Very recently, because of the outstanding contributions to the field of nano materials, Polymer Science and Engineering, Prof. Thomas has been conferred Honoris Causa (DSc) by the University of South Brittany, Lorient, France. Professor Thomas has published over 650 peer reviewed research papers, reviews and book chapters. He has co-edited 53 books published by Royal Society, Wiley, Wood head, Elsevier, CRC Press, Springer, Nova, etc. He is the inventor of 6 patents. The h index of Prof. Thomas is 124 and has more than 76,500 citations. Prof. Thomas has delivered over 300 Plenary/Inaugural and Invited lectures in national/international meetings over 40 countries. He has established a state of the art laboratory at Mahatma Gandhi University in the area of Polymer Science and Engineering and Nanoscience and Nanotechnology through external funding form DST, CSIR, TWAS, UGC, DBT, DRDO, AICTE, ISRO, DIT, TWAS, KSCSTE, BRNS, UGC-DAE, Du Pont, USA, General Cables, USA, Surface Treat Czech Republic, MRF Tyres and Apollo Tyres. Professor Thomas has several international collaborative projects with a large number of countries abroad. He has already supervised 75 PhD theses. The former co-workers of Prof. Thomas occupy leading positions in India and abroad.

For Product Safety Concerns and Information please contact our EU
representative GPSR@taylorandfrancis.com
Taylor & Francis Verlag GmbH, Kaufingerstraße 24, 80331 München, Germany

www.ingramcontent.com/pod-product-compliance
Lightning Source LLC
Chambersburg PA
CBHW082106220326
41598CB00066BA/5638